AF340680

Metabolic Profiling
Disease and Xenobiotics

Issues in Toxicology

Series Editors:
Professor Diana Anderson, *University of Bradford, UK*
Dr Michael Waters, *Integrated Laboratory Systems Inc., N Carolina, USA*
Dr Timothy C Marrs, *Edentox Associates, Kent, UK*

Adviser to the Board:
Alok Dhawan, *Ahmedabad University, India*

Titles in the Series:
 1: Hair in Toxicology: An Important Bio-Monitor
 2: Male-mediated Developmental Toxicity
 3: Cytochrome P450: Role in the Metabolism and Toxicity of Drugs and other Xenobiotics
 4: Bile Acids: Toxicology and Bioactivity
 5: The Comet Assay in Toxicology
 6: Silver in Healthcare
 7: *In Silico* Toxicology: Principles and Applications
 8: Environmental Cardiology
 9: Biomarkers and Human Biomonitoring, Volume 1: Ongoing Programs and Exposures
10: Biomarkers and Human Biomonitoring, Volume 2: Selected Biomarkers of Current Interest
11: Hormone-Disruptive Chemical Contaminants in Food
12: Mammalian Toxicology of Insecticides
13: The Cellular Response to the Genotoxic Insult: The Question of Threshold for Genotoxic Carcinogens
14: Toxicological Effects of Veterinary Medicinal Products in Humans: Volume 1
15: Toxicological Effects of Veterinary Medicinal Products in Humans: Volume 2
16: Aging and Vulnerability to Environmental Chemicals: Age-related Disorders and their Origins in Environmental Exposures
17: Chemical Toxicity Prediction: Category Formation and Read-Across
18: The Carcinogenicity of Metals: Human Risk through Occupational and Environmental Exposure

19: Reducing, Refining and Replacing the Use of Animals in Toxicity Testing
20: Advances in Dermatological Sciences
21: Metabolic Profiling: Disease and Xenobiotics

How to obtain future titles on publication:
A standing order plan is available for this series. A standing order will bring
delivery of each new volume immediately on publication.

For further information please contact:
Book Sales Department, Royal Society of Chemistry, Thomas Graham House,
Science Park, Milton Road, Cambridge, CB4 0WF, UK
Telephone: +44 (0)1223 420066, Fax: +44 (0)1223 420247
Email: booksales@rsc.org
Visit our website at www.rsc.org/books

Metabolic Profiling
Disease and Xenobiotics

Edited by

Martin Grootveld
De Montfort University, Leicester, UK
Email: mgrootveld@dmu.ac.uk

Issues in Toxicology No. 21

Print ISBN: 978-1-84973-163-8
PDF eISBN: 978-1-84973-516-2
ISSN: 1757-7179

A catalogue record for this book is available from the British Library

Published by The Royal Society of Chemistry,
Thomas Graham House, Science Park, Milton Road,
Cambridge CB4 0WF, UK

Registered Charity Number 207890

For further information see our website at www.rsc.org

Preface

This book represents the culmination of at least several years' relatively intensive work, and provides an in-depth and sometimes highly critical review of research investigations performed in the metabolomics research area and, more generally, that concerning the 'omics' fields in general (for example, proteomics and genomics, *etc.*). My major objective was primarily to provide valuable advice from my own original, basic grounding in the statistical analysis of datasets with a biomolecular focus or otherwise. However, as the volume of work progressed, it became clearer to me that more and more researchers involved in these areas are, at least some of the time, keen to experience a revelation of some kind, and are utilising the wide range of methods and techniques developed in order to achieve a rapid research impact 'hit' without bearing out the consequences of their outputs in terms of both short- and long-term applications of their often dedicated bioanalytical chemistry and multivariate (MV) data analyses work. Indeed, particularly clear is the knowledge that, despite the polynomially-increasing number of publications available in this research area, there appear to be very few which actually manifest themselves into relatively simple diagnostic tools or probes for the diagnosis of the diseases which they were originally designed to investigate and perhaps also monitor. Part of this problem arises from the apparent inabilities of researchers to transform their findings into a clinically or diagnostically significant context (and/or the professional and financial constraints associated with this process), and there remains the potential hazard that, if taken out of context, such results may serve to provide some confusion and perhaps even misinformation. A further component (if you'll excuse the poor choice of words!) is derived from the high costs of performing such multicomponent analysis and the associated valid metabolomic/statistical interpretation of datasets acquired therefrom. Moreover, an additional major barrier is provided by the severe lack of

Issues in Toxicology No. 21
Metabolic Profiling: Disease and Xenobiotics
Edited by Martin Grootveld
© The Royal Society of Chemistry 2015
Published by the Royal Society of Chemistry, www.rsc.org

statistical validation and cross-validation techniques employed by such researchers in order to evaluate the reliabilities and reproducibilities of the methods that they have developed, *i.e.* so that they may provide a sound foundational basis for the results acquired in their experiments (such concerns are rigorously discussed in Chapters 1 and 2). However, not seeing these connections directly is not the same as not realising that they might be there!

Of much critical importance to the performance of many multivariate (MV) analyses of high-dimensional, high-throughput datasets is the satisfaction of, in many cases, essential assumptions for the effective operation of such models, and in both Chapters 1 and 2 the authors provide relevant information regarding these requirements, and also demonstrate their clear violation when an experimental dataset is subjected to a series of statistical tests for their satisfaction (including those concerning assumptions for normality, homoscedasticity and also the detection of statistical outliers, albeit in a univariate context), observations which are consistent with the very few of those made available by other researchers. In this manner, researchers should always question the validity of many MV analysis techniques which are applicable to such datasets. This problem is absolutely rampant in published work available in which the researchers involved have only employed univariate analysis methods such as t-tests, or one- or two-classification ANOVA (*i.e.* completely randomised or randomised block designs, respectively, for the latter), for example their almost complete lack of consideration for the intra-sample variance homogeneity (homoscedasticity) assumption when testing for significant differences 'Between-Classifications', and which relatively simple log- or square root-transformations of the dataset would, at least in some cases, cure. Hence, we can imagine the many problems to be encountered by workers challenged by multidimensional 'omics' research problems in this manner!

In Chapter 3 I also review and provide examples of the applications of additional MV analysis techniques which are already available, but nevertheless to date have only been applied to the metabolomic profiling, metabolomics and/or genomics areas in a limited (or very limited) manner. These include canonical correlation analysis (CCorA), and both the k-means and agglomerative hierarchal (AHC) clustering techniques, which have been previously extensively employed in alternative research areas such as ecology and environmental science. Such applications serve as an adjunct to the methods commonly employed in our field of interest. Although these methodologies are not proposed to serve as the first choice of MV analysis for such multidimensional datasets, they can, however, represent valuable strategies or aids for application in particular 'omics' investigations or circumstances, for example the use of the CCorA and canonical correspondence analysis (CCA) techniques in order to explore and evaluate any significant linkages, and also the level of dimensionality, between two separate dataset tables (or, for that matter, components or factors derived therefrom, one of which may represent biofluid or tissue biopsy metabolite levels monitored with one technique, the other perhaps a series of latent, potentially related variables such as age, gender, family status, body mass index, blood pressure components, *etc.*).

Also noteworthy is the essential knowledge that many frequently employed or employable MV analysis techniques available are critically dependent on simple linear (Pearson) correlations between the 'predictor' (X) variables acquired in such model systems; however, in view of this, such models are fraught with many difficulties in view of (1) many potential non-linear, polynomial or otherwise, relationships existing between such variables (many metabolic pathway analyses involved or implicated are either clearly or conceivably of a 'non-linear' nature), and (2) corrections for the influence of further cross-correlated variables (a problem which is resolvable *via* the computation of *partial* correlation coefficients where only a small number, say 2–5, of variables are involved in simple multiple linear regression, partial correlation and discriminatory analysis models), which may exert a major influence on a critical dependent (Y) variable, binary, ordinate, continuous or otherwise. Fortunately, recent developments in the metabolomics research area have served to provide at least some viable means of overcoming these problems, specifically the independent component analysis (ICA) and Gaussian Graphical Models (GGMs) approaches (the former making allowances for potential polynomial relationships between such putative predictor variables, the latter targeted at the consideration of the most important *partial* correlations between them).

A further very important aspect of such investigations involves the consideration of potentially a multitude of interactions between variables involved in the statistical processing of MV bioanalytical datasets (such as those encountered in factorial ANOVA experimental designs), and although this is possible for relatively small numbers of *lateral* variables such as those noted above (including clinically relevant indices, where appropriate), it remains an overwhelming challenge to deal with those arising in MV datasets consisting of hundreds or even thousands of potential predictor variables! For current considerations, however, I and my co-authors merely focus on the applications of techniques (and related examples) which combat and effectively deal with the former (much simpler) task, *i.e.* those concerning the applications of the ANOVA-Simultaneous-Component-Analysis (ASCA) method (which permit exploration of ANOVA-derived orthogonal effect matrices for underlying intra-metabolite relationships and correlations), which is described in my own Chapter 3, and, in a more problem-targeted context, in Chapter 4 by Westerhuis *et al.*, the latter also involving Multi-Level Partial Least Squares-Discriminatory Analysis (ML-PLS-DA). Indeed, in Chapter 4, the authors provide valuable information regarding the development and application of this novel technique, in particular its employment for the solution of two challenging time-series metabolomics tasks, the first investigating the differential treatments applied to a plant species, the second a polyphenolic interventional study in human participants.

Since many of the complete variances of datasets acquired in frequently conducted metabolomics investigations are accountable by variations in sample-donor identities, the time-points at which samples are collected, and also a possible range of further (albeit lateral 'independent' X variables), this relatively recent advance into the metabolomics research area serves to effectively circumvent the confounding effects of such interfering variables, and hence permit researchers to focus on the significance of the main factor(s) of interest following their removal, specifically those observed 'Between-Disease or -Treatment Classifications' as appropriate. A range of researchers have focused on isolating and determining the significance of a range of variance components in complex factorial experimental designs for very many years (although perhaps only in a univariate context), and hence it is a little surprising that metabolomics researchers in general have only recently got round to the idea that it would be highly advantageous also to perform this procedure in a corresponding MV model manner!

Professor Dziuda's contribution in Chapter 5 reveals and outlines metabolomics methods available for the analysis of datasets which have larger numbers of potential predictor (X) variables than there are samples available for analysis. This consideration is of critical importance to the great majority of scientists involved in the metabolomics and further 'omics' research areas, especially those which, in view of advice provided to them (or alternatively their viewpoint), are generally limited to the applications of conventional MV analytical techniques such as PCA or PLS-DA, which are clearly restricted or limited in the context of their applications to such ($n < P$ or $n \ll P$) datasets, especially the latter method!

This contributor also discusses the application of some commonly employed and well-established data-mining methods to such cases, and also rises to this challenge in his outline and critical appraisal of some new techniques targeted at overcoming this $P \gg n$ problem encountered in many metabolomics investigations. Primarily, this author focuses on the methods and approaches which are appropriate for the analysis of high-throughput, multidimensional 'omics' datasets, and also provides much useful information regarding some common misconceptions and pitfalls in this area. He also provides guidance concerning when exactly to employ such methods. One major point of interest and importance arising from this work is the rather severe lack of considerations for biomolecular feature selection available in the current literature. Indeed, as he states, this is, after all, the most important aspect of biomarker discovery! He then further delineates the critical importance of presenting new frontiers regarding the sensible MV statistical analysis of such complex and challenging datasets, specifically those involving selected supervised 'learning' algorithms which, when coupled to powerful feature selection methods, can serve to provide a wealth of information regarding MV biomarker identification processes. This chapter also focuses on the extreme importance of considerations for the biological interpretation and significance of the biomarkers selected (together with the critical requirement for their correct validation), plus a novel data-mining technique that permits their efficient, robust, parsimonious and biologically and/or clinically interpretable discovery.

These points are also critically considered in my own Chapters 1–3, the third of which provides full details and an application example of Dr Magidson's recently developed Correlated Component Regression (CCR) technique, which can be applied to such $n \ll P$ datasets. As noted above, a further critically important reason for necessarily implementing the application (and hopefully routine future usage) of such forms of data analysis *via* the now commonly employed ^{1}H NMR or LC-MS techniques, for example, is the high cost of preforming such investigations. Indeed, for the purposes of one grant application which I recently submitted in conjunction with clinical colleagues, the rate for the collection of blood plasma samples for one particular clinical study performed at a single UK Health Service provider was approximately £200 per collection, and this without the additional costings required for the essential provision of associated high-resolution ^{1}H NMR analysis and subsequent MV explorations of the datasets acquired!

Chapter 6 by Dr Rick Dunn and co-workers outlines the diverse applications of differing mass spectrometric platforms to the biological and metabolomics research areas, and here the authors focus on the series of advantages offered by these systems, particularly those concerning their specificities, sensitivities and the established potentials and applications of these techniques for the multicomponent analysis of biofluids and tissues (linked with the capacity to classify the identities of thousands of metabolites present in a single sample). The applications of such methodologies will undoubtedly continue to expand, and may also give rise to novel discoveries relating to human health and diseases, together with the subsequent potential development of novel and challenging therapeutic interventional strategies.

Recent developments regarding the applications of data classification - algorithms, firstly those involving unsupervised PCA and cluster analysis techniques, and secondly supervised methods such as Linear Discriminant Analysis (LDA), PLS-DA, Soft Independent Modelling of Class Analogy (SIMCA), Artificial Neural Network (ANNs), SVM machine-learning and Bayesian classification systems to the detection and characterisation of the 'biomarker' roles of metabolites in both soft and hard tissues, together with biofluids collected from humans, are outlined by Kenichi Yoshida and myself in Chapter 7. Indeed, Professor Yoshida's investigations have revealed much valuable metabolic information regarding the ability of these MV analysis techniques to distinguish between healthy and cancerous tissues collected from humans. The application of ongoing technologies for the detection and identification of biomarker patterns which are distinctive for various tumours are also discussed, as is the requirement for the performance of multiple experiments for these purposes.

In Chapter 8, Professor Adamec introduces and discusses the applications of Group-Specific Internal Standard Technology (GSIST) as a newly developed, novel and highly sensitive LC-MS method that permits the analysis of biomolecules at sensitivities required for the life science research areas. Indeed, novel derivatisation reagents and methods serve to provided major benefits regarding the LC-ESI-MS analysis of metabolites, specifically those involving enhancements of detection sensitivity, attenuations of the hydrophobicities/hydrophilicities of analytes, and their retention times, and chromatographic band-spreading patterns (processes which increase the resolution and rapidity of the separation techniques involved), and also an increased efficacy of both comparative recovery and quantification processes, the latter including the employment of isotopic adducts of selected derivatisation reagents.

Uniquely, Professor Dzeja and colleagues of the Mayo Clinic (USA) outline the value of applying stable isotope ^{18}O-assisted ^{31}P NMR and mass spectrometric analyses in order to permit the simultaneous monitoring of high-energy phosphate metabolite levels and their rates of turnover in blood and tissue specimens (Chapter 9). This novel technological breakthrough has given rise to the synchronous monitoring of both ATP synthesis and its utilisation, in addition to the detection of phosphotransfer fluxes involved in the glycolytic, and adenylate and creatinine kinase pathways. Moreover, the status of mitochondrial nucleotides, which are implicated in the Krebs cycle and its dynamics, together with the glycogen turnover process therein, can also be determined. One major advantage offered by this ^{18}O-based technology is that it has the ability to monitor virtually all phosphotransfer reactions occurring within cells (including those associated with small pool signalling molecule turnovers), and also the dynamics involved in such energetic signal communications. These investigators therefore provide much valuable information concerning the phosphometabolomic/fluxomic profiling of transgenic human disease models which explore trans-systems metabolic network adaptions, and also the potential detection and monitoring of biomarkers which may be related to the effectiveness of treatments for human diseases and/or drug toxicology.

Chapter 10 by Dr Chris Silwood and myself focuses on the application of both conventional and more recently developed methods for the MV analysis of multianalyte human biofluid datasets, the latter involving the Self-Organising Maps (SOMs, both supervised and unsupervised approaches) technique, and their applications have served to provide useful information concerning the ability of an oral rinse product added *in vitro* to exert an influence on the ^{1}H NMR metabolic profile of human saliva. Indeed, these methods readily facilitated the detection of perturbations mediated by the oxidation of critical salivary biomolecule scavengers by the actions of an active oxyhalogen agent in the product tested.

With regard to the toxicology research area, in Chapter 11 Wei Tang and Quiwei Xu provide detailed descriptions of drug-induced liver injury, focusing on the current views and understandings regarding the underlying mechanisms involved in these processes. These investigators also focus on the applications of metabolomics techniques to the provision of essential biomolecular information regarding the pathogenesis of hepatotoxicity, including the seeking, identification and plausible future applications of significant biomarkers for detection, diagnosis, prevention and clinical control of this condition.

Finally, in Chapter 12 Dr Gomase evaluates the application of chemogenomic techniques in order to seek chemical (specifically drug) targets within biosystems, in this case relevant proteins. Such research work can indeed serve as a valuable aid to developments in the areas of gene discovery and presents regulation, cheminformatics and molecular signalling opportunities with respect to the potential authentication of novel therapeutic agents for the treatment of chronic human diseases such as a series of cancers. Indeed, the reliable and effective prediction of interactions between specific proteins and low-molecular-mass molecules represents one of the most important phases in our capacities to elucidate the mechanisms involved in a multitude of biological processes, and may also play a crucial role in the development of future drug-discovery systems, together with its further application to the less hazardous and practical issues associated with stem cell regeneration processes.

I would like to express my sincerest thanks to all the authors who contributed chapters to this book (who unfortunately also had to put up with a number of delays with its preparation and completion). Thanks also go to a number of my research collaborators, including those based on my own university campus, namely Victor Ruiz Rodado, Dr Sundarchandran, Prof. Katherine Huddersman, Drs David Elizondo and Dr Dan Sillence, to mention but some, and those from other universities or elsewhere, in particular Prof. Richard Brereton (formerly of the University of Bristol), Prof. Frances Platt (University of Oxford), Prof. Geoffrey Hawkes (Queen Mary, University of London) and Dr Chris Silwood, some of whom have directly or indirectly contributed towards the generation of this work (*via* the kind provision of biofluid samples for ^{1}H NMR analysis and/or clinical/clinical chemistry datasets), and sometimes also with the MV or computational intelligence analysis of datasets generated. I also wish to thank a lot of further staff at Leicester School of Pharmacy for their kind support whilst I was involved in producing this work.

Strangely, this book was written and edited, at various stages, in the USA, Brazil, Argentina, Paraguay, Crete and Spain (and sometimes also Portugal), but most especially in various regions of the UK, including North Wales, Shropshire, Manchester, London, Leicester and next to Loch Lomond in Scotland. I also wish to thank the operators of various train, plane and automobile rides which also offered ample opportunities for me to work on the manuscripts, the Black Bear pub in Whitchurch and also the (not so) Happy Friar and Fat Cat bars in Leicester, in which the bar staff did not complain too much about me writing in their 'hospitable' environments. Finally, I also thank my fantastic wife Kerry for all the help and support she provided whilst I was working on this task (amongst many others): she really had to put up with quite a lot of difficult days involved, at least some of which were unavoidable. I also sincerely thank her for typing my many scribbled revisions to this work, and also for providing invaluable suggestions for improved ones! I hope that this book will serve as a valuable aid to both scientific and clinical researchers who wish to explore such spheres of the unknown!

Contents

Chapter 1 Introduction to the Applications of Chemometric Techniques in 'Omics' Research: Common Pitfalls, Misconceptions and 'Rights and Wrongs' 1
Martin Grootveld

1.1 Introduction 1
1.2 Principal Component Analysis (PCA) 2
    1.2.1 Critical Assumptions Underlying PCA 4
    1.2.2 Number and Significance of Explanatory Variables Loading on a PC 9
    1.2.3 Number of Extractable PCs and Their Characteristics 9
    1.2.4 Total Variance of the Dataset 10
    1.2.5 What is an Adequate Sample Size for PCA and Further Forms of MV Analysis? 10
    1.2.6 Interpretability Criteria of PCs 11
    1.2.7 Varimax Rotation 12
    1.2.8 Example Case Study 13
    1.2.9 Examination of a Wider Range of Components 15
    1.2.10 Consideration of Type I (False-Positive) Errors 16
    1.2.11 Determinations of the Suitability of MV Datasets for Analysis with PCA and FA 17
1.3 Partial Least Squares-Discriminatory Analysis (PLS-DA) 18
    1.3.1 Case Study Describing an Example of PLS-DA 'Overfitting' 20
    1.3.2 Permutation Testing 22

Issues in Toxicology No. 21
Metabolic Profiling: Disease and Xenobiotics
Edited by Martin Grootveld
© The Royal Society of Chemistry 2015
Published by the Royal Society of Chemistry, www.rsc.org

1.3.3 Procedures for the Validation and Cross-Validation of PLS-DA Models   24
1.3.4 Attainment of the Final Calibration Model   28
1.3.5 Quality Evaluation Processes   28
1.3.6 Cost-Benefit Analysis (CBA)   30
Appendix I   31
Appendix II   33
Acknowledgements   33
References   33

Chapter 2   Experimental Design: Sample Collection, Sample Size, Power Calculations, Essential Assumptions and Univariate Approaches to Metabolomics Analysis   35
Martin Grootveld and Victor Ruiz Rodado

2.1 Introduction   35
2.2 Essential Considerations for Sample Collection   36
2.3 Raw Data Preprocessing Steps   39
2.4 Data Normalisation, Scaling and Dimensionality Reduction   42
2.5 Assumption of Normality   44
2.6 Analysis-of-Variance (ANOVA): Experimental Design and Analysis   50
    2.6.1 Model I: Fixed Effects   50
    2.6.2 Model II: Random Effects   53
    2.6.3 Hierarchical or 'Nested' Models   54
    2.6.4 Factorial/Multifactorial Models   54
    2.6.5 ANOVA-Simultaneous Component Analysis   57
    2.6.6 Further Considerations of Interaction Components of Variance in MV Modeling   57
2.7 Outline of the Applications of Univariate Approaches to the Analysis of Metabolomics Datasets   58
    2.7.1 More on Essential Assumptions Required   60
    2.7.2 Bonferroni Correction for Multiple Comparisons of Mean Values   62
2.8 Power (Sample Size) Computations for Untargeted, Univariate Investigations of Metabolomics Datasets   64
2.9 Sample Size Requirements and Statistical Power Computations for High-Dimensional, Metabolomic Datasets   67
2.10 Error Analysis   69
Acknowledgements   69
References   70

Chapter 3 Recent Developments in Exploratory Data Analysis and Pattern Recognition Techniques 74
Martin Grootveld

3.1 Introduction 74
3.2 Canonical Correlation Analysis (CCorA) 75
    3.2.1 CCorA Case Study 76
3.3 Classification and Regression Tree (CART) Analysis 80
3.4 Moderated t-Statistic Methods 81
    3.4.1 Significance Analysis of Microarrays (SAM) 81
    3.4.2 Empirical Bayesian Approach Modelling (EBAM) 82
3.5 Machine Learning Techniques 83
    3.5.1 Self-Organising Maps (SOMs) 83
    3.5.2 Support Vector Machines (SVMs) 85
    3.5.3 Random Forests (RFs) 86
3.6 Cluster Analysis 86
    3.6.1 Agglomerative Hierarchal Clustering (AHC) Methods 89
    3.6.2 Clustering Analysis Case Study 91
3.7 Novel Approaches to the Analysis of High-throughput Metabolomics Datasets 92
    3.7.1 Genetic Algorithms 95
    3.7.2 Gaussian Graphical Models 96
    3.7.3 Independent Component Analysis (ICA) 98
3.8 Multidimensional Data ($P > n$) Problems Encountered in MV Regression Modelling 101
    3.8.1 Regression Regularisation 102
    3.8.2 Model Tuning and Optimisation *via* an M-Fold Cross-Validation Process 102
    3.8.3 Principal Component Regression (PCR) 103
    3.8.4 Partial Least Squares Regression (PLS-R) 104
    3.8.5 Correlated Component Regression (CCR) 104
References 110

Chapter 4 Analysis of High-dimensional Data from Designed Metabolomics Studies 117
*Johan A. Westerhuis, Ewoud J. J. van Velzen,
Jeroen J. Jansen, Huub C. J. Hoefsloot and Age K. Smilde*

4.1 Introduction 117
4.2 Case Study 1: The Effect of Jasmonic Acid on the Production of Glucosinolates in *Brassicaceae oleracea* 119
    4.2.1 The ANOVA Model 121

	4.2.2	The ASCA Model	121
	4.2.3	Concluding the Glucosinolate Study	125
4.3	Case study 2: Metabolic Modifications Following Polyphenolic Intervention in Humans	125
	4.3.1	Multivariate Consequence	126
	4.3.2	The Multilevel PLSDA Model	128
	4.3.3	The Study Setup	130
	4.3.4	Analysis of Pooled Samples	130
	4.3.5	Dynamic Non-linear Analysis of the Urinary ^{1}H NMR Data	133
	4.3.6	Short Conclusion on Case Study 2	133
4.4	Conclusion	134
	Acknowledgement	135
	References	135

Chapter 5   Current Trends in Multivariate Biomarker Discovery   137
Darius M. Dziuda

5.1	Introduction	137
5.2	Common Misconceptions in Biomarker Discovery based on $p \gg N$ Datasets	138
	5.2.1	Univariate (Rather than Multivariate) Analysis	139
	5.2.2	Using Unsupervised (Rather than Supervised) Learning Algorithms	140
5.3	Feature Selection	142
	5.3.1	Search Models	143
	5.3.2	Search Strategies	143
	5.3.3	Stability of Results	144
5.4	Supervised Learning Algorithms	145
	5.4.1	Linear Discriminant Analysis	145
	5.4.2	Support Vector Machines	150
	5.4.3	Random Forests	155
5.5	Searching for Multivariate Biomarkers that are Robust and Biologically Interpretable	156
	5.5.1	Informative Set of Genes	157
	5.5.2	Modified Bagging Schema	158
	5.5.3	Identification of Parsimonious Biomarkers that are Robust and Interpretable	159
	References	160

Chapter 6   Discovery-based Studies of Mammalian Metabolomes with the Application of Mass Spectrometry Platforms   162
Warwick B. Dunn, Catherine L. Winder and Kathleen M. Carroll

| 6.1 | Introduction | 162 |

6.2   Mass Spectrometry Instrumentation   168
    6.2.1   Sample Introduction   169
    6.2.2   Ion Formation   170
    6.2.3   Mass Ion Separation According to Mass-to-charge Ratio   170
    6.2.4   Ion Detection and Data Acquisition   172
    6.2.5   Instrument Control and Data Processing   174
    6.2.6   Other Considerations   175
6.3   Sample Introduction Systems   176
    6.3.1   Direct Infusion Mass Spectrometry (DIMS)   176
    6.3.2   Gas Chromatography-Mass Spectrometry   177
    6.3.3   Comprehensive GCxGC-MS   180
    6.3.4   High Performance Liquid Chromatography-Mass Spectrometry   181
    6.3.5   Capillary Electrophoresis-Mass Spectrometry   184
6.4   Moving from Small-scale to Large-scale Metabolomic Studies   185
6.5   Concluding Remarks   188
Acknowledgements   188
References   188

Chapter 7   Recent Advances in the Multivariate Chemometric Analysis of Cancer Metabolic Profiling   199
Kenichi Yoshida and Martin Grootveld

7.1   Introduction   199
7.2   MV Chemometric Analysis of Cancer   202
    7.2.1   Infrared Spectroscopy   202
    7.2.2   Nuclear Magnetic Resonance Spectroscopy   203
    7.2.3   Mass Spectrometry   206
    7.2.4   Other Methods   207
    7.2.5   Further Considerations   208
7.3   Summary   208
References   209

Chapter 8   Group-specific Internal Standard Technology (GSIST) for Mass Spectrometry-based Metabolite Profiling   220
Jiri Adamec

8.1   Introduction   220
8.2   Basic Principles of GSIST   221
8.3   Application of GSIST   223
    8.3.1   Absolute Quantification Targeting Specific Functional Groups: Determination of Estrogens   223

8.3.2   Absolute Quantification Targeting
Multiple Functional Groups: Analysis of
Central Carbon and Energy Metabolism   227
8.3.3   Relative Quantification of Unknown
Metabolites in Complex Samples:
Determination of Triterpenoid Metabolomic
Fingerprints   234
8.3.4   Discovery of Novel Metabolites   240
8.4   Conclusion   247
Acknowledgements   247
References   248

**Chapter 9 ^{18}O-assisted ^{31}P NMR and Mass Spectrometry for
Phosphometabolomic Fingerprinting and Metabolic
Monitoring   255**
*Emirhan Nemutlu, Song Zhang, Andre Terzic and
Petras Dzeja*

9.1   Introduction   255
9.2   Methodology   259
9.2.1   Phosphometabolomic Platforms   259
9.2.2 ^{18}O Metabolic Labelling Procedure   261
9.2.3   GC/MS Analysis of ^{18}O-labelling of Metabolite
Phosphoryls   262
9.2.4 ^{31}P NMR Analysis of ^{18}O Incorporation into
Phosphoryl Metabolites   265
9.2.5   Phosphometabolite Analysis by ^{1}H-NMR   267
9.2.6   Data Analysis and Calculations of Phosphoryl
Turnover and Phosphotransfer Fluxes   267
9.2.7   Multivariate Statistical Analysis   269
9.3   Results   270
9.3.1   Phosphometabolomic Profiling of
Transgenic Animal Models   270
9.4   Conclusions   277
Abbreviations   278
Acknowledgments   278
References   279

**Chapter 10   Investigations of the Mechanisms of Action of Oral
Healthcare Products using ^{1}H NMR-based Chemometric
Techniques   287**
C. J. L. Silwood and Martin Grootveld

10.1   Introduction   287
10.1.1   High-resolution NMR Analysis of Human
Saliva   288

      10.1.2 Applications of Multivariate (MV)
            Statistical Techniques to the Interpretation
            of Salivary ^{1}H NMR Profiles   290
  10.2 Case Study: ^{1}H NMR-based Multivariate Statistical
      Analyses of Human Saliva Samples before and after
      Treatment with an Oxyhalogen Oxidant-containing
      Oral Rinse Product   293
      10.2.1 Materials and Methods   294
      10.2.2 Results   303
      10.2.3 Discussion   314
  10.3 Conclusions   316
  References   317

**Chapter 11 Metabolomics Investigations of Drug-induced
Hepatotoxicity**   323
Wei Tang and Qiuwei Xu

  11.1 Introduction   323
  11.2 Drug-Induced Liver Injury (DILI)   324
  11.3 Possible Mechanisms Underlying DILI   326
      11.3.1 Drug Metabolism and Elimination   326
      11.3.2 Direct Cytotoxicity and Immune-mediated
            Reactions   329
      11.3.3 Ambiguous Nature of Current
            Understandings of DILI   333
  11.4 Metabolomics   336
  11.5 Application of Metabolomics in Studies of DILI   339
      11.5.1 Mechanistic Investigation   339
      11.5.2 Searching for Biomarkers   342
  11.6 Summary and Closing Remarks   345
  Acknowledgements   347
  References   347

Chapter 12 Chemogenomics   357
*Virendra S. Gomase, Akshay N. Parundekar and
Archana B. Khade*

  12.1 Introduction   357
  12.2 Privileged Structures   358
  12.3 Drugs Arising from the Side-effects – SOSA
      Approach   359
  12.4 Classification of Chemogenomics   360
  12.5 Chemogenomics Screens   361
  12.6 Haploinsufficiency Profiling   362
  12.7 High-content Screening   363

12.8   Mode of Action by Network Identification   364
12.9   Current Research in Chemogenomics   364
12.10   Bioinformatics   365
12.11   Kinase Activity   365
12.12   Oncology   366
12.13   Ligand-binding Study   367
12.14   Metabolomics   367
12.15   Pharmacophore   367
12.16   Cheminformatics   368
12.17   Pharmacogenomics   368
12.18   Drug Safety   369
12.19   Evaluating Complex Signalling Networks   369
12.20   Current Trends in Chemogenomics   370
        12.20.1   Stem Cells   370
        12.20.2   Schistosomiasis   370
        12.20.3   Ligand-Enzyme Interaction   370
        12.20.4   Cytoscape Plug-ins   370
        12.20.5   Novel Screening Technologies   371
        12.20.6   Anti-HIV Drugs   371
12.21   Discussion   371
12.22   Conclusion   372
References   373

Subject Index   **379**

Introduction to the Applications of Chemometric Techniques in 'Omics' Research: Common Pitfalls, Misconceptions and 'Rights and Wrongs'

MARTIN GROOTVELD

Leicester School of Pharmacy, Faculty of Health and Life Sciences, De Montfort University, The Gateway, Leicester LE1 9BH, UK
Email: mgrootveld@dmu.ac.uk

1.1 Introduction

In this first chapter, I shall focus mainly on the two most widely employed multivariate (MV) assessment systems available in practice, specifically Principal Component Analysis (PCA) and Partial Least Squares methods, particularly Partial Least Squares-Discriminatory Analysis (PLS-DA), the first of which is an unsupervised exploratory dataset analysis (EDA) method, the second being a supervised pattern recognition technique (PRT). I have chosen to concentrate on these particular MV analysis methods here since there are numerous documented examples of the applications of these in the scientific, biomedical and/or clinical research areas in which they have sometimes been employed inappropriately, to say the least! Further details

Issues in Toxicology No. 21
Metabolic Profiling: Disease and Xenobiotics
Edited by Martin Grootveld
© The Royal Society of Chemistry 2015
Published by the Royal Society of Chemistry, www.rsc.org

regarding the principles and modular applications of these two MV analysis approaches are provided in Appendices I and II.

1.2 Principal Component Analysis (PCA)

The applications of Principal Component Analysis (PCA)[1,2] to the interpretation of MV metabolomic or chemometric datasets are manifold, and this is, perhaps, one of the most extensively applied techniques, examples of which are provided in refs 3–7, and which is sometimes employed in the first instance, if only for the detection and removal of statistical 'outlier' samples. The principles of this method involve the reduction of a large MV dataset (such as that arising from the 'bucketed' ^{1}H NMR analysis of, say, a collection of biofluid samples, tissue biopsies or their extracts, or otherwise) to a much smaller number of 'artificial' variables known as Principal Components (PCs), which represent linear combinations of the primary (raw) dataset 'predictor' variables and, hopefully, will account for at least some, if not most, of their variance. These PCs can then, at least in principle, be employed as 'predictor' or criterion (X') variables in subsequent forms of analyses. It is clearly a valuable technique to apply when at least some level of 'redundancy' is suspected in the dataset, *i.e.* when some of the X variables are correlated or highly correlated (either positively or negatively) with each another. In metabolomics experiments, it is often the case that one or more (perhaps many) biofluid metabolite concentrations (or proportionately related parameters such as a resonance, signal or peak intensity) will be significantly correlated with one (or more) others, either positively or negatively. Obviously, in such situations, many of the predictor (X) variables can be rendered redundant, and this forms the basis of the PCA technique in terms of its dimensionality reduction strategy.

PCA is a procedure that converts a very large number of 'independent' variables (more realistically described as 'interdependent' variables in view of their multicorrelational status), *i.e.* 0.02–0.06 ppm ^{1}H NMR spectral 'buckets' (which have variable frequency ranges if 'intelligently selected', and constant, uniform ones if not, the latter often being a pre-selected size of 0.04 or 0.05 ppm), many of which are correlated into a smaller number of uncorrelated PCs. Hence, a major objective of this form of multivariate analysis is to alleviate the dimensionality (*i.e.* the number of independent, possible 'predictor' variables) of the dataset whilst retaining as much of the original variance as possible. Hence, the first (primary) principal component is that which explains as much of the total variance as possible, the second as much of the remaining variance as possible, and so on with each succeeding PC until one with little or no contribution to variance is encountered; all components are, of course, orthogonal to (*i.e.* uncorrelated with) each other.

PCA can effectively delineate differing classifications within MV metabolomics datasets, and this is conducted according to the following procedure:

The data matrix is reduced to the much smaller number of PCs describing maximum variance within the dataset through decomposition of the

X predictor variable matrix (containing the integral NMR buckets) into T score (containing class information projections of sample data onto each principal component through displacement from the origin) and P loading (describing the variables that influence the scores) matrices, such that $X = t_1 \cdot p_1^T + \cdots + t_A \cdot p_A^T$, where the subscripted A value represents the total number of PCs, the residual information being included in a residual matrix E. The first PC should contain the maximum level of variance in the X matrix, such that the resulting deflated X matrix is then employed to seek a second component, orthogonal to the first, with the second highest variance contribution, and so on. PCA loadings with large values correspond to variables that have particularly high variance contributions towards them, and therefore they impart more to the total variance of the model system investigated.

However, there still remains much confusion regarding differences between the PCA and exploratory Factor Analysis (FA) techniques. Although similar in many respects (many of the stages followed are virtually identical), one of the most important conceptual differences between the two methods lies with the assumption of an underlying causal structure with FA (but not with PCA). Indeed, the FA technique relies on the assumption that covariation in the observed X variables is ascribable to the presence of one or several latent variables (or factors) that can (or do) exert a causal influence on the X variable dataset.[8,9] Indeed, researchers often use FA when they are perhaps aware of a causal influence of latent factors on the dataset (for example, the clear influence of thyroid disease status on blood plasma thyroxine levels, or a type 1/type 2 diabetes disease classification on blood plasma glucose and, where appropriate, ketone body concentrations), and this technique has been much more extensively employed in, for example, the social and environmental science areas rather than in metabolomics research; hence, an exploratory FA permits researchers to identify the nature, total number and relative influence of these latent factors.[10] Similarly, for sufficiently large MV datasets, the multiple FA (MFA) method serves to determine underlying relationships or 'signatures' between a series of causal latent variables and the MV dataset attained. In FA or MFA, we may also add the 'diagnostic' or other variables as supplementary ones rather than as latent causal factors.

For PCA, however, no prior assumptions regarding potential underlying causal latent variables are made; indeed, it is simply a dimensional alleviation technique that gives rise to a (relatively) much smaller number of (uncorrelated) PCs which account for as much of the MV dataset as possible (although the influence of or differences between such latent or explanatory variables are, of course, frequently investigated in a metabolomics sense).

Since PCs are defined as linear combinations of optimally weighted predictor (X) variables, it is possible to determine the 'scores' vectors of each one on each PC, which is considered significant (commonly determined *via* a Scree plot[11]). For example, the first PC may be primarily ascribable to selected metabolic differences between two (or more) disease classification groups, whereas the second may arise from a second series of perhaps unknown, unrelated metabolic perturbations, or alternatively a further influential (perhaps latent) variable such as dietary habit or history, or further differences between sample donors, for example those regarding gender, age, family, ethnicity status, *etc.* Figure 1.1 shows typical Scree plots arising from the metabolomic PCA of intelligently bucketed datasets arising from the ^{1}H NMR analysis of (a) human salivary supernatants (with 209 predictor variables, 480 samples and 2 oral health disease classifications) and (b) human urine (with only 22 predictor variables, 60 samples and again 2 disease classifications). For this latter example, we selected the most important bucket predictor variables *via* the prior performance of (1) model-directed repetitions (>60 times) of the logistic model of correlated component regression (CCR, as outlined in Chapter 3) with corresponding validation, cross-validation (CV) and permutation testing, and (2) selected computational intelligence techniques, again with accompanying validation, cross-validation and permutation testing. For these Scree plots displayed in Figures 1.1(a) and (b), and Tables 1.1(a) and (b), respectively, list the number of PCs with eigenvalues >1, and their corresponding eigenvalues (*i.e.*, the mean number of predictor X varoables per PC), together with the percentage of total variances accounted for by these PCs (the latter both individually and cumulative). From Figure 1.1(a) and Table 1.1(a), it can be observed that 14 PCs had eigenvalues >1, the first (PC1) with an eigenvalue of 121.66 (*i.e.* a mean value of 121.66 positively and/or negatively correlated predictor variables are responsible for it), the second 27 or so, the third 11 and the fourth 10, *etc.*; these first four PCs account for 58.2%, 12.85%, 5.3% and 4.9% of the total variance, respectively (total 81.2%). In Figure 1.1(b), however, only 8 PCs had eigenvalues >1, the first five accounting for only *ca.* 60% of the total variance. It should also be noted from Figure 1.1(b) that the Scree plot appears to have more than one simple break-point, the first after PC6, the second after PC12 (although PCs 9–12 are considered irrelevant since their eigenvalues are all <1). Therefore, for this latter example, it would appear that only PCs 1–6 should be considered as providing valuable MV information.

1.2.1   Critical Assumptions Underlying PCA

Now here's the difficult part! Indeed, this is where a lot of PCA applications to the analysis of metabolomics/chemometric datasets fall down, and hence fail or completely fail to provide satisfactory models for the diagnosis of human diseases, determinations of their severities, or responses to treatment, *etc.*

As with many alternative MV analysis techniques, the satisfactory application of PCA to the recognition of patterns or 'signatures' of metabolic biomarkers in metabolomics datasets (^{1}H NMR-derived or otherwise) is critically dependent on the satisfaction of a series of assumptions. Unfortunately, such assumptions are rarely checked, evaluated or monitored prior to the performance of PCA, and hence results acquired can hardly be considered as having a sound basis. However, as noted below, some of these

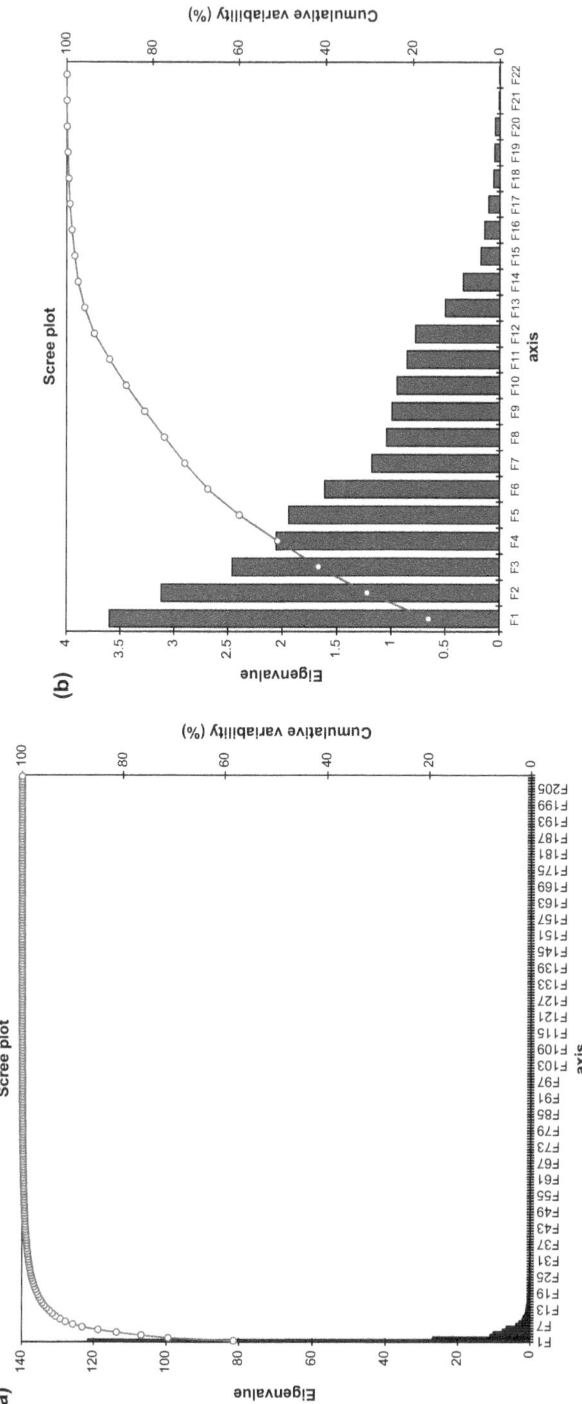

Figure 1.1　Scree plots for (a) exploratory human salivary supernatant ^{1}H NMR metabolomics dataset consisting of 2 oral disease classifications (healthy controls and clinically defined oral disease-positive), 480 samples (240 in each 'disease' group) and a total of 209 possible explanatory (X) variables, the latter comprising 'intelligently selected' ^{1}H NMR buckets with frequency ranges of 0.02 to 0.08 ppm, and (b) a human urinary ^{1}H NMR dataset (intelligently bucketed in the same manner) arising from 60 samples, 2 disease classifications (46 and 14 in each classification group) and only 22 predictor X variables (the latter were selected from a total of 222 original ones *via* a repeated and permutated correlated component regression (CCR) cross-validation process; details of the CCR technique employed are provided in Chapter 3). For the salivary supernatant dataset (a), resonance intensities were normalised to that of a specified pre-added concentration of a 3-trimethylsilyl [2,2,3,3-2H4] propionate (TSP) internal standard. Each column of the two datasets was subjected to standardisation (autoscaling) prior to data analysis. The eigenvalues of the PCs arising from these typical scree plots are listed in Table 1.1.

Table 1.1 Lists of eigenvalues, percentages of variance explained and cumulative percentage variabilities for the two intelligently bucketed ^{1}H NMR datasets specified in Figures 1.1(a) and (b), respectively.

(a)

PC	Eigenvalue	% Variance explained	% Cumulative variability
PC1	121.66	58.21	58.21
PC2	26.86	12.85	71.06
PC3	11.02	5.27	76.34
PC4	10.19	4.87	81.21
PC5	7.59	3.63	84.84
PC6	6.59	3.15	87.99
PC7	3.89	1.86	89.85
PC8	3.05	1.46	91.31
PC9	2.25	1.08	92.39
PC10	1.86	0.89	93.28
PC11	1.35	0.65	93.93
PC12	1.22	0.58	94.51
PC13	1.11	0.53	95.04
PC14	1.04	0.5	95.54

(b)

PC	Eigenvalue	% Variance explained	% Cumulative variability
PC1	3.60	16.36	16.36
PC2	3.12	14.17	30.53
PC3	2.46	11.18	41.71
PC4	2.06	9.35	51.06
PC5	1.94	8.82	59.88
PC6	1.61	7.32	67.20
PC7	1.18	5.35	72.55
PC8	1.04	4.75	77.30

assumptions are of much more importance than others, and the technique serves to be relatively robust to violations of the selected criteria required.

These assumptions are:

(1) Primarily, since PCA is conducted on the analysis of a matrix of Pearson correlation coefficients, datasets acquired should satisfy all the relevant assumptions required for this statistic.

(2) A random sampling design should be employed, and hence each biofluid, tissue or alternative sample should contribute one, and only one, value (specifically, metabolite concentration or related measure, normalised and/or standardised) towards each observed 'predictor' (X) variable; these values should ideally represent those from a random sample drawn from the population(s) investigated.

(3) All biomolecule predictor (X) variables should be evaluated on suitable concentration (or directly proportional spectroscopic or chromatographic intensity measures), concentration interval or concentration ratio measurement levels.

(4) Each predictor variable measurement (for example, concentration or signal intensity) should be distributed normally, and those that deviate from this (*i.e.* those that demonstrate a limited level of kurtosis or skewness) can, at least in principle, be appropriately transformed in order to satisfy this assumption.

(5) Each pair of predictor (X) variables in the plethora of those available in an MV dataset should conform to a bivariate normal distribution; specifically, plots derived therefrom should form an elliptical scattergram. Notwithstanding, Pearson correlation coefficients are remarkably robust against deviations from this assumption when the sample size is large (although this is often not the case in metabolomics experiments!). However, selected MV analysis techniques such as independent component analysis (ICA), which is covered in Chapter 3, also allow for quadratic or higher order polynomial relationships between the exploratory variables (although selected transformations of the dataset acquired may serve to convert such non-linear relationships to linear or approximately linear ones). An example which describes the application of a series of four such tests of normality for a large number of predictor X variables within a ^{1}H NMR multivariate 'intelligently bucketed' urinary dataset is provided in Chapter 2. Appropriate transformations for the conversion of such non-normally distributed X variable datasets include the logarithmic ($\log_{10}$- or $\log_e$-) transformation for variables in which the standard deviation is proportional to the mean value (in this case, the distribution is positively skewed); the square root transformation for variables in which the estimated variance (s^2) is proportional to the mean (which frequently occurs in cases where the variables represent counts such as the number of abnormal cells within a microscopic field, *etc.*); the reciprocal transformation for variables with standard deviations proportional to the square of the mean (this is usually applied to highly variable predictors such as blood serum creatinine concentrations); the arcsine $(\%)^{1/2}$ transformation for variables expressed as percentages, which tend to be binomially distributed (this transformation is likely to have some application to MV metabolomic datasets which have been normalised to a constant sum (say 100%) both with and without their subjection to the subsequent standardisation pre-processing step, details of which are provided in Chapter 2). Of course, the standardisation process (involving mean-centring and unit-variance scaling), will provide variables with mean values of zero and standard deviations and variance values of unity, and hence the performance of such transformations may be considered inappropriate). However, this standardisation process will certainly *not* achieve the conversion of a significantly skewed distribution into a non-skewed, symmetrical and perfectly normally distributed one!

(6) Watch out for outliers! The presence of even just one outlying data point can sometimes give rise to a strong (but overall false!) apparent

correlation between, say, two metabolite levels, even if the complete dataset has been subjected to normalisation (row operation) and standardisation (column operation) procedures. Figure 1.2 shows an example of how this might arise. In addition to checking for outlying biofluid or tissue samples, which can easily be achieved by examinations of two- or three-dimensional PCA scores plots (such samples may occur from their collection from study participants taking or

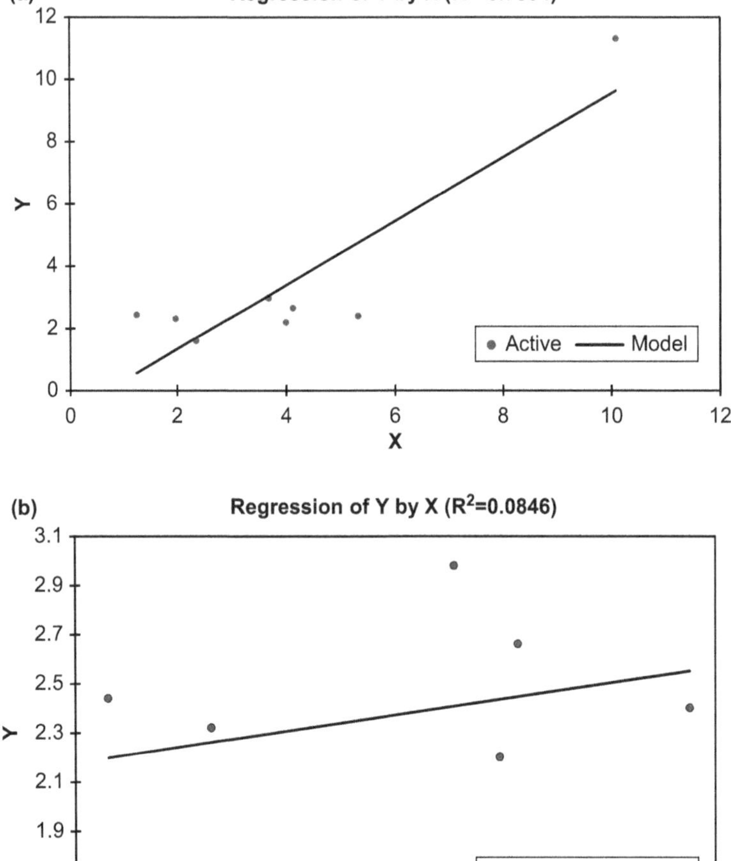

Figure 1.2 (a) Apparent correlation between Y and X arising from the effects exerted by a single 'outlying' data point. (b) Corresponding plot arising from the removal of this outlying data point; this removal substantially diminishes the R^2 value (square of the Pearson correlation coefficient) of this relationship from 0.7884 ($p = 0.0032$) to 0.0846 ($p = 0.5268$, ns).

receiving project- or clinical trial-unauthorised medication, or further programme-prohibited agents such as alcoholic beverages, for example), researchers should also endeavour to check *all* the predictor variables individually for such outlying data points, and perhaps remove them if proven necessary. In this manner, we can at least be confident that each predictor variable (column) dataset is outlier-free and will not be violating the 'no-outlier' assumption.

1.2.2 Number and Significance of Explanatory Variables Loading on a PC

When one or more explanatory variables, biomolecular or otherwise, load on a principal component, it is highly desirable for researchers to have an absolute minimum of three or so of these X variables per component; indeed, it is generally considered good practice to retain five or more of these variables per component, since some of these may be subsequently removed from the diagnostic criteria developed. However, in metabolomics datasets consisting of perhaps 200 or more of such variables (such as those generated from the high-resolution ^{1}H NMR or LC-MS analysis of selected biofluids), it is not uncommon to encounter PCs that contain as many as 100–1000 or more of these X variables, which are all correlated (positively and/or negatively), and hence have autonomy and perhaps independence regarding their contributions to successive PCs, *i.e.* those which account for less and less of the total variance encountered in the dataset.

A further important consideration is whether or not a particular potential (biomolecular) explanatory X variable significantly loads on a specified component: this is generally considered the case if its PC loading value is >0.40. It is the author's view that these loadings should be checked and monitored more closely during the MV analysis of large or very large metabolomics datasets, since this does not seem to occur very often in the extensive range of publications available which have been extensively surveyed by the author! However, if indeed this is the case, then the predictor (X) variable can be considered as one which significantly contributes to a particular PC.

1.2.3 Number of Extractable PCs and Their Characteristics

A PC is defined as a linear combination of 'optimally weighted' predictor (X) variables, and here 'optimally weighted' indicates that these variables are weighted in such a manner so that the PCs arising therefrom account for the maximal proportion of variance in the complete dataset; the 'linear combination' descriptor refers to the information that the particular scores on a component are generated by a simple simulation of those on each X variable.

In many PCAs performed on metabolomics datasets (^{1}H NMR-derived or otherwise), usually it is only the primarily extractable components (say, up to

6, but this value can often be as many as 10 to 20) which qualify for retention, further interpretation and employment in any further forms of analyses (MV or alternative methods). The remaining PCs (which can, in principle, represent a very large number from typical metabolomics datasets containing 100–200 or more X variables) are likely to account for only trivial levels of the complete X variable dataset variance, and hence may be removed from the analysis. Of particular importance is the deletion of those PCs which have eigenvalues <1, *i.e.* those with an average number of <1 predictor variable per component.

The first PC derived from the PCA of a metabolomics dataset will, of course, account for a maximal quantity of the total variance in the observed predictor (X) values, and hence it will also be significantly correlated with at least some (perhaps as many as 100 or so) of them. However, the second PC, which will account for the second largest percentage of such variance (and which was not accounted for by the first PC), is correlated with a smaller number of X variables that did not exhibit strong correlations with PC1. One major property of PC2 (and also subsequent PCs, *i.e.* PC3, PC4, PC5, *etc.*) is that it will be completely uncorrelated with PC1, *i.e.* the two PCs are *orthogonal*. Of course, the remaining PCs account for lower and lower percentages of the total X dataset variance and, again, they are all uncorrelated with each other, together with the first two (primary) PCs.

1.2.4   Total Variance of the Dataset

Since each of the observed variables is, in general, standardised during the course of PCA (although, for particular reasons, not always!), and each X variable therefore has a mean value of zero and unit variance (and hence unit standard deviation), the total variance of the dataset is therefore the sum of the observed variables' variances, and hence is equivalent to the number of X variables subjected to analysis in this manner. As an example, if 180 X variables are being considered and analysed, the total variance will be 180, and the extracted PCs effectively partition this variance, with PC1 perhaps accounting for 23 total variance units, the second PC (PC2) perhaps for 13 units, and so on, and the PCA proceeds until all the dataset variance has been accounted for (although realistically it should be terminated when one of the eigenvalues has a <1 value).

1.2.5   What is an Adequate Sample Size for PCA and Further Forms of MV Analysis?

Basic PCA theory suggests that, since the method is designed as a large or very large sample process, the minimum number of samples subjected to analysis (by ^{1}H NMR, FTIR or LC-MS techniques, for example) should be the larger of 100 or 5 times the number of 'predictor' X variables. Therefore, if we have a ^{1}H NMR dataset with 200 or so resonance intensity buckets ('intelligently selected' or otherwise), then we should, at least in principle, have a sample size of 1000 or more! This clearly has implications for many such metabolomic investigations – indeed, the author has often seen many examples in the scientific or clinical research areas where the disease or response status of a series of biofluid samples have been 'correctly' classified from datasets containing only 20–30 or so samples (rows), sometimes as few as 10–12, and the number of intensity buckets (columns) approaches or is greater than 200! This sample size problem represents a major assumptive criterion in this research area, and many researchers clearly fail to allow for this, a factor which can regularly give rise to the 'overfitting' of experimental datasets to selected models in many further forms of MV statistical analysis (particularly the supervised PLS-DA technique, which has a reputation for being 'over-eager to satisfy'!).[12,13] However, PCA is somewhat less susceptible to this problem since it is an unsupervised EDA technique.

As expected, if, in an experimental design, we select, say, 300 participants to serve as donors for a particular biofluid sample (with adequate control for the potential interference of xenobiotic agents), it is highly, if not extremely, likely that one or several of these may not be able to provide samples (or, for that matter, insufficient volumes of them), and hence they will not enter into the final analysis; a finite number of participants can always be expected to fail to provide specimens under the required pre-specified conditions of the experiment, and/or at the correct time-points, if appropriate (as specified in a Participant Information Sheet approved by the particular Research Ethics Committee involved). Therefore, it is always sensible to recruit a larger number of participants to the study (*via* its experimental design), say 350 in this case in order to allow for this.

It should also be noted that these sample size criteria only represent minimum (lower level) requirements, and some researchers have made strong arguments that they should only be applicable if, firstly, many X ('predictor') variables are expected to load on each contributory PC, and, secondly, if the variable communalities are high, specifically if a particular X variable loads substantially on at least one of the retained PCs.

1.2.6   Interpretability Criteria of PCs

For each retained PC, it is of much importance to confirm that our interpretation of them makes 'metabolic sense' regarding the nature of the explanatory X variables employed, *i.e.* those which are found to load on each component. Basic selection criteria for this include requirements for (1) an absolute minimum of three variables, each with significant loadings on a particular retained PC; (2) the variables significantly loading on a selected PC sharing the same conceptual (metabolic) interpretation, *i.e.* perhaps these loadings on a selected PC arise from or relate to a disturbance in a particular metabolic pathway (perhaps only partially)?; (3) the differing X variables loading on differing PCs to reflect differing constructs (*e.g.* if five metabolites load significantly on PC1, and four further ones load significantly on PC2, do the first five PC1-loading variables appear to reflect a construct that is, in principle, different from those loading on PC2?).

Further considerations are that we should employ the minimum eigenvalue of ≥ 1.0 for each PC (especially since if a PC has an eigenvalue of <1, then an average of <1 X variable contributes towards it, and hence it is of no significance or consequence!), and also we should realistically determine the 'break' in the curve from the Scree plot acquired (often, these are unclear!). Since there can frequently be more than one such break in a Scree plot, the consideration of more than one possible solution may be required. It is also accepted that the combined retained PCs should account for a minimum of 70% of the cumulative variance; indeed, if $<70\%$ is covered, then it is recommended that alternative models with a larger number of PCs should be considered, perhaps those also including quadratic and/or multinomial representations of one or more of the potentially very many X variables.

1.2.7 Varimax Rotation

The rotated factor (PC) pattern should demonstrate a relatively 'simple' structure, *i.e.* (1) a range of the X variables should exhibit high loadings on only one retained PC, and near-zero ones on further PCs, and (2) most retained PCs or factors should demonstrate relatively high PC loadings for some X variables, and hopefully near-zero ones for the remainder.

Both PCA and FA primarily extract a series of components (otherwise known as factors) from a dataset, and these factors are predominantly orthogonal, and their relative importance is ordered according to the percentage of the total variance of the original dataset that these components account for.

However, generally only a (small) sub-set of these components is retained for further consideration, the remaining ones being considered as either non-contributory or non-existent (for example, in ^{1}H NMR-linked metabolomics analysis, they may arise from measurement error or 'noise').

So that we can interpret the PCs/factors that are considered relevant, it is important that the preliminary selection step is succeeded by a 'rotation' of the PCs that were primarily isolated and retained. There are two major classes of rotation employed, specifically orthogonal (in which the newly constructed axes are orthogonal to each other), and oblique (in which there is no requirement for the new axes to be orthogonal to each other). Since the rotations are conducted in a sub-space (known as the component or factor space), these new axes are always explicable by a lower level of variance than the original components/factors (which are, of course, optimally computed), but the portion of variance explicable by the total sub-space following rotation remains the same as it was prior to rotation (*i.e.* only the variance partition has been modified). Since the rotated axes are not defined according to a pre-specified statistical inference, their major focus and advantage is to assist interpretation of the results acquired.

Since these rotations take place in a sub-space (specifically the retained component/factor space), it must be optimally chosen, since this sub-space selected powerfully influences results arising from the rotation. Therefore, a range of sizes for the retained factor sub-space should be explored in order to evaluate the robustness of the rotation's final interpretation.

In general, the initial matrix is not interpreted, and the PCs/factors are rotated to generate a more parsimonious solution, in which each variable has a new combination of high and low loadings across the factors involved. The interpretation of this form of PCA or FA involves an identification of what is common amongst the variables which load highly on a particular component/factor (perhaps a chemopathological disturbance in a selected metabolic pathway), and what distinguishes them from those having low loadings on that particular one.

1.2.8   Example Case Study

In this experimental PCA case study example, I attempt to relate a salivary ^{1}H NMR metabolomics dataset to a single classification model, the classification being the presence or absence of a particular oral health condition (*i.e.* healthy controls *versus* active disease qualitative classifications). The original dataset consisted of 209 'intelligently selected' ^{1}H NMR bucket variables, and from Figure 1.3(a) it can be clearly observed that there are no visually apparent classification distinctions observed in three-dimensional (3D) interactive scores plots of PC3 *vs.* PC2 *vs.* PC1. However, three further, highly correlated 'false-dummy' latent variables (with scores ranging from 0 to a maximum value of 10) were then introduced into the experimental design model (correlational details of which are provided in Table 1.2), and supplemented to the original dataset in a stepwise fashion, so that there were 210, 211 and finally 212 explanatory (X) variables in the 'revised' dataset; for these added variables, it was ensured that each one was strongly (Pearson) correlated to an assigned binary 'disease classification' score of 0 for no disease activity (*i.e.* the healthy control group) and 1 for the oral disease classification group. There was only a relatively small number of

Table 1.2   Pearson correlation matrix between the three 'false-dummy' X (predictor) variables which were sequentially introduced into the MV salivary ^{1}H NMR dataset subjected to PCA, as outlined in Figure 1.3. The correlations of these three variables with an arbitrarily assigned 'real' dummy variable (*i.e.* disease score, comprising values of 0 and 1 for healthy control and oral disease-active patients, respectively) are also provided. Each of these correlation coefficient (r) values are statistically significant at a *p* value of <0.0001.

Correlation matrix (Pearson):

Variables	X1	X2	X3	Disease score
X3	0.9238	0.9412	1	0.8908
Disease score	0.9780	0.9448	0.8908	1
X2	0.9723	1	0.9412	0.9448
X1	1	0.9723	0.9238	0.9780

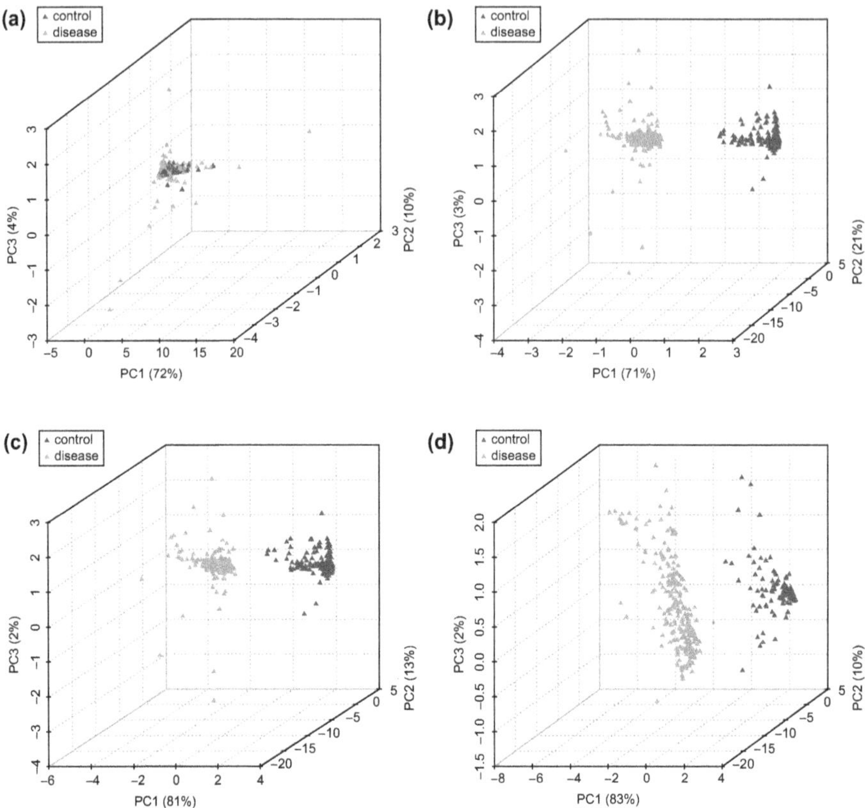

Figure 1.3 (a) PC3 *versus* PC2 *versus* PC1 scores plots of an 'intelligently bucketed' human salivary supernatant dataset comprising 2 oral disease classifications, 480 specimens (240 in each group) and 209 putative predictor X variables (the cubed-root transformation was applied to the dataset, followed by Pareto scaling prior to the performance of MV statistical analysis). Corresponding PC3 *versus* PC2 *versus* PC1 scores plots arising from the supplementation of this dataset with 1 [(b)], 2 [(c)] and 3 [(d)] highly correlated 'false-dummy' predictor variables, which were also highly correlated with an assigned 'dummy' disease score variable of 0 for healthy control and 1 for oral disease-active participants (a score which was *not* included in the potential predictor variable dataset); Pearson correlation coefficients for the relationships between these 'false-dummy' variables are provided in Table 1.2.

significant Pearson correlation coefficients between these dummy variables and those of the original, unsupplemented ^{1}H NMR bucket variables.

Figures 1.3(b)–(d) exhibit interactive 3D scores plots of the models in which there were 1, 2 and 3, respectively, of these 'false-dummy' variables added sequentially, and the classification status was either 'healthy control' or 'oral disease-positive' patients. Clearly, the introduction of these three new variables gives rise to major differences in the levels of discrimination between the two disease status classifications. Indeed, the level of 'Between-Disease

Classifications' distinction between these four datasets clearly increases with increasing number of 'false-dummy' variables included, although the inclusion of only one of them gives rise to a satisfactory level of discrimination between them.

1.2.9   Examination of a Wider Range of Components

A further important point for consideration is the knowledge that, more often than not, one or more of the PCs or factors which account for only a relatively small percentage of the overall dataset variance can be responsible for and hence reveal major distinctions between the subsequently specified supplementary PCA classifier variables, and may also serve to offer much valuable information regarding specific biomarkers available in the dataset. Indeed, many researchers involved in the metabolomics research area simply investigate and plot the first few (strongest) PCs against one another in an attempt to seek and detect any significant discriminatory potential amongst the classification groups, and, in view of this, are sometimes disappointed! The author therefore recommends that investigators should first perform dataset-constrained univariate significance testing procedures, *i.e.* t-tests and ANOVA, the latter containing and also considering as many latent sources of variation as possible, together with those ascribable to their possible first- or second-order interactions; this constraint can be implemented *via* the attainment of a Bonferroni-corrected p value for testing the significance of the source of variation of major interest, that 'Between-Disease Classifications', for instance.

In this manner, researchers may select putative metabolic biomarkers which exhibit the most highly significant differences 'Between-Classifications' or otherwise, and then search for these and their loadings on (contributions towards) PCs up to the first 10, 15 or even 20 of these PCs (linear combinations), provided that they all have eigenvalues ≥ 1, and that they all significantly contribute towards the total dataset variance, albeit in a relatively small manner. This approach, which involves a relatively unique combination of both univariate and MV analytical approaches, serves to inform us about small numbers of metabolic biomarkers which are not included as major or substantial contributions to the first few PCs (PC1, PC2 and PC3, *etc.*), and one or more of the biomolecular signals loading on which may also serve as major discriminatory indices between two or more disease classification groups.

Figure 1.4 shows an example of this, which exhibits a plot of PC9 *versus* PC8 from an experiment in which three predictor ^{1}H NMR 'intelligently selected' bucket intensities loaded substantially on PC8 and PC9 (each with loading values of *ca.* 0.40 on PC8, and percentage contributions of 12.1, 12.1 and 11.8% towards it, *i.e.* these three variables alone accounted for >35% of the total variance of this component); this experiment also involved an exploration of the metabolic classification of human saliva specimens into two classification groups (for this example, healthy control participants *versus* those with a further known oral health condition); the eigenvalues of PC8

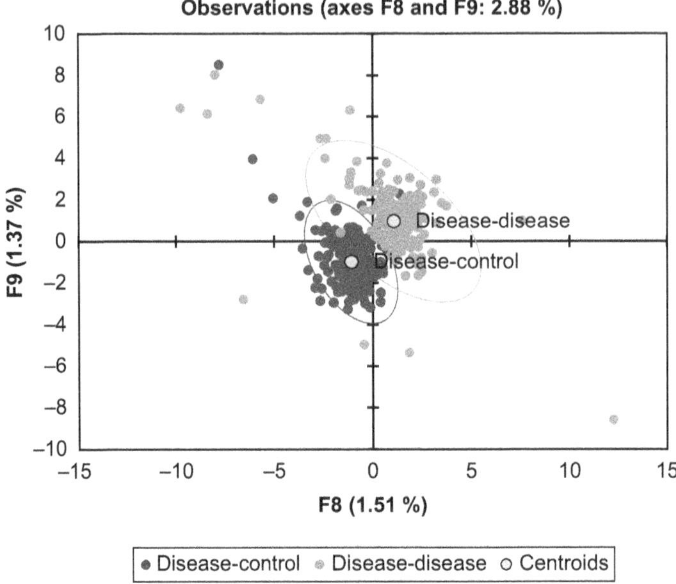

Figure 1.4 Plot of PC9 *versus* PC8 (with eigenvalues 3.20 and 2.91, respectively) for an 'intelligently-bucketed' human salivary supernatant ^{1}H NMR MV dataset (consisting of 2 classifications, 204 explanatory X variables and a total of 428 samples), which reveals discriminatory potential for 3 X predictor variables which exert a particularly high loading on PC8 in this model. Centroids and 95% confidence ellipses for the two disease classification groups are indicated (the former outlined as central circles). This TSP-normalised dataset was also subjected to autoscaling prior to data analysis.

and PC9 were 3.20 and 2.91, respectively, *i.e.* approximately three explanatory X predictor variables loaded on each one); there were 204 potential explanatory X variables and a total of 428 salivary supernatant samples involved in this model system. Allowing for the presence of a number of 'outlier' samples (as noted above, this serves as an efficient means of 'policing' clinical trials, for example the detection of samples containing exogenous agents such as drugs, oral healthcare product agents or further 'foreign' exogenous agents, for example, in participants who are not rigorously adhering to clinical trial protocols), it is clear that there is a major distinction between the two disease classification groups, with the disease one having a centroid with positive scores for PC8 and PC9, and the healthy control one with a centroid which has a negative score for both these PCs (95% confidence ellipses for these two classification groups are also exhibited).

1.2.10   Consideration of Type I (False-Positive) Errors

If, as in the PCA, PLS-DA, Partial Least Squares-Regression (PLS-R) techniques, or a range of further MV analytical methods which are based on the preliminary computation of a Pearson correlation (and covariance) matrix, we primarily generate such a matrix of, for example, $200 \times 200 = 4000$ Pearson correlation coefficients for an experimental design incorporating 200 predictor (X) metabolic variables which are generated *via* ^{1}H NMR or LC-MS analysis, and if we specify a significance level of $p = 0.01$, then we will achieve an average of 40 stunningly significant correlations purely by chance alone! Furthermore, if our p value was more liberally set to a value of 0.05, this probability-mediated number of significant correlations would escalate to no less than 200! These considerations are outlined in more detail in Chapter 2.

1.2.11 Determinations of the Suitability of MV Datasets for Analysis with PCA and FA

How exactly do we determine whether or not PCA or FA is appropriate for application to our MV metabolomics, proteomics or genomics datasets? Well, firstly we may employ the Kaiser–Meyer–Olkin (KMO) measure of sampling accuracy, and this method serves to provide essential information regarding whether or not the magnitudes of the *partial* correlations measured amongst variables are sufficiently low. If two variables share a common PC or factor with a series of further variables, their partial correlation coefficient ($\acute{r}_{ij}$) will be low, and this criterion will serve to inform us of the 'unique' variance shared between them [however, readers should note that such *partial* correlations, and their further application to the analysis of MV datasets, for example as in Gaussian Graphical Models (GGMs), are outlined in more detail in Chapter 3].

Critical considerations include whether or not the relationships existing between the predictor (X) variables are strong enough, and are we therefore confident in proceeding with the application of a PCA or FA model to the dataset? Indeed, this KMO test represents an index for comparisons of the magnitudes of the observed (Pearson) correlation coefficients to those of the *partial* ones [eqn (1), in which r and $\acute{r}$ depict the Pearson and *partial* correlation coefficients respectively, the latter equivalent to $r_{ij \bullet 1,2,3,\ldots k}$]. Hence, if $\acute{r}^{2}_{ij} \approx 0$, then the KMO statistic ≈ 1 and we may conclude that the predictor variables explored serve as representative measures of the same PC or factor, whereas if $\acute{r}^{2}_{ij} \approx 1$, then the variables involved are not considered to be expressing measurement of the same PC or factor. Hence, high values attained for the KMO statistic indicate that application of PCA or FA models to datasets acquired are acceptable approaches for their analysis (an absolute minimum value of 0.50 is preferable). Generally, such models are considered exceptional if its value is >0.90, very good if its magnitude lies between 0.80 and 0.90, good for values between 0.70 and 0.80, mediocre for values within the 0.50–0.70 range and unacceptable if <0.50.

$$\text{KMO test value} = (\Sigma\Sigma\ r^{2}_{ij})/[(\Sigma\Sigma\ r^{2}_{ij}) + (\Sigma\Sigma\ \acute{r}^{2}_{ij})] \tag{1}$$

However, a further method of determining the strength of the relationships amongst the predictor variables is Bartlett's Sphericity Test, which simply evaluates the null hypothesis that the variables present within the whole population's correlation matrix are uncorrelated, *i.e.* that the inter-correlation matrix is derived from a population in which the X variables are non-collinear (specifically an identity matrix). This test computes the determinate of the matrix of the sums of products and cross-products which generate the inter-correlation matrix. This matrix determinate is then tested for its statistical significance *via* a Chi-squared statistic.

1.3 Partial Least Squares-Discriminatory Analysis (PLS-DA)

Typical metabolomics profiling investigations involve two or more classifications of participants (human, animal, plant, cell or otherwise), and when there are only two of these, they can be divided into disease case *versus* healthy control or perhaps treatment *versus* untreated control groups. These investigations can be performed in either an exploratory or a predictive manner: the former is focused on whether the dataset acquired contains a level of information which is sufficient for us to discriminate between the two classifications, whilst the latter's objective is to determine whether or not we can predict whether an unknown sample can be successfully classified into one of these two (or more) groups, and, if so, to what level of confidence, exactly?

Partial Least Squares Discriminatory Analysis (PLS-DA) is based on the Partial Least Squares (PLS) model (Appendix I) in which the dependent (Y) variable represents membership of a particular classification (*e.g.* diseased *versus* healthy control, *etc.*), and since common metabolomics experiments contain a very large number of resonances, signals or peaks representing a multitude of biomolecules (at least some of which may serve to be valuable biomarkers of diseases and perhaps also their activities), these considerations can sometimes present many perplexing choices for mathematical modelling, validation and CV options. Indeed, as noted above for PCA, the minimum sample size required for a satisfactory model increases substantially with the number of variables monitored, and since the number of samples provided for analysis and/or sample donors is frequently somewhat or even much lower than the number of predictor (X) variables incorporated into the model, this leads to many validation challenges. These problems arise in view of the increasing likelihood of models with (apparently) effective group classifications which are generated purely by chance alone (*via* the now increasingly recognised 'overfitting' problem)!

Hence, a recommended means for the MV analysis of any metabolomics dataset is to employ a (relatively large) series of randomly classified datasets in order to establish the reliability and precision of the model's predictive capacity, and Westerhuis *et al.* (2008)[12] have provided some valuable and convincing examples of the importance of classifying metabolomic datasets according to a series of selective rules which involve the prior random permutation of disease and/or treatment classifications for PLS-DA models and, consequently, related binary score values for PLS-regression (PLS-R) ones.

Since far too many metabolomics investigations seem to involve far too many predictor (X) variables, and perhaps far too few analysed samples, these arguments are very true and valid; indeed, many of the studies reported in the relevant scientific literature can involve as many as 200–1000 or more X values, and perhaps as few as 20–40 samples for such MV analysis! Under such circumstances, employment of the above PLS models will nearly always give rise to a perfect clustering/separation of the two (or more) classifications investigated. A reputable and perhaps famous quote by Snedecor in 1956[14] is that there will be a perfect predictive fit between a single dependent variable and six 'predictor' variables in an experimental model which also contains only six samples, with measurements provided for the dependent variable and also the six independent ones. This is probably the best statement regarding the overfitting 'curse of dimensionality' that I am aware of, and, unlike many metabolomics investigations, this example only involves the fitting of six predictor variables to an equivalent number of cases. Therefore, metabolomics experimenters should carefully consider this fact when attempting to 'fit' as many as hundreds or even thousands of X variables to the bioanalytical profiles of as few as 20 or so biofluid or tissue biopsy samples collected!

The validation and CV of PLS-DA and PLS-regression (PLS-R) models is indeed an area of serious concern, and a number of pertinent reviews published have revealed that acceptable methods for processes are either lacking or not even attempted,[15–18] and have also outlined the most important problems associated with it. Indeed, as delineated above, one of the most important of these considerations is a very limited sample size; in view of the high economic cost of acquiring multicomponent bioanalytical profiles on biofluids and/or tissue sample biopsies (including the collection of the sample itself), this is very often the case!

In order to effectively evaluate the results acquired, CV processes can be performed; however, Anderssen *et al.* (2006)[19] have noted that very often the methods selected for these are either erroneous or, for that matter, not performed in correct or acceptable manners. Indices which are frequently employed to quantify the effectiveness of classification selection criteria include (1) simply the number of misclassifications, (2) the ubiquitous (and sometimes mysterious) Q^2 value, which indicates the variation of predicted values and hence the quality of prediction (the range is 0 to 1, where values of 0.50 and 0.90 are considered good and excellent, respectively), and (3) a wide range of criteria-determining sums and/or ratios of correct and false-positives and/or -negatives of what is classically known as a confusion matrix. Furthermore, it is also common for researchers involved in this area to provide the Area Under the Receiver Operating Characteristic Curve (AUROC). If this value is close to 1.0, then the classification criteria are viewed as 'good' or even 'excellent', whereas if it is close to 0.50, then the classification function employed is considered to be of very little or zero use. However, it is important to note that there remains a major problem with all of these model efficacy evaluation measures: the value corresponding to a high level of classification efficacy is unknown, and p values for the statistical significance of the discriminatory effects observed are rarely provided (in any case, such a value is critically dependent on the number of samples placed in both the 'training' and 'test' sets).

Most of us are already aware that models constructed from routine or even especially selected CV techniques can contain differing numbers of PLS components and, for that matter, differing 'significant' predictor variables with different loading coefficients for each sub-set of these models, and with the exception of a number of recent developments in this area (particularly those involving random permutation testing and determinations of the statistical significance of such evaluations), for example Westerhuis *et al.* (2008),[12] there are currently no or very limited acceptable criteria for this. This is, of course, of much significance regarding the transference of such information from the sub-set of models to the full dataset, and, more importantly, for its future application to the diagnosis, and perhaps severity determination and monitoring of the chemopathologies of disease processes and their treatment regimens. Moreover, what is the clinical significance of these models?

1.3.1 Case Study Describing an Example of PLS-DA 'Overfitting'

Here, a typical example of the 'overfitting' of datasets by PLS-DA to a model employing 222 predictor X variables (intelligently selected chemical shift bucket intensities of the ^{1}H NMR profiles of human urine, normalised and standardised *a priori*) and only 20 biofluid samples is described. This may appear to be statistically unacceptable to many readers (and of course, it is!), but this form of experimental design and MV analysis is not that uncommon in the scientific/biomedical literature!

For this experiment, the classification groups of the healthy control and disease classifications (10 in each group) were randomly permuted 30 times, and then PLS-DA was performed on each of these permuted classification status sets.

Figure 1.5 shows PLS-DA scores (t_2 *versus* t_1) plots for six of the PLS-DA sample classification permutations tested in this manner. Clearly, there are very high levels of sample classification clusterings and hence discrimination notable for each of these examples, and these results provide ample evidence for the overfitting of a very large number of predictor (X) variables (222) to a statistically small sample size (n = 20) using this technique; the acquisition of false-positive results in this manner is not that unusual in the metabolomics/scientific literature. However, out of the complete set of 30 random permutations of the sample classification status, models

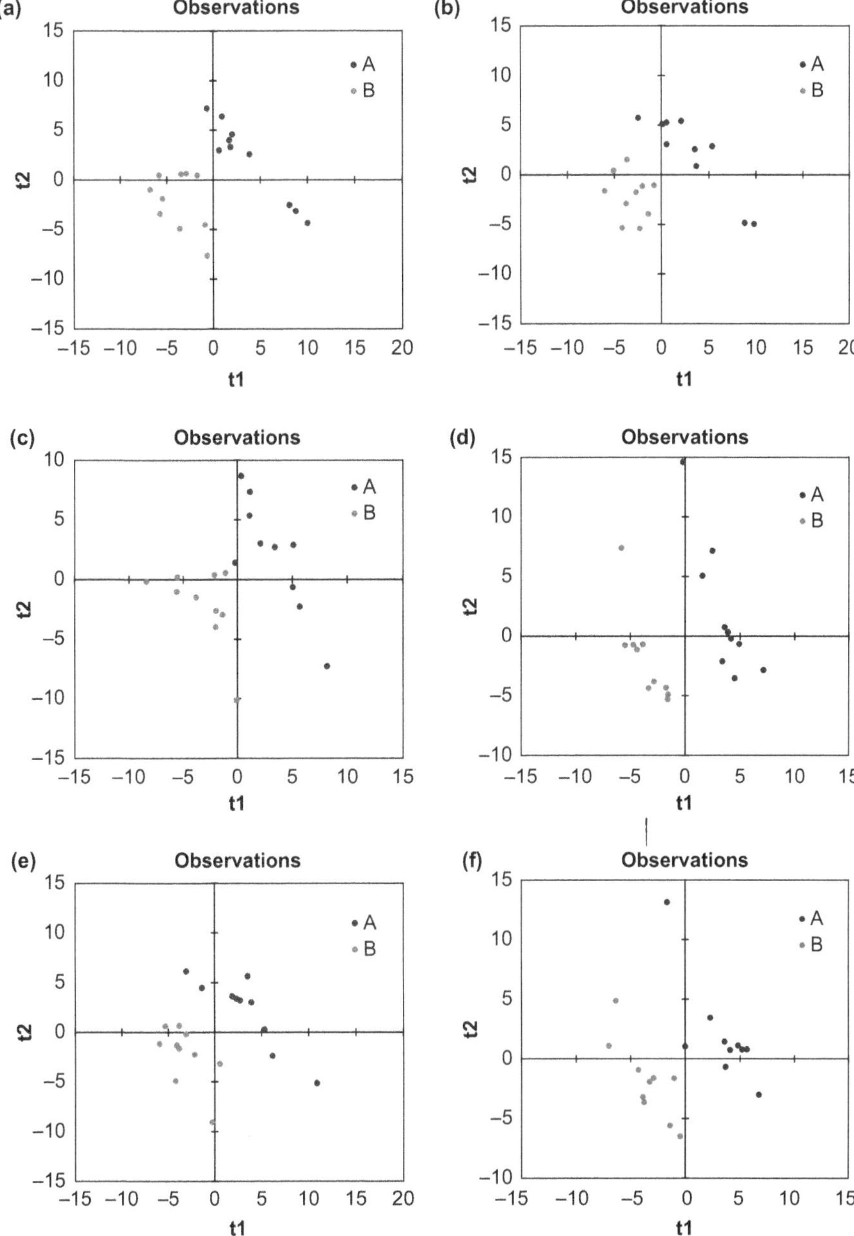

Figure 1.5   (a)–(f) Six typical examples of randomly permuted PLS-DA t_2 *versus* t_1 scores plots arising from an experimental model involving the random permutation of classification groups amongst 20 samples (10 in each group were specified) for a ^{1}H NMR urinary metabolomics dataset (intelligently selected buckets) consisting of 222 predictor X variables. A total of 30 randomly permuted classifications were performed for this experiment.

with only a single PLS-DA component were constructed in 13 cases ($Q^2 = -0.070 \pm 0.272$, mean $\pm$ SD), and those with two components were built in 14 cases ($Q^2 = -0.075 \pm 0.308$ and 0.081 ± 0.328 for the first and second components, respectively). Moreover, a further one of the sample classifications had a total of five components (with Q_1^2 and $Q_2^2 = 0.273$ and 0.403 for the first and second components, respectively), and another had as many as seven (with Q_1^2 and Q_2^2 values of 0.005 and 0.061, respectively)!

1.3.2 Permutation Testing

Similarly, for the above case study, a series of permutation tests was employed in order to explore relationships between the full set of 222 ^{1}H NMR bucket 'predictor' variables and the hypothetical disease classification status. This rigorous testing system serves to determine whether or not the disease status classifications of the study participants is significantly improved over that arising from any other random classification of these groups; the class labels of the healthy control and disease classifications are permuted, and then randomly assigned to different patients. With these 'incorrect' disease class labels, a classification model was again computed. Hence, the rationale was that for these 'incorrect' class labels, the computed model for classification purposes should be ineffective at class prediction (since the groups are generated randomly, the null hypothesis is that there are no differences between them). With repetition of this permutation test many times (2000 times individually for each of the initially randomly assigned class labels, *i.e.* an overall two-phase randomisation process), a null distribution of classifications which are expected to be insignificant was formed, and if the computed *pseudo*-F statistic lies outside at least the 95% or 99% confidence bounds of this distribution for 'real', genuine classification labels, then it could be concluded that there is a significant (linear) relationship between the X predictor variables and classification status.

For 52 out of a prior 56 randomly-permuted class labels, the *pseudo*-F value statistic was not significant (*i.e.* $p > 0.050$: Figure 1.6); the four that were significant had p values of 0.0495, 0.0375, 0.024 and 0.011). Therefore, with a significance value of 0.05, we can expect, on average, approximately 2.8 of the statistic values to be significant by chance alone, and the value of four significant values obtained here is not that far off this expected figure!

In a further PLS-DA experiment, the above random permutations were also performed in order to test the 'overfitting' of the model to a total of 10 sample donors (patients) included in the study, again with 222 explanatory X variables and only $n = 20$ samples collected therefrom. A typical result arising from this further investigation is shown in Figure 1.7; the t_2 *versus* t_1 scores plot obtained reveals that quite a high level of distinction is achievable between each of the 10 participants involved by PLS-DA overfitting in this experimental design which contains many more X variables than samples available (>10-fold in this case)!

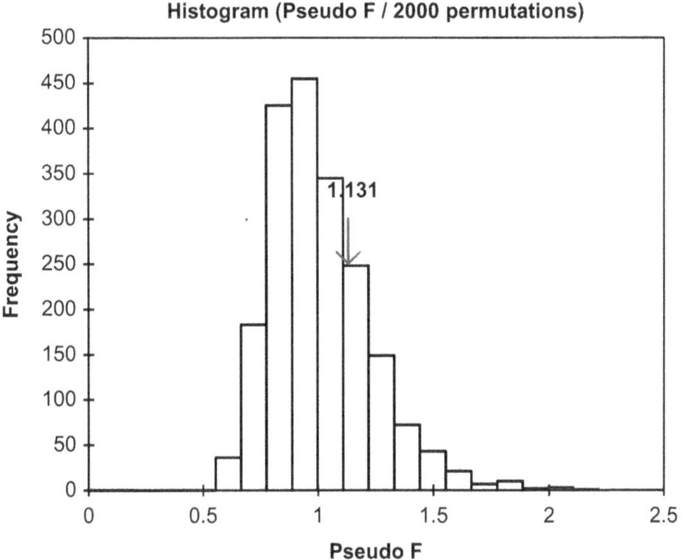

Figure 1.6 Typical permutation testing output derived from the above randomly classified ^{1}H NMR urinary metabolomics dataset analysed as described in Figure 1.5. In this case, the *pseudo*-F statistic value was 1.13; the values of this statistic computed were only significant in 4 out of 56 random permutation cases tested (2000 subsequent permutations were performed for each of the 56 classifications arising from primary permutation).

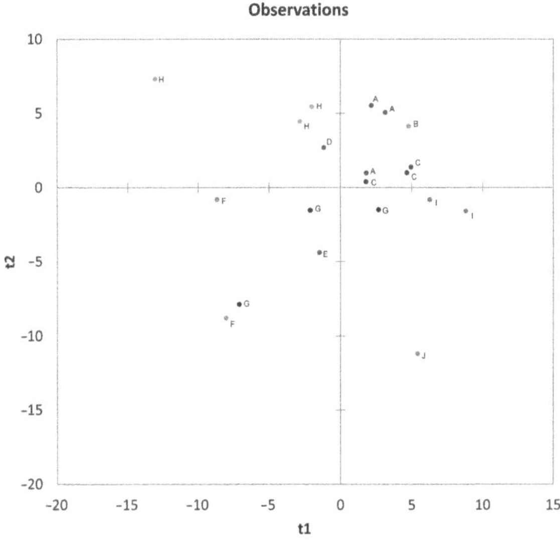

Figure 1.7 Corresponding PLS-DA scores plot acquired on the dataset explored in Figure 1.5, but in this case with family status serving as a qualitative lateral classification variable. The different letters correspond to different families from which the sample donors are derived.

1.3.3 Procedures for the Validation and Cross-Validation of PLS-DA Models

As indicated above, although used very infrequently, a critically important aspect of CV processes is permutation testing, which usually involves the analysis of a very large number (say 500–500 000, or even more) of versions of the dataset with randomly-assigned classification labels. In this manner, a random distribution for the null (H_0) hypothesis that no differences exist between the two (or more) classifications is attained, and hence we are able to test the significance of any key differences observable, MV or otherwise.

The major advantage offered by such permutation testing is that *via* the analysis of a very large number of versions of the complete dataset (say, up to 10 000 or so) with randomly assigned classification labels, a reference (null or H_0) distribution is acquired, and if our computed statistic (*e.g.* a *pseudo*-F ratio statistic value, as employed in redundancy or partial redundancy analysis) lies within this distribution without a significant p value (<0.05, <0.01 or otherwise, a parameter pre-selected by the researcher), then we can conclude that there is no evidence available for a significant departure from the null hypothesis and hence there is not a significant influence of the disease classification and/or an administered therapeutic regimen or toxic agent, the latter in the case of animal model experiments performed to investigate the toxicological insults and effects of selected agents (on target organs such as the liver or kidney, for example) on the metabolic profile of the biofluid or tissue biopsy sample evaluated in this manner.

The performance of such permutation testing has revealed that the erroneous application of CV methods can often give rise to too (and far too) optimistic classification outputs. A range of previous publications focused on this area have indicated this danger and sometimes confirmed it *via* the performance of detailed assessments, *e.g.* Westerhuis *et al.* (2008).[12] Indeed, when employed incorrectly, a result consisting of only a small number of misclassifications is obtainable. Moreover, the application of such permutation testing to the now highly utilised PLS-DA model quality parameters such as Q^2 and AUROC values, together with the misclassification rate(s), provides null hypothesis (H_0) distributions of such values that may be obtained in the case of no or near-zero differences observed between two (or more) classification criteria, and in this manner we have the ability to determine which particular CV index lies significantly outside this hypothetical permuted distribution. In this manner, we may be able to propose the application of a range of model systems which differ only in the slightest sense rather than just a single one, and as such we can derive a series of estimates of classification memberships. Indeed, Westerhuis *et al.* (2008)[12] have argued strongly that such an extensive series of these model systems should be employed as a powerful confidence measure and reassurance index for such classification membership assignment tasks.

There is a major requirement for the employment of CV models in view of the frequently (and increasingly!) small numbers of biofluid or tissue biopsy samples available for such metabolomics investigations, especially since their prior segregation into 'training', 'validation' and 'test' sets is, for a large number of studies, just not possible. Hence, selected CV techniques serve to provide a more realistic use of datasets tested in this manner (although we should, of course, note that it is required to expose the complete modelling process to CV strategies in order to yield a reliable error rate estimation). A further important consideration is that the classification index or indices predicted should *not*, under any circumstances, be employed for model development.[14,15,18] Although this particular stringent requirement has been noted in a relatively large number of publications, unfortunately it remains a very uncommon practice!

So that we may be confident with the nature of and results acquired from the CV method performed, the dataset should be divided into training, optimisation (validation) and test sets; a model is developed from the training and optimisation datasets, and the test set is then employed solely for determining the model's performance.

Repetition of such a process in a manner involving the inclusion of each sample in the test set only once allows a realistic estimate of prediction error which is representative of future samples entering the model test system. In order to ensure the complete independence of the test set, samples therein should remain exclusive to all operations involved in the model's development, including prior dataset pre-treatment systems employed for the 'training' set, for example transformations, normalisation, scaling and standardisation, *etc.*

In the single cross-validation (1CV) method, which is employed for an extensive range of systems and applications, a number of samples (or sample donors) are removed from the complete dataset and utilised as a validation set. The remaining samples which form a training set are then employed to generate a whole series of classification models with the number of PLS components ranging from 1 to perhaps 10 or 20, although the latter higher component range can sometimes be a little unlikely or unrealistic! Subsequently, a predictive capacity and prediction of all validation set members is provided (and the predictive errors of all these developmental models are stored for future use). Henceforth, a new patient or participant dataset is introduced, and subsequently this process is repeated for these up to the stage where all of them have been placed in the validation dataset once and only once, and in this manner the total predictive error for all models throughout all test samples is completed; that with the lowest predictive error then serves as the optimal one for further development and, hopefully, application to real test samples! The predictive errors acquired *via* the employment of this technique are then utilised in order to compute the Q^2 value and the misclassification rate. For this particular CV model, it should be noted that samples originally incorporated into the validation set are also utilised to determine the most effective model parameters, and therefore they do not remain completely independent, which represents an important requirement for an acceptable CV model.

Cross-model validation (*i.e.* double cross-validation, abbreviated 2CV) has been put forward as a system suitable for dealing with problems arising from the dependency existing between the prediction error for new samples and the model optimisation parameters. In this system, one series of samples is completely isolated as the 'test' set, and the remaining ones then undergo a single CV testing process, in which they are also sub-divided into training and validation sets (the single CV regimen again giving rise to an optimum number of PLS components). The optimised model derived therefrom is then employed to predict the classification status (disease class or otherwise) of those samples (biofluid, tissue biopsy, *etc.*) placed into the test set. Subsequently, the whole process undergoes a repetitive construct until all the samples have been placed in the test set once, and only once, and it should also be noted that it is of much importance to select the validation samples in a random manner in order to further optimise the inclusion of differing combinations of validation and training sets for each newly selected test set. In this manner the final model will have been constructed in the complete absence of the test set, and hence its predictive capacity remains independent of the model optimisation regimen utilised.[20]

In the quality-of-fit (FIT) model, the single CV basis of the above 1CV technique seeks the optimum number of PLS components, and a PLS-DA model is then constructed from all samples available with this optimal number. Subsequently, the classification groups of all of these samples are determined (or estimated) with this particular model; this, however, represents a re-substitution rather than an acceptable prediction process, and in this manner Q^2-FIT, the number of misclassifications, and AUROC values may be obtained.

Therefore, overall the variability of the estimated parameters, and their influence on the model's predictive capacity, are evaluated *via* the 1CV method. However, the 2CV technique offers advantages since it is also provides an assessment of the variability of meta-parameters and their overall contribution towards the predictions obtained; the classifications predicted for the samples analysed are only completely independent of the remaining dataset when this particular technique is utilised.

Figure 1.8 shows a PLS-DA analysis of a very large thyroid disease dataset comprising a series of explanatory variables, including the blood serum concentrations of thyroxine (T4) and thyroid-stimulating hormone (TSH). Results obtained revealed very clear distinctions between the three classes of disease [healthy controls (euthyroid), hypothyroid and hyperthyroid patients]; the validation process involved the prior removal of approximately one-third of the samples, and the PLS-DA model was then built on the remaining two-thirds of them. Validation of the model in this manner gave rise to an excellent agreement between the predicted sample identities and their known ones (mean classification rates of 100% for euthyroid and hyperthyroid patients, and 98.9% for the hypothyroid group).

Also shown are results derived from a Partial Least Squares-Regression (PLS-R) model in which the hypothyroid, euthyroid and hyperthyroid disease classifications were assigned arbitrary scores of -1, 0 and $+1$, respectively. Again, the model evaluated demonstrated an excellent predictive capacity.

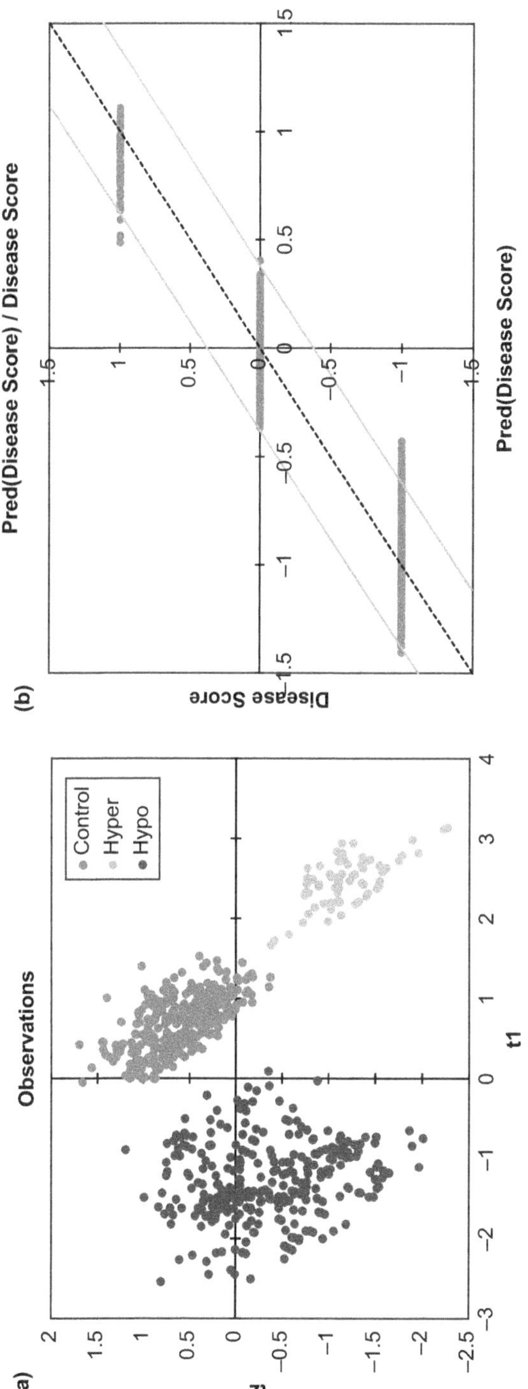

Figure 1.8 (a) PLS-DA t_2 *versus* t_1 scores plot derived from the analysis of a thyroid disease dataset consisting of a series of explanatory X variables [including the blood serum concentrations of thyroxine (T4) and thyroid-stimulating hormone (TSH)] and 300 euthyroid, 300 hypothyroid and 71 hyperthyroid patients [the dataset was mean-centred and scaled to X variable (column) standard deviations prior to performing the analysis]. Results obtained revealed very clear distinctions between the three classes of disease [healthy controls (euthyroid), hypothyroid and hyperthyroid patients]. The validation performed involved the prior removal of approximately one-third of the samples; the PLS-DA model was then built on the remaining two-thirds of them; validation performed in this manner gave rise to an excellent agreement between the predicted sample (patient identities with their known ones (this validation process was repeated five times, and mean classification rates of 100% for euthyroid and hyperthyroid patients and 98.9% for the hypothyroid classification were obtained). (b) Plot of observed disease score *versus* estimated disease score for a corresponding Partial Least Squares Regression (PLS-R) model in which the hypothyroid, euthyroid and hyperthyroid disease classifications were assigned arbitrary (dependent Y variable) disease scores of -1, 0 and $+1$, respectively. Abbreviations: Control (healthy control, euthyroid patients); hypo, hypothyroid patients; hyper, hyperthyroid patients.

1.3.4   Attainment of the Final Calibration Model

Selection of the predictive capacities of each of the separate testing systems generates a range of somewhat differing models, with differing numbers of 'biomarker' variables and also perhaps components, which arise from the random selection of some specimens into the 'training', 'validation' and, where appropriate, 'test' sets, each of which has differing contributions towards each parameter evaluated: however, this approach serves to complicate the optimisation of a final 'diagnostic' model system.[14] Indeed, the precision of such a final predictive model system should always be greater than those generated during the CV regimen (*i.e.* those developed on sample sub-sets), and hence at this stage the full applicability of the final calibration system is not required.

Notwithstanding, as an alternative to such final models, a whole series of these incorporating one per test set can be made available in order to metabolomically classify future test samples, and therefore a group of possible classifications for each one may be available, rather than a single one [such computations are likely to involve the consideration of further lateral 'predictor' variables such as gender, age, BMI and, where appropriate, length of treatment (if any), *etc.*]. Henceforth, this group can serve to provide 'mean level' predictions based on the individual predictive models developed, and therefore appropriate confidence intervals (CIs) for the overall predictive capacities can be developed, and their stabilisation and stabilities throughout research work performed. Indeed, the 'bagging' procedure of Breiman[21,22] is of much relevance here.

1.3.5   Quality Evaluation Processes

Since there is currently a range of criteria employed to aid determinations of a pre-selected classification of 'unknown' samples, including percentage classification successes based on the confusion matrix (and consisting of numbers of false-positives and -negatives, together with true positives and negatives[14]), it is necessary for us to be clear about the particular measures adopted for this purpose, and also their possible influence (facilitatory or adverse, for that matter), on the classification of samples collected from patients participating in future clinical or metabolomics investigations. For a particular class of disease-positive participants, the proportions/percentages of true positives is known as the *sensitivity*, whilst those of false-positives is referred to as the (1-*specificity*) parameter, and a combination of these two criteria gives rise to the so-called Receiver Operator Characteristic (ROC) curve. Indeed, the ROC curve comprises a plot of *sensitivity versus* (1-*specificity*), and this relationship is often employed to determine the successful (or unsuccessful) performance of a clinically-relevant MV (*e.g.* a biomolecular concentration index) dataset or alternative measurement system. Of course, *sensitivity* is defined as the number of correct (true) positives found expressed as a percentage of all the available positives (*i.e.* those with a particular disease classification or, alternatively, response to a particular treatment, *etc.*). *Sensitivity* values lie between 0 and 1, with 0 being no success whatsoever, and 1 a 100% classification rate. The (1-*specificity*) index, however, represents the number of false-positives expressed as a percentage of all such negative (disease-free) values (*i.e.* those for a 'control', healthy participant dataset). For an effective and reliable model system, *sensitivity* values should be close to 1.0, although the *specificity* should also be close to this particular extreme value, so that (1-*specificity*) remains close to 0 [the classification boundary pre-set by the investigator(s) determines the overall *specificities* and *sensitivities* of the model system tested]. A modification of these selection parameters may give rise to an elevation in the number of true positives, although the number of false-positives will also be enhanced, and *vice-versa*. Hence, the classification boundary of the model system tested determines the effectiveness of a ROC curve, which delineates both the *specificities* and *sensitivities* of models with perhaps differing classification barriers or thresholds.

Indeed, we may select values other than 0 as the classification boundary cut-off value, and the choice of a slightly lower value may increase the sensitivity of the +1 (disease-positive) group, although this is inevitably coupled to an alleviation of the (1-*specificity*) parameter. Therefore, the overall classification quality measurement utilised is the area under the ROC curve (AUROC) value, which is 1.0 for an ultimately perfect class distinction, and 0.50 if there is absolutely no separation detectable or present.

Q^2 values, however, represent predictive capacity default parameters, which are commonly employed in PLS-DA investigations, and are targeted at determining the efficacies of classification label predictions from newly derived datasets. Q^2 is defined in eqn (2), in which SS represents the fraction of the mean-corrected sum-of-squares of the Y classification codes explained for each PLS component obtained, and PRESS the sum of squared differences between the observed and predicted Y values for all biofluid or tissue biopsy specimens incorporated into the test system (further details are provided in Appendix I). As might be expected, the optimal value of Q^2 (1.0) is extremely difficult to attain in practice in view of considerations that (1) the requirement for it is that the classificational

$$Q^2 = (1 - \text{PRESS}/\text{SS}) \tag{2}$$

prediction of all such samples should be exactly equivalent to their class labels, and (2) the always present inherent variation (perhaps that 'Between-Participants', 'Between-Samples-within-Participants' and/or further latent or 'hidden' variables) nested within the same classification criterion. Therefore, the Q^2 value derivable depends not only on the 'Between-Classifications' variability, but naturally also on the 'Between-Samples-within-Classifications' one, and this renders it somewhat difficult to achieve a Q^2 value which is representative of a high classification prediction capacity, and hence it is highly recommended to employ a series of permutation tests in order to evaluate the distributional status of such model-dependent Q^2 values in the complete absence of any influential or constraining effects exerted by two (or more) classification criteria, which may (or may not) exert significant effects on this random permutation distribution.

However, since the AUROC value, and also the number of misclassifications found, reflect simple 'extent of classification' error measurements, and only serve to inform us of the numbers correctly and incorrectly classified, they are clearly of less value than a permuted distribution of values which arises from the null hypothesis of no effects exerted by the 'Between-Classification' factor or factors. Indeed, Q^2 is a prediction error measure that, perhaps fortunately, is able to distinguish between correct and incorrect classifications; for example, a class prediction value of $+0.90$ is penalised more so than one of $+0.60$ for a correct class label of 0, *i.e.* some estimated classification status values are more equal than others! Notwithstanding, the above AUROC and number of misclassification measures noted above serve to view these prediction errors as exactly the same – *i.e.* in these cases, all incorrectly classified errors are equal!

In view of the large number of variables available to classify the disease (or alternative) status of biofluid or tissue biopsy specimens, the MV metabolomics data analysis arising therefrom remains a highly complex process. Indeed, there remains a very wide range of modelling solutions available to effectively 'solve' these problems, and hence 'overfitting' is a very common example available in the scientific literature, *i.e.* the model employed appears to classify 'training' datasets very efficiently, but its application to samples collected in future, corresponding investigations has a very poor or perhaps virtually zero classification ability (please note the examples given above in Sections 1.3.1 and 1.3.2)! Clearly, such studies are opportunistic, highly presumptive and largely hypothesis-driven arguments, which eventually fail to offer the high level of merit proclaimed from the original modelling MV experiments performed. Indeed, as a highly typical example, the PLS-DA scores plots, which are documented in a very significant proportion of disease status classifications in metabolomics-based publications, may represent highly exaggerated or over-optimistic visions of such classification differences (however, they *may* reveal some level of significant 'within-classification' differences, which perhaps were unknown to the investigators prior to performing the analysis). Indeed, results arising from putative classification studies of this nature employ predictions rather than fitted values (*e.g.* PLS-DA scores) as a foundation, and the failure to perform one or more of the validation, CV and corresponding permutation monitoring of the dataset acquired will not provide researchers with a high level of confidence regarding the results acquired!

1.3.6 Cost-Benefit Analysis (CBA)

Briefly, this procedure can be performed in order to select the optimal number of 'biomarker' variables, and also to determine the diagnostic benefit of adding additional ones (although the cost of, for example, employing 30 rather than 5 biomarker variables could represent a 6-fold increase, the diagnostic benefit derived may be limited, with improvements of perhaps only a few per cent in terms of those correctly classified). Indeed, successful models may be formed on only the top 5 or so ranked explanatory (X) metabolic predictor (biomarker) variables, or even less than this number.

Appendix I

Partial Least Squares-Discriminatory Analysis (PLS-DA)

Partial least squares-discriminatory analysis (PLS-DA) represents a regression-extended class of PCA, which involves the derivation of latent variables (analogous to principal components), which maximise the co-variation between the monitored dataset(s) (*i.e.* conventional or 'intelligently selected' ^{1}H NMR spectral bucket areas) and the response variable which it/they is/are regressed against. PLS-DA represents a special form of PLS data modelling which, in the case of a significant discriminant function, has the ability to distinguish between known or established classifications of samples in a calibration set, and is focused on seeking a range of discriminatory variables and directions in a greater than bivariate (*i.e.* multivariate) space. This procedure involves the computation of an indicator matrix of potential classification (predictor X) variables for each classification group incorporated in the calibration dataset [for a two classification system, each group may be assigned a value of 0 or 1 (or -1 and $+1$) according to which particular class a study participant who provides a 'diagnostic' biofluid or tissue biopsy sample belongs].

Like PCA, Partial Least Squares (PLS) performs a dimensionality reduction of the X matrix, but also relates X variances to that of Y contained in a Y response matrix. The matrices are simultaneously decomposed, exchanging respective scores information so that the technique maximises their covariance. Components that successfully maximise any remaining covariance are then generated, the optimal number defining the model dimensionality. A PLS-DA analysis involves a Y matrix containing class information (and hence it is a supervised technique), the biomolecule concentrations or proportional NMR, LC-MS or GC-MS intensity measurements (X matrix) being related to nominal categorical codes (Y column dummy matrix) by an equivalent correlation matrix B [eqn (1)]; the

$$Y = XB \tag{1}$$

analysis can therefore maximise the correlation (or covariance) between X and Y. The X and Y matrices are converted to eqns (2) and (3), in which T and P represent the scores and

$$X = TP^T + E \tag{2}$$

$$Y = UQ^T + F \tag{3}$$

loadings matrices for X, respectively, U the corresponding Y scores matrix, Q^T the y weighting matrix, and E and F the residual matrices which accommodate information not related to X/Y correlations. The X weights w (describing the variation in X correlated to the Y class information, *i.e.* through their covariance, as well as information on the variation in X not related to Y) are also employed for calculating T [eqn (4)]. The W^* matrix is transformed from the original W matrix so that it is

$$T = XW^*$$
(4)

PLS component-independent, since the X scores T are linear combinations of the X variables, and when multiplied by P they will essentially return the original variables (with small E values). Equations (2) and (4) can then be combined to yield eqn (5) [*i.e.* a modified form of eqn (1) which allows for residuals], in order to set up the regression model according to eqn (6).

$$Y = XB + E$$
(5)

$$B = W^*Q^T$$
(6)

A range of output parameters can be generated from PLS analytical software packages, including goodness-of-fit parameters such as the fraction of the mean-corrected sum-of-squares (SS) of the Y codes explained for each generated PLS component, *i.e.* R^2 [eqn (7)], where RSS represents the

$$R^2 = (1 - RSS/SS)$$
(7)

fitted residual sum of squares, *i.e.* the sum of the squared differences between the observed and fitted y values [eqn (8)].

$$RSS = \Sigma(Y_{fitted} - Y_{actual})^2$$
(8)

The presence of many, potentially highly correlated, X predictor variables indicates the possibility of data overfitting, and hence there is a requirement to test the model's predictability for each PLS component. However, model validation through deduction of the number of significant PLS components can be determined *via* a 'leave-one-out' CV method, in which data for one sample is removed from the model, and the predicted classification groups or analogous Y value codes are then compared with those of the removed sample, the process being repeated until all samples have been left out once. The predictive residual sum of squares (PRESS) is the sum of the squared differences between the observed and predicted y values for the CV process [eqn (9)],

$$PRESS = \Sigma(Y_{predicted} - Y_{actual})^2$$
(9)

and the fraction of total variation in the Y codes that can be predicted by each PLS component is defined by Q^2 [the 'cross-validated R^2 value', eqn (10)].

$$Q^2 = (1 - PRESS/SS)$$
(10)

The number of components that cause a minimum computed PRESS value (within a limit of 5% between each subsequent component) is noted, and this number can then be pre-set in the developing model program for further computations.

Appendix II

Brief Summary of Further Forms of Discriminatory Analysis (DA) Available

There is a wide variety of approaches for discriminant analysis, and many of them are still not well established amongst the metabonomics research community. These are of two forms: One- and Two-class classifiers. One-class classifiers allow us to build models of varying complexity around each class separately (for example, between two or more disease classification groups to be examined), from, for example, their ^{1}H NMR spectral profiles, so that researchers can predict whether a patient has a disease, and/or belongs to a disease sub-group, to a given specified level of confidence. They can also permit us to determine how well modelled a particular class is. Using Receiver Operator Characteristic (ROC) curves, prediction thresholds can be computed in order to determine optimum conditions for the minimisation of false-negatives or false-positives. However, Two-class classifiers attempt to form a 'hard boundary' between two (or more) classes (samples close to the boundary are somewhat ambiguous and therefore difficult to classify), and for each sample a model stability can be determined. In metabolomics investigations, a variety of statistical methods for validation can be utilised in order to ensure that the models are sound, and methods employable include Linear Discriminant Analysis, Quadratic Discriminant Analysis, Partial Least Squares Discriminant Analysis, Learning Vector Quantisation and Support Vector Machines (SVMs), in one- or two-class formats (where appropriate).

Acknowledgements

In this work the author utilised XLSTAT2013, MetaboAnalyst 2.0, MetATT and ACD Spectrus Processor 2013 software.

References

1. R. Johnson and D. W. Wichern, *Applied Multivariate Statistical Analysis*, Pearson Prentice Hall, Upper Saddle River, NJ, 2007.
2. G. P. Quinn and M. J. Keough, *Experimental Design and Data Analysis for Biologists*, Cambridge University Press, Cambridge, 2002.
3. M. M. Beckwith-Hall, J. K. Nicholson, A. W. Nicholls, P. J. Foxall, J. C. Lindon, S. C. Connor, M. Abdi, J. Connelly and E. Holmes, Nuclear magnetic resonance spectroscopic and principal components analysis investigations into biochemical effects of three model hepatotoxins, *Chem. Res. Toxicol.*, 1998, **11**, 260–272.

4. T. R. Brown and R. Stoyanova, NMR spectral quantitation by principal-component analysis. II. Determination of frequency and phase shifts, *J. Mag. Res. Series B*, 1996, **112**, 32–43.
5. R. Madsen, T. Lundstedt and J. Trygg, Chemometrics in metabolomics-a review in human disease diagnosis, *Anal. Chim. Acta*, 2010, **659**, 23–33.
6. M. Mamas, W. B. Dunn, L. Neyses and R. Goodacre, The role of metabolites and metabolomics in clinically applicable biomarkers of disease, *Arch. Toxicol.*, 2010, **85**, 5–17, DOI: 10.1007/s00204-010-0609-6.
7. A. M. Weljie, J. Newton, P. Mercier, E. Carlson and C. M. Slupsky, Targeted profiling: Quantitative analysis of 1H NMR metabolomics data, *Anal. Chem.*, 2006, **78**, 4430–4442.
8. J. O. Kim and C. W. Mueller, *Introduction to Factor Analysis. What It Is and How To Do It*, Sage, Beverly Hills, CA, 1978.
9. J. O. Kim and C. W. Mueller, *Factor Analysis: Statistical Methods and Practical Issues*, Sage, Beverley Hills, CA, 1978.
10. R. J. Rummel, *Applied Factor Analysis*, Northwestern University Press, Evanston, IL, 1970.
11. R. B. Cattell, The scree test for the number of factors, *Multivar. Behaviour. Res.*, 1966, **1**, 245–276.
12. J. A. Westerhuis, C. Huub, J. Hoefsloot, S. Smit, D. J. Vis, A. K. Smilde, E. J. J. van Velzen, J. P. M. van Duijnhoven and F. A. van Dorsten, Assessment of PLSDA cross validation, *Metabolomics*, 2008, **4**, 81–89.
13. E. Szymańska, E. Saccenti, A. K. Smilde and J. A. Westerhuis, Double-check: validation of diagnostic statistics for PLS-DA models in metabolomics studies, *Metabolomics*, 2012, **8**(1), 3–16.
14. G. W. Snedecor, *Statistical Methods*. Iowa State University Press, Des Moines, Iowa, USA, 1956.
15. R. G. Brereton, Consequences of sample size, variable selection, and model validation and optimisation, for predicting classification ability from analytical data, *Trac-Trends Anal. Chem.*, 2006, **25**(11), 1103–1111.
16. D. I. Broadhurst and D. B. Kell, Statistical strategies for avoiding false discoveries in metabolomics and related experiments, *Metabolomics*, 2006, **2**(4), 171–196.
17. P. D. B. Harrington, Statistical validation of classification and calibration models using bootstrapped Latin partitions, *Trac-Trends Anal. Chem.*, 2006, **25**(11), 1112–1124.
18. C. M. Rubingh, S. Bijlsma, E. P. P. A. Derks, I. Bobeldijk, E. R. Verheij, S. Kochhar and A. K. Smilde, Assessing the performance of statistical validation tools for megavariate metabolomics data, *Metabolomics*, 2006, **2**(2), 53–61.
19. E. Anderssen, K. Dyrstad, F. Westad and H. Martens, Reducing over-optimism in variable selection by cross-model validation, *Chemomet. Intell. Lab. Syst.*, 2006, **84**(1–2), 69–74.
20. M. Stone, Cross-validatory choice and assessment of statistical predictions, *J. Royal Stat. Soc. B*, 1974, **36**, 111–147.
21. L. Breiman, Bagging predictors, *Mach. Learn.*, 1996, **24**(2), 123–140.
22. L. Breiman, Arcing classifiers, *Ann. Stat.*, 1998, **26**(3), 801–824.

Experimental Design: Sample Collection, Sample Size, Power Calculations, Essential Assumptions and Univariate Approaches to Metabolomics Analysis

MARTIN GROOTVELD* AND VICTOR RUIZ RODADO

Leicester School of Pharmacy, Faculty of Health and Life Sciences,
De Montfort University, The Gateway, Leicester LE1 9BH, UK
*Email: mgrootveld@dmu.ac.uk

2.1 Introduction

Multivariate (MV) metabolomics experiments are often underdetermined (*i.e.* they frequently contain many more variables than samples),[1] and this is a situation which can give rise to many challenges and problems regarding the statistical analysis of data acquired therefrom. Indeed, typical 'omics' experiments commonly involve datasets containing measurements made on several hundreds to tens of thousands of potential predictor (X) variables (for example, hundreds of metabolites determined in a metabolomics experiment, or all the genes detectable in a microarray experiment); however, in view of the expenses incurred, frequently only a relatively small number of

Issues in Toxicology No. 21
Metabolic Profiling: Disease and Xenobiotics
Edited by Martin Grootveld
© The Royal Society of Chemistry 2015
Published by the Royal Society of Chemistry, www.rsc.org

samples are collected and analysed in order to explore these multidimensional inter-relationships.

Furthermore, in many metabolomic investigations which focus on the identification and determination of the statistical and clinical significance of potential biomarkers for selected diseases, drug toxicity issues and environmental stresses, it has been recognised that the variation 'Between-Study Participants' is often larger than that observed 'Between-Classification Groups', and hence can exert a major confounding effect on results acquired from high-throughput metabolomics investigations. Indeed, in many metabolomics investigations, this frequently unknown, perhaps inaccessible, source of 'Between-Study Participants' variation is very often pressing on our major 'Between-Disease' or 'Between-Treatment' Classification focus of interest so that it exerts an unknown and recurrently confounding influence on our analysis (such influences can, of course, exert major effects on our MV analyses which remain remote from the conceptions of at least some researchers working in this area; this lack of prior consideration can sometimes give rise to a mountain range forest of confusion!).

Therefore, in this chapter we outline essential criteria regarding frequently required considerations for the collection of biofluid and/or tissue biopsy specimens from human metabolomics study participants (for example, is pre-fasting required, and what are the appropriate timings for such aspirations or alternative forms of collection?), raw dataset preprocessing stages (including 'bucketing' or 'binning' procedures for resonance, signals or peaks, together with data normalisation and scaling processes), and satisfaction of the essential assumption of normality (which is often a prerequisite for the MV analysis of these datasets). Moreover, we also outline both relatively simple and complex analysis of variance (ANOVA) experimental designs and their applications to the analysis of multidimensional datasets such as that performed using the ASCA technique.

We also discuss univariate approaches to the analysis of high-dimensional datasets (predominantly employed as an essential preliminary analysis method, but sometimes also as a variable selection one), and also provide further essential information regarding the critical assumptions required for these and subsequent MV approaches, together with the performance of power calculations (*i.e.* determinations of the minimal experimental sample sizes required) for both these forms of modelling systems. Finally, the detection of differing classes of uncertainty in bioanalytical investigations, and their experimental or statistical proliferation (known as error analysis), is also briefly described.

2.2   Essential Considerations for Sample Collection

The institution of carefully selected sample collection regimens is considered to be of major importance for the successful performance of metabolomics experiments. Indeed, there are many potential interferences or 'contaminators' of datasets acquirable which may arise during periods of biofluid or tissue biopsy sample collection, the time-points at which samples were collected, transport to the laboratory, and their storage and/or preparation, phenomena that may give rise to the generation of artifactual information. For example, the detection of microbially induced catabolites in biofluid samples which arise from their erroneous storage or pre-treatment regimens, the former perhaps for unacceptably prolonged periods of time, the latter without consideration for their prior treatment with an effective microbicidal agent such as azide, for example.

As a key example, our research group regularly performs MV metabolomics investigations of human saliva and, in order to avoid interferences arising from the introduction of exogenous agents into the oral environment, sample donors are requested to collect all saliva available, *i.e.* ('whole') saliva expectorated from the mouth, into a sterile plastic universal tube or container immediately after waking in the morning on a pre-selected day (*i.e.* following a 'sleep-fasting' period of 6–8 hours). Each participant is also requested to refrain completely from oral activities (eating, drinking, tooth-brushing, oral rinsing, smoking, *etc.*) during the short period between awakening and sample collection (*ca.* 5 min.). Moreover, each collection tube contains sufficient sodium fluoride (15 μmol) in order to ensure that metabolites are not generated or consumed *via* the actions of micro-organisms or their enzymes present in whole saliva during periods of sample preparation and/or storage. For ^{1}H NMR analysis, it is also recommended that samples are treated with a small volume of a relatively concentrated phosphate buffer solution of a selected pH value (say 7.00 or 7.40) prior to analysis.

Researchers should also be aware of the artifactual, time-dependent O_2-mediated oxidation of biofluid/tissue biopsy sample electron donors during episodes of transport, preparation and storage, *e.g.* oxidation of thiols such as L-cysteine and glutathione to their corresponding disulphides (processes which also involve the generation of superoxide anion and/or hydrogen peroxide as by-products), and also the oxidative transformation of ascorbate to dehydroascorbate and further products. Exclusion of atmospheric O_2 (*e.g.* storage of such samples in its absence, for example under N_2 or helium) can, of course, serve to circumvent such issues.

Dunn *et al.* (2008)[2] conducted a metabolomic study involving GC-TOF-MS metabolic profiling in order to evaluate the stability of human blood serum and urine samplings during 0 and 24 hr periods of storage at 4 °C prior to freezing them at −80 °C; for these experiments, >700 unique metabolite peaks (and >200 per sample) were detectable with an associated high degree of reproducibility, and these were explored utilising both univariate and MV statistical analysis methodologies in order to determine possible modifications in the metabolomes of these samples. However, no such differences in analytical variance were observed between technical replicates, nor those between samples stored at 48 °C for 0 and 24 hr time periods were found for both blood serum and urine samples. Although PCA scores and loadings plots revealed that a few samples differed metabolically for both serum and urine, corresponding univariate analysis demonstrated that these differences were associated with only a small number of metabolites and were not found to be statistically significant. Hence, these researchers concluded that there were only minimal modifications observed in the biomolecular compositions of these biofluids.

Interestingly, comparisons of the intra- *versus* inter-subject metabolic variabilities of blood serum and urine showed that the variance in the metabolome of a single subject stored at $-80\,^{\circ}$C or $48\,^{\circ}$C for 24 hr is small when expressed relative to that of the metabolomes of 40 healthy volunteers (in addition to differing genotypes, many phenotypic factors also influence the composition of the human metabolome including diet, health and lifestyle, together with diurnal and oestrus cycle effects). Hence, in well-designed metabolomic studies involving the collection of biofuid samples from a large number of participants, apparent changes in the metabolome artefactually or adventiously induced *via* storage at $48\,^{\circ}$C for 24 hr periods are minimal when expressed relative to those observed 'Between-Participants'.

In a related animal study, Schreir *et al.* (2013)[3] explored the possible influence of urine sample modification potentially arising from sample storage and preparation episodes, and also the status of animal health in the groups involved in their study. This involved an analysis of the appropriateness of the ^{1}H NMR-linked quantification of rat urine biomolecules for statistical MV data analysis (specifically, metabolomics-based sample classifications according to their criteria) so that they may attain a rapid, robust and cost-effective delivery of an acceptable level of data in order to facilitate and promote the modelling processes involved, and these researchers demonstrated that recommended means of urine sample storage (at $-20\,^{\circ}$C) could be effectively employed for periods of up to 24 months, whereas storage at only $4\,^{\circ}$C in a standard refrigerator could be utilised for only a 14-day period. Additionally, neither pH- nor ionic strength (salt, I)-induced modifications to these samples gave rise to changes in measurement accuracy, an observation indicating the feasibility of exact determinations subsequent to a single pH buffering stage, and that the successful attainment of this is achieved through the focusing of an NMR facility dedicated to this form of research investigation, this serving as an option which clearly optimises the quality of research results acquired. In this manner, the authors of this work concluded that high-resolution ^{1}H NMR spectroscopy and its associated MV analysis technologies provide a highly accurate, robust and high-throughput analytical system for biomolecule quantification and metabolomics classification strategies, the latter for diagnostic purposes or otherwise.

Briefly, it is also of much importance to note that it is well known that common clinical chemistry determinations of a range of biomolecules which serve as biomarkers for a range of diseases also have stringent requirements for careful sample collection, and also their storage prior to biochemical analysis. For example, for determinations of blood plasma lipid profiles, samples should always be collected from patients in the fasting state, since low-density-lipoprotein (LDL), and especially total triacylglycerol concentrations are substantially influenced by recent food intake. In these cases, it is now scientifically and clinically recognised that such samples should only be collected from patients who have primarily undergone a 12 hr fasting period (such samples should then be stored in the recommended manner). Therefore, it is perhaps a little surprising that at least some researchers involved in the metabolomics research area will continue to collect samples from study patients without first implementing any strict considerations for fasting periods, collection times and, where appropriate, standardisation of these processes throughout the entire participant sampling group. Moreover, the authors are aware of at least some publications in which metabolomics investigators have apparently 'discovered' the 'diagnostic' significance of a series of lipidic biomarkers in appropriate diseases without allowing for or even first considering the often significant confounding effects of failing to fast the patient cohort involved!

2.3 Raw Data Preprocessing Steps

The primary stage in most forms of metabolomics data analysis (univariate, MV or a combination of both) is processing of the 'raw' dataset, and this involves a number of sequential steps which depend on the methodologies employed.[4] This, of course, represents an extremely important consideration in data analysis. Such raw data processing procedures, and their application to the differing bioanalytical techniques utilised in this research area, have been reviewed in detail by Hansen (2007),[4] Katajamaa and Ores (2007),[5] Scalbert *et al.* (2009),[6] Schripsema (2010)[7] and Spraul *et al.* (1994),[8] and hence they will not be further outlined in detail in this work.

Generally, ^{1}H or alternative nucleus-based NMR spectral datasets encompass variations in peak widths, line-shapes and exact resonance frequencies in view of differences in solution sample matrices (ascribable to pH, ionic strength or the presence of metal ions, the latter at trace levels or otherwise), or alternatively variations in instrumental performance. Thus, it is essential that such raw datasets should be subjected to preprocessing methods in order to correct for such variations. Frequently, NMR data preprocessing techniques include corrections of linewidths *via* a line-broadening parameter (which employs a tuned exponential multiplication), Fourier transformation, a phase correction conducted with user pre-defined phase constants, together with positioning and scaling.[9] Moreover, these preprocessing techniques also include 'bucketing' or 'binning', peak-picking and spectral deconvolution.[10] Chemical shift bucketing is routinely required for NMR data processing prior to the performance of MV or alternative statistical analysis,[8,11] and is performed *via* separations of spectra acquired into multiple discrete regions (otherwise known as hixels), which are subsequently averaged and integrated. Although this process may give rise to a marked loss of information, it does achieve correction of datasets for peak shifts ascribable to small differences in physicochemical properties between samples (*e.g.* pH, ionic strength, *etc.*) A further advantage of this preprocessing step is, of course, the high level of data reduction involved, which serves to simplify exportation of datasets and their subsequent MV analysis. Although the bucket width is frequently pre-set to a value of 0.04 or 0.05 ppm, which gives rise to reductions of high-field spectral profiles (of, for example, 16–64 K data points) to a maximum of approximately 250 bucketed or binned ones, problems can arise from this 'indiscriminate' bucketing technique in view of the frequent occurrence of >1 signal per bucket, a problem which can arise from resonance overlap. Such issues can also arise during the course of other forms of multianalyte spectroscopic analysis, together with LC-MS profiles, for example. Moreover, bucketed ^{1}H NMR data are directly exportable to a wide range of statistical packages for subsequent MV statistical analysis.

Since the application of pre-fixed NMR chemical shift buckets can lead to problems regarding the incorporation of >1 metabolite signal (and some-times several or more) resonances or partial resonances per bucket or bin, some researchers have focused on the provision of potential solutions to this problem. Indeed, this complication can be overcome *via* the employment of 'Intelligent-Bucketing' processing, software for which is readily available (this 'Intelligent-Bucketing' algorithm and its applications are described in detail below). Spectral alignment and peak-picking involves the employment of alignment algorithms; these include (1) the linear fit procedure[12] and (2) the automated removal of frequency shifts in spectra *via* the application of PCA.[13] Deconvolution of spectral profiles into individual components[10] offers some advantages over alternative preprocessing techniques since it permits the simultaneous identification and quantification of individual biomolecules present.

Intriguingly, Weljie *et al.* (2006)[14] outlined a technique for the deconvo-lution of complex spectral profiles, the basis of which is the mathematical modelling of individual NMR resonances extractable from pure, authentic compound spectral libraries in order to generate a component database. Subsequently, a database search is performed in order to allow the identi-fication and quantification of these biomolecules. These researchers termed this approach 'targeted profiling', and its application was validated against a standard 'spectral bucketing' procedure; the method demonstrated a very high level of stability in PCA-based pattern recognition processes, and was insensitive to the potential influence of water resonance suppression tech-niques, scaling factors and relaxation times.

'Intelligent Bucketing' has been introduced to further enhance the bene-fits of autoprocesing to metabolic profiling and metabolomic investigations, and this technique was designed to perform 'smart' bucketing divisions (*i.e.* bucketing decisions) for complex, multicomponent ^{1}H NMR spectra such as those acquired on human saliva and urine.

Of course, modelling of such data can, at least in principle, be improved with fewer principal components when smarter, 'intelligently selected' bucketing divisions are made, such as those that are optimised to ensure that single resonances *do not* span two (or, very occasionally, more) buckets, a process which clearly segregates (shares) the biomolecular information available. Usually, a PCA will take into account a resonance that is encompassed by two buckets by placement of both of the bucket regions into the same PC. However, there are two major problems with this approach to NMR data analysis: (1) the remainder of the signals in the bucket could arise from further agents that may have been generating an independent X predictor variable (such a contribution will thus be lost in the statistical model) and (2) if the exact chemical shift location of a resonance changes between samples (spectra), however slight (potentially ascribable to small modifications in pH, ionic strength, divalent cation concentration, temperature, viscosity, *etc.* as desribed below), its relative contribution to each of the two involved bucket intensities will, of course, vary, a process decreasing the accuracy of the overall analysis, and hence will potentially confuse and confound interpretations of the results acquired. Intelligent bucketing avoids such problems and gives rise to more accurate models which can take into account the biochemical processes involved.

Hence, intelligent bucketing permits inprovements in the accuracy of biomedical NMR data modelling *via* the removal of inherent problems associated with classical (fixed chemical shift range) bucketing. Such problems arise from the inherent senstivity of NMR analysis to the molecular environment of the biofluid examined, particularly senstivities to pH, temperature and the presence of metal ion-biomolecule interactions (*i.e.* complexation reactions such as those involving the chelation of biofluid Ca^{2+} and Mg^{2+} by salivary citrate and lactate).

Intelligent bucketing represents an algorithm designed to make critical divisional decisions, *i.e.* those which define precisely where a bucket (or bin) division should be. As noted above, frequently, the edge of a bucket will be positioned in the centre of a signal, and hence its net contribution is spread over two (or occaisionally more) integral regions. Although the nature of PCA itself corrects such errors *via* a combination of two such spectral regions together into a single PC (as a consequence of their relatively strong correlation), poor and inaccurate results are acquired when the resonance is subject to pH-, temperature-, divalent metal cation-, ionic strength- (and, in some cases, viscoelastic-) controlled variations in its chemical shift value(s) between samples collected for comparative purposes and, under these circumstances, the contribution of this signal is asymetrically divided between more than one bucket integration region. Intelligent bucketing 'selects' bucketing divisions which are based on local minima, and therefore this algorithm avoids this chemometrical error. It also has the ability to perform 'bucketing' on a whole series of simultaneously overlaid spectra, and hence resonances which shift as a consequence of the above noted factors can be considered and, where required, negated.

Two-dimensional NMR techniques, including correlation spectroscopy (2D-COSY), total correlation spectroscopy (2D-TOCSY), heteronuclear single quantum and multiple bond coherences (2D-HSQC and -HMBC, respectively), J-resolved spectroscopy (2D-JRES) and high-resolution magic angle spinning (2D-HRMAS), the latter for application to intact tissue biopsy specimens, have also been employed in metabolomics investigations in order to enhance the specificity of biomolecule identification and their quantification,[10,15] but these applications will not be discussed further here.

Data processing for the mass spectrometric technique involves noise reduction, spectral deconvolution, and peak detection and integration, together with chromatogram alignments, component detection and identification and quantification.[5] Of course, both 'raw' and 'real' mass spectrometric datasets contain background and noise levels, the former representing a slowly varying spectral signal shift, the latter involving 'spikes' in signal intensities. These matters will not be further dealt with here, and readers are advised to refer to a number of excellent reviews and developments in this particular subject area.[16–20]

2.4   Data Normalisation, Scaling and Dimensionality Reduction

Prior to the performance of metabolomics data analysis, it is advisable that metabolite resonance, signal or peak intensities should be normalised in order to account for differences in metabolite recoveries during selected extraction processes, where appropriate, or alternatively systematic errors arising from variations in instrumental performance. Such normalisation processes (dataset row operations) can be performed by the employment of either single or multiple internal standards which have been pre-spiked into the sample matrix prior to analysis, or alternatively *via* the utilisation of pre-defined normalisation factors.[21]

A common and frequently employed normalisation process is to express individual resonances or signal intensities relative to that of the entire spectral or chromatographic profile (*i.e.* assignment to each signal a percentage of that of the total spectral or chromatographic features observed), allowing, of course, for the removal of particular regions which should be excluded in view of their interfering and potential confounding nature, for example the residual water and intense urea resonances in urinary ^{1}H NMR profiles, together with those of drugs and their metabolites, and further, perhaps unexpected, exogenous agents. However, as might be expected, this process may give rise to problems if a large number of detectable metabolites are increased or diminished in concentration as a consequence of a particular disease process, or, alternatively, as a response from the administration of a particular therapeutic agent or toxin. If this is the case, then expression of their intensities to that of a total spectral or chromatographic profile may render interpretations of the results acquired problematical.

It is also often desirable to 'standardise' each predictor (X) variable included *via* the now well-known autoscaling process, which involves mean-centering of each data point followed by division by the variable's estimated sample variance so that the mean and variance of each one becomes 0 and

1, respectively (*i.e.* unit variance for the latter); this process allows each measurement to be considered as equivalent, irrespective of their prior magnitudes (for example, in the ^{1}H NMR spectral profiles of healthy human urine, some metabolite signals are reproducibly of a much higher intensity than others, for example, urinary citrate has a much higher urinary concentration than those of bile acids)! In this manner, any heteroscedasticity problems with dataset X variables are removed, although it is important to note that this scaling process is unable to protect against (*i.e.* transform) outlying predictor variable data points (in a univariate sense); indeed, following scaling to mean zero and unit variance, any highly outlying predictor variable ones (and their relative, adverse 'weightings' arising in that variable) remain for subsequent MV analysis, which causes problems with the assumption that individual data points therein are sampled from a normal distribution of metabolite levels (or directly proportional spectroscopic or chromatographic measure). A further disadvantage associated with autoscaling is the deleterious inflation of measurement errors.

However, Pareto scaling is currently a highly recommended prior data treatment step, and this process (which involves mean-centring followed by division by the square root of the X variable sample standard deviation) provides a transformed variable that is in the form of somewhere between no scaling applied and autoscaling, and yields a 'variance' that is equivalent to the sample standard deviation rather than the unit variance one delivered with the autoscaling technique. Although the objectives of this scaling method are alleviations of the relative importance of high values (coupled with a partial preservation of the original dataset structure, and therefore providing sample representations which remain closer to the original metabolite level values than those derived from application of the autoscaling method), its major disadvantage is that it is still particularly sensitive to large fold-changes.

Alternative scaling processes employed by researchers include the Range (mean-centring followed by division by the range of the X variable sample group, a process highly sensitive to outlying data points), VAST (the product of the autoscaled data point and that variable's mean divided by its standard deviation, which is targeted at small metabolic variable fluctuations, but is not appropriate for high levels of induced X variable variation lacking an overall group structure), and Level scaling (the objective of which is to focus on relative responses, and is valuable for specific biomarker identification, but again has problems concerning the adverse inflation of measurement errors) approaches.

For the correct performance of MV statistical analysis of datasets acquired, it is sometimes considered desirable to further alleviate the number of potential predictor (X) variables in order to maximise the achievement of uncorrelated spectral or chromatographic biomolecular features. As outlined below in this chapter, this may be performed *via* determinations of the (univariate) statistical significance of each metabolite variable using t-tests or ANOVA, *etc.*, in addition to the computation of linear combinations of such variables with PCA, or by the employment of evolutionary algorithms, *e.g.* genetic algorithms or genetic programming. In general, evolutionary algorithms are performed in combination with a second analysis algorithm (such as partial least squares or discriminant function analysis) that seek combinations of variables which serve to demonstrate the highest level of effectiveness in the secondary algorithm, and are mediated by the principles of evolution and species selection processes (reviewed in ref. 22). Previously, Kell (2002)[23] has successfully applied such evolutionary algorithms to the analysis of multicomponent metabolomics datasets; these strategies are further discussed in Chapter 3 (section 3.7.1).

2.5   Assumption of Normality

The great majority of parametric statistical assessments and evaluations, including univariate applications such as t-tests, linear regression and analysis-of-variance (ANOVA), and in a more complex sense, MV exploratory data analysis and pattern recognition techniques, rely on the assumption of normality of the distribution from which data are sampled. Hence, it is of much importance for us to evaluate the validity of this essential assumption prior to performing such hypothesis- or non-hypothesis-driven statistical analysis. Indeed, differing shapes and parameters of probability distributions is of much importance regarding pre-defined metabolomics research objectives and, in view of this, the observed distributional frequencies in an experiment should be monitored in order to determine if they correspond to (or significantly depart from) a theoretical model normal distribution. Notwithstanding, although there are many non-parametric univariate statistical tests available which overcome requirements for such assumptions, and also MV ones which involve the application of non-parametric methods (for example, a PCA can be based on Spearman correlation coefficients rather than Pearson ones), such model systems can be considered to be less powerful than parametric ones, and hence determinations of the 'best-fit' of a continuous data distribution to that of a (perhaps simulated) theoretical normal one remains a critically important practice for the prior analysis of datasets, if only in a univariate sense. However, Mardia's test of MV normality can be applied to determine whether the observations present in MV datasets are sampled from an MV normal distribution (the effective operation of many of these techniques is critically dependent on the satisfaction of this assumption!).

Currently, there is a range of such normality/goodness-of-fit tests available, although the χ^2 goodness-of-fit one must be employed for relatively large sample sizes.[24] These tests obviously determine whether or not a particular frequency distribution, or smaller sample of size n, fits a specific distributional pattern. Of these tests that are available, one of the most employed is a modification of the Kolmogorov–Smirnov (K–S) curve fitting algorithm, *i.e.* the Lilliefors normality test,[25] which, of course, has the null hypothesis that the distribution of the error observed between the experimental (test) distribution and a normal one is normally distributed. A further means of determining such goodness-of-fit is the Anderson–Darling (A–D) test (also a modification of the K–S test), and provides a greater weight to the distributional tails than the K–S method. Since the A–D test employs a specific distribution for the computation of critical values, this permits a more sensitive test, although such critical values are required to be computed for each of these distributions. However, the Shapiro–Wilks (S–W) approach has been shown to be one of the most powerful normality tests, and also represents a significant improvement on the K–S one.[26] Each of these test systems, however, generates differing results, with some accepting and others rejecting the null hypothesis of normality, and such phenomena create a high level of confusion amongst researchers. Razali and Wah (2011)[27] evaluated the power of four such normality testing systems, with the power of each assessed by the Monte-Carlo simulation of a series of datasets produced from both alternative symmetric and asymmetric distributions. Results acquired revealed that the S–W test system had the greatest power in this context, with the A–W, Lilliefors and K–S ones having less power in that order. Notwithstanding, the power of all of these tests remains rather low for small sampling groups.

Mendes and Pala (2003)[28] made comparative evaluations of the type I errors and powers of the K–S, Lilliefors and S–W tests, and found that their effectiveness in these contexts was in the order S–W, Lilliefors $>$ K–S [an observation consistent with that of Razali and Wah (2011)[27]], and that the highest power of all these tests was observed when they were performed on datasets sampled from an exponential distribution. A further investigation performed by Oztuna *et al.* (2006)[29] disclosed that the Jarqua–Bera (J–B) test was the most powerful for detecting the normality of distributions, whereas the S–W one was the most powerful for revealing those of a non-normal character, the latter achieving a sufficient level of power for relatively small sample sizes. Interestingly, Saculinggan and Balase (2012)[30] recently investigated the power of a total of six well-established goodness-of-fit tests for normality [χ^2 goodness-of-fit, K–S, A–P, S–W, Lilliefors and the D'Agostino–Pearson (D–P)] tests for small, intermediate and large sample sizes, and also a range of dataset contamination levels; the power of each of these tests was then determined *via* Monte-Carlo simulations involving a sample size of 10 000 and a pre-selected type I error rate for each pre-generated alternative distribution. Their results demonstrated that, as expected, the power of all these testing methods was low for small sample sizes ($n < 20$), but when the n value was 20, the S–W and A–D tests were the most powerful. However, when $n = 60$, the S–W and Lilliefors tests served as the best ones available, with the former of these representing the most powerful one for large sample sizes. These researchers also concluded that for large sample sizes, the D–P test achieved the greatest power level under all conditions that were evaluated. This D–P test is known as an 'omnibus' test since it synchronously employs test statistics for both skewness and kurtosis to provide a single *p* value (DP value $= Z_{g1}^2 + Z_{g2}^2$, a parameter which follows a χ^2 distribution with 2 degrees of freedom).

Intriguingly, Seier (2002)[31] also investigated the power of selected univariate normality tests *via* computer simulations for small, moderate and large sample sizes, and also involving a series of symmetric, skewed, contaminated and mixed (including bimodal) distributions, and proposed a newly developed test system based on skewness and kurtosis. This investigator found that some regression-type tests appear to serve as the best option for testing normality in view of their power over alternative options. Indeed, Cen-Shapiro's alternative QH* test,[32] which is based on normalised spacings, performs more consistently than further Q regression tests.[33] However, for the detection of symmetric distributions with high levels of kurtosis, the employment of tests based on skewness-kurtosis parameters, *e.g.* those of D'Agostino (1990)[34] and DeCar and Hosking (1990)[35] were found to be effective, although it should be noted that Pearson's well-known measure of kurtosis $[\beta_2 = E(x - \mu)^4/[E(x - \mu)^2]^2]$ has a value of 3.0 for a range of further symmetric distributions in addition to the normal one. However, the G_ω^{2*} statistic [where $G_\omega^{2*} = [a(b_1)^{1/2}]^2 + [z_\omega]^2$, with $a = n/(n-2)[6/(n+1)]^{1/2}$ and $\beta_1^{1/2} = (1/n)\Sigma[(x_i - \bar{x})/s]^3]$ is the most efficient evaluation method for scale-contaminated normal distributions in which the standard deviation of the contaminating distribution may be greater or smaller than that of the uncontaminated, main distribution investigated. However, the manner in which kurtosis is measured contributes to the power of the test systems employed for differing classes of distributions. Indeed, selected tests fail to perform effectively, some especially when applied to distributions with kurtosis markers lower than that expected for a normal distribution (3.0).

If the sample distribution evaluated has a kurtosis value of >3.0, then it is described as leptokurtic (with central peak lower and broader, and distributional tails longer and broader than those of a perfect normal distribution); hence, a Student's t-distribution with $\nu = 4$ degrees of freedom has infinite kurtosis. However, distributions with kurtosis values <3.0, with a central peak lower and broader, and tails shorter and thinner than those of the perfect normal distribution, are known as platykurtic distributions. The ultimate distributional extreme in this case is one with only two likely outcomes (such as the results arising from the repetitive tossing of a coin). In such a case, there is no central peak and no tails, and hence the 'distribution' has the lowest possible kurtosis value (1.0).

The measurement of kurtosis has much relevance to the analysis of metabolomics data, since if a large biofluid metabolite concentration (or proportional signal or peak intensity) dataset is acquired, and there are significant or highly significant 'Between-Disease Classifications' effects observed for one or more of the biomolecules monitored (or signal or peak area directly related to this), then the (column) 'distribution' will clearly be bimodal (with two distributional 'peaks' within a column dataset), and hence kurtosis measurements will be significantly lower than those expected for a normal distribution (*i.e.* $\ll 3.0$), although they may provide us with some valuable information concerning biomolecules which may serve as valuable biomarkers for the disease process investigated.

As an example case study, we evaluated the normality of each of the 222 potential predictor (X) variables in an MV urinary ^{1}H NMR dataset comprising two disease classifications using the S–W, A–D, Lilliefors and J–B testing systems (Table 2.1). We also applied a range of dataset transformations [logarithmic, square root, $(1 + X)^{1/2}$ and arcsine $\sqrt{\%}$ ones], and also standard (sum) normalisation (to the total spectral intensities), together with standardisation procedures (the latter involving mean-centering and division by the standard deviation or the square root of the standard deviation (the latter known as Pareto-scaling) in order to investigate their influence on the fitting of these X variable datasets to a perfect (albeit theoretically-hypothesised) univariate normal distribution. From these

Table 2.1 (a) Statistical significance of the normality testing of 222 explanatory X variables in an MV ^{1}H NMR 'intelligently bucketed' urinary dataset from patients with a neurodegenerative lysosomal storage disease (the significance levels at $p < 0.05$ and < 0.01 are indicated) (n = 14). A range of dataset transformations, together with selected normalisation and standardisation processes, were evaluated by four different normality tests. (b) As (a), but for a corresponding heterozygous carrier (parental) control group classification (n = 46). Abbreviations: SW, Shapiro–Wilks; AD, Anderson–Darling; L, Lilliefors; and JB, Jarqua–Bera tests for normality. Abbreviations: ns, not significant.

(a)

Disease classification dataset		Significance level		
Data preprocessing/normalisation	Test	ns	0.05	0.01
Raw (unprocessed)	SW	83	27	112
	AD	89	22	111
	L	108	23	91
	JB	129	11	82
Square root	SW	124	16	72
	AD	127	24	71
	L	137	35	50
	JB	169	16	37
$(1 + x)^{1/2}$	SW	92	24	106
	AD	95	21	106
	L	111	23	88
	JB	135	10	77
$\ln(1 + x)$	SW	90	28	104
	AD	93	27	102
	L	114	22	86
	JB	139	15	68
Arcsine	SW	113	31	78
	AD	114	30	78
	L	131	35	56
	JB	159	22	41
Normalised	SW	125	24	73
	AD	127	20	75
	L	134	35	53
	JB	169	13	40

Table 2.1 (*Continued*)

(a)

Disease classification dataset		Significance level		
Data preprocessing/normalisation	Test	ns	0.05	0.01
Normalised/standardised	SW	125	24	73
	AD	127	20	75
	L	134	35	53
	JB	169	13	40
Normalised/Pareto-standardised	SW	125	24	73
	AD	127	20	75
	L	134	35	53
	JB	169	13	40

(b)

Control classification dataset		Significance level		
Data preprocessing/normalisation	Test	ns	0.05	0.01
Raw (unprocessed)	SW	15	13	194
	AD	22	12	188
	L	33	22	167
	JB	36	6	180
Square root	SW	44	20	158
	AD	52	18	152
	L	65	25	132
	JB	82	26	114
$(1+x)^{1/2}$	SW	25	11	186
	AD	29	15	178
	L	43	13	166
	JB	46	10	166
$\ln(1+x)$	SW	36	13	173
	AD	43	7	172
	L	49	16	157
	JB	56	8	158
Arcsine (normalised)	SW	41	19	162
	AD	50	25	147
	L	69	26	127
	JB	75	27	120
Normalised	SW	40	18	164
	AD	46	23	153
	L	72	21	129
	JB	85	18	119
Normalised/standardised	SW	40	18	164
	AD	46	23	153
	L	72	21	129
	JB	85	18	119
Normalised/Pareto-standardised	SW	40	18	164
	AD	46	23	153
	L	72	21	129
	JB	85	18	119

analyses, it was found that high or very high percentages of these sample groups exhibited significant deviations from normality at the 1% level (with fewer of these significant at the 5% one). However, these analyses also revealed that the square root transformation served as the most effective one for satisfying the normality assumptive criterion of these MV datasets, albeit only partially, followed by those which were normalised, or normalised and standardised; the rank of their effectiveness in this context then followed the sequence arcsine $\sqrt{\%} > \ln(1+X) \geq (1+X)^{1/2} >$ the raw (unprocessed and untransformed) dataset (as expected, there were no differences observed between datasets which were normalised, or normalised and then standardised, the latter *via* the auto- or Pareto-scaling routes). Therefore, these operations clearly show that the normality of these particular datasets is improved somewhat by these preprocessing treatments; however, even with the most effective transformation applied ($\sqrt{X}$), there were still many highly significantly non-normally distributed predictor (X) variables remaining!

Of further interest are the differences observed in the abilities of the four normality testing systems employed. Indeed, these results demonstrated that the S–W and A–D tests found the highest number of significant potential explanatory (X) variables in this context, followed by the Lilliefors and then the J–B tests, *i.e.* their normality testing stringencies were found to be in the order S–W$\geq$A–D$>$Lilliefors$>$J–B.

Figure 2.1 shows a distributional histogram of just a single intensity bucket from the ^{1}H NMR profiles of urine samples collected from the two

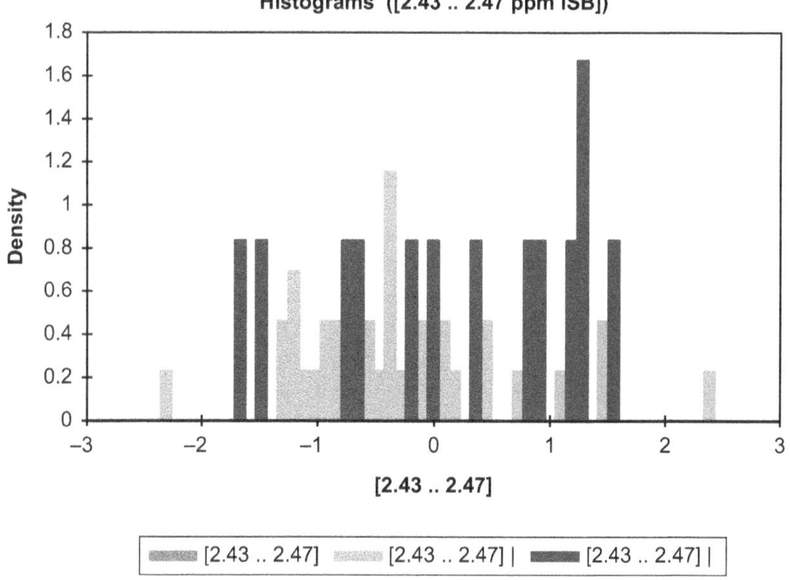

Figure 2.1 Distributional histogram of a single 'intelligently-selected' the L-Glutamine-C_4-CH_2 function intensity bucket from the ^{1}H NMR profiles of urine samples collected from two disease classifications [the complete dataset comprised 222 potential predictor (X) variables and a total of n = 60 samples].

disease classifications involved in the investigation detailed in Table 2.1. Clearly, there appear to be major deviations from an assumed normal distribution for each one!

Figure 2.2 exhibits a further example which involves the attempted fitting of a normal distribution to the distribution of blood serum thyroxine (T4) concentrations for three thyroid disease classifications, *i.e.* euthyroid, hypothyroid and hyperthyroid conditions (and involving n = 300, 300 and 71 patients, respectively). Although there were no deviations from an assumed normal distribution found for the hyperthyroid group of patients, those observed for the euthyroid and hypothyroid classification groups displayed extremely highly significant ones. Indeed, for the serum T4 levels of these two groups, it appears that the distributions plotted are multimodal, or bi-modal at the very least, an observation which may be related to the influence of further lateral variables which are not considered here.

Of the tests available for the evaluation of multinormality, Mardia's test method (which is based on MV skewness and kurtosis statistics) is one of the most popular, although unfortunately it remains very infrequently employed for the prior testing of multidimensional metabolomics datasets! This test is based on computation of the standardised third and fourth moments, and involves the construction of affine invariant test statistics, a process in which the dataset vectors are primarily standardised *via* employment of the sample mean vector and covariance matrix.

2.6 Analysis-of-Variance (ANOVA): Experimental Design and Analysis

Essentially, ANOVA serves as a methodology for isolating variance components (sources of variation) ascribable to the effects of factors from the total variation of a response measure. Hence, the dataset must contain clear information regarding any given source of variation, and also its nature (fixed or random), prior to us attempting to estimate its contribution to a particular mathematical model. Indeed, estimation of these components is best achieved from experiments which have been specifically designed for this purpose. The classification procedure to be employed in an ANOVA model will, of course, be critically dependent on the particular nature and number of the independent variance components, which, in the light of various restrictions, can be identified as possible contributors to the total variance of the dataset acquired. Indeed, such datasets can be classified with regard to each source of variation (according to a pre-specified mathematical model), and a complete classification represents an essential primary stage of the complete analysis.

2.6.1 Model I: Fixed Effects

In this model, the influences exerted by the main 'treatment' factors (or further major sources of variation investigated) are 'fixed' (or pre-defined),

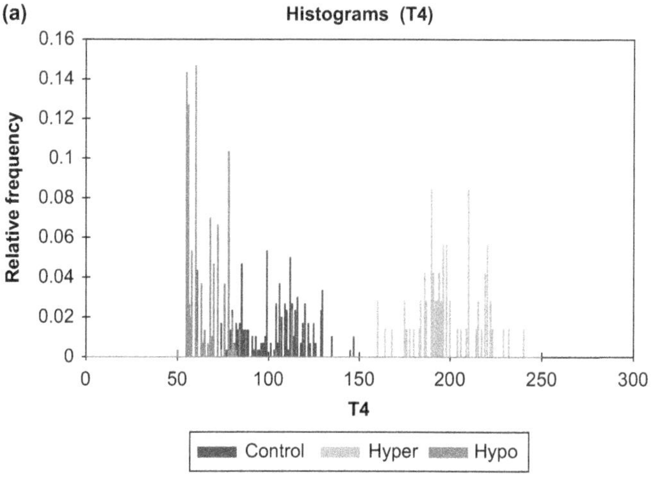

(a)
Histograms (T4)
Relative frequency
0.16
0.14
0.12
0.1
0.08
0.06
0.04
0.02
0
0   50   100   150   200   250   300
T4
Control   Hyper   Hypo

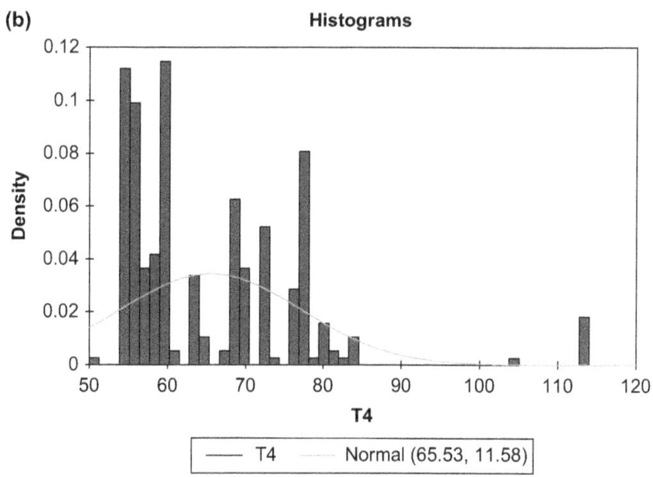

(b)
Histograms
Density
0.12
0.1
0.08
0.06
0.04
0.02
0
50   60   70   80   90   100   110   120
T4
T4   Normal (65.53, 11.58)

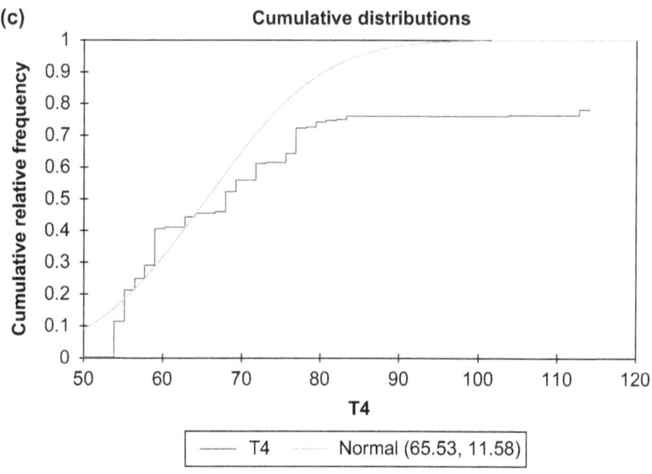

(c)
Cumulative distributions
Cumulative relative frequency
1
0.9
0.8
0.7
0.6
0.5
0.4
0.3
0.2
0.1
0
50   60   70   80   90   100   110   120
T4
T4   Normal (65.53, 11.58)

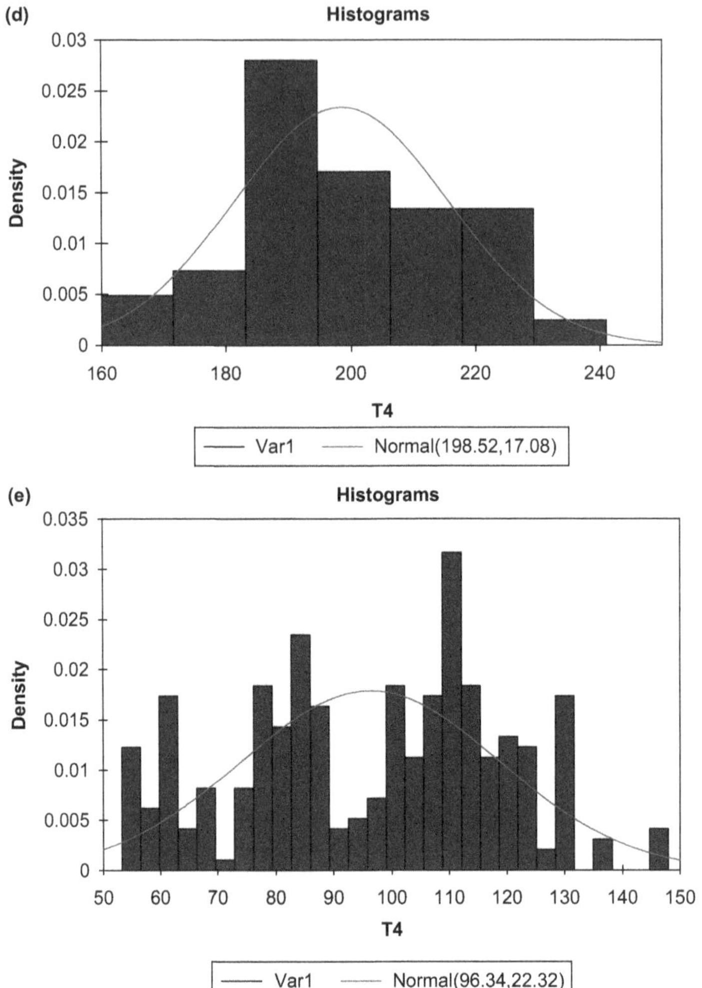

Figure 2.2 (a) Relative frequency histograms of the distributions of blood serum thyroxine (T4) concentrations of patients within three thyroid disease classifications: healthy controls (euthyroid, $n = 300$); hypothyroid ($n = 300$); and hyperthyroid ($n = 71$). The overall mean $\pm$ SD values for the entire (3 classification) dataset was 93.38 ± 42.82 ng ml^{-1}; (b) relative frequency histogram and modelled normal distribution for the hypothyroid disease classification, the latter based on the mean $\pm$ SD parameters of 65.52 ± 11.58 ng ml^{-1} for this group; (c) cumulative relative frequency diagram for the hypothyroid patient group showing clear differences between the observed dataset and the modelled normal distribution; (d) relative frequency histogram and modelled normal distribution for the hyperthyroid disease classification, the latter based on the mean $\pm$ SD parameters of 198.52 ± 17.08 ng ml^{-1} for this group; (e) relative frequency histogram and modelled normal distribution for the euthyroid classification, the latter based on the mean $\pm$ SD parameters of 96.33 ± 22.32 ng ml^{-1} for this group.

and are assumed to determine the population mean values, *i.e.* the sample means determined estimate the (putatively corresponding) 'fixed' population ones. Hence, for a very simple (randomised blocks) fixed effects ANOVA model, any observation will represent the sum of three components, specifically an 'overall' mean value, a disease- or treatment-induced deviation from this value plus a random (error) element arising from a normally distributed population of mean zero and standard deviation σ. Indeed, the mathematical model for this simple model comprising two sources of variation is that described by eqn (1), where T_I is the 'treatment' effect, and ε_{ij} is the error (residual) component, which, in a metabolomics experiment

$$X_{ij} = \mu + T_I + \varepsilon_{ij} \tag{1}$$

may represent 'within-groups' and, unless controlled for in more detailed ANOVA-based models/investigations, also 'Between-Patient or -Participant' variance within each disease or treatment classification group.

In this model, the researcher involved is interested in specific comparisons of pre-selected classification groups. For example, it may be required to determine the concentration of a particular metabolite in a biofluid by three different analytical methods, with two or more samples collected from each participant, and each method involving replicate determinations. Clearly, the 'Between-Methods' variance component, which may at least be partially attributable to potential analytical biases arising from the analytical technique or techniques employed, differs from those ascribable to sampling and analytical errors. In this model, we simply wish to compare the mean values of each analytical method, and not estimate the 'Between-Sample Donors' and 'Between-Replicates-within-Methods' variance components, which can also be considered in more complex experimental designs.

2.6.2 Model II: Random Effects

Random effects are explored *via* the sampling of (hopefully) normal populations, and 'Between-Patient-within-Disease or Treatment' groups serve as good examples of this since we may be interested in evaluating the component of variance amongst patients, together with comparisons of it with perhaps further sources of variation and the overall error (residual) component. In this manner, variance components can be viewed as a random sample from an infinitesimal population of such classifications, and hence the overall purpose of the experiment is to estimate the population variance. Indeed, repeated biofluid or tissue sample collections from a range of donors (perhaps stratified according to age, gender or further latent variables), together with repeated determinations of a metabolite concentration using the same technique, can all be regarded as random elements of total populations of such components, and all three of these ('Between-Donors', 'Between-Repeated Samples' and 'Between-Replicate Analyses') have components of variance (in this case $\sigma_D{}^2$, $\sigma_S{}^2$ and $\sigma_R{}^2$, respectively) which can be estimated. Hence, the purpose of this experiment is to successfully estimate and determine the magnitude of each of these components.

2.6.3 Hierarchical or 'Nested' Models

A typical classification in chemometrics is known as a Hierarchical or Nested classification, and an example of this would be a biochemical researcher sampling a biofluid at two or more time-points from a particular group of patients (healthy control or diseased, say n = 10) in order to measure a particular metabolite (for example, lactate or 3-D-hydroxybutyrate), and then analysis is performed in triplicate on each one. In this experiment, we actually have three variance components: firstly, that 'Between-Patients'; secondly, that 'Between-Samples-within-Patients'; and thirdly, that 'Between-Replicate Analyses-within-Samples'. A final variance estimate is that attributable to error, *i.e.* that which is unexplainable by the above three sources of variation.

2.6.4 Factorial/Multifactorial Models

ANOVA is also frequently employed to explore the effects of multiple factors simultaneously, and in cases where the experimental design involves observations made at all possible combinations of levels (or sub-groups) of each factor, it is commonly known as a factorial experiment. Indeed, such factorial experiments do, of course, exert a higher level of efficacy over a (multiple) series of single-factor designs; this efficacy increases with the number of factors considered. The major advantage offered by this experimental design is the inclusion of components of variance attributable to interaction effects, in addition to those of the main factors.

In a three-factor experimental design involving factors A, B and C, an ANOVA model will incorporate variance components arising from the effects of these three main factors (A, B and C), together with those ascribable to the first- (AB, AC and BC) and second-order (ABC) interaction components, and all of these require hypothesis testing [the mathematical model for this experimental design is depicted in eqn (2)]. Notwithstanding, the propagation of interaction components in such models does increase the risk of a type I error, *i.e.* that false-positives are generated by chance in some tests performed. However, higher-order interactions (*e.g.* that of the ABC one in this design) are rarely encountered in practice in such multifactorial experimental designs, and it is considered crucial to evaluate the presence of interactions and their significance between two (or more) experimental factors in order to enhance the interpretational status of datasets acquired. The testing of only single factors one at a time not only hides such interactions, but also generates inconsistent experimental results; for example, as indicated in Figure 2.3, if the value of a response of a dependent variable to the first factor classification is significantly or substantially influenced or altered by changes in the second factor, then clearly the factorial or

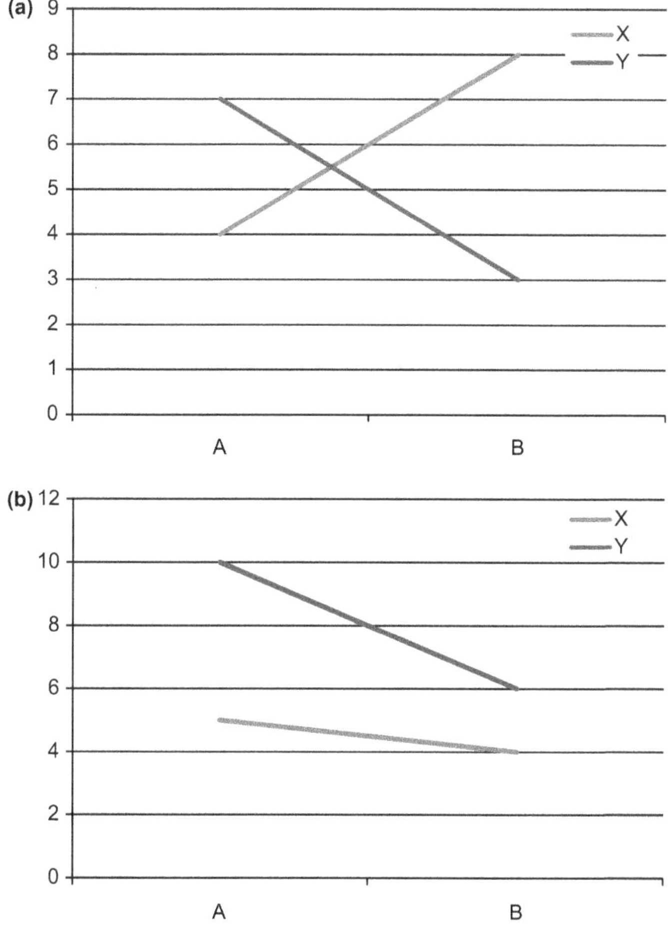

Figure 2.3 Plots demonstrating interaction in a two-factor experimental design. A and B represent two classifications or levels of the first factor, and X and Y two classifications or levels of the second. Interaction can arise from differences in (a) the direction or (b) the magnitude of the responses observed.

multifactorial experimental design is the most appropriate (albeit univariate) statistical model performable for the analysis.

$$Y_{ijk} = A_I + B_j + C_k + AB_{IJ} + AC_{Ik} + BC_{Jk} + ABC_{IJk} + e_{ijk} \qquad (2)$$

Researchers are advised to exercise much caution when significant interactions are detected, and should primarily test these interaction terms in order to evaluate their significance and effects on the Y response variable (quantitative or qualitative). Indeed, the analysis should be subsequently expanded, and, if significant, the effects of one or more of the main factors (A and B in a relatively simple two-factor design) are required to be critically

Table 2.2 Experimental design for the univariate or ASCA analysis of biomolecule concentration/intensity data acquired in a metabolomic time-series investigation, representing a combination of a completely randomised with a randomised block design: mixed model with Treatments 'nested' within each Participant ($n = 2$ per Participant). Abbreviations: $\kappa_{T(P)}{}^2$, 'Between-Treatments-within-Participants' component of variance (Fixed Effect); $\kappa_S{}^2$, 'Between-Diurnal Time-Points' component of variance (Fixed Effect); $\sigma_P{}^2$, 'Between-Participants' component of variance (Random Effect); $\sigma_{TP}{}^2$, Treatment $\times$ Participant Interaction component of variance; $\sigma_{TS}{}^2$, Treatment $\times$ Diurnal Time-Points Interaction component of variance; $\sigma_{PS}{}^2$, Participant $\times$ Diurnal Time-Point component of variance; σ^2, Error (Residual) variance.

Source of variation	Levels	Degrees of freedom (d.f.)	Nature	Parameters estimated for mixed model
Between Treatments (Active *vs.* Control) (T)-within-Participants	2	1	Fixed	$\sigma^2 + 30\sigma_{TP}{}^2 + 180\kappa_{T(P)}{}^2$
Between Participants (P)	30	29	Random	$\sigma^2 + 12\sigma_{PS}{}^2 + 12\sigma_P{}^2$
Between Diurnal Time-points (S)	6 per participant	5	Fixed	$\sigma^2 + 12\sigma_{PS}{}^2 + 60\kappa_S{}^2$
Treatment $\times$ Participant Interaction	60	29	Fixed	$\sigma^2 + 30\sigma_{TP}{}^2$
Treatment $\times$ Time-point Interaction	12	5	Fixed	$\sigma^2 + 30\sigma_{TS}{}^2$
Participant $\times$ Time-point Interaction	180	145	Fixed	$\sigma^2 + 12\sigma_{PS}{}^2$
Error (Residual)	n/a	145	n/a	σ^2
Total	n/a	359	n/a	n/a

reconsidered. However, as we might expect, one or more significant interaction terms will frequently serve to mask those exerted by the main effects.

Table 2.2 exhibits a 'cross-over' analysis-of-variance (ANOVA)-based time-series experimental design which was recently employed by the authors in a clinical trial; each of the participants recruited received both an 'active' test product treatment and a control (placebo) one. In this model, ANOVA was employed to determine the statistical significance of each component of variance for each metabolomics predictor (X) variable (monitored and determined by ^{1}H NMR analysis in this case). Indeed, the aim of this procedure was to determine the significance of the 'Between-Treatments' and 'Between-Time-Points' effects for each of the Treatment classifications (*i.e.* the 'active' product *versus* its placebo control) incorporated into the investigation, and also the further components of variances involved, specifically that 'Between-Participants' and those arising from the Treatment $\times$ Diurnal Time-Point, Treatment $\times$ Participant and Participant $\times$ Diurnal Time-Point first-order interactions. The first of these interaction components was considered to be of critical importance regarding evaluations of the relative effectiveness of the two formulations investigated. A sufficient time period (14 days) was adhered to prior to permitting the participants to 'cross-over' from the placebo to the treatment regimens or *vice-versa*.

The experimental design for this investigation is classified as a mixed-model, 3-factor system with treatments (one 'active' product tested against a corresponding placebo control) and time-points at which the determinations are made being fixed effects at 2 and 6 levels, respectively, and participants (n = 30 in total) representing a random effect. Mixed-model component analysis for each biomolecule determined comprises the three main effect factors, their associated interactions and fundamental (residual) experimental error. The 'Between-Treatments' factor is 'nested' within each Participant (Table 2.2), and in this model the dataset was routinely $\log_e$-transformed prior to statistical analysis in order to satisfy assumptions of normality and variance homogeneity.

This experimental design serves as an example for the univariate analysis of multicomponent metabolic datasets, and has also been adopted by the authors to form the basis of a model for ANOVA-Simultaneous Component Analysis (ASCA) as described below.

2.6.5 ANOVA-Simultaneous Component Analysis

ASCA represents an MV extension of the univariate ANOVA approaches described above, and this form of analysis can also incorporate experimental designs containing one or more interaction effects. For the simplest form of this complex MV model [eqn (3)], the algorithms employed primarily partition the variance of the complete dataset into those attributable to each factor (say A and B), together with the AB first-order interaction term where AB_{ij} depicts the interaction effect and e_{ij} the residual (error) term.

$$Y_{ij} = A_I + B_j + AB_{IJ} + e_{ij} \tag{3}$$

The SCA portion of the analysis involves the application of PCA to each of the A, B and AB components of variance extractable so that we are permitted to evaluate MV variations within each partition (following a pre-specification of the number of components to be employed for each model). The ASCA technique and its applications to the analysis of metabolomics datasets are discussed in more detail in Chapters 3 and 4 of this volume.

2.6.6 Further Considerations of Interaction Components of Variance in MV Modeling

It should also be noted that PLS-DA, Partial Least Squares Regression (PLS-R) and Principal Component Regression (PCR) models (the latter two discussed in more detail in Chapter 3), for example, can also incorporate sources of variation arising from interactions between one or more of the potential predictor (X) variables. Notwithstanding, much caution must be exercised regarding such applications, since in a typical metabolomics dataset containing, say, 200 separate biofluid metabolite level variables, then the number of only first-order interactions available for testing in this manner is 199^2, *i.e.* 39,601, and the inclusion of such an elevated number of potential 'predictor' variables will obviously give rise to a series of substantial overfitting problems (rather than simply those commonly encountered with the relatively simple ones commonly encountered in experimental models of insufficient sample size which do not consider interactions, at that)! Additionally, although higher order interactions such as second-order, 3-variable ones are also possible (albeit unlikely), they have not even been considered in such a complex but poorly designed, lacklustre model! However, in analytical datasets in which we incorporate only a small number of such predictors, *e.g.* up to 10 (which may be generated *via* the prior performance of an acceptable and pre-validated metabolomics dataset filtering process), or, alternatively, in situations where we may have special reasons for considering only a small number of selected interactions (say, those restricted to only first-order ones involving six or so selected X variables), then of course we may proceed with such an analysis, provided that we have a sufficient sample size to accommodate it.

2.7 Outline of the Applications of Univariate Approaches to the Analysis of Metabolomics Datasets

Despite the potentially wide range of applications, relative power and largely accepted versatility of MV statistical techniques and/or models available for the analysis of metabolomics datasets, it is important to note that such approaches are, of course, subject to what is now known as the 'curse-of-dimensionality' problem, which commonly arises when such datasets contain an excess of 'sparse' data regarding the provision of input variables employed, which frequently outnumbers the biofluid/tissue biopsy sample size available. Hence, with a selected sample size, the performance of a particular algorithm or algorithmic model will deteriorate rather than improve for systems which involve more (or many more) than the optimal number of predictor (X) variables. In this manner, attempts to facilitate the MV model's 'fit' to observed datasets can introduce substantially deleterious 'overfitting' problems which will, of course, severely limit its applicability in terms of the provision of marked errors, and hence also markedly hamper its predictive power; these problems are extensively delineated in Chapter 1, and also later on in this one. In particular, as also noted in Chapter 1, the absolute requirement for the performance of essential validation and cross-validation performance testing is a necessary pre-requisite, which is often ignored by many researchers working in this area. However, such data analysis can also be focused on univariate approaches which employ conventional/traditional statistical techniques [*i.e.* those that serve to consider single predictor (X) variables individually and independently of the remaining ones present in the MV dataset available]. In this context, it should be clearly noted that the incorporation of both univariate and MV analysis techniques into the final statistical model does *not* represent a problematical process, and often the two forms of data analysis performed harmoniously together will provide a high level of metabolomics information. Indeed, the authors are aware that in many studies performed in this manner, the univariate analysis of MV datasets can provide similar (or virtually equivalent) trends and results to those achieved *via* MV analysis of these by PCA, PLS-DA or further derivations of the latter, allowing, of course, for the statistical correction of the significance level of each X variable found for all the possible ones available (say, 200 or many more) in view of the much increased probability of the occurrence of Type I errors when such a large number of simultaneous tests are performed.

However, such univariate methods available are occasionally employed to serve as statistical 'filters' for the metabolomics analysis of multicomponent biofluid or tissue biopsy datasets in order to retain only those that are statistically significant at a pre-specified, albeit univariate, pre-selected p value ($p < 0.05$, or more realistically a Bonferroni-corrected one). In this manner, a potentially information-rich set of metabolic features is then subjected to a series of MV analyses according to recommended guidelines. However, it is, of course, important to consider that such an approach fails to recognise correlations between ^{1}H NMR bucket or LC-MS mzRT feature intensities, and hence information concerning such correlated or multicorrelated metabolite levels is not considered in these experiments (except, albeit subsequently, those between the remaining, univariately significant ones). However, the application of such MV approaches without consideration of pre-selected univariately significant metabolites/biomarkers also involves the prior acceptance of many important assumptions regarding the distributions of each predictor (X) variable, notably the normality of their distributions, homoscedasticity (homogeneity of intra-sample variances, and also those 'Between-X variables', the latter representing a markedly important point for consideration), and also the independence of samples entering the investigation, *i.e.* in a metabolomics sense, the entry or incorporation of one sample (biofluid or otherwise) into the multianalytical dataset should provide no clues or identity of information regarding the metabolite level values or classification status of others therein. Basically, for the latter consideration, one interpretation of this is that the researchers involved should not be 'selective' regarding the entry of samples into the study, especially if they are already aware of the positive (or, where appropriate, negative) classification status of such specimens.

Furthermore, when performing primary univariate statistical tests (t-tests, ANOVA, *etc.*) on MV analytical datasets, researchers should also be aware of the much increased risk of false-positive results (*i.e.* type I errors) possible; unfortunately such issues are frequently overlooked by research workers who subject their untargeted metabolic datasets to such univariate analysis, a process that can lead to many statistically compromised results!

2.7.1 More on Essential Assumptions Required

2.7.1.1 *Further Considerations Regarding the Normality Assumption*

As noted above in Section 2.5, the statistical assumptions always required for the correct performance of both the univariate and MV analysis of datasets tend to depend on basic mathematical constructs which are, unfortunately, rarely encountered or even recognised in research practice. With regard to the normality assumption, which is applicable to the testing of univariate, bivariate, trivariate and MV, *etc.* datasets (all, but especially the latter in a metabolomics context), this problem has been put into context in Section 2.5. Notwithstanding, it is important to note that for small or relatively small sample sizes, the tests available lack the power required to detect such non-normal distributions [although we should, necessarily, not neglect the bimodal distribution of two or more populations for each of the exploratory predictor (X) variables in which there are clear or very clear 'Between-Classification' differences, in which case the normality assumption clearly falls down!]. However, as the sample size increases, the normality assumption is less of a 'hindrance' in view of the Central Limit Theorem. The primary evaluation and testing of this assumption has been described by Box (1953)[36] as 'commencing a journey in a rowing boat in order to determine whether or not the launch of an ocean liner presents hazards'. Fortunately, such parametric statistical tests and evaluations remain somewhat robust against small departures from normal distributional assumptions. Moreover, as noted in Section 3.5, such tests of normality for the predictor (X) variables can be supported by descriptive statistical measures such as those for skewness or kurtosis. Whilst valuable, the employment of probability or Q-Q plots for each X variable is potentially problematic in view of the very large number of them to be examined for each metabolomic dataset (for example, several hundreds or even thousands!).

2.7.1.2 *Homoscedasticity (Homogeneity of Variances) Assumption*

This statistical assumption is essential for the testing of all parametric, univariate (and hence necessarily MV) statistical hypotheses and their future performances, and it requires that all 'Within-Classification' or 'Within-Group' variances are homogenous (*i.e.* exhibit homoscedasticity). If these variances are, however, heterogeneous (*i.e.* are heteroscedastic), then the probability of attaining a significant difference between the mean values of two or more sample classifications (assuming that there is a true difference in these mean population values) is likely to be smaller than that required under homoscedastic constraints. However, it should also be noted that with the standardisation (autoscaling) technique that is often considered essential for the analysis of MV datasets as a preprocessing stage, explanatory (X)

Table 2.3 Statistical significance of ^{1}H NMR untransformed/non-centered and standardised, intelligently bucketed MV urinary ^{1}H NMR spectral dataset containing 222 potential predictor (X) variables *via* (a) Levene's and Bartlett's univariate tests for heteroscedasticity (intra-sample variance heterogeneity) and (b) Grubb's univariate test for statistical 'outliers'.

(a)

Levene's	$p<0.01$	$p<0.05$	Bartlett's	$p<0.01$	$p<0.05$
	99	40		193	7

(b)

p-value	Number of buckets
<0.01	112
<0.05	27

variable variances (and therefore standard deviations) are standardised to a value of 1.00 (unit variance), with each sample having a mean of zero. However, this process will not prevent 'Between-Classification-within-X Variable' heteroscedasticity problems, particularly when there are clear statistically significant differences between the two mean classification values compared (in which case we will also experience a *bimodal* distribution of observational measures for each predictor variable involved). If indeed there are clear 'Between-Classification' differences present within selected X predictor variables, then each of the two (or more) distributions of biofluid or tissue biopsy samples therein may have significantly or very different variances (the unit variance parameter arising from the standardisation process therefore represents a weighted mean of the individual variances of the two sampling distributions). In any case, if this standardisation process is performed in such situations, where exactly is the statistical logic in that?

Statistical test systems available for the determination of heteroscedasticities amongst multiple predictor (X) variable datasets are Bartlett's and Levene's tests, with the latter displaying less sensitivity to departures from normality. Above we show examples of the application of both these test systems to an untransformed/non-centered and standardised, intelligently bucketed MV urinary ^{1}H NMR spectral dataset containing 222 potential predictor (X) variables (Table 2.3); each bucket intensity was primarily normalised to that of a fixed added concentration of a ^{1}H NMR chemical shift reference compound [3-trimethylsilyl-1-[2,2,3,3-^{2}H$_4$] propionate (TSP), $\delta = 0.00$ ppm], which also served as a quantitative internal standard. Clearly, there are high proportions of these variables which were significantly heteroscedastic using both of these testing systems: indeed, with Levene's test, 45 and 18% of these variables were significant at the $p = 0.01$ and 0.05 levels, respectively, whereas for Bartlett's test, as many as 87% were significant at the $p = 0.01$ level (together with 3% at the $p = 0.05$ level).

Similarly, we conducted a test for outliers (albeit, a univariate one) in order to determine how many of these 222 predictor variables contained significantly outlying data points (these tests were performed using Grubb's test). Again, the results acquired revealed that a very high proportion of the variables had outlying data points (112 and 27 of them were significant at the $p = 0.01$ and 0.05 levels, respectively). These results are displayed in Table 2.3 and Figure 2.4.

2.7.2 Bonferroni Correction for Multiple Comparisons of Mean Values

For all of the hypotheses tested in an MV metabolomics dataset consisting of perhaps 200–1000 potential predictor variables, a widely accepted index known as the family-wise error rate (FWER) is employed in order to provide important information regarding the probability of acquiring one or more false-positive values.[37] Therefore, if k independent comparisons are made, the FWER is elevated according to the formula $1 - (1 - \alpha)^k$, where α is the pre-specified probability threshold value applied to each test performed. In order to ensure the maintenance of a selected FWER value (*e.g.* 0.05 or 0.01) whilst conducting such multiple tests, obviously the α value required must be constrained in order to reflect this, and hence it must be lower or much lower than the FWER value selected. The magnitude of this decrease is proportional to the number of X variables selected (*i.e.* α itself is inversely proportional to the number of predictor variables).

However, the Bonferroni correction stands as an accepted approach for modulating the FWER rate *via* the specification of a revised α value on consideration of the number of predictor (X) variables explored for each test [eqn (4)]. For example, if, in a model system, we have 200 intelligently se-lected buckets in the ^{1}H NMR profiles of human urine samples, we should calculate a modified α value of $\alpha = 0.05/200 = 2.50 \times 10^{-4}$ for each 'in-dependent' test system in order to accept a FWER value of 0.05.

$$\alpha = \text{FWER}/k \tag{4}$$

Hence, mean differences observed between compared features which are significant at the $p = 2.50 \times 10^{-4}$ level would be considered to be statistically significant in this case. In this situation, the p (FWER) value required for attaining ≥ 1 false-positive(s) amongst the 200 hypotheses evaluated is $1 - [1 - (2.50 \times 10^{-4})]^{200}$, equivalent to 0.0488, a figure which is substantially lower than that attained if no such correction is applied, *i.e.* $\text{FWER} = 1 - [1 - (0.05)]^{200}$, which is nearly equivalent to a value of 1.0 (ac-tually, 0.999965)! Hence, application of the Bonferroni correction in the correct manner provides a marked elevation in the selectivity of univariately significant metabolite variables, but restricted by a FWER value virtually statistically equivalent to 0.05, the expected p value for a simple two-sample comparison of means test (*i.e. via* a two-sample t-test). As such, this critically constraining method for the univariate evaluation of multiple X predictor variable-containing datasets markedly diminishes the occurrence of type I

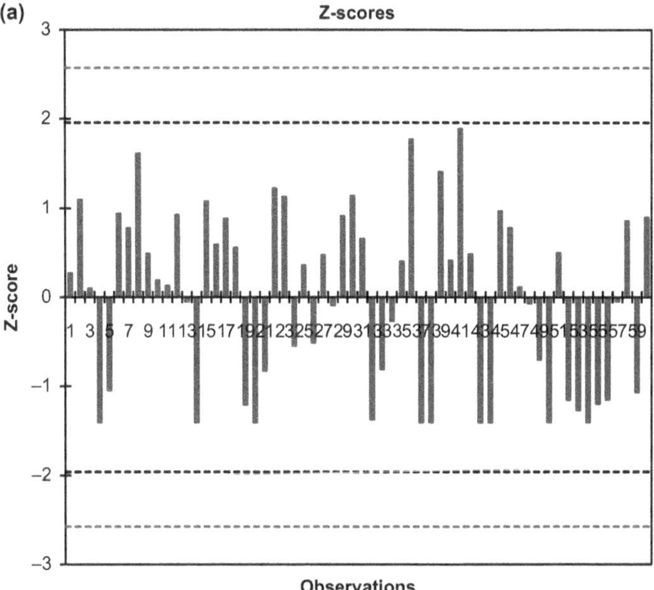

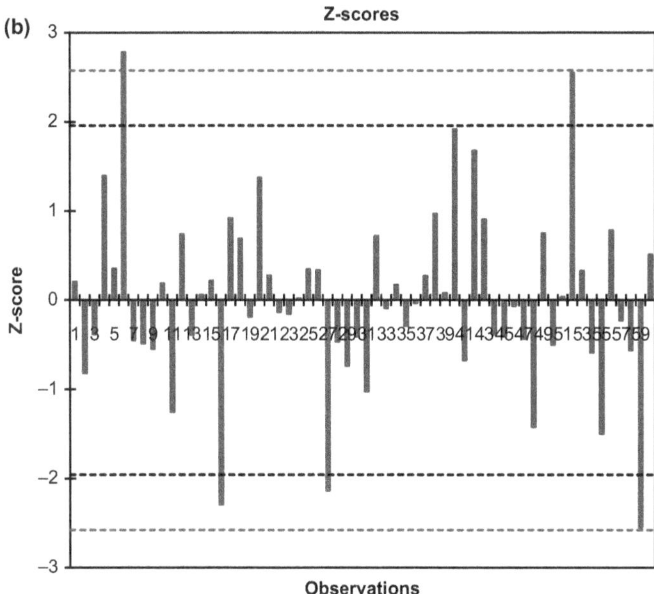

Figure 2.4 Examples of Z scores arising from the application of Grubb's test for outliers to the intensities of two intelligently-selected urinary ^{1}H NMR dataset variables (which were normalised and autoscaled prior to analysis). Results from typical buckets (a) without and (b) with significant outliers are shown (the latter contains two observations which were significant at the $p = 0.05$ level, and three at the $p = 0.01$ level).

(false-positive) errors, although it should be noted that this is at the cost of an increased level of type II (false-negative) ones, and the latter problem may give rise to the exclusion of potential biomarkers of some level of diagnostic or prognostic significance (and, where appropriate, results derivable from such 'untargeted' metabolomics investigations).

A less stringent form of the Bonferroni correction process is that which employs the 'step-down' (Holm) procedure, and this method involves a prior ranking of each variable from the most to the least significant (and correspondingly from the smallest to the largest p values); the smallest p value is then multiplied by the total number of predictor (X) variables available X_{total}, and if this product remains lower than a value of 0.05, then this first X variable is considered significant. Subsequently, the second-lowest p value is multiplied by the total number of X variables minus 1 ($X_{total} - 1$), and if this product is also <0.05, then this variable is also considered significant, and this sequence is repeated until all the primarily significant variables have been tested in this manner, and one has been found to be insignificant at the 'corrected' significance level.

A further means of correcting such multiple testing errors is the False Discovery Rate (FDR), which serves to remedy the type II error problem associated with the Bonferroni correction method.[38] In this proposed method, a probability-corrected q-value is computed for each spectral, chromatographic or metabolic feature expected; for this parameter, the expected proportion of false-positives is also considered when the statistical significance of each (univariately tested) predictor (X) variable is varied throughout acceptable zones or ranges. Therefore, a metabolite which expresses a q-value as small as 0.01 reveals that 1% of all such variables with corresponding p values of 0.01 are present as false-positives. Hence, a useful consideration is that for a p value of 0.01, 1% of all univariate tests performed will give rise to false-positive features, whereas a q-value of 0.01 provides evidence that only 1% of all the *significant* tests will comprise false-positives.

2.8 Power (Sample Size) Computations for Untargeted, Univariate Investigations of Metabolomics Datasets

The predictable sample size required for each classification status or group represents an extremely important consideration for all such investigations of this nature, and the great majority of bodies (governmental, research council, charitable, industrial or otherwise) involved in the provision of funding for proposed clinical, biomedical and metabolomics research programmes now insist on or require that grant funding applicants make an important provision for essential pilot data, together with a consideration of statistical power calculations regarding the recruitment of a sufficient number of participants/patients to such studies in order to achieve a satisfactory level of statistical significance, univariate or otherwise. Indeed, in order to achieve the successful, ambivalent attraction of such research funding for both clinically- and metabolomically- (or chemometrically-)related research funding, the provision of essential pilot data with associated (and realistic!) power calculations is considered essential and beyond reproach. The authors are also aware that bodies available for the funding of non-clinical research also demand an appropriate, study-targeted evaluation of statistical power calculations for the prospective undertaking of such future investigations.

Clearly, the statistical magnitude of a too-small sample size gives rise to a lack of reproducibility (precision) of data acquired, and hence cannot provide reliable evidence with the metabolomics hypothesis or hypotheses explored.

Conversely, an inappropriately selected too large sample size may give rise to an unnecessary overspend on resources for only a limited or very limited further informational gain. Moreover, this also represents an ethical consideration, since in investigations which involve an unnecessarily large sample (human participant) size, researchers may be adversely exposing at least some of the participants involved to selected risk factors or hazards. As might be expected, *a priori* determinations of an acceptable sample size for multifeature, MV datasets is of a high level of complexity and, with the exception of the investigations performed by Guo *et al.* (2010),[39] together with those delineated below, there is little or nothing yet available in this area for such applications. Moreover, in 2011 Hendricks *et al.*[40] concluded that at that time there were no methods available for the prior estimation of sample sizes required for the exploration of multicollinear, MV datasets.

However, classical power analysis techniques conceptualise the estimation of sample sizes required for univariate analysis. Indeed, the sensitivity (or power) of a statistical test system is represented as $1 - \beta$, where β is defined as the probability of obtaining a false-negative or type II error in such hypothesis testing procedures. Furthermore, the statistical power of a test is the probability of it permitting the detection of significant differences over and above a selected level of confidence. Of course, traditional power calculations for a given potential (univariate) biomarker molecule predictor (X) variable requires a full consideration of sample means and variances, together with the effect of sample sizes on the attainment of significant mean differences at the desired *p* level.

Notwithstanding, for MV datasets potentially containing at least several significant biomarker (X) variables, it is necessary for such power calculation estimations to be substantially reconsidered. In such cases, the average/mean power is employed (rather than simple power), and the significance level defined is required to incorporate multiple testing criteria. For example, multiple values of sample standard deviations and effect sizes require a high level of consideration!

Of much importance to this research area, Ferreira *et al.* (2006a, 2006b)[41,42] applied power analysis calculations to such high-dimensional, MV datasets *via* univariate methods coupled to appropriate multiple testing criteria corrections, and these researchers explored a pilot microarray dataset in order to estimate statistical power, the distribution of effect sizes and also the minimum sample size required according to these criteria. An adaption of this technique by van Iterson *et al.* (2009)[43] forms a component of the BioConductor SSPA package, and this involves the treatment of multidimensional datasets as a series of multiple univariate feature responses; however, correlations between these predictor variables are neglected. However, this method can serve to provide effective guidelines for the ratification of experimental design options on the basis of pilot data acquired *a priori*. Moreover, as previously noted, it is, of course, of much importance to note that the major determinants of the number of samples available for such MV classificational status determinations or estimations are limited by both financial and/or ethical constraints.

Recently, Vinaixa *et al.* (2012)[44] evaluated the performance of the SSPA package with respect to the estimation of effect sizes and the performance of relevant power calculations in a series of untargeted metabolomics datasets. As noted by these researchers, a bimodal density of statistical frequencies is, of course, expected when the classification status gives rise to significant or relatively significant differences between the mean values of the two classifications selected for comparison. Therefore, in cases in which differences between the two (or more) classification criteria are apparently unimodally represented, their effects are poorly described in terms of the effect sizes required (as indeed might be expected!). Therefore, these researchers recommended that the incorporation of a minimum sample size of n = 10 is the best option required in order to facilitate and perhaps promote the statistical power of univariate test systems when indeed such a difference between the two classification groups is present.

A major consideration regarding untargeted MV ^{1}H NMR and LC-MS metabolomics datasets is, of course, the attribute that they contain many multiple correlations between their spectral and chromatographic features (*i.e.* ^{1}H NMR buckets or bins, and mzRT variables for LC-MS datasets), specifically multicollinearity. For ^{1}H NMR datasets, obviously clear correlations will be observed between resonances arising from the same biomolecule, for example those between the -CH$_3$ and -CH group resonances of lactate [$\delta = 1.33(d)$ and $4.13(q)$ ppm, respectively] and the α-CH$_2$, β-CH$_2$ and γ-CH$_3$ signals of *n*-butyrate [$\delta = 2.14(t)$, $1.55(tq)$ and $0.90(t)$ ppm, respectively], in addition to 'Between-Metabolite' ones. Indeed, the latter correlations arise from metabolic inter-relationships, *i.e.* many biomolecules participate in a wide range of inter-connected enzymatic reactions in metabolic pathways (for example, as co-factors, substrates and products), and hence modulate and/or attenuate such reactions (*e.g. via* feedback inhibition processes). As such, all targeted or untargeted MV metabolomics datasets are multidimensional and hence multicollinear and multicorrelated, and this obviously causes problems for conventional multiple regression ordinary multiple linear regression (OMLR) and simple discriminatory (DA) analyses, and even for some further forms of high-dimensional analysis such as canonical correlation analysis (CCorA).

2.9 Sample Size Requirements and Statistical Power Computations for High-Dimensional, Metabolomic Datasets

Since MV datasets generated *via* a wide range of 'omics' investigations (including metabolomics) are of a very high dimensional (*i.e.* multidimensional) nature, in which the number of predictor (X) variables can often exceed the number of samples monitored (sometimes substantially so!), there are major constraints and considerations for implementation of the minimal requirements for sample sizes required for such explorations, together with the associated statistical power calculations involved. Of the very few developments in this area, Guo *et al.*[39] evaluated the relative performance of *k*-Nearest Neighbour clustering techniques, Microarray Prediction Analysis, Random Forests and Support Vector Machines (SVMs) in multidimensional omics datasets, most especially the wide range of signal-to-noise dataset ratios, imbalances in classification distributions, and metric selections available for the classifier parameters derived therefrom (the major contributory characteristics of datasets profiled from a number of human and animal studies employing high-content mass spectrometric and multiplexed immunoassay techniques were provided).

Interestingly, these investigators concluded that the mean contribution of effect size in human studies was substantially lower than in that observed in experimental animal ones, and that datasets acquired from the former group were particularly characterised by a greater biological variation and a higher incidence of outlying data points or whole samples. Indeed, simulation experiments revealed that classification prediction analysis for microarrays (PAMs) exhibited the highest statistical power, but only when the classification-conditional feature distributions were Gaussian, and those of the outcomes balanced. However, in circumstances involving skewed feature distributions and unbalanced classificational ones, application of the Random Forests (RFs) technique was optimal. The researchers involved further concluded that no single classification method exhibited optimal performance characteristics under all conditions in which they were tested, although valuable guidance for the design and performance of such biomedical investigations involving the MV analysis of multidimensional datasets was provided.

This work focused on determinations of the minimal sample size required for elucidating whether or not a particular algorithm designed for classification purposes is significantly more effective than random choices made on the sample classification criteria. However, although a selected sample size number may give rise to a high level of statistical power, the classification algorithm may not, however, include all possible biomarkers which play significant roles in distinguishing between control and disease (case) samples. Moreover, biomarker classifiers which exhibit a poor level of extension to and application in further, comparable investigations may result from this particular process. Indeed, the investigations performed by Guo *et al.* (2010)[39] were considered appropriate only for investigations conducted during the primary stages of biomarker discovery, such as those focused on clinical settings for which the clinical value of the discriminatory dataset is unknown. The subsequent attainment of a 'diagnostic' level of scientific proof for these preliminary studies can then lead to more carefully designed and performed investigations, coupled with more stringent validation and cross-validation (CV) investigations, and focused on the achievement of a set of reliable biomarkers which display high levels of accuracy and precision regarding the classification of a particular clinical condition and/or its severity. However, as noted by Ein-Dor *et al.* (2006),[45] even when the newly developed biomarker set is sufficiently reproducible, the sample size required can amount to thousands! The simulations conducted by Guo *et al.*[39] for the comparison of a series of classifiers can be subjected to a series of rigorous constraints in order to ensure that a robust biomarker set has a sufficient level of (MV) statistical power. However, further research work is required in order to provide discriminatory diagnostic classification criteria for multiple (*i.e.* >2) groups, continuous [quantitative dependent (Y) variable(s)] and/or censored parameters.

Intriguingly, Nicholson *et al.* (2011)[46] recently performed a highly detailed investigation which involved an exploration of the contributions of human sources of variation to biofluid metabolite concentrations. Their approach involved a ^{1}H NMR-based exploration of the biomolecular profiles of blood plasma and urine samples collected from both identical and non-identical twins longitudinally; variations in the metabolite concentrations of these samples were subsequently decomposed into familial (both common-environmental and genetic), individual-environmental and longitudinally unstable components. These researchers deduced that the 'stable' variance components (specifically, those ascribable to the influences of familial and individual-environmental sources) were responsible for, on average, 60 and 47% of 'biological variation' in plasma and urine, respectively, with regard to metabolites detectable by ^{1}H NMR analysis. Therefore, these researchers concluded that since clinically predictive variation in the metabolite levels of these two biofluids is likely to be 'nested' within these biologically stable components of variance, their results are of a high level of significance with respect to the design and MV statistical analysis of future biomarker discovery studies. These investigators also presented a unique power calculation method for ^{1}H NMR-linked metabolome-wide association studies (MWASs), and from these discovered that sample sizes of several thousand will be required for the attainment of a sufficient level of statistical precision in order to reliably detect ^{1}H NMR-responsive biomarkers with the ability to quantify the predisposition of individuals to disease. Therefore, this observation should clearly be of a very high level of concern to metabolomics researchers who involve only restricted or highly restricted numbers of participants, and multianalyte-containing biofluid samples derived therefrom, in their investigations!

2.10 Error Analysis

Error analysis involves the detection, identification and determination of various classes of uncertainty associated with bioanalytical measurements, and also the proliferation of these errors *via* mathematical and statistical processes. Of course, this terminology has a tendency to associate such *error* with analytical precision rather than accuracy and problems arising from experimental mistakes! Therefore, in the metabolomics research area, such error analysis serves to provide researchers with much valuable information regarding their ability to delineate the levels of confidence that they have (or may have) in results derived or derivable from MV metabolomics datasets. Indeed, as might be expected, the importance of error analysis is now considered critical with respect to the extremely large numbers and often heterogeneities of measurements available in now commonly acquired high-throughput classes of 'omics'-type experiments.

In view of the markedly elevated range of biomolecules commonly determinable in such metabolomics experiments (for example, low-molecular-mass biomolecules, proteins, polysaccharides, DNA, RNA, *etc.*), which are regularly monitored in their hundreds or even thousands, there is obviously a very high degree of heterogeneity associated with the possible analysis solutions for such metabolomics problems. Fundamentally, Moseley (2013)[47] very recently described and reviewed the involvement and employment of error analysis in MV metabolomics explorations as an improvement in overall experimental design (which are generally poorly accepted or implemented in many published investigations), and hence the prior consideration of appropriate statistical methods for their analysis (which should, of course, include validation and cross-validation models *via* permutation techniques, where relevant), essential quality control monitoring of the laboratory experiments performed and finally determinations of our confidence (and hence potential uncertainties) in the results acquired. Indeed, this researcher describes and discusses a variety of current approaches for monitoring error propagation in MV metabolomics studies, and in his review analytical derivation and approximation methodologies are included, as are Monte Carlo error analysis approaches, and also those which are involved in the potential solution of inverse metabolic challenges. Of course, the many key assumptions associated with such MV statistical techniques applicable to the analysis of metabolomics datasets serve as major constraints to the generation of effective solutions to these problems, and the violation of one or more of these readily enhances the proliferation of statistical errors!

Acknowledgements

In this work the authors employed XLSTAT2013, MetaboAnalyst 2.0, MetATT and ACD Spectrus Processor 2013 software.

References

1. I. S. Kohane, A. T. Kho and A. J. Butte, *Microarrays for Integrative Genomics*, A Bradford book, MIT Press, Cambridge, 2003.
2. W. B. Dunn, D. Broadhurst, D. I. Ellis, M. Brown, A. Halsall, S. O'Hagan, I. Spasic, A. Tseng and D. B. Kell, GC-TOF-MS study of the stability of serum and urine metabolomes during the UK Biobank sample collection and preparation protocols, *Int. J. Epidemiol.*, 2008, **37**, i23–i30, DOI: 10.1093/ije/dym281.
3. C. Schreir, W. Kremer, F. Huber, S. Neumann, P. Pagel, K. Lienermann and S. Pestel, Reproducibility of NMR analysis of urine samples: Impact of sample preparation, storage conditions, and animal health studies, *Biomed. Res. Int.*, 2013, Article ID 878374 (19 pages). http://dx.dot. org/ 10.1155/2013/878374.
4. M. A. E. Hansen, *Metabolome Analysis: An Introduction*, ed. S. G. Villas-Boas and U. Roessner, Wiley, Hoboken, NJ, USA, 2007, pp. 146–187.
5. M. Katajamaa and M. Orešič, Processing methods for differential analysis of LC/MS profile data, *BMC Bioinf.*, 2005, **6**, 179.
6. A. Scalbert, L. Brennan, O. Fiehn, T. Hankemeier, B. S. Kristal, B. van Ommen, E. Pujos-Guillot, E. Verhej, D. Wishart and S. Wopereis, Mass-spectrometry-based metabolomics: Limitations and recommendations for future progress with particular focus on nutrition research, *Metabolomics*, 2009, **5**, 435–458.
7. J. Schripsema, Application of NMR in plant metabolomics: Techniques, problems and prospects, *Phytochem. Anal.*, 2010, **21**, 14–21.
8. M. Spraul, P. Neidig, U. Klauck, P. Kessler, E. Holmes, J. K. Nicholson, B. C. Sweatman, S. R. Salman, R. D. Farrant, E. Rahr, C. R. Beddell and J. C. Lindon, Automatic reduction of NMR spectroscopic data for statistical and pattern recognition classification of samples, *J. Pharm. Biomed. Anal.*, 1994, **12**, 1215–1225.
9. A. Lommen, J. M. Weseman, G. O. Smith and H. P. J. M. Noteborn, On the detection of environmental effects on complex matrices combining off-line liquid chromatography and 1HNMR, *Biodegradation*, 1998, **9**, 513–525.
10. J. Schripsema, Application of NMR in plant metabolomics: Techniques, problems and prospects, *Phytochem. Anal.*, 2010, **21**, 14–21.
11. B. M. Beckwith-Hall, J. K. Nicholson, A. W. Nicholls, P. J. Foxall, J. C. Lindon, S. C. Connor, M. Abdi, J. Connelly and E. Holmes, Nuclear magnetic resonance spectroscopic and principal components analysis investigations into biochemical effects of three model hepatotoxins, *Chem. Res. Toxicol.*, 1998, **11**, 260–272.
12. J. Vogels, A. C. Tas, J. Venekamp and J. VanderGreef, Partial linear fit: A new NMR spectroscopy preprocessing tool for pattern recognition applications, *J. Chemom.*, 1996, **10**, 425–438.
13. T. R. Brown and R. Stoyanova, NMR spectral quantitation by principal-component analysis. II. Determination of frequency and phase shifts, *J. Magn. Reson., Ser. B*, 1996, **112**, 32–43.

14. A. M. Weljie, J. Newton, P. Mercier, E. Carlson and C. M. Slupsky, Targeted profiling: Quantitative analysis of ^{1}H NMR metabolomics data, *Anal. Chem.*, 2006, **78**, 4430–4442.
15. C. Ludwig and M. R. Viant, Two-dimensional J-resolved NMR spectroscopy: Review of a key methodology in the metabolomics toolbox, *Phytochem. Anal.*, 2010, **21**, 22–32.
16. B. L. LaMarche, K. L. Crowell, J. Navdeep, V. A. Petyuk, A. R. Shah, A. D. Polpitiya, J. D. Sandoval, G. R. Kiebel, M. E. Monroe, S. J. Callister, T. O. Metz, G. A. Anderson and R. D. Smith, MultiAlign: a multiple LC-MS analysis tool for targeted omics analysis, *BMC Bioinf.*, 2013, **14**, 49.
17. C. A. Hastings, S. M. Norton and S. Roy, New algorithms for processing and peak detection in liquid chromatography/mass spectrometry data, *Rapid Commun. Mass Spectrom.*, 2002, **16**, 462–467.
18. A. Savitzky and M. J. E. Golay, Smoothing and differentiation of data by simplified least squares procedures, *Anal. Chem.*, 1964, **36**, 1627–1639.
19. H. P. Chen, H. J. Liao, C. M. Huang, S. C. Wang and S. N. Yu, Improving liquid chromatography-tandem mass spectrometry determinations by modifying noise frequency spectrum between two consecutive wavelet-based low-pass filtering procedures, *J. Chromatogr. A*, 2010, **1217**, 2804–2811.
20. K. R. Coombes, S. Tsavachidis, J. S. Morris, K. A. Baggerly, M. C. Hung and H. M. Kuerer, Improved peak detection and quantification of mass spectrometry data acquired from surfaceenhanced laser desorption and ionization by denoising spectra with the undecimated discrete wavelet transform, *Proteomics*, 2005, **5**, 4107–4117.
21. M. Sysi-Aho, M. Katajamaa, L. Yetukuri and M. Oresic, Normalization method for metabolomics data using optimal selection of multiple internal standards, *BMC Bioinf.*, 2007, **8**, 93, DOI: 10.1186/1471-2105-8-93.
22. C. A. Pena-Reyes and M. Sipper, Evolutionary computation in medicine: An overview, *Artif. Intell. Med.*, 2000, **19**, 1–23.
23. D. B. Kell, Metabolomics and machine learning: Explanatory analysis of complex metabolome data using genetic programming to produce simple, robust rules, *Molec. Biol. Rep.*, 2002, **29**, 237–241.
24. P. E. Greenwood and M. S. Nikulin, *A Guide to Chi-Squared Testing*, Wiley, New York, 1996. ISBN 0-471-55779-X.
25. G. E. Dalal and L. Wilkinson, An analytic approximation to the distribution of Lilliefors' test statistic for normality, *Am. Stat.*, 1986, **40**(4), 294–296.
26. S. Keskin, Comparison of several univariate normality tests regarding type I error rate and power of the test in simulation based small samples, *J. Appl. Sci. Res.*, 2006, **2**(5), 296–300.
27. N. Razali and Y. Wah, Power comparison of Shapiro-Wilk, Kolmogorov-Smirnov, Lilliefors and Anderson Darling tests, *Journal of Statistical Modelling and Analytics*, 2011, **2**(1), 21–33.
28. M. Mendes and A. Pala, Type I Error rate and power of three normality tests Pakistan, *J. Inf. Technol.*, 2003, **2**, 135–139.

29. D. Oztuna, A. H. Elhan and E. Tuccar, Investigation of four different normality tests in terms of type I error rate and power under different distributions, *TuBITAK Turk. J. Med. Sci.*, 2006, **36**(3), 171–176.

30. M. Saculinggan and E. A. Balase, Empirical power comparison of goodness of fit tests for normality in the presence of outliers, *iCAST: Contemporary Mathematics, Mathematical Physics and Their Applications*, 2013, **435**, 012041, DOI: 10.1088/1742-6596/435/1/012041.

31. E. Seier, http://interstat.statjournals.net/YEAR/2002/abstracts/0201001.php, *East*, 2002, **42**(20), 1–17.

32. L. Cen and S. Shapiro, An Alternative test for normality based on normalized spacings, *J. Stat. Comput. Simul.*, 1995, **53**, 269–287.

33. P. Zhang, Omnibus test of normality using the Q statistic, *J. Appl. Stat.*, 1999, **26**, 519–528.

34. R. B. D'Agostino, A. Belanger and R. B. D'Agostino Jr., A suggestion for using powerful and informative tests of normality, *Am. Stat.*, 1990, **44**, 316–322.

35. J. R. M. Hosking, L-moments: analysis and estimation of distributions using linear combinations of order statistics, *J. Roy. Stat. Soc., Ser. B Med.*, 1990, **52**, 105–124.

36. G. E. P. Box, Non-normality and tests on variances, *Biometrika*, 1953, **40**, 318–335.

37. J. D. Storey, A direct approach to false discovery rates, *J. Roy. Stat. Soc. B Med.*, 2002, **64**, 479–498.

38. Y. Benjamini, D. Drai, G. Elmer, N. Kafkafi and I. Golani, Conflicting the false discovery rate in behaviour genetics research, *Behav. Brain Res.*, 2001, **125**, 279–284.

39. Y. Guo, A. Garber, R. N. McBurney and R. Balasubramanian, Sample size and statistical power considerations in high-dimensionality data settings: a comparative study of classification algorithms, *BMC Bioinf.*, 2010, **11**, 447, DOI: 10.1186/1471-2105-11-447.

40. M. M. W. B. Hendricks, F. A. Eeuwijk, R. H. Jellema, J. A. Westerhuis, T. H. Reijmers, H. C. J. Hoefsloot and A. K. Smilde, Data-processing strategies for metabolomics studies, *TrAC-Trend Anal. Chem.*, 2011, **30**, 1685–1698.

41. J. A. Ferreira and A. Zwinderman, Approximate sample size calculations with microarray data: an illustration, *Stat. Appl. Genet. Mol. Biol.*, 2006, **5**, Article 25.

42. J. A. Ferreira and A. H. Zwinderman, Approximate power and sample size calculations with the Benjamini-Hochberg method, *Int. J. Biostat.*, 2006, **2**(10), Article 8.

43. M. van Iterson, P. 't Hoen, P. Pedotti, G. Hooiveld, J. den Dunnen, G. van Ommen, J. Boer and R. Menezes, Relative power and sample size analysis on gene expression profiling data, *BMC Genom.*, 2009, **10**, 439, DOI: 10.1186/1471-2164-10-439.

44. M. Vinaixa, M. A. Rodriguez, S. Samino, M. Díaz, A. Beltran, R. Mallol, C. Bladé, L. Ibañez, X. Correig and O. Yanes, Metabolomics reveals

reduction of metabolic oxidation in women with polycystic ovary syndrome after Pioglitazone-Flutamide-Metformin polytherapy, *PloS One*, 2011, **6**, e29052.

45. L. Ein-Dor, O. Zuk and E. Domany, Thousands of samples are needed to generate a robust gene list for predicting outcome in cancer, *PNAS*, 2006, **103**, 5923–5928.

46. G. Nicholson, M. Rantalainen, A. D. Maher, J. V. Li, D. Malmodin, K. R. Ahmadi, J. H. Faber, I. B. Hallgrímsdóttir, A. Barrett, H. Toft, M. Krestyaninova, J. Viksna, S. G. Neogi, M.-E. Dumas, U. Sarkans, The MolPAGE Consortium, B. W. Silverman, P. Donnelly, J. K. Nicholson, M. Allen, K. T. Zondervan, J. C. Lindon, T. D. Spector, M. I. McCarthy, E. Holmes, D. Baunsgaard and C. C. Holmes, Human metabolic profiles are stably controlled by genetic and environmental variation, *Mol. Syst. Biol.*, 2011, **7**, 525, DOI: 10.1038/msb.2011.57.

47. N. B. Moseley, Error analysis and propagation in metabolomics data analysis, *Comput. Struct. Biotech. J.*, 2013, **4**(5), e201301006, http://dx.doi.org/10.5936/csbj.201301006.

Recent Developments in Exploratory Data Analysis and Pattern Recognition Techniques

MARTIN GROOTVELD

Leicester School of Pharmacy, Faculty of Health and Life Sciences,
De Montfort University, The Gateway, Leicester LE1 9YH, UK
Email: mgrootveld@dmu.ac.uk

3.1 Introduction

This chapter will focus on up-to-date and very recent developments regarding the analysis of MV datasets arising from either multidimensional metabolomics or genomics experiments, and in particular this will provide essential clues and supporting information regarding applications of the methods/techniques described to the provision of potential solutions to such statistical problems. Particularly noteworthy is the still novel development and application of methods available for datasets which contain larger (or much larger) numbers of potential predictor variables than samples available in the complete dataset: in the 1970s and 1980s, many statisticians, and further researchers involved in the analysis of biomedical data, would be completely baffled and perhaps even shocked by this prospect! However, recent developments have facilitated and, in many cases, permitted this, albeit with a major sense of caution, and also with concern for major

Issues in Toxicology No. 21
Metabolic Profiling: Disease and Xenobiotics
Edited by Martin Grootveld

Published by the Royal Society of Chemistry, www.rsc.org

requirements for (1) the satisfaction of essential assumptive criteria and (2) the essential performance of a reliable and reproducible series of validation, cross-validation and further model testing and evaluation systems. PCA was never really originally meant to be applied to hundreds or even thousands of possible explanatory variables!

Methods available involve machine-learning (computational intelligence) techniques (including Self-Organising Maps, Support Vector Machines and Random Forests, to mention but some), which have been recently applied to the analysis of metabolomics and genomics (microarray) datasets, Gaussian Graphical Models (GGMs) and Independent Component Analysis (ICA), the latter two representing statistically acceptable improvements on more traditional approaches available such as PCA and PLS-DA. After all, when is an apparent correlation not a *real* correlation? Answer: when it is derived from an extremely large covariance matrix reflecting many thousands of Pearson correlation coefficient cells! Although this answer will not be correct for many such correlations where there is a clear or already established *linear* relationship between two variables, it certainly will be so for what may be a large proportion of them, and some researchers often utilise a concatenation of these methods without any apparent consideration for this. Fortunately, the Gaussian Graphical Models (GGMs) and Independent Component Analysis (ICA) techniques described here serve as major advances towards the provision of solutions to such problems (these developments are further outlined and detailed in this chapter), but further major analytical advances are required in this area, as indeed they are in many n $<$ or $\ll$ X variable or even less demanding situations!

In the early stages of this chapter, information is provided regarding the application of techniques which, although frequently employed in the environmental and ecological research areas, today are still not very commonly applied to the MV analysis of metabolomics or genomics datasets. These techniques for analysis, specifically Canonical Correlation Analysis (CCorA) and Classification and Regression Trees (CARTs), have been applied here to the investigation of a particular clinical index/blood serum biomarker dataset for the benefit of the reader.

Also discussed is the metabolomics potential of 'clustering' techniques such as k-means and agglomerative hierarchal clustering (the latter abbreviated as AHC); their applications to the analysis of a clinical biomarker dataset is illustrated (although, admittedly, one with many more biofluid samples and participants involved than the number of potential 'predictor' variables!).

3.2 Canonical Correlation Analysis (CCorA)

CCorA is a process which permits us to explore relationships between two sets of variables.[1–3]

As an example, a clinical researcher may wish to explore the simultaneous relationship between several (correlated or uncorrelated) physiological measures (for example, systolic and diastolic blood pressure values, together with the mean articular pressure) and perhaps a series of particular blood plasma or urinary metabolite levels. In cases such as these (in which the researcher may be focused on investigating relationships between two sets of variables), CCorA would serve as the appropriate method of choice for the analysis of such datasets. In addition to determining if the series of blood plasma or urinary metabolite concentrations relate to these blood pressure parameters, CCorA can also provide valuable information regarding exactly how many dimensions (canonical variables) are required in order for us to understand the association between the two sets of variables. Details regarding this particular example are provided below.

The eigenvalues isolated *via* the employment of this technique are interpreted in terms of the proportion of variance accounted for by the correlation between the respective canonical variates, and are determined *via* an extraction of canonical roots. This proportion should only be expressed relative to the canonical variate variance, *i.e.* it represents the weighted sum scores of the two series of variables; hence, they do not provide information regarding the level of variability explained in either (isolated) set. The number of eigenvalues computable is equivalent to the number of canonical roots, specifically the minimum number of variables in either of the two datasets. As expected, successive factor (component) eigenvalues are of a diminishing magnitude; primarily, the weights which maximise the correlation between the two sum scores are computed, and following the extraction of this first root, the weights generating the second largest correlation between these sum scores are determined (with the constraint that this sum score set does not correlate with the primary one), *etc.*

A canonical correlation coefficient is simply the square root of the eigenvalue and relates only to the canonical variates (of course, such correlations between successfully extracted canonical variates diminish with increasing variate number). The largest correlation (*i.e.* that for the first factor or root) therefore serves as an overall measure of the canonical correlation between the two classes of variables, although it should be noted that further canonical variates may also be correlated in a form that makes scientific and interpretable sense.

The statistical significance of each of the canonical correlations (roots) is usually evaluated by the performance of multivariate ANOVA (MANOVA) and the Wilks–Lambda test, although Pilai's and the Hotelling–Lawley traces, and Roy's greatest root serve as further methods employable to test this.

3.2.1 CCorA Case Study

This section delineates a CCorA case study which involves an examination of the inter-relationships between two datasets: the first comprising a series of (correlated 'dependent') human blood pressure indices, the second a series of (correlated, 'independent', possibly explanatory) lateral X variables, which consisted of two blood serum thyroid disease biomarkers, an arbitrary thyroid disease score, together with a number of further blood pressure-determining variables.

In view of multiple problems associated with multicollinearity effects, the CCorA model involved an analysis of the scores vectors of pre-isolated orthogonal Principal Components (PCs) arising from PCA of both datasets; each variable was autoscaled in both datasets prior to performing the primary PCA stage of the CCorA.

The first (dependent) variable set contained clinical systolic blood pressure (SBP), diastolic blood pressure (DBP), mean arterial pressure (MAP) and the SBP:DBP ratio indices, the second ('independent') one comprising the blood serum concentrations of the thyroxine (T4) and thyroid-stimulating hormone (TSH) thyroid biomarkers, an arbitrarily assigned disease score index (*i.e.* -1, 0 and $+1$ for hypothyroid, euthyroid and hyperthyroid patient participants, respectively), and further blood pressure-determining one lateral variables, specifically patient age, BMI and gender, the latter expressed as a score of 0 for males and $+1$ for females (prior to autoscaling).

In summary, CCorA was conducted on the corresponding scores vectors of two sets of orthogonal (uncorrelated) PCA components (*i.e.* PCs) obtained from the two separate series of variables to investigate canonical correlations between them, and also their dimensionality; these canonical dimensions serve as representative 'latent' variables (corresponding to factors in factor analysis), and maximise the correlation between the two sets of variables.

The prior PCA stage of the analysis demonstrated that there were three major PCs derivable from the (independent, lateral variable) thyroid biomarker and disease score/age/BMI/gender score dataset, each with eigenvalues greater than or very close to 1; the first PC (PC1) comprised a linear combination of inversely correlated serum T4 and TSH concentrations, and also the arbitrary disease score assigned (the latter positively and negatively correlated with blood serum T4 and TSH levels, respectively), and had an eigenvalue of 2.77 (46% of variance explained), with loadings of 0.87, -0.86 and 0.97, respectively. The second PC (PC2) contained a combination of positively correlated age and BMI values (eigenvalue 1.28, with 21% of variance explained), with corresponding loadings of 0.80 and 0.57, respectively, whilst the third PC (PC3) was found to result from gender score only (eigenvalue 0.92, with 15% of the variance explicable), and this had a loading of 0.82. The squared cosines of these variables' contributions towards their respective PCs were all greater than 0.63, with the exception of that of BMI, which contributed to PC2 (0.33), and the Kaiser–Meyer–Olkin measure of sampling adequacy ranged from 0.52 to 0.77 for all variables included, an observation confirming that the dataset is satisfactory for the application of PCA.

For the series of (dependent) blood pressure variables, the preliminary PCA performed revealed that the dataset was segregatable into two clear PCs, each with eigenvalues >1, the first (PC1) comprising a linear combination of the positively correlated SBP, DBP and MAP parameters, with corresponding loadings of 0.91, 0.72 and 0.98, respectively (eigenvalue 2.72, with 68% of the variance explained), and the second (PC2) contained only the potentially diagnostic SBP:DBP ratio variable, with a powerful PC2 loading of 0.94 (eigenvalue 1.30, with 32% of variance explained). The squared cosines of these variables were within the 0.72–0.98 range.

Subsequently, the CCorA model was applied in order to explore the canonical variates and their dimensionality between the dependent scores vectors of the blood pressure PCs (the first consisting of a linear combination of SBP, DBP and MAP parameters, the second only the SBP:DBP ratio), and those constructed from the 'independent' thyroid disease biomarkers/ disease score (PC1), age and BMI (PC2) and lastly gender score (PC3). Application of this technique to the analysis of these two sets of PC scores vectors revealed that there were two dimensions between these two series of variables, and the first of these was found to be very highly significant ($p < 0.0001$, Wilks' lambda test). Furthermore, these two dimensions derived therefrom corresponded to significant canonical correlations of 0.32 for the first, and 0.18 for the second, and were responsible for 76 and 24%, respectively, of the total canonical variance.

The canonical correlation plot of the second *versus* the first 'factor of factor scores vectors' arising from this form of CCorA (Figure 3.1) clearly indicates that the first 'independent' variable PC containing positively loading serum T4 and negatively loading TSH concentrations, and also the positively loading disease score, very highly significantly contributed to the first CCorA dimension, and was strongly and positively related to the second BP PC arising from the SBP:DBP ratio variable alone. Thus, the markedly increased blood serum T4 and decreased TSH concentrations, which are employed as biomarkers for hyperthyroidism (with has a disease score of $+1$ in this model system), are strongly correlated to the SBP:DBP ratio value.

Also notable is the inter-relationship between the first BP dataset PC scores (*i.e.* that incorporating strongly and positively correlated SBP, DBP and MAP indices) and those of the second 'independent' latent variable one containing the age and BMI predictors (PC2), an observation expected in view of the known correlations between the variables incorporated in each of these. However, the third (gender status) PC developed from this lateral variable dataset was found not to exert a significant influence on orthogonal PCs arising from the blood pressure dataset.

Standardised canonical coefficients arising from this canonical correlation analysis of uncorrelated PCs were $+0.82$ for the relationship between the first lateral variable dataset PC (which contains both serum biomarker concentration and the thyroid disease score variable) and the first 'factor of factors' ('PC of PCs'), and $+0.67$ for the association of the second lateral variable dataset PC and the second 'factor of factors'. Likewise, the standardised canonical coefficient for the first blood pressure variable dataset PC (arising from a linear combination of SBP, DBP and MAP variables) and the second 'factor of factors' was 0.80, with an equivalent value for that between the second blood pressure PC (derived from the SBP:DBP ratio only). Therefore, we may conclude that the dependent BP variables are indeed

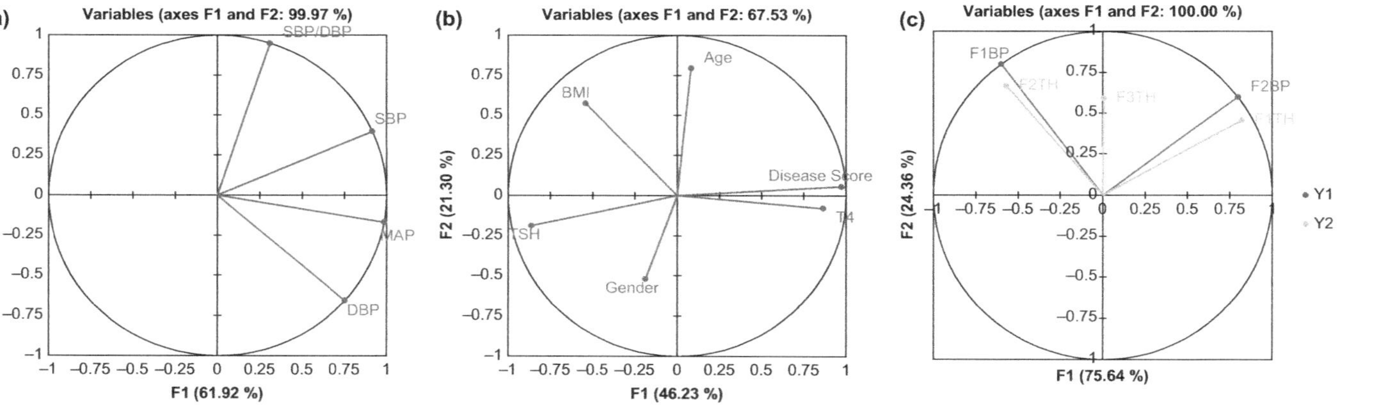

Figure 3.1   (a) Primary Principal Component Analysis (PCA) conducted on the 'dependent' BP Y variable dataset consisting of SBP, DBP, MAP and SBP:DBP ratio parameters; (b) as (a), but for the 'independent' X variable dataset containing thyroid biomarker concentrations (those of blood serum T4 and TSH levels), an arbitrary thyroid disease score (-1 for hypothyroid, 0 for euthyroid and $+1$ for hyperthyroid patients), BMI, age and gender score (specifically 0 and $+1$ for males and females, respectively); (c) plot of factor 2 (F2) *versus* factor 1 (F1) for canonical correlation analysis (CCorA) of orthogonal principal component (PC) scores vectors arising from the prior PCA analysis of the blood pressure parameter dataset (Y1) consisting of SBP, DBP and MAP values (all significantly augmenting PC1, abbreviated F1BP) and the diagnostic SBP:DBP ratio (the only significant variable significantly contributing to PC2, abbreviated F2BP), and the second dataset (Y2) comprising thyroid disease biomarkers (blood serum T4 and TSH levels) and disease score (all significantly contributing to PC1, abbreviated F1TH), together with participant age and BMI values (accounting for PC2, abbreviated F2TH) and gender score (responsible for PC3 alone, which is abbreviated F3TH). Canonical correlations (*i.e.* standardised canonical coefficients) between the first BP dataset input PC vectors were -0.599 (first BP PC) and 0.800 (second BP PC) for F1, and 0.800 (first BP PC) and 0.599 (second BP PC) for F2. Canonical correlations between the second dataset input vectors (three orthogonal PCs arising from linear combinations of blood serum thyroid disease biomarker concentrations/thyroid disease score, age/BMI values and gender score) were 0.823 (first lateral variable PC), -0.568 (second lateral variable PC) and 0.007 (third lateral variable PC) for F1, and 0.455 (first lateral variable PC), 0.667 (second lateral variable PC) and 0.590 (third lateral variable PC) for F2. For this example, the complete thyroid status dataset was analysed. Both the X and Y datasets were mean-centred and scaled to unit variance (*i.e.* standardised) prior to conducting the prior PCA analysis. Further abbreviations: in (c), Y1 represents the two BP PCs, and Y2 the three thyroid T4 and TSH biomarker/disease score, age/BMI and gender PCs. The thyroid status dataset comprised the above parameters, and biomarkers were determined on 300 euthyroid (healthy control), 300 hypothyroid and 71 hyperthyroid patients in total.

significantly linked to the thyroid disease score value and their corresponding condition-dependent T4 and TSH concentrations.

A related CCorA approach has been previously applied by Doeswijk *et al.*[4] to the analysis of multiple sensory-directed metabolomics data bocks; these researchers concluded that highly correlating partial least squares regression (PLS-R) scores vectors reveal data block portions that are closely related, and that examination of the relevant loading vectors arising from such an analysis serves to identify metabolites of interest and hence analytical value.

3.3   Classification and Regression Tree (CART) Analysis

The basis and objectives of Classification and Regression Trees (CARTs)[5] is to partition MV datasets recursively into strata of diminishing sizes according to improvements in the 'fit 'of such predictor X variables to a Y (dependent) variable, the latter of which may be qualitative or quantitative. CARTs serve to partition the 'sample space' into a set of rectangular projections and a model is then fitted to each one (primarily, this sample space is split into two such regions). Henceforth, optimal split is then sought and determined for all possible X variables at all potential 'split points', and this process is then subsequently repeated for each of the two (or more) prospective regions generated (a process which gives rise to the 'recursive portioning' term). However, the main selective operations of the CARTs method are the 'rules' employed for selection and termination. Indeed, the 'selection rule' determines exactly which stratification process to instigate at every stage, whilst the 'termination rule' provides essential information regarding the final strata that are generated. However, subsequent to creation of the strata, their 'impurities' are determined; the 'node impurity' term refers to the degree of heterogeneity of the outcome classifications or quantitative values within a particular stratum.

As noted above, CARTs are readily applicable to many types of categorical or classification variables, these including continuous indicator, ordinal and non-ordinal ones (*i.e.* they are not restricted to the estimation of classification outcomes with only two categories).

Notwithstanding, CARTs are typically 'model free' in their foundation, and to date there is only a very limited application of these techniques to the metabolomics research area (although a model-based statistical criterion can be employed for the purpose of 'splitting'). In principle, CARTs can offer a number of advantages when applied to the analysis of MV datasets (metabolomics or otherwise), specifically when they are of a non-linear and non-parametric nature. Moreover, they do not require the satisfaction of any distributional assumptions, the data generation process is treated as 'unknown', the predictors do not rely on the creation of a functional form, and there is also an assumption of additivity of predictor variables, an advantage which permits the exploration of complex interactions between these X variables.[6] Therefore, although conceptually facile, CARTs methods are powerful and can, at least in principle, provide optimal solutions to classification and deterministic problems. The differing approaches offered by the CARTs technique include AID (Automatic Interaction Detection) trees,[7] CHAID (Chi-square Automatic Interaction Detection)[8,9] and the more recently developed QUEST (Quick, Efficient, Statistical Tree) approaches.[10]

Receiver Operating Characteristic (ROC) curves may be generated from the application of the CARTS technique to models which have only two dependent variable (Y) classification groups. For a ('hypothetical') perfect model separation between them, the AUROC value is equivalent to 1.00, whilst it is 0.50 for a system with absolutely no discriminatory potential. In general, a model is considered effective when the AUROC value is >0.7, although it should be noted that for highly discriminating model systems, this value should be within the 0.87–0.90 range; a model with an AUROC value >0.90 is considered to be exceptional!

Figure 3.2 exhibits the application of the QUEST CARTs method to the (relatively simple) discrimination between the three classifications of thyroid patients explored above in Section 3.2. Clearly, this analysis gave rise to a very high level of distinction between the three classifications explored. Moreover, a validation procedure involving the prior, random removal of *ca.* one-third of the sampling population as a test set, and generation of a model with the remaining two-thirds, gave rise to disease classification distinctions which were very similar to those computed from the overall dataset. The results acquired regarding classification of the euthyroid dataset were very similar to those already employed as normal reference population values of blood serum TSH and T4 concentrations, *i.e.* 0.17–4.05 IU ml^{-1} and 55–135 ng ml^{-1}, respectively.

3.4 Moderated t-Statistic Methods

3.4.1 Significance Analysis of Microarrays (SAM)

The significance analysis of microarrays (SAM) is now a well-established statistical method for the identification of differentially expressed genes in the analysis of microarray datasets. Indeed, during the performance of multiple tests on multidimensional data, the SAM technique has the ability to address the false discovery rate (FDR), and also can provide a significance score to each predictor (X) variable in view of its modification, which is expressed relative to the standard deviation of repeated measurements. Moreover, for X variables which have scores higher than a specified threshold value, relative differences are evaluated with reference to a distribution generated from the performance of random permutations of the sample classification labels. Hence, this analytical system serves to circumvent the false discovery rate (FDR) problem associated with conducting multiple tests on such high-dimensional datasets.[11]

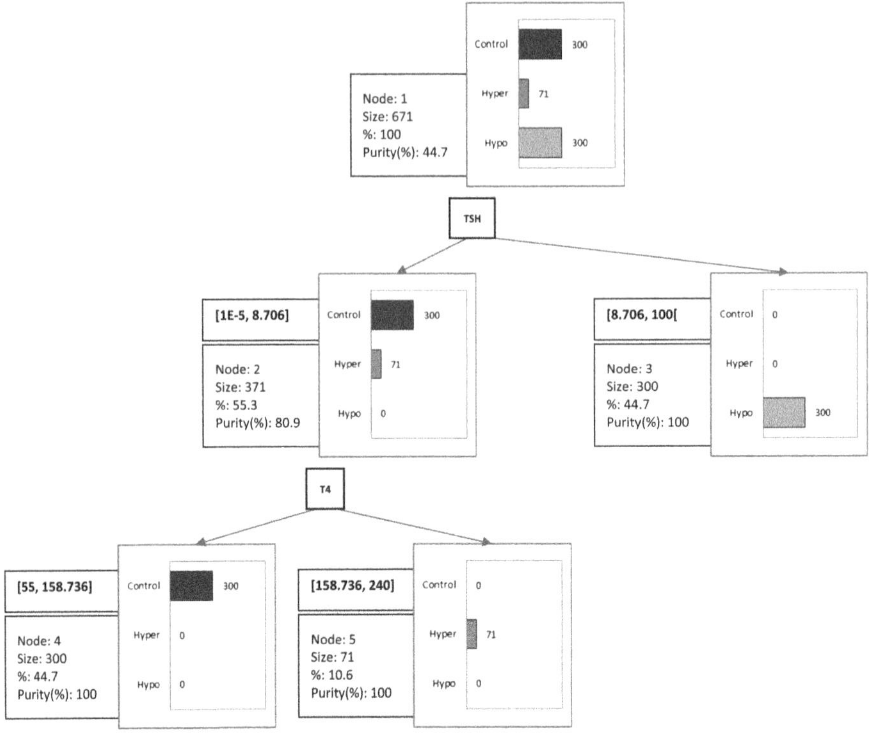

Figure 3.2 QUEST regression tree demonstrating clear distinctions between three thyroid disease classifications [specifically, euthyroid (healthy control), hypothyroid and hyperthyroid patients (n = 300, 300 and 71, respectively)] on the basis of their blood serum thyroid biomarker variables [thyroxine (T4) and thyroid-stimulating hormone (TSH) concentrations]. For this analysis, the maximal tree depth was 10, the number of intervals also 10 and the significance level was set at $p = 0.05$. Normal (euthyroid) reference population values for the blood serum concentrations of T4 and TSH are 55–135 ng ml^{-1} and 0.17–4.05 IU ml^{-1}, respectively. The values for these reference ranges determined from this analysis were 55–159 ng ml^{-1} for T4, and 0.0001–8.71 IU ml^{-1} for TSH.

3.4.2 Empirical Bayesian Approach Modelling (EBAM)

However, the EBAM technique is an empirical Bayesian approach which utilises moderated t-statistic values as its basis, and employs a model consisting of a two-classification mixture, *i.e.* those for null and differentially expressed genes or, alternatively, further potentially predictive (X) variables such as metabolic biomarkers;[12] both prior and density parameters are determined from the dataset. Hence, an X variable is considered to be significantly different (or, correspondingly, a gene differentially expressed) if the posterior value calculated for it is greater than a pre-specified δ index (with regard to microarray experiments, genes which are not differentially expressed will not have higher test scores).

3.5 Machine Learning Techniques

3.5.1 Self-Organising Maps (SOMs)

Self-organising maps (SOMs) represent a neural network-based algorithmic approach that can powerfully visualise relationships between multi-bioanalyte-containing human biofluid or tissue biopsy samples, *i.e.* they have the ability to identify any major trends available in such highly dimensional datasets. Indeed, they can be utilised for exploring 'self-similarities' between ^{1}H NMR profiles, and also visualising separations between SOM 'clusters' arising from each of the sources of variation incorporated into an experimental model (*e.g.* those arising from 'Between-Disease Classification Groups', 'Between-Participants', 'Between-Families' or 'Between-Sequential Time-Points' sources). In this manner, self-similarities between the spectral profiles can be readily detected and subsequently evaluated. The SOMs technique's foundation is based on the concept of a grid containing interconnected nodes, each of which comprises a model. Primarily, these models commence as random values, but during performance of the iterative training process, they are re-evaluated in order to represent differing sub-sets of a training set (the X and Y dimensions of the grid are required to be primarily specified by researchers).

The development and practical applications of SOMs were originally demonstrated by Kohonen in the 1980s,[13,14] and, to date, they have been widely employed for the visualisation of relationships between classification groups or samples. These methods represent a valuable alternative to more traditional MV analysis techniques such as PCA but, currently, in view of their computationally-intensive nature, they are not commonly employed in areas such as analytical chemistry or metabolomic profiling. Notwithstanding, SOMs are now much more feasible for the solution of many 'real-life' problems in the bioanalytical chemistry/biochemistry research areas.

Previously, SOMs have been employed for exploratory data analysis purposes;[15] however, this technique can also be applied in a supervised mode. Supervised SOMs[16] have been proposed for classification purposes in which a further vector of class information is incorporated into the training process, and this introduces an additional factor that serves to organise the map. Since the extent to which the class information exerts an influence on the map can be regulated, Wongravee *et al.*[17] introduced a class weight that can be adjusted according to how far the class membership information is employed in training it: a low value gives rise to a map that is close to an unsupervised one, whereas a high value may 'overfit' the dataset acquired. However, the researchers involved developed a means for the optimisation of this parameter. Supervised SOMs also provide opportunities to explore sources of variation with a low contribution to that of the complete dataset *via* organisation of the maps on the basis of these variations. A novel discriminatory index (SOM-DI) for purposes of identifying significant biomarkers arising from the supervised SOM analysis (and representative of selected sources of variation) was also proposed.[17]

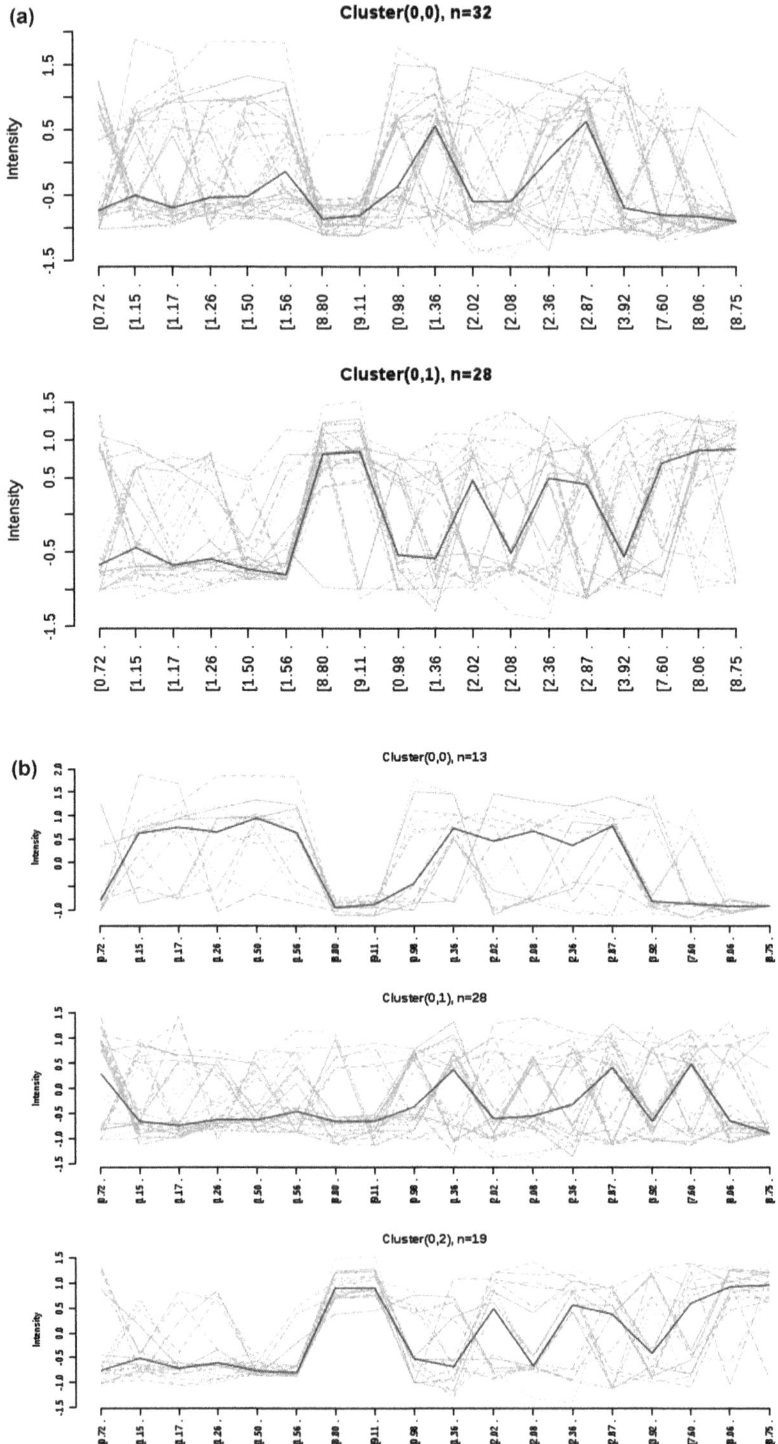
(a)
Cluster(0,0), n=32
Intensity
(b)
Cluster(0,0), n=13
Intensity
Cluster(0,1), n=28
Intensity
Cluster(0,1), n=28
Intensity
Cluster(0,2), n=19
Intensity

Although in PLS-DA, PLS weights and regression coefficients can be employed as indicators of significance, MV analysis by the SOMs technique has limitations since the classification and variable information are weighted with an equivalent level of significance; such a weighting is often undesirable, and the implementation of supervised SOMs outlined[17] permits researchers to label the classifier(s) and experimental data with pre-selected levels of significance. Despite the extension of PLS-DA to the MV analysis of multiple classification groups, such implementations can be of a high level of complexity.

The classification of a relatively small ($n = 60$) bucketed urinary ^{1}H NMR dataset containing only 20 ANOVA-selected 'predictor' variables into two or three 'self-similar' classification groups is shown in Figure 3.3.

3.5.2 Support Vector Machines (SVMs)

The SVM classification algorithm seeks a non-linear decision function in the input space *via* a mapping of the dataset into a higher dimensional feature space, and then performs separations on it through the use of a maximum margin hyperplane.[18] This process can be conducted *via* a recursive feature selection, and a sample classification procedure involving a linear kernel[19] (metabolites or alternative features are selected according to their relative contribution to the classification, a process evaluated by the consideration of cross-validation error rates). In this manner, the explanatory (X) variables of least importance are removed during subsequent stages, and the entire process generates a whole series of SVM models. The X variables utilised by the most effective model created (and therefore viewed as important) are then ranked *via* their model selection frequencies (Figure 3.4).

Figure 3.3 (a) and (b), Self-Organising Map (SOM) clustering classifications of an intelligently bucketed urinary ^{1}H NMR dataset according to (a) two and (b) three pre-specified groupings. The dataset comprised $n = 60$ samples of two disease classifications (46 in the healthy control and 14 in the disease-active group) and 20 ANOVA-selected bucket intensity values, and was creatinine-normalised, cubed root-transformed and Pareto-scaled prior to the performance of SOM analysis. The x-axes represent ^{1}H NMR bucket features, and the y-axes their relative intensities. The blue lines show the median intensities of each corresponding cluster. The X-axis corresponds to creatinine (Cn)-normalised intelligently-selected ^{1}H NMR bucket intensities. The dark lines represent the median intensities of each cluster. For the first (2-classification) model, the first cluster has upregulated 1.36–1.41, 1.56–1.58 and 2.87–2.89 ppm bucket values, whereas the second one has upregulated 2.02–2.08, 2.36–2.40, 2.87–2.89, 7.60–7.66, 8.06–8.12, 8.75–8.80, 8.80–8.86 and 9.11–9.16 ones. For the second (3-classification) strategy employed, the first cluster has upregulated 1.15–1.17, 1.17–1.22, 1.26–1.32, 1.50–1.56, 1.56–1.58, 0.98–1.03, 1.36–1.41, 2.02–2.08, 2.36–2.40 and 2.87–2.89 ppm buckets, the second upregulated 0.72–0.76, 1.36–1.41, 2.87–2.89 and 7.60–7.66 ppm ones, and the third upregulated 8.80–8.86, 9.11–9.16, 2.02–2.08, 2.36–2.40, 2.87–2.89, 7.60–7.66, 8.06–8.12 and 8.75–8.80 ppm ones.

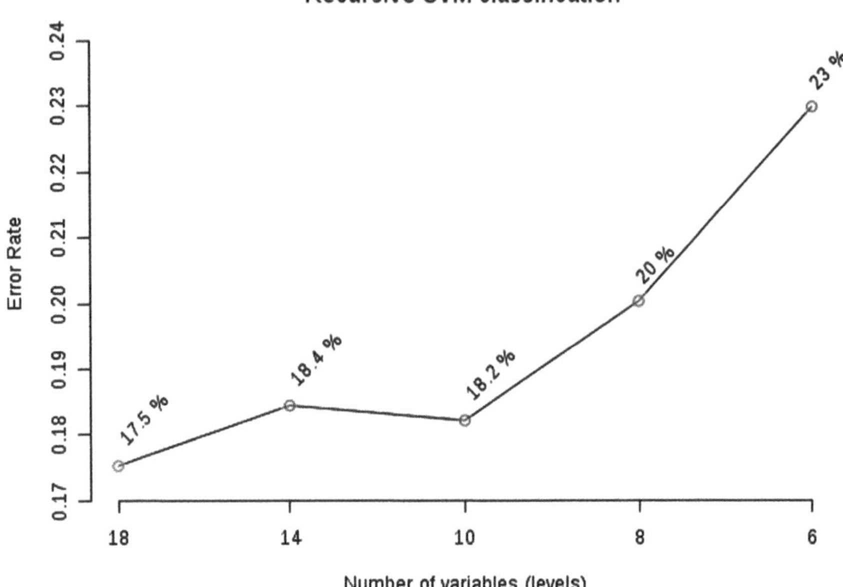

Figure 3.4 Computationally Intensive Recursive Support Vector Machine (SVM) classification of the intelligently-bucketed ^{1}H NMR dataset analysed by SOMs as described in Figure 3.3. For this analysis, five levels (or models) were constructed with the employment of 6, 8, 10, 14 and 18 ^{1}H NMR bucket features (the SVM classification was performed *via* recursive predictor variable feature selection). Clearly, the model incorporating 18 predictor variables displays the smallest error rate (17.5%).

3.5.3 Random Forests (RFs)

Random Forests (RFs)[20] represent a supervised learning algorithm which is appropriate for the analysis of multidimensional datasets. A 'forest' of classification trees is employed, each of which is generated *via* random feature selection from a 'bootstrap' sample located at each branch. The prediction of classification status arises from the majority of the ensemble.

This methodology also routinely serves to generate further valuable information, including out-of-bag (OOB) error and variable importance measures. Indeed, during tree growth, *ca.* one-third of the overall sampling group is removed from the 'bootstrap' sample, and these so-called OOB data are subsequently employed as a 'test' sample in order to acquire an unbiased estimate of the classification (OOB) error. Variable importance is evaluated by measuring the increase of the OOB error when it is permuted. Some modules available also have features which permit outlier detection (Figure 3.5).

3.6 Cluster Analysis

Cluster analysis includes techniques for the combination of similar explanatory (X) variables into clusters or classifications according to

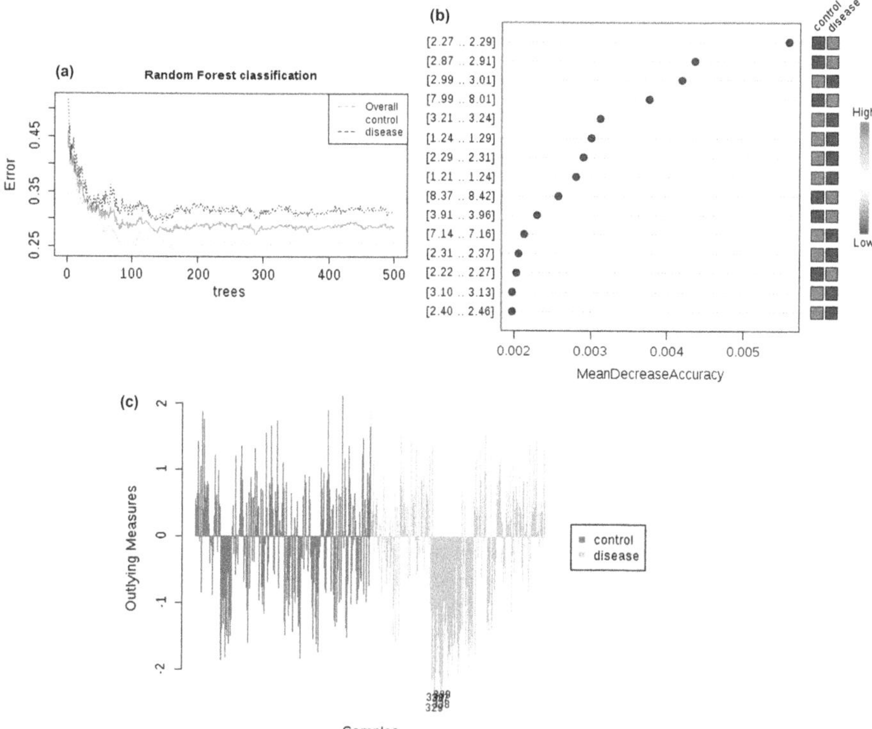

Figure 3.5 (a) Cumulative error rates computed in a Random Forest (RF) classification of an intelligently selected bucket ^{1}H NMR salivary supernatant dataset with 198 putative explanatory variables, two classification groups (healthy control *versus* oral disease) and 480 samples (240 in each classification). For this investigation, there were only 74 and 69% classification success rates for the healthy control and oral disease-active classifications, but this is often the case in many metabolomics investigations; however, at least some valuable biomarker data was still extractable form the dataset. The error rate for the complete dataset is shown as the black line, whilst the red and green lines display the error rates for each classification explored. (b) Significant spectral features ranked by the RF model for the healthy control and oral disease-active classifications (the *y*-axis represents the top 15 ^{1}H NMR chemical shift bucket features, which are ranked *via* the mean decrease in classification accuracy when permuted). (c) Detection of outliers by the RF technique (only the five most highly significant of these are labelled). The dataset was normalised to sample medians, cubed root-transformed and Pareto-scaled prior to analysis.

their similarities or dissimilarities, and these are then usually depicted in a tree-like ordination diagram which is classically known as a dendogram.[21] These techniques are now frequently employed by biologists working in the taxonomic and phylogenetic research areas in which 'dissimilarity' monitoring measures are often classified as genetic and/or molecular differences between organisms and species, *etc.*, or morphological, and hence the dendograms arising therefrom may be representative of possible evolutionary sequences. However, to date there remains only a limited application of such analysis to the metabolomics field of research. In principle, and if correctly applied, these methods are very likely to provide a high level of valuable linkage information if adapted to the classification of human diseases, their severity status and putative prognostic outcomes, together with the responses of patients to particular therapeutic avenues or regimens. Moreover, investigators may also apply these procedures to investigate the (inverted) clustering of predictor X variable columns, and hence acquire valuable information concerning which putative predictor biomolecules are clustered together in the context of their inter-relationships to disease status and severity, *etc.* Indeed, in the latter case, pre-identified or identifiable ^{1}H NMR or LC-MS mzRT values may be linked together in subsequent pathway analysis techniques, which may serve to be representative of amino acid, nucleotide, fatty acid and/or methylamine metabolism, for example, together with the Krebs cycle, and therefore may indicate which biochemical pathways are involved in the disease process investigated, and may also reveal information regarding which biomolecular routes are featured in its pathogenesis. In the case of ^{1}H NMR-selected buckets or bins (intelligently selected or otherwise), of course we would expect major 'similarities' between individual resonances arising from the same biomolecule (for example, strong intensity correlation linkages between the two $-CH_2-$ group resonances of 2-oxoglutarate, $^-O_2C.C\underline{H}_2C\underline{H}_2.CO.CO_2{}^-$), although such similarities would also be expected for metabolites exhibiting significant, strong or very strong correlations between each other, as might be expected between those arising from pathway-related biomolecules.

Agglomerative hierarchal and k-means clustering approaches are now common and frequently employed clustering techniques, and these methods are also complementary. Agglomerative hierarchical clustering (AHC) is an iterative classification method which is focused on dissimilarities between the 'objects' (perhaps patients or participants recruited to a metabolomics investigation) to be grouped together. Indeed, a class of dissimilarity can then be selected which is based on the subject matter explored and the dataset nature. From the dendogram acquired [which exhibits the progressive grouping(s) of the dataset], it is then possible to acquire much valuable information regarding a suitable number of classes into which the dataset can be grouped. However, in the k-means clustering technique (a non-hierarchical clustering strategy which commences *via* the creation of k clusters), an 'object' may be assigned to one class or grouping during one iteration, then change class during operation of the subsequent iteration (which is not possible with AHC for which such assignment is irreversible). Indeed, several potential solutions may be explored.

K-means clustering is a non-hierarchical clustering technique which commences *via* the creation of k clusters in accordance with a pre-specified process. Primarily, this approach first computes the means of each of the clusters, and if it transpires that one of the observations is closer to the centroid of another one, then that observation becomes a member of that alternative cluster. This process is repeated until none of the observations are re-assignable to a differing cluster.

In Section 3.6.1, the applications and potential limitations associated with the AHC technique are focused upon.

3.6.1　Agglomerative Hierarchal Clustering (AHC) Methods

The AHC technique serves to join and cluster individual explanatory (X) variables, and subsequently variables and classification strata, together until all such variables occur in one large group. Predominantly, AHC analysis algorithms commence with an overall matrix of pairwise similarities or dissimilarities (d_{hi}) between the X predictor variables, and therefrom this process involves (1) generation of a primary cluster between two variables with the smallest dissimilarity level, (2) recomputation of the dissimilarities existing between that particular pre-formed cluster and the remaining X variables, (3) production of a second cluster between the primary one generated and the X variable which is most similar to it, and (4) continuation of the process until finally all X variables are associated within such clusters.

The cluster analysis graphical depiction reveals connectivities between the classification groups, the line lengths indicating dissimilarities. Notwithstanding, if, as in a common metabolomics experiment, there are many predictor (X) variables, then the standard dendogram can be very complex and hence difficult to view in a single diagram. Alternatives to this include so-called polar dendograms in which the X variables are circularly arranged, and their distance from the circle reflects dissimilarities between variables and groups of them (although for the latter, the interpretations are somewhat subjective!).

Differences notable between the series of available AHC techniques are ascribable to exactly how the dissimilarities observed between separate clusters and variables are recomputed. Three frequently employed 'linkage' methods available for this purpose are (1) single ('nearest-neighbour') linkage, in which the dissimilarities observed between two clusters is determined *via* the minimal dissimilarity between all two-variable combinations (one from each clustering classification); (2) 'furthest-neighbour' (complete) linkage, in which the dissimilarity features observed between two clusters are determined *via* the maximal level of dissimilarity computed between all combinations of two variables (one from each cluster); and (3) group mean or average linkage, in which the dissimilarity between two clusters is monitored by the average of all such dissimilarities between two variables, again with one from each cluster [an unweighted pair-groups method involving arithmetic means (UPGMA) frequently represents a recommended approach for this].

There are a range of advantages and disadvantages associated with both the UPGMA and further available linkage methods (reviewed in refs 21–23), the additional ones including a weighted version of UPGMA (WPGMA), in which the original dissimilarities are weighted differentially, and an unweighted clustering analysis (UPGMC), which is focused on centroid indices rather than mean values. If the MV dataset demonstrates striking dissimilarities, then each of these methods will generate similar dendograms. However, for datasets with only a weak clustering structure, each of these differing linkage approaches may give rise to very different patterns.[22]

However, AHC analysis also has a number of disadvantages associated with its application to MV datasets and, as described above, these are primarily associated with dendogram interpretations. Indeed, once a cluster is generated from ≥ 2 X variables, then, as noted above, that particular cluster

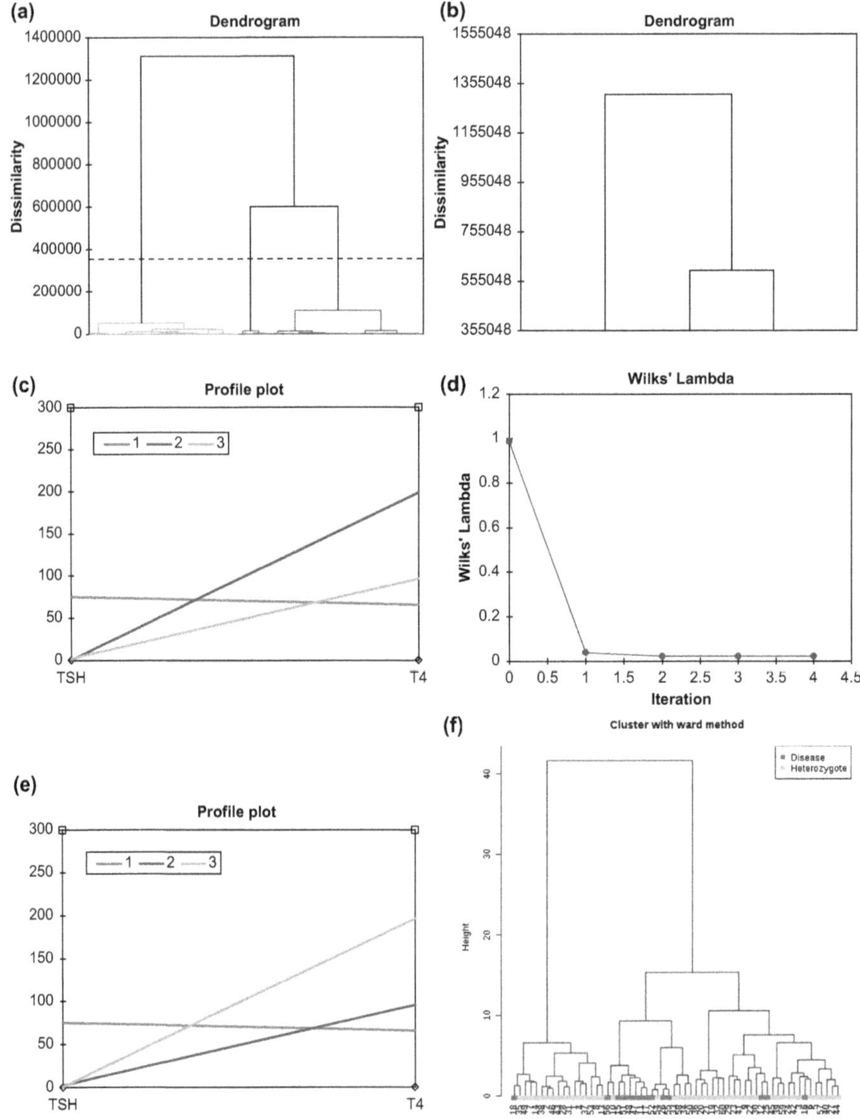

cannot be subsequently decomposed. Consequently, the representation of the dendogram does not include *all* pairwise dissimilarities between the predictors incorporated, unlike those observable in the multidimensional scaling (MDS) technique. Therefore, a misleading or potentially misleading clustering generated during the primary phases of the process will unfortunately exert an influence on all the remaining clusters formed therefrom. Moreover, as with many metabolic analysis techniques, some inexperienced researchers tend to focus too highly on the clusterings observed without first exploring the particular reasons for (*i.e.* the metabolic variable contributions towards) the dissimilarities detectable.

3.6.2 Clustering Analysis Case Study

Figure 3.6 shows an example of such a clustering analysis applied to the exploration of blood serum thyroid disease biomarker concentrations [specifically those of thyroxine (T4) and thyroid stimulating hormone (TSH)] in

---

Figure 3.6 (a)–(c) Agglomerative Hierarchal Clustering (AHC) analysis applied to a dataset consisting of 671 matched blood serum thyroxine (T4) and thyroid-stimulating hormone (TSH) levels in three thyroid disease classification groups [300 healthy control (euthyroid), 300 hypothyroid and 71 hyperthyroid patients], *i.e.* only two biomarker concentration variables. (a) and (b) Complete and simplified classification dendograms, respectively, for the distinction of the three thyroid disease classifications; in (a), the green-, violet- and brown-coloured classifications represent the hypothyroid, hyperthyroid and euthyroid conditions, respectively. (c) Profile plot for the three disease classifications [in this diagram, classes 1 (red), 2 (blue) and 3 (green) represent the hypothyroid, hyperthyroid and euthyroid disease classification groups respectively]. In this model, the dissimilarity index was the Euclidean distance, and Ward's agglomerative method was employed (the truncation method was automatic). This methodology gave rise to an exceptional discrimination between the three disease classifications (100% for each case). Furthermore, from (a), two or more major subclassifications of the euthyroid and hypothyroid classifications are also detectable, phenomena which may be ascribable to further clinical criteria (*i.e.* selected lateral variables). (d)–(e) k-means clustering applied to the dataset analysed by AHC. (d) Plot of the determinant of W *versus* number of iterations (this determinant, which is pooled within the covariance matrix, represents a criterion which is much less sensitive to the effects of scale than the corresponding W trace one). (e) Profile plot for the three thyroid disease classifications [for this plot, classes 1 (red), 2 (blue) and 3 (green) represent the hypothyroid, euthyroid and hyperthyroid disease classification groups, respectively]. For this analysis, 10 000 iterations and a convergence of 10^{-5} were employed; data were not centred and reduced prior to analysis, the initial partition was random and 10 000 repetitions were performed. The classification success levels were 99.7% for the euthyroid and 100% for both the hypothyroid and hyperthyroid disease classifications (although it should be noted that 3 out of a total of 300 euthyroid participants were incorrectly classified as hypothyroid). (f) Application of AHC to the MV analysis of two disease classifications in a urinary ^{1}H NMR dataset.

euthyroid (healthy control, *i.e.*, those already explored by CCorA in Section 3.2.1; the normal reference population range of T4 and TSH concentrations are 55–135 ng ml^{-1} and 0.17–4.05 IU ml^{-1}, respectively), hypothyroid (diminished T4 and elevated TSH levels) and hyperthyroid (elevated T4 and reduced TSH concentrations) [Figures 3.6(a)–(c), respectively]. Clearly, this methodology is very successful in distinguishing between clusters arising from each of the three separate thyroid disease patient classification groups, and hence this approach serves as a valuable means of discriminating between them. This example is particularly noteworthy in view of the consideration that only two (albeit key) biomarker variables were required to achieve these results, specifically those commonly employed in clinical practice. However, the incorporation of further (lateral) X predictor variables may serve to improve the discriminatory potential observed between the three thyroid disease classifications investigated, and may also provide evidence for the presence of sub-clusters detectable therein. Figures 3.6(e) and (f) show a corresponding k-means clustering analysis, which also reveals an excellent classification of the three thyroid disease classifications, and Figure 3.6(f) displays results acquired from the AHC analysis of a urinary ^{1}H NMR dataset.

3.7 Novel Approaches to the Analysis of High-throughput Metabolomics Datasets

Classical data analysis methods serve to evaluate the significance of 'Between-Group' responses to perhaps a disease process or, alternatively, a treatment applied, either in a univariate sense, *i.e.* systems involving the testing of single potential predictor variables individually (*via* t-tests, ANOVA or corresponding non-parametric methods applied in this manner), or through the application of MV techniques such as MV ANOVA (MANOVA), ASCA or further multidimensional testing systems, PCA or PLS-DA, for example.

Although commonly rejected on consideration that *univariate* differences will, in general, not provide a similar contributory discriminatory 'picture' in a multidimensional space or even hyperspace, such univariate analysis techniques can be employed in order to alleviate the problem of a potentially very large ^{1}H NMR or GC-MS dataset containing many non-discriminatory variables (*i.e.* those which do not contribute to the metabolomic distinction potentially arising between two or more classification groups, for example), and in this manner diminish it to a smaller or much smaller number consisting of those which are contributory, at least in a univariate sense. In this manner, this smaller dataset will reveal biomolecules which demonstrate the most powerful responses to the conditions set by the experimental design hypothesis, and therefore experimenters may perform 'variable selection' modelling, evaluations and validation and cross-validation techniques in a subsequent MV fashion.

Two examples in which such univariate methods have been employed for the analysis of metabolomic datasets include the application of a two-way (randomised blocks) ANOVA design to assess therapeutic agent-induced modifications to selected biomolecules/metabolites,[24] and a further study which has involved a combined non-parametric Wilcoxon rank sum test/ ANOVA evaluation of differing rates of cancer progression (for pathologies varying from benign prostate to metastatic disease).[25] As noted above, such methods can be valuable if there are highly or very highly significant differences between the particular pre-selected biomolecule levels of the two (or more) criteria of classification involved, but their inclusion in multi-component, hyper-dimensional evaluations of their discriminatory potential may also serve to represent them as important classifiers, either directly as contributions to components (usually orthogonal) or, alternatively, as 'suppressor' variables.

However, the author is also aware of many situations in which the univariate analysis of such MV metabolomics datasets yields similar or very similar results to those acquired from PCA or PLS-DA explorations, for example, and in these cases the prior or single performance of t-tests or ANOVA for single criteria of classification or factors can provide valuable statistical disease classification data, but only at the *univariate* level: no allowance for multiple correlations is made, although factorial experimental designs (such as that illustrated below) can also involve the incorporation of a wide range of factors, including both fixed and random effects (for which estimated population components-of-variance are derivable for the latter), 'nested' effects, together with a range of possible first-, second- and even third-order factor interactions (*i.e.* multifactor experiments). This consideration is of much importance, since a large or even substantial proportion of the variance observed in MV metabolomics datasets (*e.g.* those comprised of ^{1}H NMR buckets or bins) is accountable for by further lateral variables, in addition to the 'Between-Disease Group Classification' factor, which is of major interest to investigators. Indeed, the statistical significance of random effects such as those arising from 'Between-Study Participants' and 'Between-Samples-within-Participants (the latter as a 'nested' effect for investigations in which more than one sample is collected per participant), perhaps in the absence of further lateral variables available, *e.g.* time-points in time-series ones) are readily determined from such experimental designs, and it is now widely accepted amongst metabolomics researchers that effects such as these are responsible for much of the total variance in MV metabolomics datasets, especially that 'Between-Participants'. Indeed, unless they are taken into account, these effects can mask and confound the testing of the major ones of interest to the study.

Furthermore, quantitative lateral variables can also be incorporated or accounted for by the application of Analysis-of-Covariance (ANCOVA) models (*e.g.* participant age and BMI, relevant cell counts, *etc.*), although it should be noted that, not unlike many MV experimental models, these systems assume that there are *linear* relationships between these (X) covariables and the dependent (Y) variable of interest, and also that there are no multi-collinearities between the X variables (if there is more than a single one of the latter), as in simple multiple regression designs with three or so 'independent' variables. Methods available for overcoming such problems include (1) transformations to linearity (*e.g.* logarithmic, exponential, power, square root, cube root, *etc.*) and (2) polynomial ANCOVA models, the latter of which permits fitting of the response variables to quadratic or higher-order relationships to the quantitative covariates. Additionally, a range of possible first-, second- or even higher-order interactions between the effects of the putative qualitative and quantitative putative explanatory (X) covariables can also be evaluated and tested for their significance.

ANOVA Simultaneous Component Analysis (ASCA) is one relatively recent development which serves to overcome the potential confounding effects exerted by one or more of such lateral variables in MV analysis models,[26] and represents a combination of PCA and ANOVA techniques. Hence, the applications of ASCA are predominantly and principally focused on the analysis of relatively complex experimental designs which may incorporate the potential influential effects of 'Between-Participant' and/or 'Between-Samples-within-Participant' effects (both random ones), for example, although one major application of this particular technique lies with the analysis of MV time-series datasets in which the MV influence of the (often highly significant) 'fixed' effect of time is explored; this procedure may involve the isolation of a component of variance ascribable to a 'Between-Sampling Time-Points-within-Participants' effect. Specifically, the significance of the Time-Point$\times$Treatment interaction effect (MT_{ij}) is determined also commonly in such models [eqn (1)], in which M_i represents the treatment effect (perhaps a drug or alternative agent investigated), T_j the effects of the fixed time-point effect factor and e_{ij} the (unexplained) residual error term. ^{1}H NMR-linked metabolomic investigations which employ this particular MV model are provided in detail by

$$Y_{ij} = \mu + M_i + T_j + MT_{ij} + e_{ij} \qquad (1)$$

Westerhuis *et al.* in Chapter 4 of this volume.

Since it has the ability to incorporate a range of simultaneously monitored covariates, it permits researchers to directly evaluate the effects of many experimental design factors on variation observed in MV datasets. One well-cited example of this process involves exploration of the influence of an oral rinse formulation on the ^{1}H NMR metabolomic profiles of human saliva,[27] but since this technique and its applications are presented in much more detail later in this volume (Chapter 4), it will not be considered further here.

A further series of unsupervised MV techniques is generally described under the 'cluster analysis' descriptor, and these predominantly include Self-Organising-Maps (SOMs),[13,28] k-means clustering[29,30] and AHC[31,32] analysis techniques as outlined in Sections 3.5 and 3.6. In summary, such clustering methods serve to provide visualisation profiles of samples incorporated into the analytical model according to intrinsic 'self-similarities' in the MV datasets acquired, and their pre-defined classification status. However, with these methods, it has come to light that there are a number of issues associated with their applications to MV metabolomic datasets, including poor levels of reproducibility of the primarily detected clusterings, complications with interpretability criteria and also the propagation of errors.[33–35]

Partial Least Squares Regression (PLS-R) represents one of a class of supervised linear mixture models, and its focus is to seek an optimal predictor (X) variable dataset decomposition process when investigators are provided with a pre-specified matrix of possible responses. Hence, like other supervised techniques available for the analysis of MV datasets, its objective is to unravel inherent patterns therein, specifically clear metabolic ones that are perhaps strongly related to the pre-specified classification status of the dataset obtained. This strategy is further described in Section 3.8.4.

However, a relatively recent extension of the PLS models available [the orthogonal-PLS (OPLS) technique[36]] has been established, its main objective being to segregate dataset variation into that of major interest, which in turn is related to the response variable and also a portion ascribable to an orthogonal noise component [the latter of course not being associated with the response (Y) one]. Therefore, application of this process gives rise to a more facile means of interpreting the results acquired, and permits investigators to evaluate the 'Within-Classification' variance, in addition to that 'Between-Classifications'.[37–39] Indeed, a wide range of classificational applications of this technique have been reported, including those regarding the prognostic monitoring of kidney transplant patients,[40] and molecular epidemiology,[41] together with alternative medicine.[42] However, as with the PLS-DA technique discussed in detail in Chapter 1, a major problem with this methodology is the potential hazard of 'overfitting',[43] which again rather commonly arises from the incorporation of a too small (or much too small) experimental sample size in such systems, and also the unintended inclusion of statistical 'noise' into the model employed, such as that arising from the excessive 'learning' performed on a 'training' dataset. Notwithstanding, there are a number of validation methods available for such investigations, such as bootstrapping[44] or cross-validation processes,[45] which may serve to counter this problem.

3.7.1 Genetic Algorithms

Genetic Algorithms (GAs) have been demonstrated to have a high level of effectiveness regarding the selection of important and 'real' biomarker variables from multidimensional datasets (reviewed in ref. 46), and represent a class of evolutionary algorithms in which numerical optimisation techniques are employed. Such techniques have a 'biological' inspiration and their descriptive noun has analogies in biological mutation and selection processes. In a GA, a population consisting of randomly generated testing solutions (known as 'chromosomes') is assessed in order to generate a 'model fitness' criterion, and subsequently newer 'generation' solutions are constructed *via* a reproductive process (the fitness function provides an indication of the likelihood of any individual chromosome reproducing). This procedure is thence re-iterated for a succeeding series of generations up until the point when a satisfactory solution 'evolves'.

Therefore, researchers may employ GAs to 'chromosomally' seek and identify relatively small sub-sets of signals or peaks in biofluid or tissue biopsy profiles (chemical shift buckets in the NMR context) which have the ability to collectively discriminate between two or more criteria of sample classifications (*e.g.* healthy control *vs.* disease-active participants); the 'goodness-of-fit' criterion is determined by the classification success rate in a two group linear discriminant analysis which has been subjected to a cross-validation process. However, for the application of this technique, a series of parameters is required to be pre-set by the operator, and these include sub-set size, 'mutation' rate, 'chromosome' number, fitness and convergence measures, *etc.*, all of which influence GA performance and/or the rate at which it converges!

The complete GA routine (known as an 'epoch') can be repeated many times (say, up to 1000 repeats), and the outcome of each epoch is markedly affected by the random basis of the primary 'chromosomes', and also the options selected and hence occurring during the GA analysis. In view of this, the best solution or solutions which arise from each epoch are retained in the model system. However, unfortunately GAs are also subject to the 'curse of dimensionality', for example the larger the sub-set size selected, the more easily GA erroneously 'overfits' the dataset, and this is a very important consideration for researchers choosing to employ this methodology. Indeed, in the many (or very many!) examples in which there are larger or much larger numbers of X variables than there are samples, even the application of cross-validation methods offers only limited security against this overfitting problem. However, the application of alternative validation models to the dataset acquired, such as its partitioning into training, tuning and in-dependent test sets, is to be recommended when there are sufficient num-bers of samples available, and in this manner will, in general, avoid the adverse selection of spurious X predictor variables that adventitiously fit the 'structure' of the training dataset.

Despite these potential problems, GAs serve to effectively perform multiple epochs with completely randomised primary 'chromosomes' in order to identify those which serve as biomarkers in each classification (*i.e.* those sampled from two differing populations, if indeed there are metabolomic or genomic differences between them in terms of one or more of the X variables monitored).

3.7.2 Gaussian Graphical Models

Gaussian Graphical Models (GGMs) serve to remove indirect interactions *via* the 'conditioning' of simple two-variable correlations between the explanatory (X) variables against all the remaining ones, and have their foundation in *partial* correlation coefficients rather than simple Pearson ones, which are rather infrequently employed in all areas of the biomedical and clinical sciences, but their major applications lie within the analysis of correlations between two such potential explanatory (biomarker concentration) variables, whilst also allowing for the correlating or anti-correlating effects of further variables available within the experimental design. Indeed, simple partial correlation models have previously involved perhaps only up to five or six variables, the number of variables to be tested in this manner being also critically dependent on the requirement for a critical minimum sample size! Moreover, many commonly-employed MV analysis techniques such as PCA and PLS-DA critically depend on a matrix of simple two-variable Pearson correlations between an extremely large number of predictor (X) variables, and also the covariance matrix derived therefrom. However, with regard to the now common model fitting of large or very large MV datasets (*e.g.* those containing 200 or more potential explanatory variables) to selected experimental design classifications, it is important to note that one or more of a multitude of such simple (Pearson) correlations between two such variables may easily be induced *via* one or more 'confounding' ones which is (latently) responsible for that observed [such a phenomenon may also arise from the activities of 'suppressor' variables which are rationally considered with reference to the Correlated Component Regression (CCR) technique discussed in detail in Section 3.8.5 below].

A GGM therefore comprises an undirected graphical system in which each node is represented by a random (X) variable, and an 'edge' between two such nodes is constructed if those concerned are conditionally correlated whilst allowing for the effects of all further X variables.[47] Such GGM model systems have recently been applied to the analysis of metabolomics datasets,[48,49] and previously to the transcriptomics analysis field.[50,51] Notwithstanding, of critical importance to such investigations is the knowledge that the full computation of all possible (*i.e.* full-order) partial correlations requires a very much larger number of available samples available than there are predictor variables, and this requirement is, of course, frequently not met or even addressed in many metabolomic investigations. However, a number of alternative estimation algorithms which employ only low-order partial correlation coefficients (*i.e.* those between a maximum of three or even four of the most highly partially-correlated X variables for each predictor considered),[52] shrinkage estimation[53] or, alternatively, bootstrap resampling.[54]

Krumsiek *et al.*[55] recently explored the applications of GGMs to the analysis of metabolic datasets, and revealed that these techniques had the capacity to recover important metabolic inter-relationship data in such matrices derived from human blood plasma. Primarily, differing computer-simulated reaction systems were employed to produce *in silico* metabolomics datasets and, as expected, these model systems demonstrated that GGMs offer substantial advantages over those which simply rely on standard (Pearson) correlation networks. These advantages arise from the ability of GGMs to recover correct, more focused correlational metabolic information underlying the structure of the response network, and therefore their application in the presentation of efficient solutions to a range of MV analytical or bioanalytical problems. Intriguingly, Bartel *et al.* (2013)[56] recently examined a series of genuine metabolomic datasets from a population cohort, and applied GGMs to established and available metabolic pathway databases. They discovered that the high partial correlation coefficients attained in their investigations corresponded to known metabolic pathway reactions, and also that a number of novel possibilities for pathway interactions could be determined. Moreover, these findings were confirmed *via* application of the GGM technique to further metabolic datasets, and also generally confirmed their applications to biomarker-identification.[57]

Interestingly, Jourdan *et al.*[58] employed GGMs in order to explore and establish a connectivity between fat-free mass index and a small number of blood serum biomolecules, and some researchers have proposed the utilisation of undirected partial correlation information to directional network inferences, for example those with strategies based on directed partial correlation coefficients,[59] partial variance,[60] or the d-separation principle.[60] Briefly, the d in the d-separation and d-connection terms is an abbreviation for dependence; hence, if two variables (X and Y) are d-separated when expressed relative to a series of variables T in a directional graphical system, then in all the probability distributions that the graph can depict, they are said to be independently conditional on T. If information regarding X provides no additional information regarding Y when supplied with knowledge of Z, then X and Y are independently conditional upon T. Therefore, when the values of each T variable are known, X will provide no further information on Y. Clearly, a path is considered as active if it carries dependency information, and two variables X and Y could be connected *via* a range of graphical paths, of which all, only a proportion or none whatsoever are active. However, X and Y are d-connected if there exists *any* active path between them, but they are considered d-separated if *all* the paths that connect them are *inactive* or, equivalently, if no path between them has activity.

Table 3.1 lists the Pearson and corresponding *partial* correlation coefficients for putative relationships between the 'false-dummy' predictor (X) variables employed in the MV data analysis shown in Figure 1.3 and Table 1.2 of Chapter 1. Clearly, there are major differences between these values; indeed, computation of the partial correlation coefficients for these 'apparent' relationships shows that at least several of the extremely highly significant Pearson ones substantially diminish to either insignificant or virtually zero values! This clearly demonstrates the high value that the GGMs technique available has to offer.

3.7.3   Independent Component Analysis (ICA)

Although they have a lot of merit regarding the MV analysis of metabolomic datasets, one major limitation of the PCA, PLS-DA and even GGM methods is their reliance on second-order (*i.e.* linear) statistical dependencies

Table 3.1 (a) Pearson and (b) corresponding *partial* correlation coefficients for putative relationships between the 'false-dummy' predictor (X) variables employed in the MV data analysis shown in Table 1.2 of Chapter 1. In (a), all Pearson correlation coefficients were significant at the $p < 0.0001$ level, whereas in (b) only X1 has a significant (positive) partial correlation with disease score.

(a)
Correlation matrix (Pearson):

Variables	*X3*	Disease score	*X2*	*X1*
X1	1	0.8908	0.9412	0.9238
X2	0.8908	1	0.9448	0.9780
X3	0.9412	0.9448	1	0.9723
Disease score	0.9238	0.9780	0.9723	1

(b)
Proximity matrix (Partial Correlation Coefficient):

	X1	*X2*	*X3*	Disease score
X1	1	0.5422	0.1563	0.7785
X2	0.5422	1	0.4702	−0.0568
X3	0.1563	0.4702	1	−0.1127
Disease score	0.7785	−0.0568	−0.1127	1

(covariances) between the explanatory (X) variables. In view of the regular occurrence of higher-order dependencies, which may arise from non-linear metabolic systems, the almost blatant neglect of such relationships by at least some metabolomics researchers using relatively simple, conventional MV analysis techniques is somewhat concerning! Indeed, many inter-relationships between two (or more) explanatory variables may be of a curvilinear, quadratic or even higher polynomial nature rather than a simple linear one, although it should be noted that it may be possible to transform such non-linear relationships to linear ones, for example *via* $\log_{10}$-, reciprocal or alternative transformations of one or both of the correlated X variables involved. Furthermore, the linearity of such relationships between determined metabolite concentrations (or directly proportional spectroscopic or chromatographic measures) represents an approximation which is correct only for normally distributed populations from which biofluids or alternative bioanalytical matrices are sampled (*i.e.* a bivariate normal distribution as outlined in Chapter 1). Krumsick *et al.* (2011, 2012)[55,61] have revealed that the prior subjection of MV datasets to a logarithmic transformation process failed to satisfy the normality distributional assumption required for a high proportion of metabolic predictor (X) variables incorporated therein, an observation confirming that obtained by Grootveld and Ruiz Rodado in Chapter 2 of this volume (in both cases this transformation process was performed so that a log-normal distribution could primarily be assumed). However, these researchers also note that the employment of Spearman (and principally also Kendall) rank correlation coefficients, or, alternatively, mutual information, serves as a potentially valuable means of overcoming this problem,[62] although it should also be noted that such possible solutions do, of course, give rise to a diminished level of statistical power!

A recently developed technique which has the ability to recognise and capture the above-noted higher-order dependencies is Independent Component Analysis (ICA),[63] which extends the conceptual attributes of standard correlations to statistical dependencies. Indeed, to date, this methodology has been applied to the areas of functional magnetic resonance imaging (MRI),[64] molecular biology for cancer class determinations,[65] cellular proliferation explorations,[66] electroencephalographic (EEG) neurobiological monitoring[67] and, more recently, metabolomics analysis, for example the investigation of colitis in a mouse model.[68] The major difference between this method and those of PCA-based classifications is based on considerations of the particular nature of inherent relationships existing between the isolated components. However, the ICA method further advances the component generation (decorrelation) stage of PCA to statistical independence, *i.e.* it converts pre-acquired MV metabolomics profiles to statistically independent components known as ICs. Indeed, combinations of metabolic pathways (representing corresponding biological processes), each of which differentially contribute towards the overall metabolic patterns of biofluids and tissues, provides a rationale for this. In concept, the ICA method attempts to resolve the profiles of these specimens into meaningful information which relates to the individual pathways which give rise to it, and the analysis involves a compartmentalisation of the dataset matrix (A) into source and mixing matrices (S and M, respectively). However, this approach permits differing interpretational choices with respect to these matrices. Indeed, a defined metabolic pathway which 'mixes' up to the complete metabolic profile (^{1}H NMR-defined or otherwise) can serve to be representative of each row in S, and A reflects how powerfully each process is activated in an investigational sample dataset. One current major source of debate is focused on the estimation of an acceptable number of K components, and one approach to this is the employment of heuristic methods,[69] although it should be noted that there are no allowances made for the inclusion of prior (explanatory or latent) sources of variable information.

However, the above concerns may be effectively solved *via* the utilisation of a Bayesian ICA technique,[70] which has been employed to determine the optimal number of full ICs on which the MV dataset can be based and deciphered. Intriguingly, in 2002 Hojen-Sorensen *et al.*[71] used a mean-field Bayesian ICA approach in order to establish a non-negativity constraint for both of the above matrices when applied to MV metabolomics datasets.[72] Indeed, these researchers proposed that such non-negative constraints are biologically more acceptable than arbitrary values in view of the knowledge that metabolite levels cannot, of course, be negative, and also that biological process activities are either zero or positive. The dataset employed consisted of 218 pre-determined metabolites for 1764 blood serum samples (arising from the German KORA F4 cohort); the biomolecules determined corresponded to a range of metabolic pathways. On performing comparative evaluations of results arising from the applications of ICA to those derived from a standard PCA and k-means clustering approaches, the researchers involved demonstrated that the ICA technique out-performed the latter two more conventional analytical methods in the context of a more acceptable decomposition of the dataset.

Specifically, the ICs obtained revealed a powerful enrichment of distinct metabolic pathways, whereas application of PCA gave rise to only an inconsistent metabolite distribution. Similarly, application of ICA to the investigation of a gene expression dataset also demonstrated that this technique exhibited a more powerful enrichment than those arising from PCA and k-means clustering approaches.[73]

Interestingly, ICs were correlated to high-density-lipoprotein (HDL) concentrations in human blood plasma samples, a biomarker system which demonstrated a powerful relationship to a particular IC. Since this lipoprotein has strong links to a series of biological processes, including the transportation of triacylglycerols, cholesterol and cholesterol esters, these results offer a high level of clinical research potential.[74,75] Further investigation of the IC involved also demonstrated a marked contribution of branched-chain amino acids (BCAAs), an observation which may provide evidence for a previously undiscovered relationship between these biomolecules and blood plasma HDL concentrations.

3.8 Multidimensional Data (P > n) Problems Encountered in MV Regression Modelling

When the number of explanatory X variables (P) approaches or exceeds the patient or participant sample size (n), which is very often the case in ^{1}H NMR-based metabolomics investigations in which there may be several hundred or more chemical shift buckets (of fixed or intelligently selected variable size), and the number of samples investigated is often lower (or substantially lower) than this variable size, traditional multiple regression or discriminatory analysis techniques available become unstable and cannot be employed in view of multicollinearity problems (*i.e.* singularity of the covariance matrix). These high correlations observable between two or more predictor variables renders them redundant in such statistical models. Indeed, such collinearities or multicollinearities (spurious or otherwise) can also give rise to the well-known *overfitting* phenomenon, which is outlined in detail in Chapter 1.

Moreover, in metabolomics datasets with very large numbers of predictor (X) variables, the number of spurious correlations (particularly those arising purely by chance alone in a correlation matrix containing very nearly 200^2 Pearson r values) also increases substantially with increasing size of P, even if we pre-set an (uncorrected) significance level (*p* value) of only 0.01. Further spurious multicollinearities (perhaps many more) may arise from the presence of outlying data points. Therefore, as noted in Chapter 1, a high level of caution is recommended for researchers attempting to fit one or more of the many forms of supervised MV analysis methods (particularly PLS-DA) to such datasets, especially those with $P \gg n$ features.

In the simplest forms of MV analysis applicable (*e.g.* logistic or ordinary multiple/least squares regression analysis), a perfect separation between, for example, healthy 'control' or 'untreated', and 'diseased' or 'treated' classifications, respectively, is frequently observable in view of this overfitting problem. Of course, such correlations between the predictor (X) variables increase as their number approaches or exceeds that of the sample size. Indeed, when P is equivalent to n, any predictor variable may be expressed as an exact linear combination of the further predictors (*i.e.* perfect multi-collinearity), even if there is absolutely no correlation between them in the population, and hence a range of traditional multiple regression or even more advanced MV analysis approaches are completely unable to even begin to explore such problems.

3.8.1   Regression Regularisation

Regularisation involves the imposition of one or more model restrictions or assumptive criteria in MV analysis methods, and these may alleviate predictor variable error variances; if these restrictions are valid, then no bias is created in the model. However, if not, the variance may still be diminished despite the bias introduced, and this gives rise to a net reduction in prediction error (traditionally known as a 'bias-variance trade-off'). These forms of regularisation include (1) the traditional setting of one or more regression coefficients to zero, which reduces the number of X variables (P) directly–indeed, removal of one or more extraneous predictor variables (with actual regression coefficients of zero or approaching zero) reduces variance, maintains unbiasedness and hence reduces prediction error; (2) penalised regression, in which the magnitude of the regression coefficients is restricted, and biases them towards zero, and hence variance is diminished (known as the 'Ridge Regression' technique); and (3) component or dimensional reduction strategies, in which the influence of higher dimensions is set to a value of zero, a procedure also reducing variance [these methods include the Principle Component Regression (PCR), Partial Least Squares-Regression (PLS-Regression) and Correlated Component Regression (CCR) approaches].

3.8.2   Model Tuning and Optimisation *via* an M-Fold Cross-Validation Process

For this purpose, the dataset is divided into a total of a recommended 5–10 (M) equivalent group folds, and the modelling process is applied M times, each one omitting one fold; notably CCR employs two 'tuning' parameters, k components and P predictor (X) variables to be included in the model. Subsequently, the performance criteria (loss of function) from biofluid/tissue biopsy sample potential biomarker predictor variables present in the omitted folds is computed, *i.e.* the mean cross-validated (CV)-R^2 value based on all M omitted folds is calculated, and the best performing tuning parameters are then selected (*i.e.* those with the smallest errors).

Moreover, we are also able to estimate the standard error value for the CV-R^2 value, and this is based on M rounds of M-fold CV. Of course, the mean CV-R^2 value is computable from separate estimates of this parameter.

In reality, the complete predictor variable (X) dataset is likely to include more or many more extraneous or completely irrelevant ones (*i.e.* those with population coefficients of zero) than those which are of value for this purpose. Therefore, selected 'sparse' approaches, which employ methods for the exclusion of these 'redundant' variables, are of high value in such metabolomics analysis.

3.8.3 Principal Component Regression (PCR)

Principal Component Regression (PCR) features three major stages. Firstly, PCA is performed on the table of explanatory (X) variables (candidate predictors); secondly, an ordinary least squares (OLS) regression process is performed on selected components (PCs) arising therefrom; and thirdly, a computation of the model parameters that correspond to the input (X) variables is conducted. The PCA stage permits the transformation of an X predictor variable table with n observations into an S principal component table containing n scores vectors described by k components, where k $\leq$ P, and each k value represents a weighted sum of all of the X predictor variables. The most effective k components are then included as 'predictors' in the model (*i.e.* those which explain the highest level of predictor variable variance).

PCR's advantages include its ability to incorporate information on a wider range of candidate predictor variables. Although there may be only k ($\ll$ P) predictor variables (actually PCs) included in the final model, each component isolated takes into account information provided by a multitude of X variables, and therefore this method could, at least in principle, serve to provide an improved estimation of the dependent variable (Y), or an assigned binary score value for particular classifications. Moreover, since the components derived therefrom are not correlated (*i.e.* they are orthogonal), any problems arising from adverse multicollinearity effects are effectively removed. Notwithstanding, the components arising therefrom do not necessarily have a predictive capacity towards the (Y) dependent variable, and therefore may not give rise to an improved level of prediction when expressed relative to that provided by, for example, stepwise linear regression techniques. Indeed, the first component derivable (PC1) may have no relevance to prediction of the Y values. Furthermore, a complete dataset of all P predictions arising from all the X (explanatory) variables is a pre-requisite for the applications of this particular model.

Supervised PCR, however, selects only the k components that serve as significant (and direct) predictors of the (Y) dependent variable(s), and this offers some advantages. However, a major and consequently metabolomically significant disadvantage of this system is that it excludes components that may serve as 'suppressor' variables,[76] and therefore the method may provide a poorer predictive capacity than that obtained with both the PLS-Regression and CCR techniques. Again, datasets including measurements made on P of the original X variables available are required in order to complete this task, *i.e.* it is a 'non-sparse' technique.

In PCR, the most valuable and relevant coefficients relate the dependent variable scores (typically 0 for healthy control, and 1 for a positive disease-active classification) to the predictor (X) variables themselves, and not the components derivable therefrom. Indeed, we can readily obtain coefficients for each of the significant predictor X values *via* a substitution process, since each component represents a weighted sum of its X value contributors.

3.8.4 Partial Least Squares Regression (PLS-R)

PLS-R models are employed to predict, hopefully with a reliable means of confidence, a quantitative dependent (Y) variable from a series of many correlated or uncorrelated predictor (X) variables (again, in typical metabolomics experiments, the latter can often exceed several hundreds or even thousands of metabolite or potential biomarker concentrations in number). The Y variable may represent a disease severity or a pain intensity score, for example (say, 1–5 as in a Likert scale), or a related physiological parameter such as blood pressure, body mass index or, for that matter, a further particular, perhaps disease-specific biofluid metabolite concentration (*e.g.* blood plasma glucose concentration in investigations involving the study of type-1 or -2 diabetes).

The concept of this technique involves replacement of the P predictor variables with $k \leq P$ orthonormal predictive components, *i.e.* $v_1, v_2, \ldots v_k$ (the components derived therefrom are orthogonal and standardised with a unit variance; both the Y and X values should be mean-centred).

3.8.5 Correlated Component Regression (CCR)

Selected recent developments regarding the analysis of datasets of a high dimensionality status have, however, revealed the attainment of reliable MV predictions when the number of putative explanatory X variables available (P) exceeds the sample size (n). Correlated Component Regression (CCR) incorporates a 'step-down' algorithm for diminishing the number of potential predictor variables.[76–79] Indeed, the powerful ability of the CCR method to 'capture' the effects of 'suppressor' variables in MV metabolomics datasets (Figure 3.7) provides an explanation for its very high predictive capacity.

The CCR technique employs k correlated components, each of which represents a composite of the original predictor (X) values, and may be employed in order to determine a disease's nature or classification, or its status. The first (primary) component (k_1) incorporates the effects of predictors which express a direct effect on disease classification, and represents

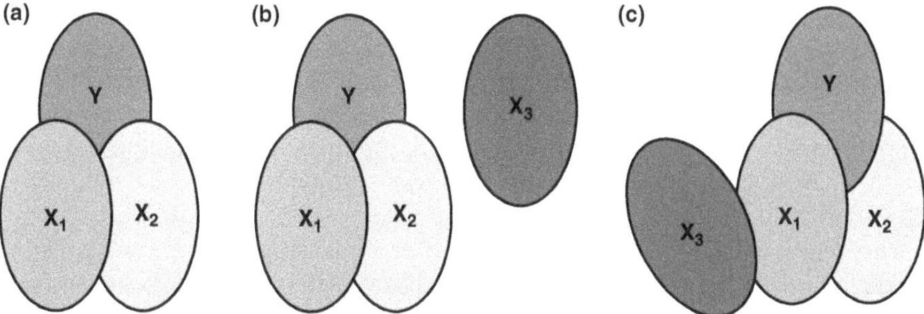

Figure 3.7 Diagrams displaying models with (a) two valid (relevant) predictor variables (X_1 and X_2) in an MV dataset (X_1, X_2 and Y are all correlated) and (b) two valid (X_1 and X_2) and one irrelevant (X_3) predictors (X_3 is not correlated with either X_1, X_2 or Y). Y represents the dependent variable (binary, ordinal or otherwise). (c) Diagram illustrating the influence exertable by a classical 'suppressor' variable (X_3 in this case) on X_1, the latter serving as a valid predictor variable [adapted with permission from Magidson *et al.* (2010)]. Such suppressor variables are potentially very common in multidimensional, metabolomics datasets.

a weighted mean of all these direct predictor (X) variable influences. However, the second component (k_2), which is correlated with k_1 (*i.e.* they are not orthogonal), generally captures the effects exerted by one or more 'suppressor' variables[78] (Figure 3.7), and this consideration can serve to improve the predictive capacity of the model *via* the removal of extraneous variation arising from one or more of the k_1-containing (direct) predictive variables. In a biochemical or metabolomics sense, we can envisage a situation in which finite fluctuations in the level of one (or more) metabolite(s) involved in a particular metabolic pathway which, although not a final product derived from this process, may exert a substantial (although 'masked') effect on the biofluid or tissue concentrations of a biomolecule that is, the latter representing the one (although perhaps not the only one) that researchers monitor as a biomarker for a particular disease process. Similarly, in genomics, although 'proxy genes' do not exert *direct* influences, they do, however, markedly enhance the predictive capacities of models by significantly influencing the effects of genes which do exert such effects directly (otherwise known as 'prime' genes).[76] Such suppressor variables frequently occur in gene expression and further high dimensionality datasets, and can sometimes also feature as the most valuable predictor (X) variables. The CCR technique effectively serves to 'capture' the effects of one or more suppressor variables, and in this manner improves a model's predictive ability *via* the removal of such extraneous variation from one or more of the predictor X variables which do indeed exert direct effects.

Overall, the Correlated Component Regression (CCR) technique can employ four classes of regression methodologies, and these employ rapid CV processes in order to determine the level of regularisation required to produce reliable predictions from data with P correlated explanatory (X) variables (in which multicollinearity is likely to be a confounding factor) and P is often greater than the sample size n. These techniques involve generalised linear models (GLMs), and one option is to activate a CCR 'step-down' algorithm in order to remove any irrelevant X (predictor) variables. The linear portion of the model serves as a weighted average of k predictive components [$k = (k_1, k_2, \ldots\ldots, k_K)$], each one representing a linear combination of the explanatory (X) variables.

The regression methods selectable differ according to the assumptions made regarding the scale type of the dependent variable Y (continuous *versus* dichotomous or binary), and the distributions (if any) assumed regarding the predictor (X) variables. Currently, there are four possible methods available for the application of the CCR technique to MV metabolomics datasets. For a continuous dependent Y variable, the CCR-linear model (CCR-LM) approach is available, in which the components arising therefrom are (as noted above) permitted to be correlated, rather than the non-correlated component option provided by PLS-R. Moreover, the CCR-LM technique is not influenced by standardisation of the predictor (X) variables, unlike the PLS-R approach, which gives rise to differing results subsequent to the application of this preprocessing step to the dataset.

However, for cases in which a binary Y dependent variable is involved (as indeed it would be if we label two distinct clinical classifications, to which we may assign scores of 0 for healthy control subjects and 1 for disease-active patients, for example), the CCR-Linear Discriminant Analysis (CCR-LDA) and CCR-Logistic Regression (CCR-Logistic) options are available. Although the former of these requires satisfaction of the assumption that the potential predictor (X) variables are concordant with an MV normal distribution within each Y variable classification [with differing group mean values and homogenous (common) variances and covariances], the latter logistic regression approach is not constrained by any distributional assumptions.

3.8.5.1   CCR Case Study

Figure 3.8 shows results arising from the application of the CCR-Logistic model to the analysis of an ANOVA-selected urinary ^{1}H NMR metabolomics dataset (containing 20 potential 'predictor' variables and a total of 60 urine samples) for the purpose of distinguishing between patients with a particular disease classification and their corresponding heterozygous (parental) carrier controls (CV component and step-down plots, together with that of the standardised predictor variable coefficients, and a ROC curve are shown). Results acquired revealed that for this model system, 4 or 5 correlated components and 12 explanatory X variables were optimal; clearly, this model gave rise to a very high level of distinction between the two classification groups: a typical AUROC value obtained was 0.995 (further details are provided in the Figure 3.8 legend). However, the area under the curve (AUC) and accuracy (ACC) vales obtained for a model with a single (k_1) component were almost as effective as those developed with 4 or 5 of these [Figure 3.8(a)].

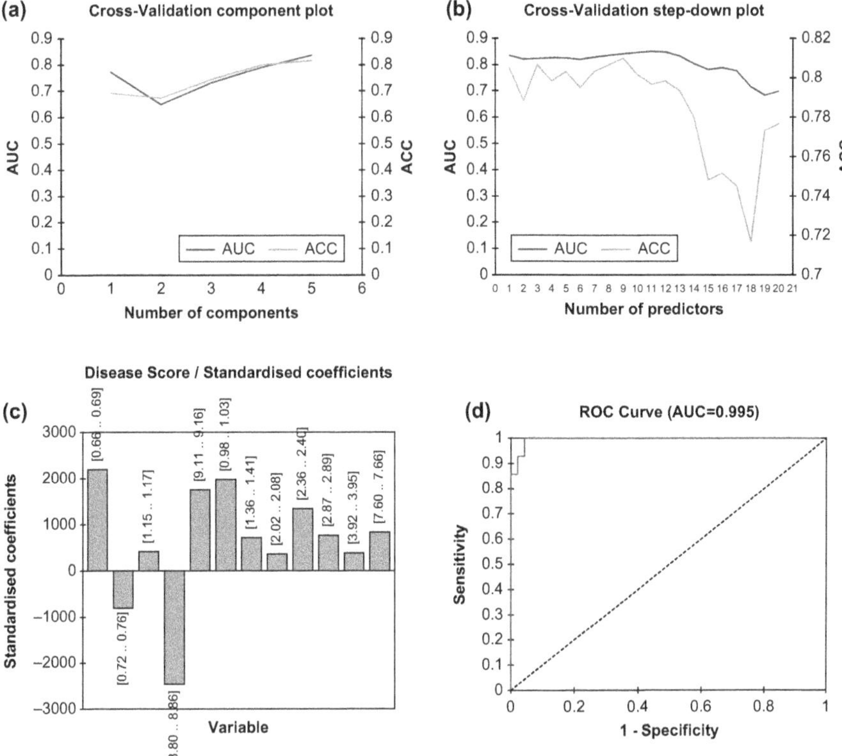

Figure 3.8 Correlated Component Regression (CCR) analysis performed on an ANOVA-selected urinary dataset containing 20 predictor (X) variables (intelligently selected ^{1}H NMR chemical shift buckets), 60 samples and two classification groups [disease-active *versus* their heterozygous (parental) controls, 14 and 46 samples in these groups, respectively]. (a) Cross-Validation (CV) component plot displaying the influence of the number of components incorporated on the area under the curve (AUC) and accuracy (ACC) values obtained; (b) Cross-Validation 'step-down' plot of AUC and ACC *versus* the number of predictors modelled; (c) diagram of standardised coefficients selected for the 12 ^{1}H NMR buckets incorporated into the model; (d) corresponding typical receiver operating characteristic (ROC) curve acquired for this analysis. The classification success rates for the heterozygous carrier controls and the disease-active participants were 95.65 and 100%, respectively. The step-down algorithm was applied to a maximum number of 20 predictor variables, 10 iterations, a cut-point for the disease scores (0 for the heterozygous controls, 1 for the disease active patients) of 0.50, and 10 rounds and 10 folds performed for the CV, the latter with stratification. The ^{1}H NMR-bucketed intensity dataset was normalised to that of urinary creatinine concentration and autoscaled prior to analysis.

Similarly, application of the PLS-DA method to this particular dataset was also found to provide a high level of valuable information regarding the selective metabolomics-based diagnosis of this condition (Figure 3.9).

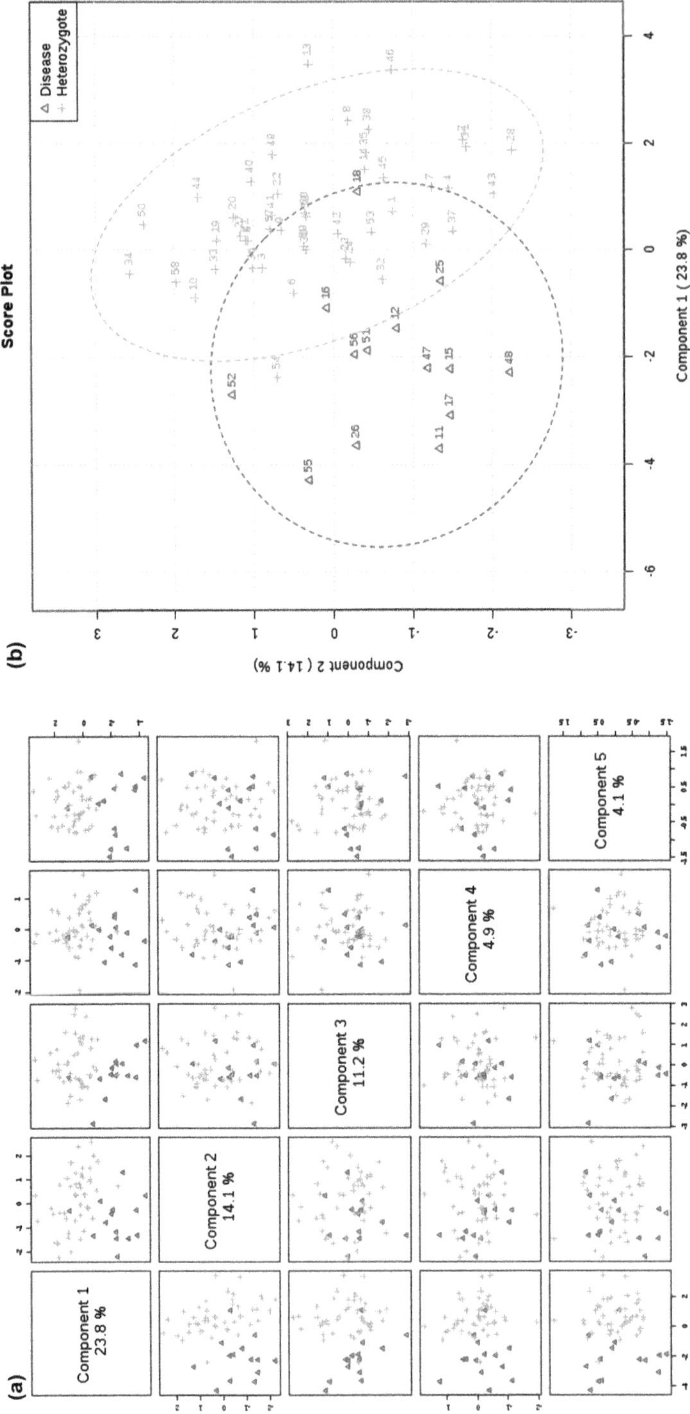
Score Plot
△ Disease
+ Heterozygote
Component 1 (23.8 %)
Component 2 (14.1 %)
Component 1
23.8 %
Component 2
14.1 %
Component 3
11.2 %
Component 4
4.9 %
Component 5
4.1 %

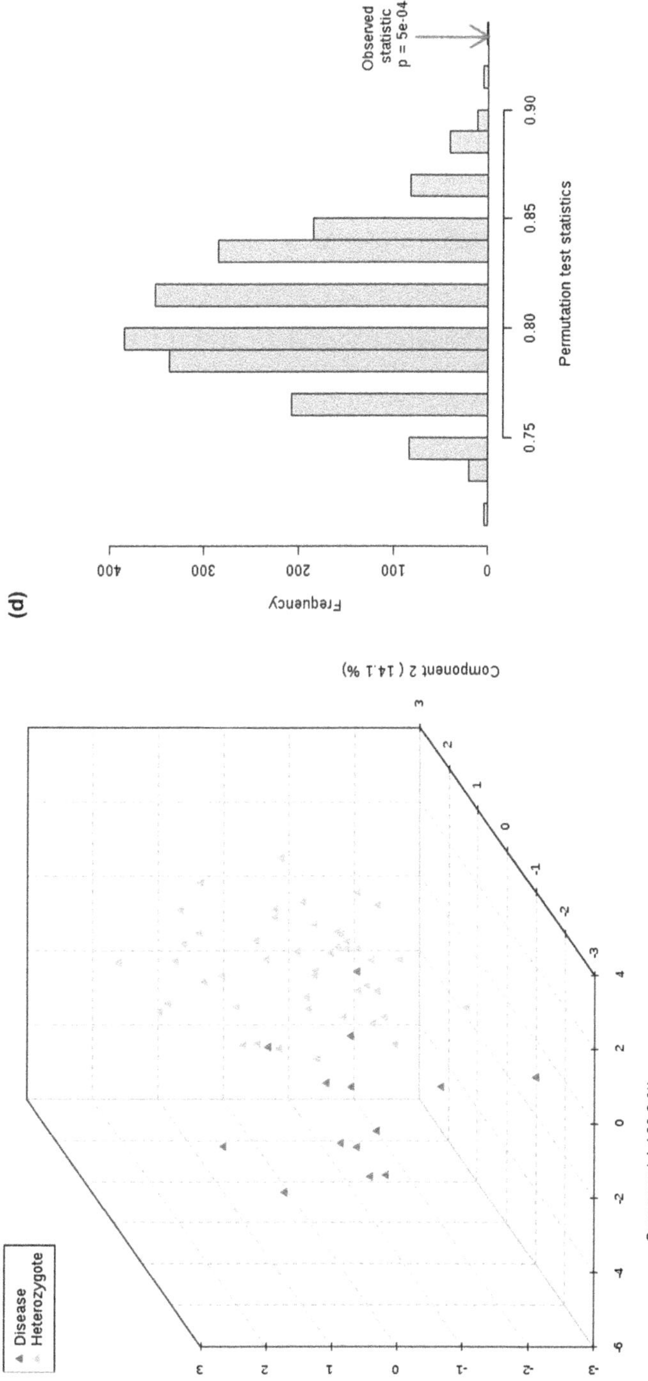

Figure 3.9 Results acquired from the corresponding analysis of the dataset explored in Figure 3.8 by PLS-DA. (a) Pairwise scores plots for the first five PCs (1–5); (b) PC2 *versus* PC1 scores plot with 95% confidence ellipses; (c) three-dimensional (3D) scores plot of PC3 *versus* PC2 *versus* PC1; (d) results arising from the CV permutation testing system performed (1000 permutations, $p < 0.0005$, a value based on prediction accuracy). The dataset was normalised to urinary creatinine concentration, cubed root-transformed and Pareto-scaled prior to analysis.

Indeed, cross-validation permutation testing of the dataset revealed a very high level of distinction between the two classifications $p < 0.0005$).

Therefore, in this example, both Partial Least Squares-Discriminatory (PLS-DA) and Logistic Correlated Component Regression (CCR-Logistic) analysis performed on creatinine-normalised intelligently selected ^{1}H NMR chemical shift buckets gave rise to high classification success rates for both patients with this disease and their heterozygous (parental) controls (90–100%). A series of MV permutation validation tests were also performed both with and without the consideration of sample donor families as a 'conditioning' variable, and a very strong relationship between the intelligently-selected ^{1}H NMR metabolic bucket predictor (X) variables and the disease classification status, *i.e.* disease-active patients *vs.* heterozygous controls, was found ($p < 0.0001$). Furthermore, two neural network non-probabilistic classification methods were also applied to analysis of the complete urinary ^{1}H NMR dataset (containing >200 bucket variables), specifically Support Vector Machines (SVMs) and Linear Discriminant Analysis (LDA), in conjunction with Genetic Algorithms (GAs). The disease classification prediction accuracies of these methods were excellent (97–98% success rate), with a very high level of reproducibility ($\pm 3\%$). From the above MV analysis pilot investigation, biomolecules which significantly contributed to the determination of classification status for this particular disease included selected amino acids and their degradation products, pyrimidine catabolites, nicotinate and nicotinamide pathway intermediates and products, and bile acids.

Interestingly, in view of the CCR component loadings observed for them, several of these metabolites appeared to serve as 'suppressor' variables, *i.e.* biomolecules which did not themselves directly contribute to the disease classification score value (*i.e.* 0 for heterozygote controls and $+1$ for disease-active patients), but nevertheless were correlated with one or more of those which were effective in this context.

References

1. M. S. Bartlett, The Statistical Significance of Canonical Correlations, *Biometrika*, 1941, **32**, 29–38.
2. W. R. Dillon and M. Goldstein, *Multivariate Analysis: Methods and Applications*, Wiley, New York, 1984.
3. P. E. Green and J. Douglas Carroll, *Mathematical Tools for Applied Multivariate Analysis*, Academic Press, New York, 1978.
4. T. G. Doeswijk, J. A. Hageman, J. A. Westerhuis, Y. Tikunov, A. Bovy and F. A. van Eeuwijk, Canonical correlation analysis of multiple sensory directed metabolomics data blocks reveals corresponding parts between data blocks, *Chemometr. Intell. Lab. Syst.*, 2011, **107**, 371–376.
5. L. Breiman, J. H. Friedman, R. A. Olshen and C. J. Stone, *Classification and Regression Trees*, Wadsworth, Pacific Grove, CA, 2nd edn, 1984.

6. T. Hastie, R. Tibshirani and J. Friedman, *The Elements of Statistical Learning*, Springer, New York, 2nd edn, 2009.
7. J. N. Morgan and J. A. Sonquist, Problems in the analysis of survey data and a proposal, *J. Am. Statist. Assoc.*, 1963, **58**, 415–434.
8. G. V. Kass, An exploratory technique for investigating large quantities of categorical data, *Appl. Stat.*, 1980, **20**(2), 119–127.
9. D. Bigss, B. Ville and E. Suen, A method of choosing multiway partitions for classification and decision trees, *J. Appl. Stat.*, 1991, **18**(1), 49–62.
10. W. Y. Loh and Y. S. Shih, Split selection methods for classification trees, *Stat. Sinica*, 1997, **7**, 815–840.
11. V. G. Tusher, R. Tibshirani and G. Chu, Significance analysis of microarrays applied to the ionizing radiation response, *Proc. Natl Acad. Sci. USA*, 2001, **98**, 5116–5121.
12. B. Efron, R. Tibshirani, J. D. Storey and V. Tusher, Empirical Bayes analysis of a microarray experiment, *J. Am. Statist. Assoc.*, 2001, **96**, 1151–1160.
13. T. Kohonen, Self-organized formation of topologically correct feature maps, *Biol. Cybernetics*, 1982, **43**, 59–69.
14. T. Kohonen, *Self-Organizing Maps*, Springer, Berlin, 2000.
15. F. Marini, A. L. Magrìa, R. Buccia and A. D. Magrìa, Use of different artificial neural networks to resolve binary blends of monocultivar Italian olive oils, *Anal. Chim. Acta*, 2007, **599**, 232–240.
16. U. Siripatrawan, Self-organizing algorithm for classification of packaged fresh vegetable potentially contaminated with foodborne pathogens, *Sensor. Actuat. B-Chem.*, 2008, **128**, 435–441.
17. K. Wongravee, G. R. Lloyd, C. J. Silwood, M. Grootveld and R. G. Brereton, Supervised Self Organizing Maps (SOMs) for classification and variable selection: illustrated by application to NMR metabolomic profiling, *Anal. Chem.*, 2010, **82**(2), 628–638.
18. C. J. C. Burges, A tutorial on Support Vector Machines for pattern recognition, *Data Min. Knowl. Dis.*, 1998, **2**, 121–167.
19. X. Zhang, X. Lu, Q. Shi, X. Q. Xu, H. C. Leung, L. N. Harris, J. D. Iglehart, A. Miron, J. S. Liu and W. H. Wong, Recursive SVM feature selection and sample classification for mass-spectrometry and microarray data, *BMC Bioinformatics*, 2006, **7**, 197, DOI: 10.1186/1471-2105-7-197.
20. L. Breiman, Random forests, *Mach. Learn.*, 2001, **45**, 5–32.
21. P. Legendre and L. Legendre, *Numerical Ecology*, Elsevier Science BV, Amsterdam, 2nd English edn, 1998.
22. J. R. Ludwig and J. F. Reynolds, *Statistical Ecology: a Primer on Methods and Computing*, John Wiley and Sons, New York, 1988.
23. M. Kent and P. Coker (ed.), *Vegetation Description and Analysis*, Belhaven Press, London, 1992.
24. E. Altmaier, S. L. Ramsay, A. Graber, H.-W. Mewes, K. M. Weinberger and K. Suhre, *Endocrin.*, 2008, **149**, 3478–3489. http://dx.doi.org/10.1210/en.2007-1747 PMid:18372322.
25. A. Sreekumar, L. M. Poisson, T. M. Rajendiran, A. P. Khan, Q. Cao, *et al.*, Metabolomic profiles delineate potential role for sarcosine in prostate

cancer progression, *Nature*, 2009, **457**, 910–914. http://dx.doi.org/10.
1038/nature07762 PMid:19212411 PMCid:272474625.

26. A. K. Smilde, J. J. Jansen, H. C. J. Hoefsloot, R.-J. A. N. Lamers, J. Van der
    Greef, *et al.*, ANOVA-simultaneous component analysis (ASCA): a new
    tool for analysing designed metabolomics data, *Bioinformatics*, 2005, **21**,
    3043–3048.

27. A. Lemanska, M. Grootveld, C. J. Silwood and R. G. Brereton, Chemo-
    metric variance analysis of ^{1}H NMR metabolomics data on the effects of
    oral rinse on saliva, *Metabolomics*, 2012, **8**, 64–80. http://dx.doi.org/10.
    1007/s11306-011-0358-4.

28. V. P. Mäkinen, P. Soininen, C. Forsblom, M. Parkkonen, P. Ingman,
    P. Ingman, K. Kaski, P.-H. Groop, M. Ala-Korpela, on behalf of the
    FinnDiane Study Group, ^{1}H NMR metabonomics approach to the dis-
    ease continuum of diabetic complications and premature death, *Molec.
    Syst.   Biol.*,   2008,   **4**,   168.   http://dx.doi.org/10.1038/msb4100205.
    PMid:18277383 PMCid:2267737.

29. J. A. Hageman, R. A. Van Den Berg, J. A. Westerhuis, H. C. J. Hoefsloot
    and A. K. Smilde, Bagged K-means clustering of metabolome data,
    Crit. Rev. Anal. Chem., 2006, **36**, 211–220. http://dx.doi.org/10.1080/
    10408340600969916.

30. X. Li, X. Lu, J. Tian, P. Gao, H. Kong, *et al.*, Application of fuzzy c-means
    clustering in data analysis of metabolomics, *Anal. Chem.*, 2009, **81**, 4468–
    4475. http://dx.doi.org/10.1021/ac900353t PMid:19408956.

31. O. E. Beckonert, M. Bollard, T. M. Ebbels, H. C. Keun, H. Antti, *et al.*,
    NMR-based metabonomic toxicity classification: hierarchical cluster
    analysis and k-nearest-neighbour approaches, *Anal. Chim. Acta*, 2003,
    490, 3–15. http://dx.doi.org/10.1016/S0003-2670(03)00060-6.

32. E. Holmes, R. L. Loo, J. Stamler, M. Bictash, I. K. S. Yap, *et al.*, Human
    metabolic phenotype diversity and its association with diet and blood
    pressure, *Nature*, 2008, **453**, 396–400. http://dx.doi.org/10.1038/nature06882
    PMid:1842511032.

33. R. Goodacre, S. Vaidyanathan, W. B. Dunn, G. G. Harrigan and D. B. Kell,
    Metabolomics by numbers: acquiring and understanding global me-
    tabolite data, *Trends Biotechnol.*, 2004, **22**, 245–252. http://dx.doi.org/10.
    1016/j.tibtech.2004.03.007 PMid:15109811.

34. D. Jiang, C. Tang and A. Zhang, Cluster analysis for gene
    expression data: A survey, *IEEE Trans. Knowl. Data Eng.*, 2004, **16**, 1370–
    1386.

35. B. Andreopoulos, A. An, X. Wang and M. Schroeder, A roadmap of
    clustering algorithms: finding a match for a biomedical application,
    Brief Bioinform., 2009, **10**, 297–314. http://dx.doi.org/10.1093/bib/bbn058
    PMid:19240124.

36. T. Kind, V. Tolstikov, O. Fiehn and R. H. Weiss, A comprehensive
    urinary metabolomic approach for identifying kidney cancer, *Anal. Bio-
    chem.*, 2007, **363**, 185–195. http://dx.doi.org/10.1016/j.ab.2007.01.028
    PMid:17316536.

37. J. Trygg and S. Wold, Orthogonal projections to latent structures (O-PLS), *J. Chemomet.*, 2002, **16**, 119–128. http://dx.doi.org/10.1002/cem.695.

38. M. Bylesjö, M. Rantalainen, O. Cloarec, J. K. Nicholson, E. Holmes, *et al.*, OPLS discriminant analysis: combining the strengths of PLS-DA and SIMCA classification, *J. Chemometr.*, 2006, **20**, 341–351. http://dx.doi.org/10.1002/cem.1006.

39. S. Wiklund, E. Johansson, L. Sjostrom, E. J. Mellerowicz, U. Edlund, *et al.*, Visualization of GC/TOF-MS-Based Metabolomics Data for Identification of Biochemically Interesting Compounds Using OPLS Class Models, *Anal. Chem.*, 2008, **80**, 115–122. http://dx.doi.org/10.1021/ac0713510 PMid:18027910.

40. H. Stenlund, R. Madsen, A. Vivi, M. Calderisi, T. Lundstedt, *et al.*, Monitoring kidney-transplant patients using metabolomics and dynamic modeling, *Chemomet. Intell. Lab. Syst.*, 2009, **98**, 45–50. http://dx.doi.org/10.1016/j.chemolab.2009.04.013.

41. E. Holmes, R. L. Loo, J. Stamler, M. Bictash, I. K. S. Yap, Q. Chan, T. Ebbels, M. De Iorio, I. J. Brown, K. A. Veselkov, M. L. Daviglus, H. Kesteloot, H. Ueshima, L. Zhao, J. K. Nicholson, P. Elliott, Human metabolic phenotype diversity and its association with diet and blood pressure, *Nature*, 2008, **453**, 396–400. http://dx.doi.org/10.1038/nature06882 PMid:18425110.

42. J. Kang, M.-Y. Choi, S. Kang, H. N. Kwon, H. Wen, C. H. Lee, M. Park, S. Wiklund, H. J. Kim, S. W. Kwon, S. Park, Application of a ^{1}H nuclear magnetic resonance (NMR) metabolomics approach combined with orthogonal projections to latent structure-discriminant analysis as an efficient tool for discriminating between Korean and Chinese herbal medicines, *J. Agricult. Food Chem.*, 2008, **56**, 11589–11595. http://dx.doi.org/10.1021/jf802088a PMid:19053358.

43. D. I. Broadhurst and D. B. Kell, Statistical strategies for avoiding false discoveries in metabolomics and related experiments, *Metabolomics*, 2006, **2**, 171–196. http://dx.doi.org/10.1007/s11306-006-0037-z.

44. R. Wehrens, H. Putter and L. M. Buydens, The bootstrap: a Tutorial, *Chemomet. Intell. Lab. Syst.*, 2000, **54**, 35–52. http://dx.doi.org/10.1016/S0169-7439(00)00102-7.

45. J. A. Westerhuis, C. Huub, J. Hoefsloot, S. Smit, D. J. Vis, A. K. Smilde, E. J. J. van Velzen, J. P. M. van Duijnhoven, F. A. van Dorsten, Assessment of PLSDA cross validation, *Metabolomics*, 2008, **4**, 81–89. http://dx.doi.org/10.1007/s11306-007-0099-6.

46. R. Leardi, Genetic algorithms in chemometrics and chemistry: a review, *J. Chemometr.*, 2001, **15**(7), 559–569.

47. S. L. Lauritzen, *Graphical Models*, Oxford University Press, Oxford. Available at: http://books.google.de/books?hl = de&lr = &id = mGGQWkx4guhAC&oi = fnd&pg = PA1&dq = Lauritzen + SL. + Graphical + Models&ots = 2IgpudGqZe&sig = zdDURuy8ZDSh3SE92apflfppkLM.

48. T. Çakır, M. M. W. B. Hendriks, J. A. Westerhuis and A. K. Smilde, Metabolic network discovery through reverse engineering of

metabolome data, *Metabolomics*, 2009, **5**, 318–329. http://dx.doi.org/10.1007/s11306-009-0156-4 PMid:19718266 PMCid:2731157.

49. J. J. B. Keurentjes, J. Fu, C. H. R. De Vos, A. Lommen, R. D. Hall, *et al.*, The genetics of plant metabolism, *Nature Genet.*, 2006, **38**, 842–849. http://dx.doi.org/10.1038/ng1815 PMid:16751770.

50. A. De La Fuente, N. Bing, J. Hoeschele and P. Mendes, Discovery of meaningful associations in genomic data using partial correlation coefficients, *Bioinformatics*, 2004, **20**, 3565–3574. http://dx.doi.org/10.1093/bioinformatics/bth445 PMid:15284096.

51. P. M. Magwene, J. Kim, *et al.*, Estimating genomic coexpression networks using first-order conditional independence, *Genome Biol.*, 2004, **5**, R100.   http://dx.doi.org/10.1186/gb-2004-5-12-r100   PMid:15575966 PMCid:545795.

52. A. De La Fuente, N. Bing, I. Hoeschele and P. Mendes, Discovery of meaningful associations in genomic data using partial correlation coefficients, *Bioinformat.*, 2004, **20**, 3565–3574. http://dx.doi.org/10.1093/bioinformatics/bth445 PMid:15284096.

53. J. Schäfer and K. Strimmer, A shrinkage approach to large-scale covariance matrix estimation and implications for functional genomics, *Stat. Appl. Genet. Mol. Biol.*, 2005, **4**, Article 32.

54. J. Schäfer and K. Strimmer, An empirical Bayes approach to inferring large-scale gene association networks, *Bioinformatics*, 2005, **21**, 754–764. http://dx.doi.org/10.1093/bioinformatics/bti062 PMid:15479708.

55. J. Krumsiek, K. Suhre, T. Illig, J. Adamski and F. Theis, Gaussian graphical modeling reconstructs pathway reactions from high-throughput metabolomics data, *BMC Syst. Biol.*, 2011, **5**, 21. http://dx.doi.org/10.1186/1752-0509-5-21 PMid:21281499 PMCid:3224437.

56. J. Bartel, J. Krumsiek and F. J. Theis, Statistical methods for the analysis of high-throughput metabolomics data, *Comput. Struct. Biotechnol. J.*, 2013, **4**(5), e201301009. http://10.5936/csbj201301009.

57. K. Mittelstrass, J. S. Ried, Z. Yu, J. Krumsiek, C. Gieger, *et al.*, Discovery of sexual dimorphisms in metabolic and genetic biomarkers, *PLoS Genet.*, 2011, **7**, e1002215. http://dx.doi.org/10.1371/journal.pgen.1002215 PMid:21852955 PMCid:3154959.

58. C. Jourdan, A.-K. Petersen, C. Gieger, A. Döring, T. Illig, R. Wang-Sattler, C. Meisinger, A. Peters, J. Adamski, C. Prehn, K. Suhre, E. Altmaier, Gabi Kastenmüller7, R.-M. Werner, F. J. Theis, J. Krumsiek, H.-E. Wichmann, J. Linseisen, Body Fat Free Mass Is Associated with the Serum Metabolite Profile in a Population-Based Study, *PLoS One*, 2012, **7**. Available at: http://www.ncbi.nlm.nih.gov/pmc/articles/PMC3384624/. http://dx.doi.org/10.1371/journal.pone.0040009.

59. Y. Yuan, C.-T. Li and O. Windram, Directed Partial Correlation: Inferring Large-Scale Gene Regulatory Network through Induced Topology Disruptions, *PLoS One*, 2011, **6**, e16835. http://dx.doi.org/10.1371/journal.pone.0016835 PMid:21494330 PMCid:3071805.

60. R. Opgen-Rhein and K. Strimmer, From correlation to causation networks: a simple approximate learning algorithm and its application to high-dimensional plant gene expression data, *BMC Syst. Biol.*, 2007, **1**, 37, PMid:17683609 PMCid:1995222.

61. J. Krumsiek, K. Suhre, A. M. Evans, M. W. Mitchell, R. P. Mohney, *et al.*, Mining the Unknown: A Systems Approach to Metabolite Identification Combining Genetic and Metabolic Information, *PLoS Genet.*, 2012, **8**, e1003005. http://dx.doi.org/10.1371/journal.pgen.1003005 PMid:23093944 PMCid:3475673.

62. R. Steuer, J. Kurths, C. O. Daub, J. Weise and J. Selbig, The mutual information: Detecting and evaluating dependencies between variables, *Bioinformatics*, 2002, **18**, S231–S240. http://dx.doi.org/10.1093/bioinformatics/18.suppl_2.S231 PMid:12386007.

63. A. Hyvärinen, J. Karhunen and E. Oja, *Independent Component Analysis*, Wiley-Interscience, Chichester, 1st edn, 2001. http://dx.doi.org/10.1002/0471221317.

64. P. Gruber, A. Meyer-Bäse, S. Foo and F. J. Theis, ICA, kernel methods and non-negativity: New paradigms for dynamical component analysis of fMRI data, *Eng. Applic. Art. Intell.*, 2008, **22**, 497–504. http://dx.doi.org/10.1016/j.engappai.2008.11.010.

65. A. E. Teschendorff, M. Journée, P. A. Absil, R. Sepulchre and C. Caldas, Elucidating the Altered Transcriptional Programs in Breast Cancer using Independent Component Analysis, *PLoS Comput. Biol.*, **3**, 2007. Available at: http://www.ncbi.nlm.nih.gov/pmc/articles/PMC1950343/. http://dx.doi.org/10.1371/journal.pcbi.0030161 PMid:17708679 PMCid:1950343.

66. D. Lutter, P. Ugocsai, M. Grandl, E. Orso, F. Theis, E. W. Lang, G. Scmitz, Analyzing M-CSF dependent monocyte/macrophage differentiation: Expression modes and meta-modes derived from an independent component analysis, *BMC Bioinformatics*, 2008, **9**, 100. http://dx.doi.org/10.1186/1471-2105-9-100 PMid:18279525 PMCid:2277398.

67. S. Makeig, A. J. Bell, T. P. Jung and T. J. Sejnowski, Independent component analysis of electroencephalographic data, *Adv. Neur. Inform. Proc. Syst.*, 1996, 145–151.

68. F. Martin, S. Rezzi, D. Philippe, L. Tornier, A. Messlik, G. Holzlwimmer, P. Baur, L. Quintanilla-Fend, G. Loh, M. Blaut, S. Blum, S. Kochhar and D. Haller, Metabolic Assessment of Gradual Development of Moderate Experimental Colitis in IL-10 Deficient Mice, *J. Proteome Res.*, 2009, **8**, 2376–2387. http://dx.doi.org/10.1021/pr801006e.

69. I. R. Keck, F. J. Theis, P. Gruber, E. W. Lang, K. Specht, G. Fink, A. Tome and C. Puntonet, Automated clustering of ICA results for fMRI data analysis, *Proc. CIMED*, 2005, 211–216.

70. G. Schwarz, Estimating the dimension of a model, *Ann. Stat.*, 1978, **6**, 461–464.

71. P. A. Højen-Sørensen, O. Winther and L. K. Hansen, Mean-field approaches to independent component analysis, *Neural. Comput.*, 2002, **14**,

889–918. http://dx.doi.org/10.1162/089976602317319009 PMid:11936966. http://dx.doi.org/10.1214/aos/1176344136.

72. J. Krumsiek, K. Suhre, T. Illig, J. Adamski and F. J. Theis, Bayesian Independent Component Analysis Recovers Pathway Signatures from Blood Metabolomics Data, *J. Proteome Res.*, 2012, **11**, 4120–4131. http://dx.doi.org/10.1021/pr300231n PMid:22713116.

73. S.-I. Lee and S. Batzoglou, Application of independent component analysis to microarrays, *Genom. Biol.*, 2003, **4**, R76. http://dx.doi.org/10.1186/gb-2003-4-11-r76. PMid:14611662 PMCid:329130.

74. J. F. Desforges, D. J. Gordon and B. M. Rifkind, High-density lipoprotein-the clinical implications of recent studies, *New Eng. J. Med.*, 1989, **321**, 1311–1316. http://dx.doi.org/10.1056/NEJM198911093211907. PMid:2677733.

75. A. Von Eckardstein, Y. Huang, G. Assmann, *et al.*, Physiological role and clinical relevance of high-density lipoprotein subclasses, *Curr Opin. Lipidol.*, 1994, **5**, 404. http://dx.doi.org/10.1097/00041433-199412000-00003. PMid:7712045.

76. J. Magidson, Correlated Component Regression: A Prediction/Classification Methodology for Possibly Many Features, *Proc. Am. Stat. Assoc.*, 2010. Available for download at http://statisticalinnovations.com/technicalsupport/CCR.AMSTAT.pdf.

77. J. Magidson, Correlated Component Regression: A Sparse Alternative to PLS Regression, 5th ESSEC-SUPELEC Statistical Workshop on PLS (Partial Least Squares) Developments, 2011, Available for download at http://statisticalinnovations.com/technicalsupport/ParisWorkshop.pdf.

78. J. Magidson and K. Wassmann, The Role of Proxy Genes in Predictive Models: An Application to Early Detection of Prostate Cancer. *Proc. Am. Stat. Assoc.*, pp. 2739–2753, 2010, Available for download at: http://statisticalinnovations.com/technicalsupport/Suppressor.AMSTAT.pdf.

79. M. Tenenhaus, Conjoint use of Correlated Component Regression (CCR), PLS regression and multiple regression, *5th ESSEC-SUPELEC Statistical Workshop on 'PLS (Partial Least Squares) Developments*, 2011.

Analysis of High-dimensional Data from Designed Metabolomics Studies

JOHAN A. WESTERHUIS,*[a,b] EWOUD J. J. VAN VELZEN,[a,c]
JEROEN J. JANSEN,[a,†] HUUB C. J. HOEFSLOOT[a] AND
AGE K. SMILDE[a]

[a] Biosystems Data Analysis, Swammerdam Institute for Life Sciences,
University of Amsterdam, The Netherlands; [b] Centre for Business,
Mathematics & Informatics, North West University, Potchefstroom,
South Africa; [c] Unilever Research and Development, Vlaardingen,
The Netherlands
*Email: J.A.Westerhuis@uva.nl

4.1 Introduction

In functional genomics many different approaches are used to understand the changes to an organism as a function of an applied treatment.[1] These changes can be defined in many ways, but popular functional genomics tools are transcriptomics, proteomics and metabolomics measurements.

Metabolomics is the 'systematic study of the unique chemical 'fingerprints' that specific cellular processes leave behind' – specifically, the study of their small-molecule metabolite profiles.[2-4] The metabolome represents

†Present address: Institute for Molecules and Materials, Analytical Chemistry, Radboud University Nijmegen, Toernooiveld 1, 6525 ED, Nijmegen, The Netherlands.

Issues in Toxicology No. 21
Metabolic Profiling: Disease and Xenobiotics
Edited by Martin Grootveld
© The Royal Society of Chemistry 2015
Published by the Royal Society of Chemistry, www.rsc.org

the collection of all metabolites in a biological cell, tissue, organ or organism, which are the end-products of cellular processes.[5] However, recently it has been understood that measuring the 'whole' metabolome leads to a too-large analytical challenge and more targeted approaches are used (*e.g.* as in lipidomics).[6]

The promise of functional genomics is to obtain an understanding of the dynamic properties of an organism at the cellular and/or organismal levels. This would provide a more complete picture of how biological function arises from the information encoded in a genome. To study the impact of a treatment on a biological system in a systematic manner, it is important that the treatment is employed in a standardised approach using an appropriate experimental design. The use of such a design is of importance to provide sufficient variation in the biological response of the system. Furthermore, it is also important to be able to estimate the independent effect of the treatment factors on the system with as little bias by other 'uncontrolled factors' as possible. A typical experimental design in a functional genomics study comprises the combination of a specific treatment and the time after treatment in which the response is studied.[7] In particular, the interaction of these two factors is of interest since it reflects the difference in the effects of two (or more) treatments throughout a specified time domain.

The default approach to study the effect of different treatments is the Analysis of Variance (ANOVA) approach.[8] ANOVA separates the contributions of the different treatment factors to the total variance of the data. Subsequently, different post–hoc tests (parametric as well as non–parametric) can be used to estimate the significance of differences between selected or particular treatment factors.

However, ANOVA is only applicable on a single biological response. When many response variables are measured simultaneously, as is the case in metabolomics studies, analysing each response variable by ANOVA ignores the correlation between the response variables. MANOVA is a multivariate extension to ANOVA that can deal with multiple biological response variables simultaneously.[9] It uses the covariance structure of the response variables, but breaks down when the number of variables becomes too large (larger than the number of experiments) as is usually the case in most metabolomics studies.

Notwithstanding, using only a multivariate data analysis method such as Principal Component Analysis (PCA) for exploring the data from a functional genomics experiment ignores the specific experimental design underlying the study.[10] The observed treatment effects in PCA will therefore describe an unknown confounding mix between all treatment effects, hampering the interpretability of the model.

Recently, a number of methods have appeared that combine ANOVA with PCA to overcome the drawbacks of both approaches. This has led to a range of methods, specifically ANOVA Simultaneous Component Analysis (ASCA),[7,11] Principal Response Curves (PRC),[12] Geometric Trajectory Analysis[13,14] and ANOVA-PCA.[15,16] The main difference between these methods is the specific definition of the ANOVA model and how the PCA method is used. Functional genomics studies with an underlying experimental design (*e.g.* cross-over and parallel designs, time-resolved experimentation, *etc.*) can greatly benefit from these new methods. Whenever an ANOVA model can be defined for a single response variable of the study, then these methods can be applied.

In the remainder of this chapter we will discuss two case studies in which a biological system was disturbed in a systematic manner and where multiple systems-response variables were analysed. The combination of ANOVA and a multivariate analysis method will be used to explore the dataset while focusing on the specific treatment effects.

4.2   Case Study 1: The Effect of Jasmonic Acid on the Production of Glucosinolates in *Brassicaceae oleracea*

The first example comprises the study of how plants react to a herbivore attack at their leaf and root sites. Plants in the *Brassicaceae* family (*e.g.* cabbage) produce very specific defence compounds when under attack by herbivores. These glucosinolates come in about 120 different species, varying only in the side group R (see Figure 4.1).[17] In two important groups of glucosinolates, the side group R is derived from methionine (aliphatic glucosinolates, AGs) or from tryptophan (Indole Glucosinolates, IGs).

To simulate a herbivore attack and thus induce the plant to synthesise glucosinolates, the plant hormone Jasmonic acid (JA) was administered to either the roots (root-induced) or the leaves (shoot-induced) of feral cabbage plants (*B. oleracea*). Subsequently, the glucosinolate levels were measured 1, 3, 7 and 14 days following treatment (this measurement was destructive, *i.e.* different plants were analysed at all time-points). Eleven different glucosinolates were determined in the plants and Table 4.1 shows the 11 glucosinolates that were monitored during the study.[18]

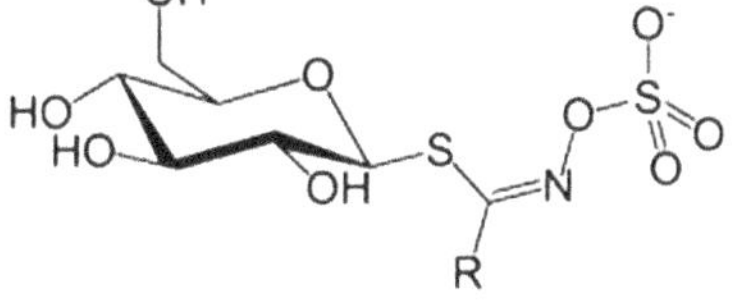

Figure 4.1   Structural formula of glucosinolates.

Table 4.1   Eleven glucosinolates determined in the study.

1	2	3	4	5	6	7	8	9	10	11
PRO	RAPH	ALY	GNL	GNA	4OH	GBN	GBC	4MeOH	NAS	NEO

The main goal of the study was to explore the impact of the treatment with Jasmonic acid on the glucosinolate composition of the plant, and whether these modifications vary with time. For all treatments (root-induced, shoot-induced and controls) a measurement of the 11 glucosinolates is performed for a group of plants after 1, 3, 7 and 14 days post-treatment. Different plants were analysed at the successive time points. (This represents an important implication that will become clearer in the remainder of this chapter.)

Figure 4.2 shows the measured intensity levels of some of the glucosinolates (PRO, NAS, 4MeOH and NEO) at time-points of 1, 3, 7 and 14 days after shoot- or root-induced treatments with Jasmonic acid (or not, as in the control situation). From Figure 4.2 it can be observed that PRO shows a concentration difference between the control and the root-induced treatment after 7 and 14 days. 4MeOH also shows a concentration difference with the control after 7 and 14 days, but only for the shoot-induced treatment with Jasmonic acid. However, NEO exhibits a distinct concentration difference between both root- and shoot-induced treatment for most of the time trajectories whilst NAS shows no statistically significant differences between the control group and either treatment mode.

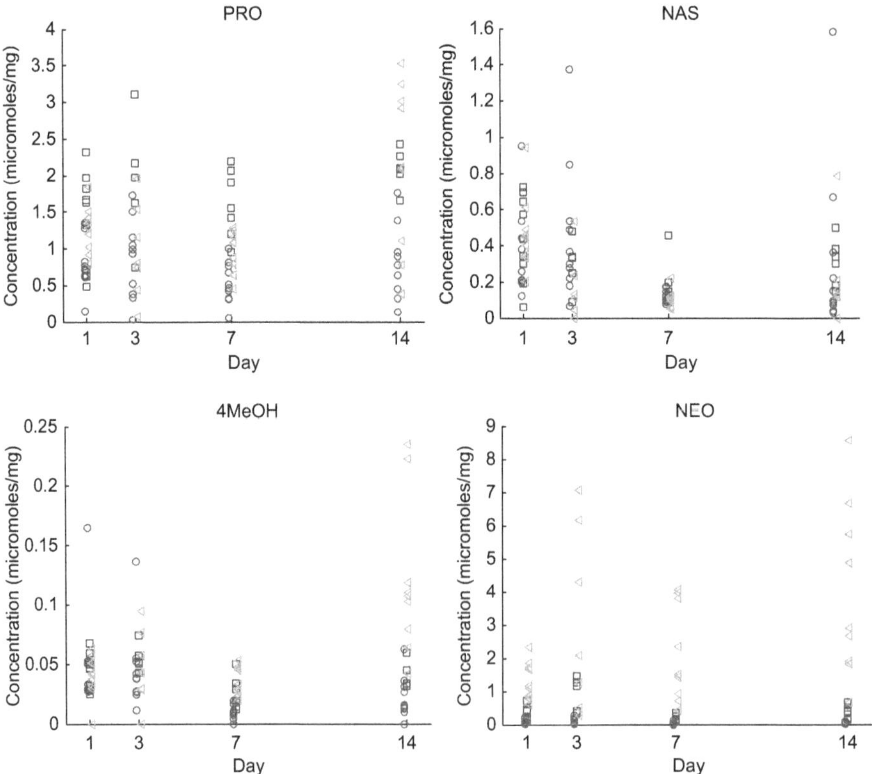

Figure 4.2 Four of the measured glucosinolates for the control (circles), root-induced (squares) and shoot-induced (triangles) after 1, 3, 7 and 14 days.

4.2.1 The ANOVA Model

ANOVA considers only a single glucosinolate at a time and estimates the effect of the treatment(s) applied on the levels of glucosinolates expressed. The concentration of a glucosinolate (*e.g.* NEO) measured at a specific time-point k for plant i_h (in treatment group h) will be denoted by the scalar x_{hki_h},

$$x_{hki_h} = \mu + \alpha_k + (\alpha\beta)_{hk} + (\alpha\beta\gamma)_{hki_h} \tag{1}$$

where μ represents the overall mean over all samples (in the absence of contributions from each further component-of-variance) and α_k represents the effect of the time factor at level k that is equal for all plants over all treatments at this time level.[7] Since α_k is an average expressed over all samples at time k, common to all treatments, it represents an ageing affect that is not of great interest to this study. The term $(\alpha\beta)_{hk}$ represents the interaction of the treatment with time, whilst $(\alpha\beta\gamma)_{hki_h}$ represents the plant-specific contribution. In this study we are mostly interested in the time effect of the treatments as deviations from the common time effect, which is represented by $(\alpha\beta)_{hk}$. The overall time effect β_h is not separately estimated but it is considered together with the time-treatment interaction in the $(\alpha\beta)_{hk}$ term. Finally, the individual plant contribution (*i.e.* how each specific plant deviates from the other plants in the same treatment group) represented by $(\alpha\beta\gamma)_{hki_h}$ is expected to be low, but it can also be used for significance testing. Note that here the measurement error is ignored as each sample was only measured once and then the individual plant contribution $(\alpha\beta\gamma)_{hki_h}$ and the measurement error cannot be distinguished.

Equation (1) shows an ANOVA model that can be used for estimating the effect of the treatments over time. This ANOVA model as defined for NEO should be repeated for each glucosinolate in order to estimate the overall time effect and the treatment $\times$ time interaction for each specific glucosinolate. This will give rise to a virtually incomprehensible table with many estimated parameters and, most importantly, the relationship between the different glucosinolates will be completely ignored in this manner. Therefore, in this investigation we will combine the ANOVA estimates with PCA.

4.2.2 The ASCA Model

Since all the J response values x_{hki_h} are obtained as multivariate measurements, they can be collected into a matrix $\mathbf{X}$ of dimensions $N \times J$, where $N = (I*K*H)$ is equivalent to the total number of samples collected in the experiment. In this study 7 plants were monitored at 3 treatment levels at 4 time-points, and thus N equals $7*3*4 = 84$. Similarly, all estimates of the ANOVA parameters on the right-hand side of eqn (1) can be collected into matrices, where the

$$\mathbf{X} = \mathbf{X_m} + \mathbf{X_\alpha} + \mathbf{X_{(\alpha\beta)}} + \mathbf{X_{(\alpha\beta\gamma)}} \tag{2}$$

$\mathbf{X_m}$ component contains the means for each response variable over all N samples, $\mathbf{X}_\alpha$ contains the estimates for the overall time effect for all samples for all responses, $\mathbf{X}_{(\alpha\beta)}$ contains the estimates of parameters $(\alpha\beta)_{hk}$, which represents the interaction of the treatment with the time, and the matrix $\mathbf{X}_{(\alpha\beta\gamma)}$ contains the $(\alpha\beta\gamma)_{hki_h}$ parameters, which represents the plant-specific contribution as a difference with the mean of each specific treatment group (the rows of matrices $\mathbf{X_m}$, $\mathbf{X}_\alpha$ and $\mathbf{X}_{(\alpha\beta)}$ are highly structured). All rows of $\mathbf{X_m}$ are exactly the same, and the rows related to one time level in $\mathbf{X}_\alpha$ are equivalent and, analogously, all rows of $\mathbf{X}_{(\alpha\beta)}$ that correspond to the combined group with levels k and h for factors α and β, respectively, are also equivalent.

The parameters of eqns (1) and (2) can be estimated in many different manners, but often the 'usual constraints'[19] are applied, which make the column spaces of the matrices in eqn (2) mutually orthogonal (*e.g.* $\mathbf{X}_\alpha^{\mathrm{T}}\mathbf{X}_{(\alpha\beta)} = 0$). This approach allows the separation of variances ascribable to the treatment effects [eqn (3)],

$$\|\mathbf{X}\|^2 = \|\mathbf{X}_m\|^2 + \|\mathbf{X}_\alpha\|^2 + \|\mathbf{X}_{(\alpha\beta)}\|^2 + \|\mathbf{X}_{(\alpha\beta\gamma)}\|^2 \tag{3}$$

where $\|\mathbf{X}\|^2$ denotes the sum-of-squares of the elements in $\mathbf{X}$. Equation (3) shows that by imposing the usual constraints, the variation in $\mathbf{X}$ can be split into independent components. This approach allows us to independently compute the contribution of each factor and interaction effect in the ANOVA model to the total variation in the dataset.

Since with this model all the variation in the dataset has been split into the corresponding effect matrices, we can use PCA to explore the high-dimensional relationships in these effect matrices. For each effect matrix, a separate PCA model is defined. This approach of separate PCA models on related data matrices is often referred to as Simultaneous Component Analysis, and hence the term ASCA is defined.

$$\mathbf{X} = \mathbf{1m}^{\mathrm{T}} + (\mathbf{T}_\alpha\mathbf{P}_\alpha^{\mathrm{T}} + \mathbf{E}_\alpha) + (\mathbf{T}_{(\alpha\beta)}\mathbf{P}_{(\alpha\beta)}^{\mathrm{T}} + \mathbf{E}_{(\alpha\beta)}) + (\mathbf{T}_{(\alpha\beta\gamma)}\mathbf{P}_{(\alpha\beta\gamma)}^{\mathrm{T}} + \mathbf{E}_{(\alpha\beta\gamma)}) \tag{4}$$

Note that the $\mathbf{X_m}$ matrix, which contains the mean of all data $(\mathbf{m}^{\mathrm{T}})$ over all samples at each row, can be decomposed as a column of ones (1) times the mean vector $\mathbf{m}^{\mathrm{T}}$. Each of the other matrices is decomposed into scores ($\mathbf{T}$) and loadings ($\mathbf{P}$) and residuals. Each PCA-model consists of a pre-defined (optimal) number of principal components indicated by R_α, while $\mathbf{E}_\alpha$ contains the residuals that are not explained in the PCA model $\mathbf{T}_\alpha\mathbf{P}_\alpha^{\mathrm{T}}$. The scores and loadings of the effects matrices represent the systematic variation in the data caused only by the specific factor applied (these scores and loadings will be explored in the remainder of this study).

We will first investigate the PCA model of effect matrix $\mathbf{X}_\alpha$. This matrix contains the general time effect independent of treatment. Two principal components were found to describe the systematic variation in this matrix, and Figure 4.3 shows the scores and loadings of the PCA model for $\mathbf{X}_\alpha$.

This PCA model shows that especially day 14 deviates from the other days involved, a difference which is mainly caused by NEO, PRO and GBN.

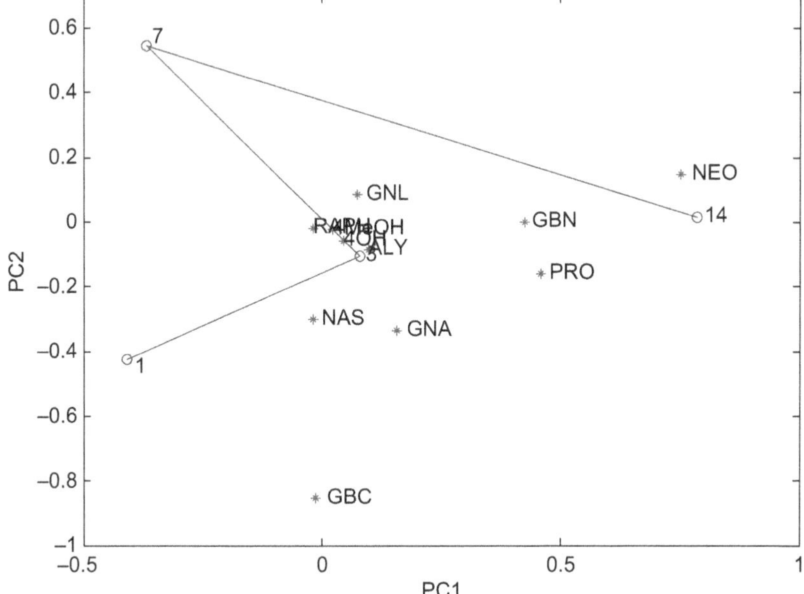

Figure 4.3   A two-component PCA model of the overall time effect. Day 14 is rather different from the rest mainly caused by variation in NEO, GBN and PRO. A second component describes the deviating levels of GBC at day 7.

The other glucosinolates only have a minor effect on the first PC. The second PC shows that day 7 is somewhat different from the other days (especially day 1), and this difference is mainly attributable to the concentration of GBC. Since the information in $\mathbf{X}_\alpha$ is of limited interest to the experimental question, the biological relevance of its interpretation is limited. However, removing the information in matrix $\mathbf{X}_\alpha$ places the focus of the remaining model (specifically that of $\mathbf{X}_{\alpha\beta}$) more on the differences observable between the individual treatment groups, which represents the original experimental question.

The second (more important) effect matrix $\mathbf{X}_{\alpha\beta}$ is also decomposed into two components, and Figure 4.4 shows the resulting scores and loadings.

For the first PC we see a clear difference between the shoot-induced treatment (green) when compared to the root-induced (blue) one and the control group (red). This effect is mainly caused by NEO and GBC for shoot-induced plants, and is also visible for root-induced plants to a considerably lower extent. The second component shows that the root-induced treated plants (blue) are clearly different from the shoot-induced and the control groups. This effect becomes more severe on days 3, 7 and 14, and is mainly ascribable to GBN and PRO.

Finally, the variation between the individual plants which are contained in the $\mathbf{X}_{(\alpha\beta\gamma)}$ PCA model can be examined: Figure 4.5 shows the single principal component derived from this matrix.

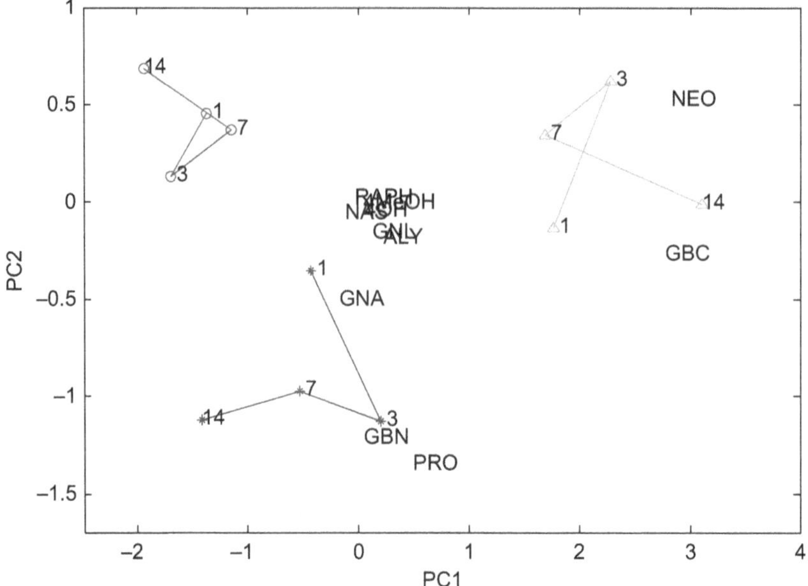

Figure 4.4 Scores and loadings of the PCA model of matrix $X_{\alpha\beta}$. The time-points (1, 3, 7 and 14) are indicated for each of the treatments (shoot-induced (triangles), root-induced (asterisks) and control (circles)).

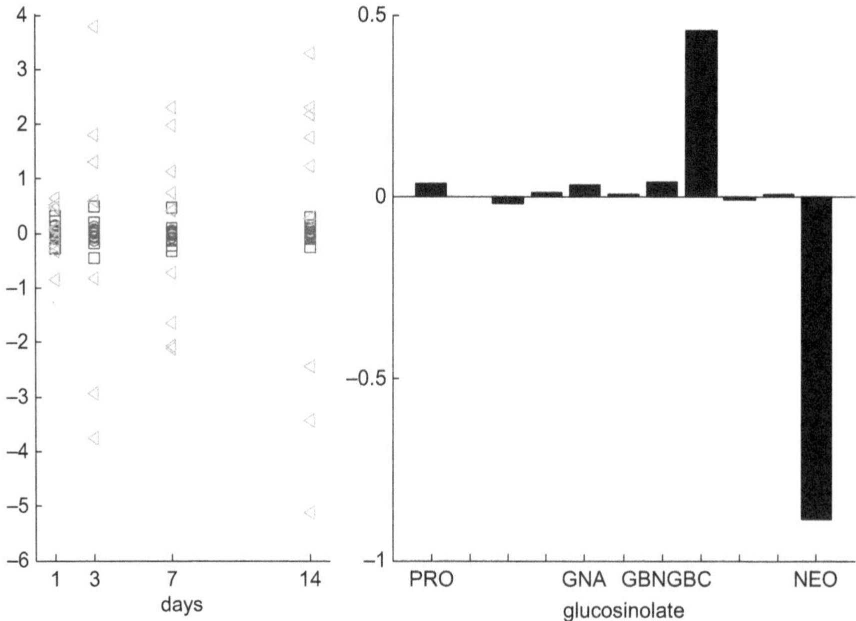

Figure 4.5 PCA of $X_{(\alpha\beta\gamma)}$. On the left the scores of the PCA model and on the right hand side the corresponding loadings of the individual plant matrix effect (shoot induced (triangles), root induced (squares) and control (circles)).

Shoot-induced plants clearly show a much higher variation in glucosino-lates, a phenomenon that starts three days after the treatment and is mainly caused by the NEO and GBC plant levels. Plants that have higher levels of NEO tend to have lower levels of GBC. This was observed from the loadings in the individual plant model. This may seem to contradict the observations in Figure 4.4, but the model of $X_{(\alpha\beta\gamma)}$ is in fact superimposed on the $X_{(\alpha\beta)}$ model. The model of $X_{(\alpha\beta)}$ shows that all shoot-induced plants have higher levels of NEO and GBC, whilst the model of $X_{(\alpha\beta\gamma)}$ reveals that shoot-induced plants with higher relative levels of NEO have less GBC, and that the same negative relationship does not occur in root-induced plants.

4.2.3   Concluding the Glucosinolate Study

In this first application, an ANOVA model for each response variable was developed. The total variation in the dataset could therefore be separated into components that are ascribable to the different experimental factors considered in the study. The multivariate analysis of the separate effect matrices therefore leads to a much better interpretation of the ecologically relevant complexity within metabolomics datasets.

4.3   Case study 2: Metabolic Modifications Following Polyphenolic Intervention in Humans

The second example in this chapter comprises metabolomics measurements performed on human urine before and after consumption of polyphenol-rich black tea. Polyphenols are plant secondary metabolites, ubiquitous in fruits, vegetables, cereals and chocolate, and beverages, such as tea, coffee or wine,[20] that have a potential role in the prevention of and/or protection against *e.g.* cardiovascular diseases, malignancies, neurodegenerative dis-orders, metabolic syndrome, *etc.*[21–23] When polyphenols enter the human body they are converted into fermentation products *via* catabolic actions of the gut microbiota. The extent and rate of these gut-mediated conversions strongly depend on the diversity of the microbiota in the gut, and the differing degradation pathways/mechanisms that are involved. The bio-availability of the polyphenol degradation products is monitored by meta-bolomics experiments conducted on plasma and urine; indeed, these experiments provide an overwhelming source of metabolites present.[24,25]

In human nutritional studies, the effect of the nutritional treatment is often small and is often much smaller than the variation observed between individuals. This is rather different from the previous plant study since the plants were bred in a controlled manner in order to minimise variation between them. The differences between participants in a human nutritional study can give rise to two important data analysis problems. Firstly, the small treatment effect can easily be overlooked in view of the large biological variation between the individuals involved. The second problem is that the effect of the treatment within the test-population differs 'between-individuals' not only in biomolecule or related signal intensity/intensities, but also in the overall metabolic profile. An average treatment effect may not be the most relevant measure in studies where sub-sets of subjects respond differently to a dietary intervention.

If the variation between individuals is considered large compared to the estimated treatment effect, a parallel study design in which different individuals are used for the differing treatments is inappropriate. In such a case, the subjects should be used as their own control by monitoring the treatment effect within each individual; this can be achieved by measuring a biofluid (*e.g.* blood or urine) metabolic profile both before and after the intervention or by monitoring the individuals for a longer time period in a longitudinal study. Another often used solution applied in such studies is a cross-over design, in which each individual undergoes a placebo treatment and a verum treatment in a random order. Thus each individual acts as its own control. Data obtained from such cross-over studies have a paired structure.

The analysis of data with a paired structure is usually performed using a paired t-test (or a repeated measures ANOVA) in the case of a single measured variable. Depending on the ratio between the inter-individual variation and the average effect magnitude of the treatment, a paired t-test is advantageous over a normal t-test since it gives improved statistical power. By using the paired data analysis method, statistical significance can be obtained for much smaller treatment effects.[26]

4.3.1 Multivariate Consequence

When data obtained from the study contains information from a series of measured metabolites (as is the case in this study), we are especially interested in the correlation between the metabolite concentration levels. For example, is the treatment effect similar for a large group of metabolites within the same metabolic pathway, or is there a second association with metabolites of a different pathway? To answer such a question, a multivariate data analysis method should be used.

Similar to the ASCA model described in the first application in this chapter, we will first introduce the ANOVA model for paired data from the cross-over study for a single metabolite. Subsequently, the model will be combined into a high-dimensional version of the ANOVA model specifically suitable for such paired data.

Consider data from a study in which I ($I = 1...I$) individuals are measured at H ($h = 1....H$) differing treatments. In our example, H is 2 since we have a placebo treatment and a tea treatment. Each measurement x_{hi} can be explained partly by an overall mean value μ, the group effect α_h and a residual e_{hi}, which cannot be explained by the treatment.

$$x_{hi} = \mu + \alpha_h + e_{hi} \qquad (5)$$

This is a simple one-way ANOVA model. The top plot in Figure 4.6 represents this situation with two groups in a one-way ANOVA situation. Now the group effect for occasion h that is estimated from the dataset is the mean treatment effect over all individuals. Actually, we estimate how the average individual of group $h = 2$ deviates from the average individual of group $h = 1$. In the case of a cross-over design where the same individuals are measured on more than one occasion, we can add an individual effect to eqn (5), as shown in eqn (6).

$$x_{hi} = \mu + \alpha_{hi} + \beta_i + f_{hi} \tag{6}$$

The individual contribution to x_{hi}, β_i, is obtained as the mean of all values for individual $\left(\beta_i = \frac{1}{H}\sum_{h=1}^{H}(x_{hi} - \mu)\right)$. The bottom plot in Figure 4.6 shows this situation in which the circles on the right represent the average value

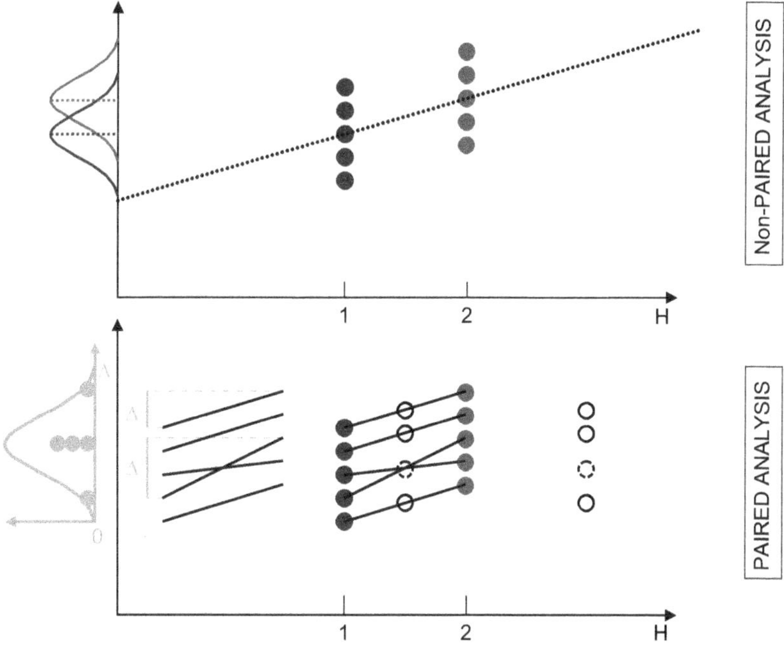

Figure 4.6 Difference between paired and non-paired analysis. On the x-axis the treatment level (H) and on the y-axis the measured value. Top figure shows a traditional analysis where the average of the $H = 1$ group is compared to the average of the $H = 2$ group. The dotted line represents the estimated effect. The distributions show quite some overlap and in this case no significant difference could be found for the two groups. Bottom figure shows the same data but now the measurements of the two occasions for the same individual are connected. The slope of the connection line represents the individual treatment effect. Note that the average of these individual treatment effects is the same as the average treatment effect from the top figure.

(β_i) for each individual. The effect of the treatment on each individual α_{hi} is not constant (as can be noted from the distribution of individual effects on the y-axis of the bottom plot, indicated by Δ), but its average value is equivalent to the estimated treatment α_h of eqn (5). The individual treatment effect α_{hi} can be obtained by removing the overall mean value μ and the individual mean β_i from the data x_{hi}

$$x_{hi} - \mu - \beta_i = \alpha_{hi} + f_{hi} \tag{7}$$

In this situation, where each individual undergoes the treatments only once, a residual f_{hi} cannot be distinguished from the individual treatment effect α_{hi}.

In summary, β_i provides information of the overall differences between individuals, while α_{hi} provides information on the variation within each individual attributable to the treatment.

4.3.2　The Multilevel PLSDA Model

The high-dimensional extension of the paired data ANOVA model of eqn (6) can be performed in a similar manner to that for the ASCA model of eqn (2). Since all the J response values x_{hi} are obtained as high-dimensional measurements, they can be collected into a matrix $\mathbf{X}$ of dimensions $N \times J$, where $N = (I*H)$ is equal to the total number of samples collected in the experiment.

$$\mathbf{X} = \mathbf{X_m} + \mathbf{X_\alpha} + \mathbf{X_\beta} + \mathbf{X_f} \tag{8}$$

where each row in $\mathbf{X_m}$ contains the mean of all the data, and $\mathbf{X_\alpha}$ contains the individual treatment effect. Each row in $\mathbf{X_\beta}$ corresponding to the same individual is equivalent and contains the overall mean of that individual. $\mathbf{X_f}$ is empty in our case, where the treatments are only applied once to each individual.

Table 4.2 demonstrates the situation for the case depicted in Figure 4.7, in which only a single variable was measured. Here $\mathbf{X_\beta}$ contains overall information about the difference between the individuals, whilst $\mathbf{X_\alpha}$ contains information regarding the treatment effect for each individual. The latter is the specific information we are aiming for in this study, *i.e.* how did the treatment affect the individuals? Thus the total variation of the dataset ($\mathbf{X}$) consisted of two levels: the 'between-individual' level estimated by $\mathbf{X_\beta}$ and the 'within-individual' treatment level estimated by $\mathbf{X_\alpha}$. The method described above was able to separate these two components of variance. Therefore, it is known as a multilevel analysis.

To develop a method closely related to Partial Least Squares-Discriminant Analysis (PLS-DA),[27] we extended the multilevel analysis to a multilevel PLS-DA model (MLPLS-DA). This method is particularly reserved for studies in which the same individuals underwent all treatments involved.

Table 4.2 Example of the separation of the variances according to eqn (6).

Individual	Treatment (H)	X	X_m	X_α	X_β
1	1	3	8.5	-1.5	-4
2	1	5	8.5	-2.5	-1
3	1	7	8.5	-0.5	-1
4	1	9	8.5	-1.5	$+2$
5	1	11	8.5	-1.5	$+4$
1	2	6	8.5	$+1.5$	-4
2	2	10	8.5	$+2.5$	-1
3	2	8	8.5	$+0.5$	-1
4	2	12	8.5	$+1.5$	$+2$
5	2	14	8.5	$+1.5$	$+4$

In the standard PLS-DA model the original data **X** is associated with a class label of the different treatments in order to find the systematic difference between them.

$$\text{PLS-DA:} \quad \begin{bmatrix} \mathbf{X}_{h=1} \\ \mathbf{X}_{h=2} \end{bmatrix}, \begin{bmatrix} 0 \\ 1 \end{bmatrix}$$

The PLS-DA model is based on the original data **X** and the class label **y** in which all samples corresponding to the placebo group $h=1$ receive the (dependent variable) score 0, whilst the samples corresponding to the treatment group $h=2$ receive the score 1.

In MLPLS-DA, the 'between-individual' variation is not considered since we are only interested in the effect of the treatment represented by the 'within-individual' variation $\mathbf{X}_\alpha$. It is possible to separate the 'between-individual' variation from the dataset because each individual was monitored for both treatments. Therefore, in MLPLS-DA, $\mathbf{X}_\alpha$ is related to the class labels.

$$\text{MLPLS-DA:} \quad \begin{bmatrix} \mathbf{X}_{\alpha,h=1} \\ \mathbf{X}_{\alpha,h=2} \end{bmatrix}, \begin{bmatrix} 0 \\ 1 \end{bmatrix}$$

In terms of a simple analogy, this method is comparable to the paired t-test, whereas the PLS-DA technique can be compared with the normal (2-sample) t-test. Hence, the MLPLS-DA shows an increased power for estimations of the treatment effect.[24,26]

Besides the search for differences ascribable to the treatments in the MLPLS-DA analysis of $\mathbf{X}_\alpha$, it is also of importance to study the 'between-individual' variation. This will provide additional knowledge on how the different individuals in the study are related, *i.e.* whether the individuals are 'clustered' in groups, or whether there are some outlying individuals. Therefore, these $\mathbf{X}_\beta$ data were also explored using a PCA model, and this form of analysis provided additional details regarding the individuals in the study that are not treatment-related.

4.3.3 The Study Setup

The case study we will explore here is an intervention study in which 20 healthy, non-smoking male subjects received black tea solids containing 800 mg polyphenols in a double-blind placebo-controlled cross-over study. The subjects were 18–40 years of age and their Body Mass Index (BMI) was between 19 and 29 kg m^{-2}. During each of the intervention periods, the subjects visited the study facility of Mediscis (Berchem Antwerp, Belgium) where they were maintained on a low-polyphenol diet for 4 days. Between the interventions, a 10-day 'wash-out' period was included, during which the subjects were free to consume their normal diet without any restrictions. The volunteers were also requested to follow a similar dietary and lifestyle pattern for the duration of the study. On the morning of the third day, the volunteers consumed a capsule containing 2500 mg dried black tea extract powder or a placebo (sucrose), both with an adequate amount of water (>200 ml). The tea extract was prepared from a spray-dried aqueous extract of Lipton Yellow Label (code LYL640, US blend) and contained 800 mg polyphenols, expressed as gallic acid equivalents.

Urine samples were collected after spontaneous urination, at non-equidistant time points, during 48 hours after administration of the black tea or placebo capsule. The weight of all urine samples produced was measured and a small volume of concentrated hydrochloric acid was added to adjust the pH value to between 3 and 4. From each acidified urine sample, an aliquot of approximately 10 ml was stored at −20 °C before analysis. ^{1}H-NMR spectra of the urine samples were obtained from a Bruker Avance 600 MHz NMR spectrometer at 300 K. The NMR spectra were bucketed into 0.00225 ppm buckets and correlation-optimised warping was used to correct for line-broadening effects and positional shifts that remained in view of pH and ion strength differences in the urine samples. A detailed description of this study has been previously documented.[25]

4.3.4 Analysis of Pooled Samples

Pooled samples were produced by collecting all urine samples from the same individual. The first step in analysing the ^{1}H NMR data is to separate the variation attributable to the nutritional treatment from the variation arising from differences between the individuals according to eqn (8). The 'within-individual' variation $\mathbf{X}_\alpha$ was centred and scaled by the square root of the standard deviation of each column of $\mathbf{X}_\alpha$ (Pareto scaling) in order to improve the weight of metabolites with low intensity in the multivariate analysis.

Figure 4.7 shows the results of an MLPLS-DA model of the 'within-individual' variation, which had two significant components. This clearly indicates that the intervention effect was not equivalent for all individuals, and that two different treatment effects can be observed. In the left column the scores plot with double cross-validated scores revealed a separation between the two intervention groups.[28] Note that double-cross validated scores

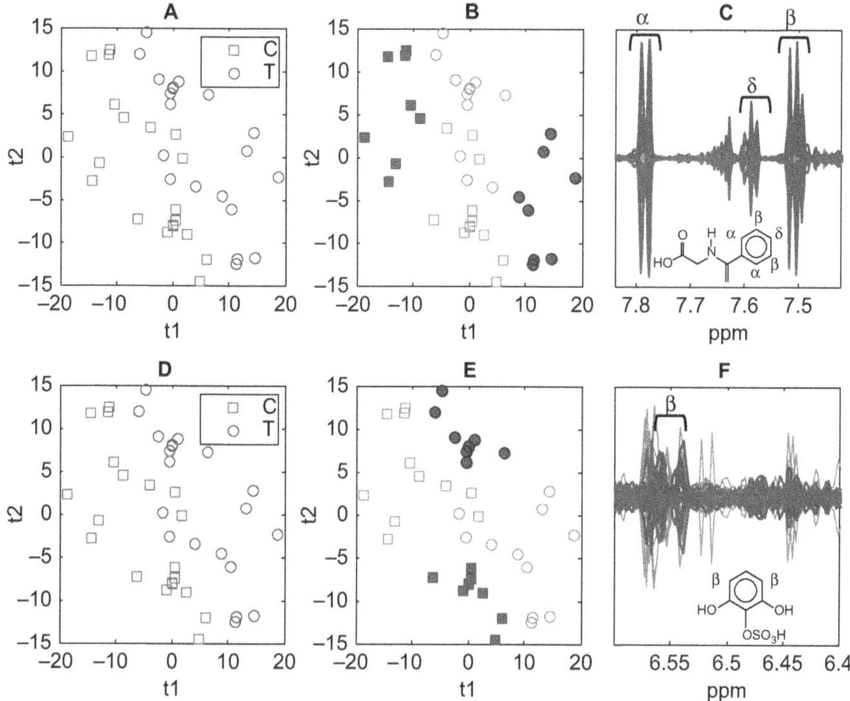

Figure 4.7 Multilevel PLS-DA double cross-validated scores which represent the urinary NMR spectra of 20 subjects after black tea intake. The (A and D) scores on the first two components (t_1,t_2) reflect the 'within subject' variation in the control period (squares) and the treatment period (circles). Two different treatment effects could be identified. The (B) first effect along the first component point towards (C) increasing hippuric acids levels and increasing 1,3-dihydroxyphenyl-2-*O*-sulfate levels. The (E) second effect along the second component is basically described by (F) 1,3-dihydroxyphenyl-2-*O*-sulfate, whereas the increase of hippuric acid is less pronounced.
With kind permission from Springer Science + Business Media: J. A. Westerhuis *et al.*, Multivariate paired data analysis: multilevel PLSDA *versus* OPLSDA, *Metabolomics*, **6**, 2010, 119–128, Figure 6.

are unbiased towards class separation and hence these scores can be interpreted for their class separation. In the second column we highlighted some individuals with large score values on the first latent variable. In the right column the corresponding loading vector is shown, and this indicates which urinary metabolites differ between the placebo and tea intervention occasions of the intervention study. Plot C shows that, in particular, hippurate differed for the individuals highlighted in plot B. In the second row, some other individuals that have a relative low value on the first latent variable but a large value on the second latent variable are highlighted. The second latent variable is dominated by another metabolite, 1,3-dihydroxyphenyl-2-*O*-sulfate. Thus, we observe that as a consequence of the single polyphenol tea intervention, a combination of two treatment effects can be discovered amongst the individuals in the study with varying excretion levels of 1,3-dihydroxyphenyl-2-*O*-sulfate and hippurate. The underlying reason why an individual has different excretion levels of these two metabolites produced during the study is not simply explicable. However, it is likely that gut bacteria play an important role in the specific degradation of the polyphenols towards these urinary end-metabolites.

Besides the 'within-individual' variation, the 'between-subject' variation was also explored. The combination of both (multilevel) analyses will then allow a comprehensive interpretation of all major sources of variation in the dataset. As shown in Figure 4.8A, the scores of four subjects on the second principal component t_B2 appear to be different from the other subjects. Whereas the first principal component (Figure 4.8B, black profile) is a generic representation of all ^{1}H NMR signal intensities, different variations amongst the NMR resonances were observed on the second principal component (Figure 4.8B). The loadings show that the 'between-subject' variation particularly depends on the ratio between the NMR signals of hippurate (δ 7.78 ppm, d; δ 7.59 ppm, t and δ 7.50 ppm, t) and the NMR signals of an unknown aromatic compound, U (δ 7.17 ppm, s; δ 7.24 ppm, s and δ 7.31 ppm, s). This unknown compound was observed in a spectral

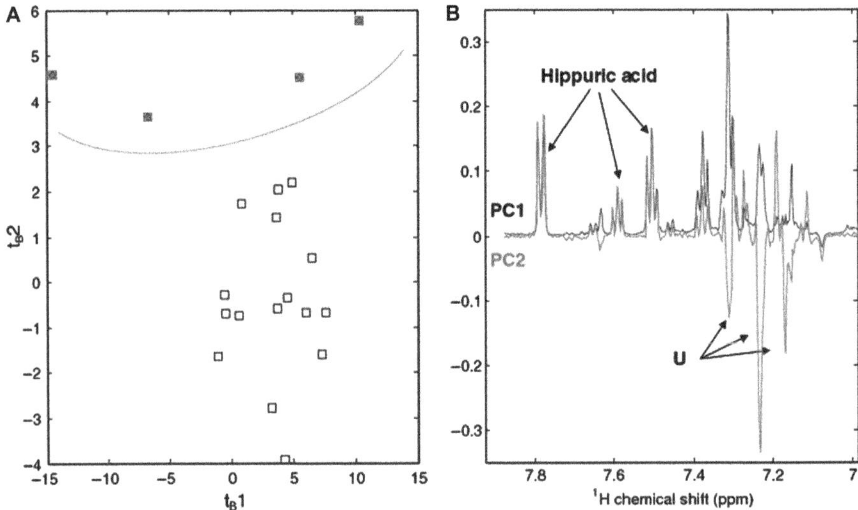

Figure 4.8 PCA analysis of the 'between individual' variation of the tea study. In the left plot the scores of the PCA analysis show a homogeneous group of individuals as well as four outlying individuals with large value on the second principal component. On the right hand side the loading of the second component shows a clear contribution of an unknown compound (U).
With kind permission from Springer Science + Business Media: J. A. Westerhuis *et al.*, *Metabolomics*, Multivariate paired data analysis: multilevel PLSDA *versus* OPLSDA, **6**, 2010, 119–128, Figure 7.

region where several other resonance patterns of aromatic amino acids, (conjugated) polyphenolic acids, (indole) alkaloids *etc.* arise together. For now, this complicates a straightforward identification of component U. Four subjects appear to have a higher signal ratio between U and hippuric acid than the remainder of the subjects in the study population.

4.3.5 Dynamic Non-linear Analysis of the Urinary ^{1}H NMR Data

The urinary output of 1,3-dihydroxyphenyl-2-*O*-sulfate was highly variable following both placebo and tea treatments. From examination of the cumulative levels of this metabolite after placebo (lighter curve) and tea treatment (black curve) in Figure 4.9, it is clear that a large difference exists between the levels of this metabolite for the different individuals. Note also that Figure 4.9 depicts the cumulative levels, and thus a small constant level of this metabolite during the placebo period leads to a linearly increasing grey curve. The slope of the placebo curve is more or less constant, an observation suggesting that the basal level during the placebo period is rather constant for all individuals. The difference in offset relates to time differences between the start of the placebo intake and the preceding urination.

The clear increase in the urinary levels of 1,3-dihydroxyphenyl-2-*O*-sulfate subsequent to tea treatment becomes apparent after 4–8 hr, and time depending per individual. Furthermore, we observed a large variation in the total excretion of this metabolite after 48 hr, *e.g.* individual 1 (s1) has the highest level after tea treatment, but only a mean level following the placebo. Thus the tea treatment was most effective for this individual. Very interestingly, we can also observe the non-responding individuals s12 and s2. These individuals show equivalent concentrations after tea and placebo treatments, *i.e.* the tea polyphenol intervention did not lead to an increased urinary output of 1,3-dihydroxyphenyl-2-*O*-sulfate for these subjects. A possible explanation for this observation is that the administered test product was directly excreted *via* the faeces without being absorbed in the human host. Another explanation may be that alternative (and less common) pathways and mechanisms were involved in (gut microbial) degradation of tea polyphenols.

4.3.6 Short Conclusion on Case Study 2

In this second case study we have shown that if the variation between the individuals is relatively large, and when these individuals are measured for all treatments, a multilevel approach can be used to separate the variation in metabolomics data. Two sources of variation can be distinguished, *i.e.* (I) variation 'between-individuals' and (II) variation within an individual that is ascribable to the treatment. This situation was rather different from the first case study where new plants were used for each experimental stage.

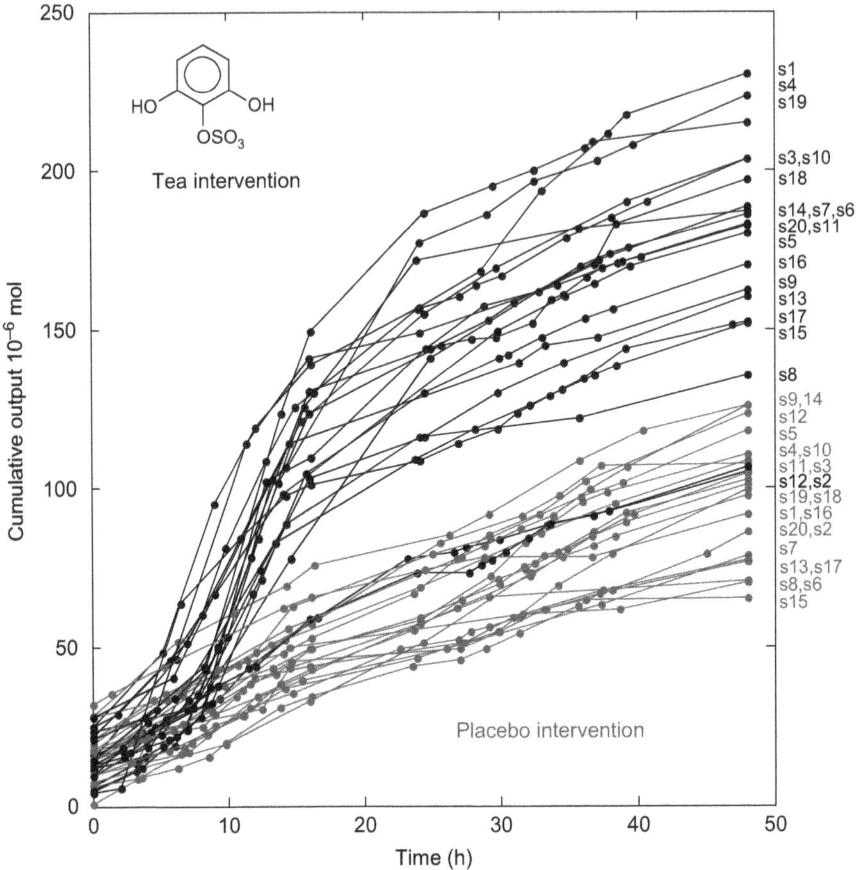

Figure 4.9 Cumulative excretion levels of 1,3-dihydroxyphenyl-2-*O*-sulfate in urine after placebo and tea intervention for all 20 individuals.
With kind permission from E. J. J. van Velzen *et al.*, *Journal of Proteome Research*, Phenotyping tea consumers by nutrikinetic analysis of polyphenol end-metabolites, **8**, 2009, 3317–3330, Figure 5.

4.4   Conclusion

The complex high-dimensional data structures that arise from functional genomics studies benefit from new methods that combine ANOVA models with high-dimensional explorative ones such as PCA, or supervised models such as PLS-DA. In view of the separation of the total variation into different effect matrices, the interpretability of the high-dimensional analysis is greatly improved. Two different case studies in which different classes of ANOVA models were combined with different data analysis methods were used to demonstrate the very broad application area where these methods can be of much value.

Acknowledgement

The *Brassicaceae oleracea* data collection was partially funded by NWO, the Netherlands Organization for Scientific Research VIDI grant, no. 864-02-001. Copyright permission for Figures 4.7 and 4.8 was granted under number 2631920034181 and for Figure 4.9 under number 2638130169823.

References

1. S. E. Calvano, W. Xiao, D. R. Richards, R. M. Felciano, H. V. Baker, R. J. Cho, R. O. Chen, B. H. Brownstein, J. P. Cobb, S. K. Tschoeke, C. Miller-Graziano, L. L. Moldawer, M. N. Mindrinos, R. W. Davis, R. G. Tompkins and S. F. Lowry, A network-based analysis of systemic inflammation in humans, *Nature*, 2005, **437**(7061), 1032–1037.
2. B. Daviss, Growing pains for metabolomics, *The Scientist*, 2005, **19**(8), 25.
3. O. Fiehn, Metabolomics – the link between genotypes and phenotypes, *Plant Mol. Biol.*, 2002, **48**(1–2), 155–171.
4. R. Goodacre, S. Vaidyanathan, W. B. Dunn, G. G. Harrigan and D. B. Kell, Metabolomics by numbers: acquiring and understanding global metabolite data, *Trends Biotechnol.*, 2004, **22**(5), 245–252.
5. K. W. Jordan, J. Nordenstam, G. Y. Lauwers, D. A. Rothenberger, K. Alavi, M. Garwood and L. L. Cheng, Metabolomic characterization of human rectal adenocarcinoma with intact tissue magnetic resonance spectroscopy, *Dis. Colon Rectum*, 2009, **52**(3), 520–525.
6. S. H. Lee, M. V. Williams, R. N. DuBois and I. A. Blair, Targeted lipidomics using electron capture atmospheric pressure chemical ionization mass spectrometry, *Rapid Comm. Mass Spectrom.*, 2003, **17**(19), 2168–2176.
7. A. K. Smilde, J. J. Jansen, H. C. J. Hoefsloot, R.-J. A. N. Lamers, J. van der Greef and M. E. Timmerman, ANOVA-simultaneous component analysis (ASCA): a new tool for analyzing designed metabolomics data, *Bioinformatics*, 2005, **21**(13), 3043–3048.
8. R. Sokal and F. Rohlf, *Biometry*, W. H. Freeman and Company, 1998.
9. L. Stahle and S. Wold, Multivariate-Analysis of Variance (MANOVA), *Chemometr. Intell. Lab. Syst.*, 1990, **9**(2), 127–141.
10. I. T. Joliffe, *Principal Component Analysis*, Springer Verlag, New York, 2002.
11. J. J. Jansen, H. C. J. Hoefsloot, J. van der Greef, M. E. Timmerman, J. A. Westerhuis and A. K. Smilde, ASCA: analysis of multivariate data obtained from an experimental design, *J. Chemometr.*, 2005, **19**(9), 469–481.
12. P. J. Van den Brink and C. J. F. Ter Braak, Principal response curves: Analysis of time-dependent multivariate responses of biological community to stress, *Environ. Toxicol. Chem.*, 1999, **18**(2), 138–148.
13. M. E. Bollard, H. C. Keun, O. Beckonert, T. M. D. Ebbels, H. Antti, A. W. Nicholls, J. P. Shockcor, G. H. Cantor, G. Stevens, J. C. Lindon, E. Holmes and J. K. Nicholson, Comparative metabonomics of

differential hydrazine toxicity in the rat and mouse, *Toxicol. Appl. Pharmacol.*, 2005, **204**(2), 135–151.

14. H. C. Keun, T. M. D. Ebbels, M. E. Bollard, O. Beckonert, H. Antti, E. Holmes, J. C. Lindon and J. K. Nicholson, Geometric trajectory analysis of metabolic responses to toxicity can define treatment specific profiles, *Chem. Res. Toxicol.*, 2004, **17**(5), 579–587.

15. P. D. B. Harrington, N. E. Vieira, J. Espinoza, J. K. Nien, R. Romero and A. L. Yergey, Analysis of variance-principal component analysis: A soft tool for proteomic discovery, *Anal. Chim. Acta*, 2005, **544**(1–2), 118–127.

16. J. R. de Haan, R. Wehrens, S. Bauerschmidt, E. Piek, R. C. van Schaik and L. M. C. Buydens, Interpretation of ANOVA models for microarray data using PCA, *Bioinformatics*, 2007, **23**(2), 184–190.

17. J. W. Fahey, A. T. Zalcmann and P. Talalay, The chemical diversity and distribution of glucosinolates and isothiocyanates among plants, *Phytochemistry*, 2001, **56**(1), 5–51.

18. J. J. Jansen, N. M. van Dam, H. C. J. Hoefsloot and A. K. Smilde, Crossfit analysis: a novel method to characterize the dynamics of induced plant responses, *BMC Bioinformatics*, 2009, **10**, 425.

19. S. R. Searle, *Linear Models*, John Wiley and Sons Inc., New York, 1971.

20. G. R. Beecher, *Overview of Dietary Flavonoids: Nomenclature, Occurrence and Intake*, 2003.

21. A. Scalbert, C. Manach, C. Morand, C. Rémésy and L. Jiménez, Dietary polyphenols and the prevention of diseases, *Crit. Rev. Food Sci. Nutr.*, 2005, **45**(4), 287–306.

22. C. Manach, G. Williamson, C. Morand, A. Scalbert and C. Rémésy, *Bioavailability and Bioefficacy of Polyphenols in Humans. I. Review of 97 Bioavailability Studies*, 2005.

23. G. Williamson and C. Manach, *Bioavailability and Bioefficacy of Polyphenols in Humans. II. Review of 93 Intervention Studies*, 2005.

24. E. J. J. van Velzen, J. A. Westerhuis, J. P. M. van Duynhoven, F. A. van Dorsten, H. C. J. Hoefsloot, D. M. Jacobs, S. Smit, R. Draijer, C. I. Kroner and A. K. Smilde, Multilevel data analysis of a crossover designed human nutritional intervention study, *J. Proteome Res.*, 2008, 7(10), 4483–4491.

25. E. J. J. van Velzen, J. A. Westerhuis, J. P. M. van Duynhoven, F. A. van Dorsten, C. H. Grün, D. M. Jacobs, G. S. M. J. E. Duchateau, D. J. Vis and A. K. Smilde, Phenotyping tea consumers by nutrikinetic analysis of polyphenolic end-metabolites, *J. Proteome Res.*, 2009, **8**(7), 3317–3330.

26. J. A. Westerhuis, E. J. J. van Velzen, H. C. J. Hoefsloot and A. K. Smilde, Multivariate paired data analysis: multilevel PLSDA *versus* OPLSDA, *Metabolomics*, 2010, **6**(1), 119–128.

27. M. Barker and W. Rayens, Partial least squares for discrimination, *J. Chemometr.*, 2003, **17**(3), 166–173.

28. J. A. Westerhuis, H. C. J. Hoefsloot, S. Smit, D. J. Vis, A. K. Smilde, E. J. J. van Velzen, J. P. M. van Duijnhoven and F. A. van Dorsten, Assessment of PLSDA cross validation, *Metabolomics*, 2008, **4**(1), 81–89.

CHAPTER 5

Current Trends in Multivariate Biomarker Discovery

DARIUS M. DZIUDA

Department of Mathematical Sciences, Central Connecticut State
University, New Britain, CT, USA
Email: dziudadad@ccsu.edu

5.1   Introduction

Current high-throughput 'omic' technologies generate datasets consisting
of many hundreds or thousands of variables, p, and usually much fewer
biological samples, N. Biomedical studies based on such $p \gg N$ datasets are
required to effectively and efficiently deal with the *curse of dimensionality*.
Although a paradigm shift from univariate to multivariate approaches is
clearly visible in the body of recent publications, limiting biomarker discovery studies to univariately based analysis still remains one of the common misconceptions. Nonetheless, it has to be stated that no single method,
multivariate or not, works optimally in all situations. Even experienced researchers may be puzzled by the fact that some of the well-established data
mining approaches fail in $p \gg N$ situations. In response to this challenge,
new data mining and statistical methods are continually being developed.

In this chapter I will discuss methods and approaches that are appropriate
for the biomarker discovery studies based on high-dimensional 'omic' data.
Firstly, I will take a look at common misconceptions in biomarker discovery,
and provide clear guidance on when to use (and when to avoid) which
methods and why. Then I will discuss feature selection, which, while still

Issues in Toxicology No. 21
Metabolic Profiling: Disease and Xenobiotics
Edited by Martin Grootveld
© The Royal Society of Chemistry 2015
Published by the Royal Society of Chemistry, www.rsc.org

underestimated by some studies, is the most important aspect of biomarker discovery. Next, I will present selected supervised learning algorithms, which, when coupled with appropriate feature selection techniques, can be used as the cores of efficient methods for multivariate biomarker discovery. I will also stress the importance of the (often neglected) biological interpretation of biomarkers, as well as the necessity for their proper validation. This chapter will be concluded with the description of a novel data mining method that allows for the identification of multivariate biomarkers that are parsimonious, robust and biologically interpretable.

Although multivariate biomarker discovery is already very important in areas such as medical diagnosis, prognosis or drug discovery, it will soon represent one of the most important aspects of personalised medicine. Tailoring therapy to the condition of a patient or evaluating the risk of adverse drug reactions are examples of personalised approaches, where many new biomarkers representing characteristic molecular profiles may or will play a crucial role. To focus our attention, we will discuss studies utilising gene expression data (such as data generated by the quite mature high-throughput microarray technologies, which allow for the direct and simultaneous measurement of gene expression at the whole-genome level). However, all of the considerations and methods presented in this chapter apply equally well to proteomic, metabolomic and other studies based on high-dimensional datasets.

5.2   Common Misconceptions in Biomarker Discovery based on $p \gg N$ Datasets

Since the $p \gg N$ biomedical datasets generated by the latest high-throughput 'omic' technologies represent a relatively new phenomenon, there is still too large a body of publications that try to stretch out the old 'one-variable-at-a time' paradigm. When one is not sure how to overcome the *curse of dimensionality*, it may seem reasonable to ignore it and apply either univariate or univariately biased approaches. These approaches may work in the situations where the differentiated phenotypic classes are very easy to separate (thus, any, or almost any, method would work). However, in general such approaches should be avoided since in most situations they may yield inferior results, or may even fail to identify the most important discriminatory information.

Another still too common misconception is the use of unsupervised methods (such as cluster analysis or principal component analysis) to 'preprocess' training data. The goal of such 'preprocessing' is to reduce the dimensionality of the problem by replacing the original variables by their subsets, or their combinations. However, by the very definition of the unsupervised approach, there is no sensible way to determine whether the identified dimensions have anything to do with the most discriminatory dimensions. Hence, we do not know how much of the important discriminatory information is removed by such unsupervised dimensionality reduction. We will now take a closer look at these misconceptions.

5.2.1 Univariate (Rather than Multivariate) Analysis

When the univariate approach is applied, each of the variables is evaluated independently of all of the other variables, and hence all correlations and interactions amongst variables are ignored. A further consequence is that such an approach is valid only when the variables are uncorrelated or when, for whatever reason, we want to focus on individual variables in isolation from the others. Neither of these is appropriate in biomarker discovery investigations based on gene or protein expression data.

A variable that is univariately insignificant may be very important when combined with other variables (see Figure 5.1). Therefore, a study that is limited to the genes that are at the top of a univariately ordered list of genes is in danger of removing from consideration important (or maybe even the most important) discriminatory information. Only multivariate approaches are capable of finding truly multivariate biomarkers, complementary expression patterns that can significantly separate the discriminated classes.

Amongst published studies, there are those that claim a multivariate approach, although they apply multivariate analysis *after*, and only *after*, limiting the number of variables to some number of major univariately identified genes (that is, top genes from a list of genes ordered by increasing *p-values* of such univariate tests as t or ANOVA F-tests). If the number of

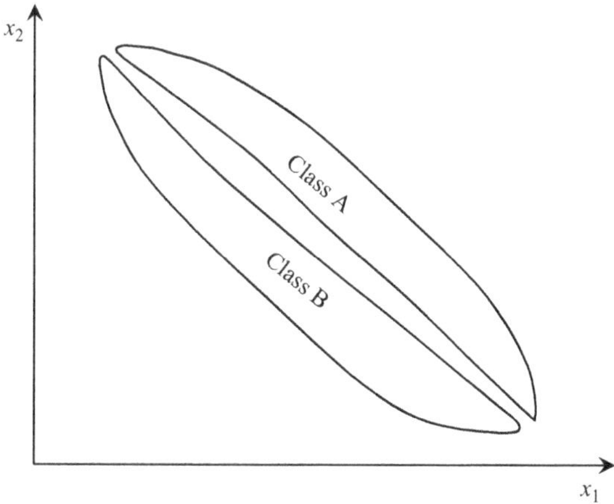

Figure 5.1 An example of a set of two variables (x_1, x_2) that can perfectly separate the two classes. However, neither of the variables is univariately significant for the class differentiation. Such combinations of two or more variables – especially when the training dataset consists of thousands of variables – can be identified only by multivariate feature selection methods.

selected top genes is small (say, a hundred or so), then the multivariate analysis is applied after the 'harm' has already been done to the dataset, and the important discriminatory information removed from consideration. In spite of the claim that multivariate analysis has been applied, such studies obviously have a strong univariate bias.

Nevertheless, one may try to defend such univariate 'preprocessing' by claiming that the goal of this 'preprocessing', is to eliminate 'noise'. Although this would not be the most appropriate method for 'noise' elimination, such a claim may have merit, but only when we can assume that the most important discriminatory information has not been eliminated. In practice, it could indicate that if we started with a dataset with, say, 20 000 variables, then this 'preprocessing' should retain some 10 000 or so of them rather than only a hundred or so. Even so, one would still remain on shaky ground, for such an approach neither efficiently deals with the *curse of dimensionality*, nor is it a recommended method for 'noise' elimination.

Let us continue playing *advocatus diaboli* and argue that it is possible for a univariate or univariately biased study to give rise to an efficient biomarker. Yes, it is indeed possible; for example, if the differentiated phenotypic classes are easily separable, then it may be possible for nearly *any* method to find an efficient biomarker. However, non-trivial phenotypic differences are most often associated with simultaneous changes in several biological processes, and it is quite likely that none of these changes are individually significant. Biomarkers which efficiently separate such phenotypic classes can only be identified by truly multivariate approaches.

5.2.2 Using Unsupervised (Rather than Supervised) Learning Algorithms

Unsupervised learning algorithms are a perfect choice for studies focused on new taxonomic knowledge. However, they should *not* be used as primary methods in biomarker discovery studies. Unfortunately, methods such as cluster analysis or principal component analysis are so popular (predominantly in view of their many excellent software implementations) that they are used quite indiscriminately, even for the studies with goals that can only be achieved by the application of supervised methods.

To identify a parsimonious multivariate biomarker, we need to perform a supervised feature selection, the goal of which is to seek and detect a small subset of variables that will allow for efficient separation of the considered phenotypic classes. As a criterion of class separation, we may use such metrics as the ratio of the variance between classes to that within classes, or the margin of a separating hyperplane.

Unsupervised methods can reduce the dimensionality by identifying such a subset (or a combination) of the original variables that preserves the most variance in the dataset. Thus, the goal of unsupervised dimensionality reduction is very different from the goal of supervised feature selection.[1]

Driving a biomarker discovery study by unsupervised methods may not only lead to inferior results, but in some situations may yield the worst possible solutions.[2]

Let us take a look at principal component analysis (PCA), a well-known unsupervised dimensionality reduction technique. PCA identifies such linear combinations of the original variables that can explain most of the data variance. However, the directions associated with most of the variance may be very different from the most discriminatory directions. Hence, using unsupervised dimensionality reduction as a 'preprocessing' step before biomarker discovery may result in discarding important (or maybe even the most important) discriminatory information. Figure 5.2 illustrates this using a simple two-dimensional example. The direction that best separates the two classes (DA) is very different from the direction of the first principal component (PC_1). If the dataset is reduced to the linear combination of the original variables identified as PC_1 (representing the direction that preserves most of the data variance), almost all of the discriminatory information will be lost. Of course, since we have only two original variables in this example, adding the second principal component will preserve entire variation in the data, but this will neither decrease the dimensionality nor identify the most discriminatory direction.

Other attempts in applying unsupervised approaches to reduce the dimensionality of data used for biomarker discovery involve cluster analysis. Genes are clustered by similarity of their expression patterns, and then a

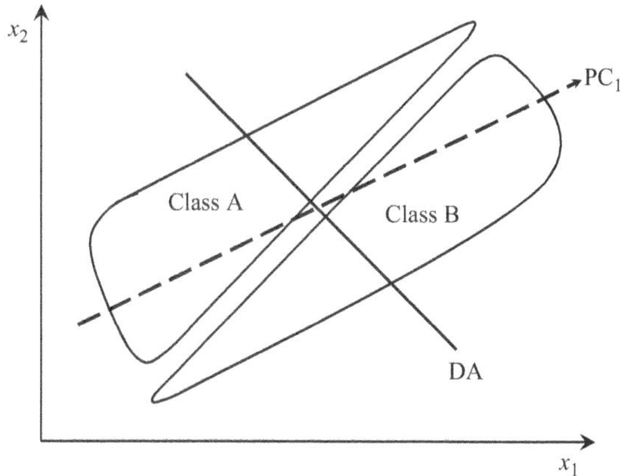

Figure 5.2 An example of two phenotypic classes (Class A and Class B), for which the direction that best separates the classes (DA) is very different from the direction of the first principal component (PC_1). PC_1 has been identified by an unsupervised method, and it represents the direction that preserves the most variation in the data. However, this direction does not have to be in any way related to the most discriminatory direction – the goal of biomarker discovery. The most discriminatory direction DA can be identified only by supervised methods (such as discriminant analysis).

study is limited only to the genes selected to represent the clusters. This approach could have merit only in the situation when the genes belonging to a cluster were either perfectly or nearly perfectly correlated. In practice, however, the genes assigned to the same cluster often share a relatively small amount of common variation (sometimes even less than 50%). The assumption that such genes carry similar discriminatory information is not only unrealistic, but also unverifiable in the unsupervised environment.

We may also look at this approach from a different point of view. The same dataset may be used in different biomarker discovery studies. The patients represented by their biological samples may be assigned to different sets of phenotypic classes, and the goals of these various studies may be to find biomarkers differentiating these diverse sets of classes. For example, one study may differentiate amongst classes representing various responses to a treatment, whilst another study may aim at predicting sub-types of the disease represented by the data, and still another study may differentiate between classes with different risks of relapse, and so on. Would it be reasonable to assume that the assignment of genes to clusters has anything to do with any of these supervised problems? Would it be reasonable to assume that the genes within a cluster carry similar discriminatory information relevant to any of these supervised goals? If so, to which of these different supervised goals? One has to realise that the results of unsupervised grouping of gene expression variables do not need to be in any way related to a particular supervised problem. As with the PCA preprocessing, this cluster-based dimensionality reduction may remove very important discriminatory information.

5.3   Feature Selection

Feature selection is the most important aspect of multivariate biomarker discovery. It is more important than selection of the learning algorithm to be used in order to build a classification system. After a properly performed and successful feature selection, which results in a parsimonious multivariate biomarker, many learning algorithms may provide efficient classification systems.

By feature selection we are referring to the identification of a small subset of variables, which together, as a set, represent the multivariate 'pattern' that can be used to differentiate amongst the phenotypic classes of interest. Please note that the identified parsimonious set of variables *is* the *multivariate biomarker*. None of the elements of this set should be called a *biomarker* (if any single variable of the set were representing an efficient biomarker, then we would not require the multivariate biomarker). Hence, such phrases as 'a set of biomarkers' should be avoided in this framework.

Although taxonomy of feature selection methods does not depend on a domain of their application, some methods and approaches that can be successfully applied to typical business data may be inappropriate for bio-medical research based on $p \gg N$ datasets. Generally, feature selection methods can be classified by their search model (filter, wrapper, hybrid and embedded models), by their search strategy (for example, exhaustive, complete or heuristic searches), by their learning approach (supervised or unsupervised), and whether they take into account interactions between variables (multivariate or univariate methods). As explained above, univariate and unsupervised approaches should not be used for multivariate biomarker discovery.

5.3.1 Search Models

Filter models perform feature selection independently of the learning algorithm to be used for a classification system. They evaluate the discriminatory power of each of the considered subsets of variables by using intrinsic characteristics of the 'training' data. Even though we should consider only multivariate filter models, some of the popular filter models often classified into this category would be better described as single-variable-centred multivariate models. *Shrunken Centroid* filters and *Correlation-based Feature Selection* are examples of the methods that include a multivariate component, yet univariately evaluate the relevance of each variable. A truly multivariate filter model may be represented by a heuristic approach that drives the search for an optimal subset of variables using a multivariate metric of class separation.

Wrapper models incorporate classification-learning algorithms into the feature selection process, that is the feature selection process is *wrapped* around the classifier. Since the classifier is employed to evaluate each of the considered subsets of features, wrapper models tend to provide more accurate classification systems tailored to the classifier's learning algorithm. However, they are usually more computationally expensive than filter models.

Some search models can be considered *hybrid models*, for they attempt to incorporate the strengths of the filter and wrapper models. For example, they may first use the filter approach to identify some number of potential biomarkers of different cardinalities, and then select one of them by using the wrapper approach utilising a specific learning algorithm.

Search models in which the feature selection process is incorporated into the learning algorithm are called *embedded models*. Usually, they use a metric of multivariate importance (of each variable) that is intrinsic to the learning algorithm. Examples of embedded search models will be discussed in the sections describing *support vector machines* and *random forests* learning algorithms.

5.3.2 Search Strategies

Although there are search strategies that guarantee finding the global optimum within the search space, they are unfeasible for datasets with a large number of variables. One such strategy is an *exhaustive search* that seeks and finds the global optimum by evaluating all possible subsets of variables. Another is a *complete search*, one that is capable of finding the global optimum without evaluating all possible subsets. For example, a complete search may be implemented by using the *branch and bound* method.[2] As stated, neither of these search strategies can be used for typical expression data with thousands of variables. Furthermore, there are indications that solutions represented by global optima may be more prone to overfitting (the training data) than those associated with local optima.[3,4] Hence, even if we could find the global optimum, there might be no reason to do so.

To efficiently deal with $p \gg N$ data, we may use *heuristic searches* that trail some 'good' path in the multidimensional search space. Since such searches evaluate only the subsets of variables that are on their search path, they result in local optima. However, properly designed heuristic searches are capable of finding local optima associated with efficient multivariate biomarkers.

Heuristic sequential searches (also known as *greedy* or *hill-climbing* strategies) may implement stepwise forward or backward selections. *Stepwise forward selection* starts with the empty set. Then each consecutive step adds one variable, the variable whose addition maximises discriminatory power of the set. *Stepwise backward selection* (also known as backward elimination) starts with all variables, and then at each consecutive step the variable with the least multivariate importance is removed. Generally, backward elimination may provide better results than forward selection, for the latter cannot evaluate a variable in the context of variables that are not included in the current set. However, some metrics of class separation (such as the metrics based on the ratio of the variance between classes to that within classes) cannot be calculated for subsets in situations where the number of variables is greater than the number of observations. This limits the use of backward elimination.

The best results may be achieved with *stepwise hybrid selection*, which incorporates both forward and backward search strategies. At each step, variables may be added or removed until discriminatory power (of a subset of the currently considered cardinality) cannot be further increased. This strategy (unlike forward or backward selections) results in subsets that are not necessarily nested, *i.e.* an optimal subset of $m + 1$ variables does not have to include an optimal subset of m variables.

To avoid solutions associated with inefficient local optima, elements of randomness may be incorporated into the feature selection process. For example, stepwise hybrid selection may start with a randomly selected variable, or feature selection may be performed many times using randomised versions of the training data.

5.3.3 Stability of Results

When various feature selection methods are applied to the same dataset, the identified multivariate biomarkers may consist of different sets of variables;

that does not necessarily indicate unstable solutions. The stability of such biomarkers has to be considered in the context of the biological processes they are hypothesised to represent. If the biomarkers tap into a common set of biological processes, then such seemingly diverse results may represent a stable and coherent solution to the class separation problem. However, identification of these biological processes underlying class differences may be a non-trivial task. For this reason, we may consider stability of biomarkers in terms of the primary expression patterns associated with class differences (see Section 5.5.3).

5.4 Supervised Learning Algorithms

Although the feature selection process may be independent of any learning algorithm (filter models), biomarker discovery studies are often performed by tailoring both the feature selection and classification stages to a specific supervised learning algorithm. Many learning algorithms can be successfully employed by such studies. The methods that make specific assumptions about the dataset (for example, assumptions about the distribution of variables, or about the independence of biological samples) are called *parametric* learning algorithms. Those that make no such assumptions are known as *nonparametric* methods.

Three learning algorithms will be described in this section. They can be used for the efficient analysis of $p \gg N$ data, and also for the identification of small multivariate biomarkers, especially when coupled with appropriate feature selection methods and used in a framework advancing stable and interpretable solutions (such as the novel framework presented in Section 5.5). *Linear discriminant analysis* is a powerful classical method that represents parametric learning algorithms. *Support vector machines* and *random forests* represent two nonparametric methods. Support vector machines are newer, but already serve as popular learning algorithms capable of delivering linear or nonlinear classifiers. Random forests belong to recent ensemble-based methods.

5.4.1 Linear Discriminant Analysis

Linear discriminant analysis (LDA) is a supervised learning algorithm that makes the following major assumptions:

- The independence of biological samples.
- Multivariate normal distribution of the variables.
- Homogeneity of the variance-covariance matrices (if the variance-covariance matrices for the differentiated classes were heterogeneous, this would give rise to *quadratic* discriminant analysis).

Let us note, however, that LDA is quite robust to some violations of the multivariate normality and homogeneity assumptions. This fact is important for at least two reasons. Firstly, with thousands of variables, it would be quite impractical to comprehensively test the assumption of their multivariate normal distribution. Secondly, for the better stability of results, it is recommended that the homogeneity assumption is made for $p \gg N$ datasets, even when the variance-covariance matrices are heterogeneous.[5]

LDA has a very good track record and belongs to learning algorithms that should be in the portfolio of any bioinformatician. Good performance of LDA may be partly explained by the fact that datasets often can only support simple linear boundaries between the differentiated classes. Furthermore, models based on parametric distributional assumptions tend to be stable.[6]

5.4.1.1 LDA Learning Algorithm

Assume that we are interested in differentiating J phenotypic classes, and that our training dataset includes n_j biological samples in each class j, where $j = 1, \ldots, J$. Hence, $N = \sum_{j=1}^{J} n_j$ is the total number of biological samples (data points) in the training set. Assume further that each of these data points is represented by p variables $x_1, \ldots, x_p$ (such as expression levels of p genes). Hence, a p-dimensional vector $\mathbf{x}_{ji} = \left[x_{1ji}, \ldots, x_{pji} \right]^T \in \Re^p$ can represent training data point i from class j (note that $i = 1, \ldots, n_j$ is used as an independent index for each class j).

For each class j, we can estimate its variance-covariance matrix as

$$\mathbf{S}_j = \frac{1}{n_j - 1} \sum_{i=1}^{n_j} \left(\mathbf{x}_{ji} - \bar{\mathbf{x}}_j \right) \left(\mathbf{x}_{ji} - \bar{\mathbf{x}}_j \right)^T, \tag{1}$$

where $\bar{\mathbf{x}}_j$ is the mean vector for training data points of class j (an unbiased estimator of the population mean vector $\boldsymbol{\mu}_j$ for class j). Let us also define $\bar{\bar{\mathbf{x}}}$ as the mean vector for all N training data points (the estimator of the overall mean vector $\boldsymbol{\mu}$).

Assuming homogeneity of variance-covariance matrices for the J populations, their common variance-covariance matrix Σ can be estimated by

$$\mathbf{S} = \frac{1}{N - J} \sum_{j=1}^{J} (n_j - 1)\mathbf{S}_j. \tag{2}$$

In the LDA context, finding the maximal separation of classes is equivalent to maximising the ratio of the variation between classes to that within classes. As a metric of class separation we can use the *Lawley–Hotelling trace statistic* T^2.[7,8] This T^2 trace criterion, which can be interpreted as the multivariate discriminatory power of a set of p variables, is defined as

$$T^2 = T^2(x_1, \ldots, x_p) = tr\left(\mathbf{HE}^{-1}\right), \tag{3}$$

where $\mathbf{H}$ is the $p \times p$ matrix representing variability amongst the classes,

$$\mathbf{H} = \sum_{j=1}^{J} n_j \left(\bar{\mathbf{x}}_j - \bar{\bar{\mathbf{x}}} \right) \left(\bar{\mathbf{x}}_j - \bar{\bar{\mathbf{x}}} \right)^T, \tag{4}$$

and $\mathbf{E}$ is the $p \times p$ matrix describing within-class variability,

$$\mathbf{E} = \sum_{j=1}^{J} \sum_{i=1}^{n_j} \left(\mathbf{x}_{ji} - \bar{\mathbf{x}}_j \right) \left(\mathbf{x}_{ji} - \bar{\mathbf{x}}_j \right)^T. \tag{5}$$

Although the exact distribution of the T^2 statistic is unknown, we can use one of its approximations. For example, the following approximation of the T^2 distribution,[5,9]

$$F = \frac{t(N - J - p - 1) + 2}{t^2 b} \, tr(\mathbf{H}\mathbf{E}^{-1}) \tag{6}$$

has an F distribution with bt and $t(N - J - p - 1) + 2$ degrees of freedom, where $t = \min(p, J - 1)$, and $b = \max(p, J - 1)$.

In order to detect and determine an optimal multivariate marker, that is, a small set of variables that sufficiently separates the classes, we can perform heuristic feature selection using the T^2 metric to evaluate the discriminatory power of each considered subset of variables. Since we are interested in truly multivariate – hence, parsimonious – biomarkers, the fact that the T^2 metric can be calculated only for the subsets including fewer than $N - J - 1$ variables imposes no practical limitations to such searches. A method implementing such a heuristic search is discussed in the next section.

Let us assume here that the feature selection process resulted in a small set of, say, $p = 10$ variables.[†] If we used this multivariate biomarker to build a classifier, classification would be performed in a p-dimensional space of the marker variables, and each biological sample would be represented by a p-dimensional point $\mathbf{x} = [x_1, \ldots, x_p]^T$ in this discriminatory space. To classify a new sample to one of the J classes, we could calculate Mahalanobis distances (D_j) between the sample point $\mathbf{x}$ and the class centroids $\bar{\mathbf{x}}_j, j = 1, \ldots, J$,

$$D_j = \sqrt{(\mathbf{x} - \bar{\mathbf{x}}_j)^T \mathbf{S}^{-1} (\mathbf{x} - \bar{\mathbf{x}}_j)}, \tag{7}$$

and assign the sample into the class with the smallest D_j value. To account for different prior probabilities q_j, we would assign the sample into the class with the minimum value of $D_j^2 - 2 \ln q_j$.[5]

Even if our optimal multivariate biomarker is parsimonious, it often includes more than three variables. When $p > 3$, neither the entire p-dimensional discriminatory space of biomarker variables, nor the complete classification results, can be presented graphically. However,

---

[†]Please note that p is used to denote any number of variables in a set. Thus, for the original training data it could represent thousands of variables, whilst for the optimal biomarker it usually represents no more than ten variables.

LDA allows us to decrease the dimensionality of the discriminatory space from p to $t = \min(p, J - 1)$. Since we usually differentiate fewer than five classes,[‡] the entire discriminatory information for most classification problems can be visualised in a space with three or fewer dimensions.

To facilitate such low-dimensionality visualisation, we solve the following generalised eigenproblem

$$\mathbf{Hv} = \lambda \mathbf{Ev}, \tag{8}$$

which has t non-zero eigenvalues $\lambda_1 \geq \lambda_2 \geq \ldots \geq \lambda_t$, and – associated with them – t normalised eigenvectors $\mathbf{v}_1, \mathbf{v}_2, \ldots, \mathbf{v}_t$. The eigenvalue-eigenvector pairs define the transformation of the p-dimensional space of the original biomarker variables into a t-dimensional space of t *features*. These features are defined by t discriminant functions $f_1, \ldots, f_t$ that are linear combinations of the p biomarker variables. For example, $f_1 = \mathbf{v}_1^T \mathbf{x}$, which is associated with the largest eigenvalue λ_1 and its eigenvector $\mathbf{v}_1 = [v_{11}, \ldots, v_{p1}]^T$, transforms a p-dimensional point $\mathbf{x}$ into the single dimension represented by the first feature, this feature being defined by the discriminant function f_1.

Thus, a biological sample in the p-dimensional space of p biomarker variables that was represented by a vector $\mathbf{x} \in \Re^p$ is now represented by a t-dimensional vector $\mathbf{w} = \mathbf{V}^T \mathbf{x} \in \Re^t$ in the new discriminatory space of t features. Columns of the $p \times t$ matrix $\mathbf{V}$ are the t eigenvectors $\mathbf{v}_1, \mathbf{v}_2, \ldots, \mathbf{v}_t$, *i.e.* $\mathbf{V}$ is the matrix of the weights associated with all linear discriminant functions $f_1, \ldots, f_t$.

Since T^2 is invariant for linear and regular transformations, the t features defining our new discriminatory space have the same discriminatory power as the p original biomarker variables. In addition, the discriminatory power of each feature is equal to its eigenvalue. Thus,

$$T^2(w_1, \ldots, w_t) = T^2(x_1, \ldots, x_p) = \sum_{k=1}^{t} \lambda_k. \tag{9}$$

Furthermore, the features have a unitary variance-covariance matrix and hence they are uncorrelated, the differentiated classes can be represented by t-dimensional hyperspheres and the Euclidean distance is equal to the Mahalanobis distance in this t-dimensional discriminatory space.

To classify a new sample, we may calculate Euclidean distances between the sample vector $\mathbf{w} = \mathbf{V}^T \mathbf{x}$ and J class centroids $\overline{\mathbf{w}}_j = \mathbf{V}^T \overline{\mathbf{x}}_j, j = 1, \ldots, J$, and assign the sample into the class with the nearest centroid. If we are required to account for different prior probabilities q_j, we would assign the sample into the class with the minimum value of the statistic $(\mathbf{w} - \overline{\mathbf{w}}_j)^T (\mathbf{w} - \overline{\mathbf{w}}_j) - 2 \ln q_j$.[10]

If, for a given significance level α, we wish to evaluate the sample membership in each of the J classes, we may check whether the sample vector $\mathbf{w}$ is within the hyperspheres representing constant density boundaries enclosing

---

[‡] A recommended way of dealing with multiclass problems is to design a multistage classification schema.[1]

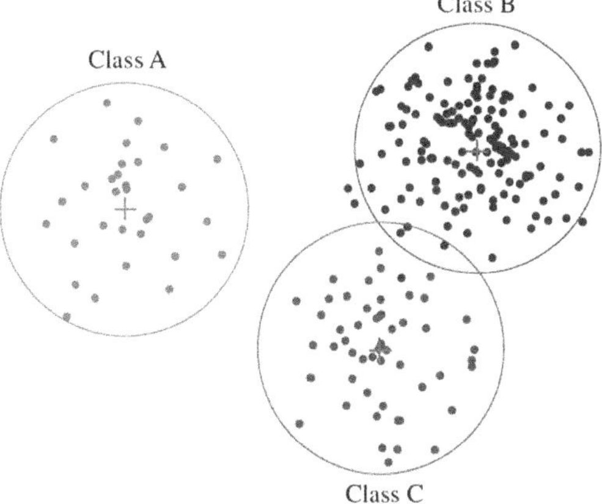

Figure 5.3   An example of discriminatory space defined by a multivariate biomarker consisting of five variables ($p = 5$). The biomarker differentiates among three phenotypic classes ($J = 3$), thus the discriminatory space is two-dimensional, $t = \min(p, J - 1) = 2$. The circles (which can be seen as two-dimensional hyperspheres) correspond to constant density boundaries enclosing 95% of the probability for each class. The points represent training data.
(Graphics from the *MbMD* biomarker discovery software, www.MultivariateBiomarkers.com.)

$(1 - \alpha) \cdot 100\%$ of the probability for each class (Figure 5.3). The radius R_j of hypersphere j can be calculated as

$$R_j = \sqrt{F_\alpha \cdot \frac{n_j + 1}{n_j} \cdot \frac{t(N - J)}{N - J - t + 1}}, \tag{10}$$

where F_α is taken from an F distribution with t and $N - J - t + 1$ degrees of freedom.[11] From such a classification, the sample may be assigned to one class, to more than one class or to none of the discriminated classes.

5.4.1.2   Feature Selection with T^2

In this section, we will describe a stepwise hybrid feature selection driven by the Lawley–Hotelling T^2 metric of discriminatory power.[1] The feature selection process commences with the empty set. The first variable added to this set may be either the single most discriminatory variable or, alternatively, a randomly selected one. Then, at consecutive steps, this heuristic search considers both forward and backward selections in order to maximise the T^2 metric of a set of p variables ($p = 2, 3, 4, \ldots$); each step starts with forward selection. From the pool of variables not included in the current set, the variable selected is that which maximises the T^2 discriminatory power of a set of p variables. Then, if $p > 2$ and if the elimination of any one of the currently selected p variables gives us a set with T^2 greater than the T^2 of the previously identified best set of $p-1$ variables, the subset of $p-1$ variables that provides the greatest T^2 becomes the current selection. The variable that was eliminated is 'sent back' to the pool to be available at later steps. This forward and backward selection continues until, for a given cardinality p, no further increases in T^2 can be achieved.

An optimal marker is required to satisfy a combined criterion of minimal size and maximal discriminatory power. The implementation of this algorithm in the *MbMD* data mining system, which is specialised biomarker discovery software (www.MultivariateBiomarkers.com) uses two user-adjustable parameters defining the search stopping criteria: *stop_T^2* and *stop_p*. The search stops when the T^2 metric of discriminatory power exceeds *stop_T^2*, or when the current set includes *stop_p* variables.

As with all heuristic searches, this search results in a biomarker associated with a local optimum. To further optimise the search, *i.e.* to avoid being trapped in an inefficient local optimum, we may repeat the search (once or more than once) starting with a different variable randomly selected in the same manner as the first one. However, a better optimisation method (utilising the *Informative Set of Genes* and *Modified Bagging Schema*) will be presented in Section 5.5.

5.4.2 Support Vector Machines

5.4.2.1 *SVM Learning Algorithms*

Introduced in the 1990s,[12,13] support vector machines (SVMs) quickly became popular in bioinformatics. They are supervised learning algorithms solving binary classification problems[§] by determining the optimal hyperplane separating the classes. SVMs designed for linearly separable training sets are called *hard-margin support vector machines*. Classifiers that allow misclassification of some of the training data points (when training data are not linearly separable) are called *soft-margin support vector machines*. Both hard- and soft-margin classifiers search for solutions that can separate classes (with or without training errors) by a linear boundary in the input space of the original variables. In cases where boundaries between classes are inherently nonlinear, the input space is mapped into a usually higher-dimensional (even infinite-dimensional) *feature space*, in which classes are linearly separable.

If training data for a binary classification problem ($J=2$) includes N biological samples characterised by p variables, then each training data point can be represented by a vector $\mathbf{x}_i = [x_{1i}, \ldots, x_{pi}]^T \in \Re^p$ and a class label $y_i \in \{+1, -1\}$, where $i = 1, \ldots, N$.

A hyperplane $\mathbf{w}^T\mathbf{x} + b = 0$ separating two classes is defined by the p-dimensional vector $\mathbf{w}$ that determines its orientation, and the scalar b that

---

[§]There are extensions of SVMs to multiclass problems.

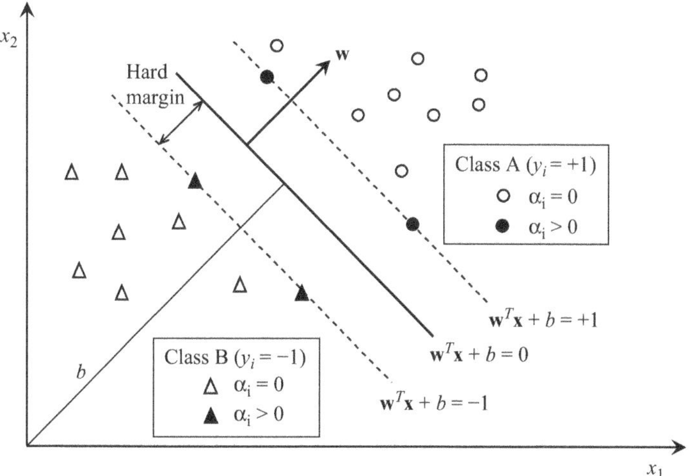

Figure 5.4   Discriminatory space of a hard-margin SVM classifier. A separating hyperplane $\mathbf{w}^T\mathbf{x} + b = 0$ is defined by the vector $\mathbf{w}$ that determines its orientation, and the scalar b that represents its offset from the origin. The training data points that lie on the support hyperplanes ($\mathbf{w}^T\mathbf{x} + b = +1$ or $\mathbf{w}^T\mathbf{x} + b = -1$) are support vectors.

represents its offset from the origin (see Figure 5.4). The *optimal separating hyperplane* is the one that maximises the *geometric margin* γ,

$$\gamma = \min_i \frac{1}{\|\mathbf{w}\|} y_i \left(\mathbf{w}^T\mathbf{x}_i + b \right), \tag{11}$$

defined as the minimal Euclidean distance between the separating hyperplane and training points of either class.[¶] This defines two *support hyperplanes*, which are parallel and equidistant to the separating hyperplane. Since $m = \min_i y_i(\mathbf{w}^T\mathbf{x}_i + b)$ can be interpreted as the functional margin of the separating hyperplane (corresponding to the value of the function $y_i(\mathbf{w}^T\mathbf{x}_i + b)$ at the points $\mathbf{x}_i$ positioned on the support hyperplanes), and since rescaling the hyperplane will not change the classification function, the optimal separating hyperplane may be found by maximising $1/\|\mathbf{w}\|$ (after rescaling *via* a setting of the functional margin m to 1). Furthermore, instead of maximising $1/\|\mathbf{w}\|$, we may minimise $\|\mathbf{w}\|$ or $\|\mathbf{w}\|^2$. Hence, the optimal separating hyperplane is the solution to the following convex optimisation problem:

$$\underset{\mathbf{w},b}{\text{minimise}} \ \|\mathbf{w}\|^2 \tag{12}$$

$$\text{subject to } y_i\left(\mathbf{w}^T\mathbf{x}_i + b\right) \geq 1 \quad \text{for all } \mathbf{x}_i, \ i = 1, \ldots, N.$$

This *primal optimisation problem* is solved with respect to $p + 1$ variables, that is, the p-dimensional vector $\mathbf{w}$ and the scalar b. Its linear constraints dictate

---

[¶] $\|\mathbf{w}\|$ is the Euclidean norm of the vector $\mathbf{w}$.

that no training point may lie between the support hyperplanes $\mathbf{w}^T\mathbf{x} + b = \pm 1$.

It has been shown that by introducing an N-dimensional vector $\alpha = (\alpha_1, \ldots, \alpha_N)^T$ of non-negative Lagrange multipliers α_i associated with training data points $\mathbf{x}_i$, $i = 1, \ldots, N$, the primal optimisation problem [eqn (12)] can be converted into the following *dual representation*:[14-16]

$$\text{maximise } W(\boldsymbol{\alpha}) = \sum_{i=1}^{N} \alpha_i - \frac{1}{2} \sum_{i,j=1}^{N} \alpha_i \alpha_j y_i y_j \mathbf{x}_i^T \mathbf{x}_j$$

$$\text{subject to } \sum_{i=1}^{N} y_i \alpha_i = 0,$$

$$\alpha_i \geq 0, \quad \text{for } i = 1, \ldots, N. \tag{13}$$

This dual optimisation problem is solved with respect to only N variables, which for $p \gg N$ data permits a very significant reduction in the number of the problem's variables. The solution to eqn (13) has to satisfy the Karush–Kuhn–Tucker, or KKT, complementarity conditions:

$$\alpha_i[y_i(\mathbf{w}^T\mathbf{x}_i + b) - 1] = 0 \quad \text{for } i = 1, \ldots, N. \tag{14}$$

The orientation of the optimal separating hyperplane is calculated as

$$\mathbf{w} = \sum_{i=1}^{N} \alpha_i y_i \mathbf{x}_i. \tag{15}$$

From the KKT conditions [eqn (14)], it is obvious that Lagrange multipliers α_i can be positive only when $y_i(\mathbf{w}^T\mathbf{x}_i + b) = 1$. Hence, the orientation of the optimal separating hyperplane depends only on the training data points that lie on the support hyperplanes $\mathbf{w}^T\mathbf{x}_i + b = \pm 1$. These training points (the only points with $\alpha_i > 0$) are known as *support vectors*.

To calculate the offset of the optimal separating hyperplane, we can solve the KKT condition for any of the support vectors,

$$b = y_i - \mathbf{w}^T\mathbf{x}_i. \tag{16}$$

Once the optimal separating hyperplane $\mathbf{w}^T\mathbf{x} + b = 0$ is identified, we can use the following classification function $f(\mathbf{x})$,

$$f(\mathbf{x}) = sign(\mathbf{w}^T\mathbf{x} + b)$$

$$= sign\left(\sum_{i=1}^{N} \alpha_i y_i \mathbf{x}_i^T \mathbf{x} + b \right), \tag{17}$$

to assign any sample $\mathbf{x}$ to the class labelled $+1$ if $f(\mathbf{x}) > 0$ and to the class labelled -1 if $f(\mathbf{x}) < 0$. It is clear from the dual representation of $f(\mathbf{x})$ that any unknown sample may be classified by using only the support vectors, that is, the training points $\mathbf{x}_i$, for which $\alpha_i > 0$. This is the reason why such classifiers are known as *support vector machines*.

Soft-margin SVMs allow misclassification of some of the training data points. Hence, optimisation is based on the 'trade-off' between maximising the margin and minimising the cost of misclassification. The optimisation problem can be formulated as

$$\underset{\mathbf{w},b,\xi}{\text{minimise}}\,\|\mathbf{w}\|^2 + C\sum_{i=1}^{N}\xi_i \tag{18}$$

$$\text{subject to } y_i\left(\mathbf{w}^T\mathbf{x}_i + b\right) \geq 1 - \xi_i \quad \text{for all } \mathbf{x}_i,\ i = 1,\ldots,N,$$

where ξ_i are non-negative slack variables that quantify the margin violation for training data points $\mathbf{x}_i$, $i = 1,\ldots,N$, and C is a regularisation parameter controlling overlap between the classes. Since $C\sum\xi_i$ represents the total cost of margin violations, increasing C will decrease the margin (which may lead to overfitting), whilst decreasing C will increase the margin (which may provide more regularised solutions).

The dual form of the soft-margin optimisation problem,

$$\text{maximise } W(\boldsymbol{\alpha}) = \sum_{i=1}^{N}\alpha_i - \frac{1}{2}\sum_{i,j=1}^{N}\alpha_i\alpha_j y_i y_j \mathbf{x}_i^T \mathbf{x}_j$$

$$\text{subject to } \sum_{i=1}^{N} y_i\alpha_i = 0, \tag{19}$$

$$0 \leq \alpha_i \leq C, \quad \text{for } i = 1,\ldots,N,$$

adds C as the upper bound on the values of Lagrange multipliers α_i. KKT conditions now include the following two constraints:

$$\alpha_i[y_i(\mathbf{w}^T\mathbf{x}_i + b) - 1 + \xi_i] = 0 \quad \text{for } i = 1,\ldots,N, \tag{20}$$

$$\xi_i(\alpha_i - C) = 0 \quad \text{for } i = 1,\ldots,N. \tag{21}$$

The training data points that have $\xi_i = 0$ do not violate their class margin. Those of them that have $0 < \alpha_i < C$ are positioned on one of the support hyperplanes, and are known as *unbounded support vectors*. The training points with $\xi_i > 0$ violate the margin of their class, and their Lagrange multipliers have to be $\alpha_i = C$ (bounded by the parameter C). They also affect the solution, and are called *bounded support vectors*. Those of them that have $\xi_i > 1$ are on the wrong side of the separating hyperplane and hence are misclassified. The ones with $\xi_i = 1$ lie on the separating hyperplane and cannot be classified.

Although the solution to the soft-margin optimisation problem can still be represented by eqns (15) and (16), and the classification function still has the form described by eqn (17), the solution differs from that for the hard-margin case – it depends not only on the support vectors positioned on the support hyperplanes, but also on the bounded support vectors.[||]

[||]To calculate the offset b we solve the constraint (eqn (20)) for any of the unbounded support vectors.

Neither hard- nor soft-margin SVMs work in situations where boundaries between classes are intrinsically nonlinear. To solve such problems, we map the input space $\Re^p$ into such a higher-dimensional *feature space* $\Re^s$, in which the classes can be linearly separable. Such a mapping (Φ) associates p-dimensional training points $\mathbf{x}_i$ with their s-dimensional images $\mathbf{z}_i$,

$$\Phi\colon \mathbf{x}_i = \left[x_{1i}, \ldots, x_{pi}\right]^T \in \Re^p \rightarrow \mathbf{z}_i = [z_{1i}, \ldots, z_{si}]^T \in \Re^s, i = 1, \ldots, N. \tag{22}$$

Solving the dual optimisation problem in the feature space leads to the classification function

$$f(\mathbf{x}) = sign\left(\sum_{i=1}^{N} \alpha_i y_i \mathbf{z}_i^T \mathbf{z} + b\right), \tag{23}$$

which is expressed in terms of the inner product $\mathbf{z}_i^T \mathbf{z}$. Given that $\mathbf{z}_i = \Phi(\mathbf{x}_i)$ is the image of one of the training points $\mathbf{x}_i$, and $\mathbf{z} = \Phi(\mathbf{x})$ is the image of the point to classify $\mathbf{x}$, by defining a function $K(\mathbf{x}_i,\mathbf{x}) = \Phi(\mathbf{x}_i)^T \Phi(\mathbf{x})$, we can re-write the classifier [eqn (23)] as

$$f(\mathbf{x}) = sign\left(\sum_{i=1}^{N} \alpha_i y_i K(\mathbf{x}_i, \mathbf{x}) + b\right). \tag{24}$$

Since both the classifier and the optimisation problem can now be expressed in terms of the function $K(\mathbf{x}_i,\mathbf{x})$, whose arguments are vectors in the original input space, we can solve the nonlinear problem without explicitly mapping data into the feature space. Functions that calculate the inner product in the feature space by performing computations in the input space are called *kernels*. The use of kernel functions to avoid explicit mapping into the feature space is known as *the kernel trick*. Amongst commonly used kernels are:

- *Polynomial kernel:* $K(\mathbf{x}_i, \mathbf{x}_j) = (\mathbf{x}_i^T \mathbf{x}_j + c)^d$, $d > 0$.
- *Radial basis function kernel:* $K(\mathbf{x}_i, \mathbf{x}_j) = \exp\left(-\dfrac{\|\mathbf{x}_i - \mathbf{x}_j\|^2}{\sigma^2}\right)$.
- *Sigmoid (hyperbolic tangent)** kernel:* $K(\mathbf{x}_i, \mathbf{x}_j) = \tanh(\beta \mathbf{x}_i^T \mathbf{x}_j + \gamma)$.

5.4.2.2 Recursive Feature Elimination

Some of the SVMs implementations include an embedded feature selection algorithm called *Recursive Feature Elimination*.[3] This algorithm follows the sequential backward selection approach. It starts with the training data including all variables, identifies an optimal classifier, calculates the multivariate importance of each variable and eliminates the variable with the least importance. Given that the p-dimensional vector $\mathbf{w}$ (that determines the

---

**This function is a kernel only for some combinations of parameters β and γ (*e.g.* for $\beta = 2$ and $\gamma = 1$).

orientation of the optimal separating hyperplane) can be interpreted as a vector of weights w_k, $k = 1, \ldots, p$, the absolute value of the weight $|w_k|$ can be employed as the multivariate importance of each variable k.

At consecutive steps, all elements of this process are repeated. The multivariate importance of the variables remaining in the training data is recalculated, and the currently least important one is eliminated. This recursive variable elimination stops when all of the variables are eliminated (or when some stopping criterion, such as the specified size of the current set of variables, is achieved). Since this is a sequential search, the results are 'nested' subsets of variables presented in the form of a list having the last eliminated variable at the top. Thus the best subset of p variables is composed of those from the top of the list.

5.4.3 Random Forests

5.4.3.1 Random Forests Learning Algorithm

The *Random Forests* learning algorithm[17] is an example of the ensemble classifier approach. It implements *bagging* – bootstrap aggregating[18] – to generate a large number of randomised training sets, and then combines decision tree classifiers built from these training sets. Efron's *nonparametric bootstrap*[19] is utilised to generate bootstrap training sets by random sampling of the original training dataset with replacement. The bootstrap sets are of the same size as the original training set, that is, they include N training data points (biological samples) and p variables. In view of drawing with replacement, each of the bootstrap training sets includes, on the average, about $0.632*N$ unique samples. The remaining (about $0.368*N$) samples are not used to train a particular classifier; they constitute its 'out-of-bag' (OOB) samples.

Each of the bootstrap training sets is used to build a decision tree classifier. At each node, m variables ($m \ll p$) are randomly selected. First, the Gini impurity index is used to identify the best split for each of these m variables. Then, the node is split based on the best selected from the identified m best splits.

Hundreds or thousands of tree classifiers are built without 'pruning'. They constitute an *ensemble classifier* that may be used to classify new samples by the plurality vote of all its classifiers. Even though, in view of their hierarchical structure, decision tree classifiers are inherently unstable, by averaging over multiple classifiers based on bootstrap training sets selected from the same distribution, the ensemble approach can improve the prediction of unstable methods by reducing the variance. Using each of the tree classifiers to classify its own OOB samples, and then aggregating the results over all classifiers, yields the OOB estimate of the misclassification error rate for the ensemble classifier. This estimate is believed to be as accurate as estimates based on an independent test set of the same size N.

5.4.3.2   Feature Selection with Random Forests

To use random forests for the identification of parsimonious multivariate biomarkers, feature selection may be embedded in the learning algorithm. It may be implemented as an iterative process similar to recursive feature elimination. At each step, a random forests ensemble classifier is built, the variable importance is calculated and the least important variable (or variables) are eliminated. Thus, a sequence of 'forests' is built. Then, one of the forests (and its set of variables – a biomarker) is selected using such criteria as the number of variables and the OOB estimate of the forest misclassification rate.

Random forests utilise metrics of variable importance that are based either on permutation experiments or, alternatively, on the decrease in the node impurity when a node is split by a variable.[17,20] The following permutation-based metric is amongst the most popular ones. To calculate the importance of variable k, we start with classifying OOB samples and counting, for every tree in the forest, the number of correctly classified samples. Subsequently, we randomly permute the variable k data, and classify such prepared OOB samples. For each tree, the difference in the number of correctly classified OOB samples when classifying the original data and when classifying the permuted data is calculated. Averaging this difference over all trees in the forest gives us the importance of variable k.

Please note, however, that permutation experiments are performed independently for each variable, and that variables have the same probability of being selected for consideration at a node. Hence, as an example, two redundant variables may have high importance scores, whilst only one of them would be required in a multivariate biomarker. Therefore, in order to account for such interactions amongst variables, and also to identify a truly multivariate biomarker, we may need an additional step of feature selection which lies outside the random forest frame.

5.5   Searching for Multivariate Biomarkers that are Robust and Biologically Interpretable

By utilising a supervised learning algorithm coupled with a heuristic approach to multivariate feature selection, we can identify a small subset of variables, which may constitute a multivariate biomarker, that is, may separate the differentiated phenotypic classes. However, we have to remember that another aspect of the *curse of dimensionality*, especially important for datasets with small numbers of biological samples, is the sparsity of the multidimensional representation of the data. Therefore, it may be possible to find good separation of training samples even for data representing randomly generated noise. Therefore, a single heuristic search may result either in an efficient biomarker (when the identified subset of variables can properly classify independent test data), or in a subset that overfits the training data (and therefore is of no value for the classification of new samples).

It should be mentioned that some researchers argue that to resolve this problem we should use only the variables that are known to have associations with the differentiated phenotypes. Although this may work in some situations, in general it would be an extremely limiting and ineffective approach. Firstly, the solutions limited to already known phenomena would have little chance to discover new biomedical knowledge. Secondly, this approach may lead to inferior results, or no valid or realistic results in situations where efficient ones could have been identified had appropriate methods been used.

Nevertheless, even when a single heuristic search results in an efficient biomarker, such a single biomarker rarely provides sufficient insight into biological processes associated with class differences. In this section, I present a novel method which allows for the identification of generalisable and interpretable biomarkers. Firstly, I introduce two concepts, the *Informative Set of Genes* and the *Modified Bagging Schema*, and then show how to combine them into a biomarker discovery method that leads to robust multivariate biomarkers that have the best chance for plausible biological interpretation.

5.5.1 Informative Set of Genes

The *Informative Set of Genes* has been defined as a set of genes whose expression data contain all of the information significant for the differentiation of phenotypic classes represented in the training dataset.[1] To identify the *Informative Set of Genes*, we repeatedly perform heuristic feature selection in order to generate a sequence of multivariate biomarkers. After the first parsimonious multivariate biomarker is found, its variables are removed from the training dataset, and the second, alternative, biomarker of the same size is identified. Subsequently, its variables are also removed, and the next alternative biomarker is identified. The process continues until the remaining training data no longer contain any significant discriminatory information. The variables selected into the first biomarker, together with those selected into the alternative biomarkers that have satisfactory discriminatory power, constitute the *Informative Set of Genes* (see Figure 5.5).

To facilitate the biological interpretation of class differences, we search for characteristic expression patterns amongst the genes selected into the *Informative Set of Genes*. Since the *Informative Set of Genes* includes all of the significant discriminatory information, we may assume that it also includes all of the characteristic gene expression patterns associated with the biological processes underlying class differences. To identify these characteristic patterns, we may use clustering methods, such as self-organising maps. Please note that the unsupervised approach is applied here to the *Informative Set of Genes*, which is the result of multivariate-supervised analysis and includes only the genes whose expressions have already been associated with the class differences.

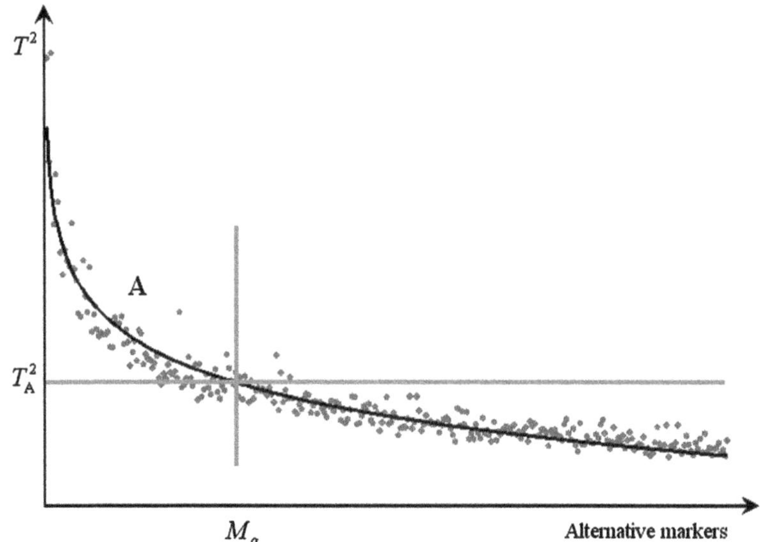

Figure 5.5   Alternative multivariate biomarkers and the selection of the *Informative Set of Genes* (based on the original implementation of the method in the *MbMD* biomarker discovery software, www.MultivariateBiomarkers.com). The points represent the T^2 measure of discriminatory power of the subsequently identified multivariate biomarkers. A logarithmic trend line approximates a decreasing tendency of this discriminatory power. The *Informative Set of Genes* is defined as the set of variables included in these of the first M_α alternative markers whose discriminatory power is not less than T_A^2 (that is, the alternative markers represented by the points in area A). The T_A^2 cut-off level of discriminatory power is adjusted for each training dataset by examining the distributions of training data points in relation to the distributions of the differentiated classes delineated by linear discriminant analysis. The alternative biomarker M_α is defined by the intersection of the trend line with the T_A^2 level of discriminatory power.[1]

5.5.2   Modified Bagging Schema

Although the *Informative Set of Genes* includes all of the significant discriminatory information, it is possible that some of the genes included in some of the alternative biomarkers were selected by random chance. To be able to find robust biomarkers, we must identify the gene expression patterns that are associated with the most important biological processes underlying class differences. The method we use is based on randomisation of the training dataset, and also on the ensemble classifier approach.

Bagging (bootstrap aggregating) is a popular method of combining classifiers built from randomised versions of the training dataset.[18] Typically, bagging utilises Efron's nonparametric bootstrap,[19,21,22] which makes no assumption about the underlying populations, selects with replacement and generates randomised (bootstrap) training sets of the same size as the original training dataset. However, sampling with replacement may cause problems for feature selection methods (applied to each of the bootstrap training sets) that require independence of the training observations.

Heuristic feature selection methods compare, at each step, the discriminatory power of subsets that consist of the same number of variables. Thus, external cross-validation cannot be used for this purpose. However, the internal cross-validation process, which uses the same set of variables and the same data for training and for validation, is unreliable in $p \gg N$ situations, to say the least. We recommend methods of subset evaluation that are based on a metric of class separation (such as the Lawley–Hotelling T^2 criterion that maximises the ratio of the variance between classes to that within classes). However, such metrics usually require that the training set observations are independent. To assure this independence, we modify bagging in a manner that the randomised training sets are generated without replacement.

The *Modified Bagging Schema* has been defined as an algorithm that selects bootstrap training sets by utilising stratified random sampling without replacement.[1] The parameter γ_{OOB} driving this selection is a proportion of 'out-of-bag' samples, that is, the proportion of the training observations that are not selected into a bootstrap training set. When using the *Modified Bagging Schema*, we generate hundreds or thousands of bootstrap training sets. Each of them includes a specified proportion of observations randomly selected without replacement from the original training dataset. For each of the bootstrap training sets, we perform an independent feature selection, and then build a classifier. For example, if $\gamma_{OOB} = 0.8$, then each classifier will be built on 80% of randomly selected training observations, with the remaining 20% of them constituting the 'out-of-bag' samples that can be used to test the performance of a particular classifier.

5.5.3 Identification of Parsimonious Biomarkers that are Robust and Interpretable

By combining the randomisation and ensemble approach (represented here by the *Modified Bagging Schema*) with the analysis of expression patterns within the *Informative Set of Genes*, we have designed a method that can be interpreted as a regularisation of feature selection leading to biomarkers that are more robust than a single multivariate biomarker selected from the entire training dataset.[1]

After the *Informative Set of Genes* is identified, its variables are clustered into groups of genes with similar shapes of their expression patterns. Following this, we limit our training dataset to include only the variables of the *Informative Set of Genes*. The *Modified Bagging Schema* is used to generate a large number of classifiers (an ensemble) based on bootstrap training sets selected from this new base training set.

By examining the distribution of each cluster's genes amongst these classifiers, we identify the clusters, the genes of which are most frequently selected into the ensemble's classifiers. Alternatively, we may limit our investigation only to those of the ensemble classifiers that perfectly (or nearly perfectly) classify their 'out-of-bag' samples. In any case, clusters with genes that are most frequently used by the classifiers of the ensemble are deemed to be the primary clusters. We assume that they represent the *primary expression patterns*, that is the patterns representing the most important biological processes associated with differences amongst the investigated phenotypic classes.

Furthermore, it is reasonable to assume that not all genes of the primary clusters are equally important for the class differentiation. The genes of the primary clusters that are most frequently selected into these classifiers that perfectly or nearly perfectly classify their 'out-of-bag' samples are called the *frequent primary genes*. They constitute the best starting point for elucidation of biological processes underlying class differences. They are also the ones that are least likely to be selected by chance into the alternative markers constituting the *Informative Set of Genes*. Therefore, if we now perform heuristic feature selection based only on the *frequent primary genes*, the identified multivariate biomarker is, of course, more likely to be robust, as well as to have a plausible biological interpretation.

References

1. D. M. Dziuda, *Data Mining for Genomics and Proteomics: Analysis of Gene and Protein Expression Data*, Wiley, Hoboken, NJ, 2010.
2. D. J. Hand, H. Mannila and P. Smyth, *Principles of Data Mining*, MIT Press, Cambridge, MA, 2001.
3. I. Guyon, J. Weston, S. Barnhill and V. N. Vapnik, Gene selection for cancer classification using support vector machines, *Mach. Learn.*, 2002, **46**(1–3), 389.
4. H. Liu and H. Motoda, Less is more, in *Computational Methods of Feature Selection*, ed. H. Liu and H. Motoda, Taylor & Francis, Boca Raton, FL, 2007, p. 3.
5. C. J. Huberty and S. Olejnik, *Applied MANOVA and Discriminant Analysis*, Wiley, Hoboken, NJ, 2006.
6. T. Hastie, R. Tibshirani and J. H. Friedman, *The Elements of Statistical Learning, Second Edition: Data Mining, Inference, and Prediction*, Springer, New York, 2009.
7. H. Hotelling, A generalized T test and measure of multivariate dispersion, *Proceedings of the Second Berkeley Symposium on Mathematical Statistics and Probability (July 31–August 12, 1950)*, Berkeley, CA, University of California Press, 1951, p. 23.
8. D. N. Lawley, A generalization of Fisher's z test, *Biometrika*, 1938, **30**(1/2), 180.
9. A. C. Rencher, *Methods of Multivariate Analysis*, Wiley, New York, 2002.

10. R. O. Duda, P. E. Hart and D. G. Stork, *Pattern Classification*, Wiley, New York, 2001.
11. D. Dziuda, Specialized PC software package for creation of computer systems supporting partial diagnostics based on numerical results of medical examinations, *Medical Informatics (London)*, 1990, **15**(4), 319.
12. B. E. Boser, I. Guyon and V. N. Vapnik, A training algorithm for optimal margin classifiers, *Fifth Annual Workshop on Computational Learning Theory*, Pittsburgh, ACM, 1992, p. 144.
13. V. N. Vapnik, *Statistical Learning Theory*, Wiley, New York, 1998.
14. S. Abe, *Support Vector Machines for Pattern Classification*, Springer, London, 2005.
15. N. Cristianini and J. Shawe-Taylor, *An Introduction to Support Vector Machines and other Kernel-based Learning Methods*, Cambridge University Press, Cambridge, New York, 2000.
16. V. N. Vapnik, *The Nature of Statistical Learning Theory*, Springer, New York, 2000.
17. L. Breiman, Random forests, *Mach. Learn.*, 2001, **45**(5), 5.
18. L. Breiman, Bagging predictors, *Mach. Learn.*, 1996, **24**, 123.
19. B. Efron, Bootstrap methods: another look at the jackknife, *Ann. Stat.*, 1979, **7**(1), 1.
20. L. Breiman and A. Cutler, *Random Forests: classification/clustering*, http://www.math.usu.edu/~adele.
21. B. Efron and R. Tibshirani, *An Introduction to the Bootstrap*, Chapman & Hall, New York, 1993.
22. M. R. Chernick, *Bootstrap Methods: a Guide for Practitioners and Researchers*, Wiley, Hoboken, NJ, 2008.

Discovery-based Studies of Mammalian Metabolomes with the Application of Mass Spectrometry Platforms

WARWICK B. DUNN,*[a,b] CATHERINE L. WINDER[b] AND KATHLEEN M. CARROLL[b]

[a] Centre for Advanced Discovery & Experimental Therapeutics (CADET), Central Manchester NHS Foundation Trust and School of Biomedicine, University of Manchester, Manchester Academic Health Sciences Centre, York Place, Oxford Road, Manchester, M13 9WL, UK; [b] Manchester Centre for Integrative Systems Biology and School of Chemistry, Manchester Interdisciplinary Biocentre, University of Manchester, 131 Princess Street, Manchester, M1 7DN, UK
*Email: w.dunn@bham.ac.uk

6.1 Introduction

Mass Spectrometry is focused on the measurement of the physical property of mass, of either atomic or molecular species. More specifically, the mass-to-charge ratio (m/z) is determined. However, metabolites, by definition, are of a low molecular weight (MW) (typically less than 1500 Da,) and are, in general, only capable of sustaining a single ionic charge; therefore, their mass (m) is directly measured. In contrast to other analytical platforms employed in metabolomic studies, mass spectrometry does not utilise the

Issues in Toxicology No. 21
Metabolic Profiling: Disease and Xenobiotics
Edited by Martin Grootveld
© The Royal Society of Chemistry 2015
Published by the Royal Society of Chemistry, www.rsc.org

absorption of electromagnetic radiation for mass measurements and is therefore described as mass spectrometry rather than mass spectroscopy. A range of technical terminologies are applied in mass spectrometry and are further discussed in this chapter. These are listed with definitions in Table 6.1.

Table 6.1 Terminologies and definitions applied in mass spectrometry and of relevance to discussions in this chapter.

1. Mass	A physical property that defines the amount of matter (including sub-atomic particles protons, neutrons and electrons) contained within an object. Sub-atomic particles have unique accurate atomic masses and each elemental isotope has a distinctive combination of each of these particles. Elemental isotopes have a unique atomic mass and combinations of these isotopes form molecular species, distinguishable by their accurate masses. The mass of isotopes is normalised to carbon-12 (^{12}C), which has a mass of 12.000000 Da. A proton (^{1}H) has approximately one-twelfth the mass of ^{12}C (1.003465). The table below describes the accurate atomic masses and isotope abundances of the most common elements.

Element	Isotope	Atomic weight (Da)	Natural abundance (%)
Hydrogen (H)	^{1}H	1.007825	99.99
Carbon (C)	^{12}C	12.000000	98.93
	^{13}C	13.003355	1.07
Nitrogen (N)	^{14}N	14.003074	99.64
	^{15}N	15.000109	0.36
Oxygen (O)	^{16}O	15.994915	99.76
	^{17}O	16.999132	0.04
	^{18}O	17.999161	0.21
Phosphorus (P)	^{31}P	30.973762	100.00
Sulfur (S)	^{32}S	31.972071	94.99
	^{34}S	33.967867	4.25
Sodium (Na)	^{23}Na	22.989769	100.00
Potassium (K)	^{39}K	38.963707	93.26
Silicon (Si)	^{28}Si	27.976927	92.22
	^{29}Si	28.976495	4.69
	^{30}Si	29.973770	3.09

Accurate atomic masses and relative abundances of commonly occurring isotopes (for further details see http://physics.nist.gov/PhysRefData/Compositions/index.html).

2. Isotope	Many chemical elements have more than one isotope, and each isotope contains the same number of protons and electrons but a different number of neutrons. Isotopes of the same chemical element have different masses normally separated by approximately one mass unit. The natural abundance of each isotope and the isotopic pattern is characteristic of the chemical element and can be used to assist metabolite identification. Natural abundance isotopic compositions are shown in the table above.

Table 6.1 (*Continued*)

3. Nominal and monoisotopic mass	The nominal mass of a metabolite is the sum of all elemental masses calculated using the integer mass of the most abundant isotope of an element, though other isotopes can also be used. For example, the nominal mass of glucose ($C_6H_{12}O_6$) is 180 Da. The monoisotopic mass of a metabolite is the sum of all elemental masses calculated using the nominal or accurate mass of the lightest stable isotope of an element. The nominal monoisotopic mass of glucose is 180 Da and the accurate monoisotopic mass is 180.0634 Da.
4. Mass resolution	Several definitions of mass resolution are used, depending on the instrument, but it is commonly (though not exclusively) defined as resolution (FWHM) (http:// goldbook.iupac.org/R05318.html), where: Resolution (FWHM) = peak width measured at 50% of its maximum height/mass. Low mass resolution indicates that all metabolites of the same nominal mass cannot be isolated by mass and are detected as one mass peak. High mass resolution indicates that metabolites of the same nominal mass but different accurate masses can be resolved according to mass and detected. High mass resolution allows accurate mass measurements to be achieved, and in so doing increases the confidence or reduces the number of false-positives during the metabolite identification process. The number of metabolites resolved and detected can theoretically increase as mass resolution increases, though resolution is inversely proportional to mass on a range of instruments, so the mass at which the resolution is measured should be described. For example, glutamine and lysine can both be detected by electrospray ionisation instruments in positive ion mode with masses of the protonated ions of 147.0767 and 147.1130, respectively. These two mass ions are not distinguishable using a low resolution mass analyser such as a quadrupole ($R = 100$) but are resolved using a medium resolution mass analyser such as a TOF ($R = 5000$) or a high resolution mass analyser such as a Orbitrap instrument ($R > 60\,000$).
5. Accurate mass measurements	The accurate atomic mass of each elemental isotope is unique. Therefore combinations of different elemental isotopes form molecules of unique accurate monoisotopic masses, with the exception of metabolites of the same empirical formula (structural isomers including leucine/isoleucine and glucose/fructose), which have identical accurate masses and are referred to as isobaric. Accurate mass measurements of metabolites can only be performed on instruments with high mass resolution (generally >4000 FWHM) and they enable the reduction in the number of possible theoretical empirical formulae for a given mass. Appropriate mass calibration protocols should be followed to ensure mass accuracy, and an internal lock-mass or reference mass(es) is often included to increase mass accuracy still further.

Table 6.1   (*Continued*)

6. Mass accuracy	Mass accuracy is defined as the difference between the theoretical and experimentally measured monoisotopic masses, normally calculated with units of ppm. Mass accuracy (ppm) = [(experimentally derived mass − theoretical mass)/theoretical mass] × 1 000 000 High mass accuracy ensures that empirical formulae can be calculated and metabolites putatively annotated with high confidence. In combination with a range of different rules (termed the seven golden rules and which include relative isotope abundances[125,126]) the number of false-positives can be reduced.
7. Chromatography	A range of physical separation methods used to separate metabolites in complex mixtures before on-line coupling to a detection system, commonly a mass spectrometer. Separation is dependent on the interaction of sample components with a stationary phase and a gas or liquid mobile phase. The stationary phase is constrained within a column through which the mobile phase passes and the metabolites are eluted according to their chemical properties, *e.g.* hydrophobicity or boiling point. The differential affinities to each phase for different metabolites provide a mechanism for separation. A range of chromatography techniques are available commercially, including gas and liquid chromatography. Raw data files are constructed of scan number (or retention time) and the Total Ion Current (TIC), from which the mass spectra are derived (for the latter, *m/z* are plotted against ion current for each scan), and Extracted Ion Chromatograms (based on the mass) if required. These data can be visualised in chromatograms.
8. Metabolite Feature	A detected feature comprising two or more reported parameters (accurate mass, retention time, retention index, EI mass spectrum). A single metabolite can be reported as multiple features and therefore the application of the term metabolite feature allows the discrimination between a metabolite and its multiple products which are detected. In GC-MS, chemical derivatisation applying trimethylsilyl derivatisation reagents can lead to the synthesis of multiple derivatisation products for a single metabolite. In HPLC-MS with electrospray ionisation, a metabolite can form multiple different ion types (for example, protonated, deprotonated, adducts, fragments, see[127]) with each having the same retention time but different accurate mass. Typically, a greater number of metabolite features are detected than the number of metabolites present in a sample.

Metabolites are composed of the most frequently observed elements carbon, hydrogen, nitrogen, oxygen, phosphorus and sulfur. The array of differing combinations of elements (referred to as the molecular or empirical formula), and their associated masses combined with their different atomic arrangements in space, all provide a complex collection of potential metabolites. Mammalian biofluids, cells and tissues contain 100 s–1000 s of selected metabolites. Mass spectrometry is a powerful analytical platform applied to investigate these metabolomes, and offers many advantages (depending on the experimental strategy applied), including high sensitivity coupled with the ability to detect 100 s–1000 s of metabolites in a single sample, high selectivity and specificity and multiple routes to identify the chemical structures of biomolecules therein (accurate mass, gas-phase fragmentation experiments, relative isotopic abundance measurements plus others).

Advances in a range of scientific disciplines have markedly progressed in view of technological advances, including the invention of mass spectrometry (as described recently[1]) and its subsequent development. These include the interfacing of chromatography to mass spectrometers *e.g.* gas chromatography (GC)[2] and subsequently high-performance liquid chromatography (HPLC),[3,4] improvements in ion source and mass analyser design (for example electrospray ionisation,[5,6] TOF mass analysers[7,8] and the Orbitrap mass analyser[9]), and increases in mass resolution and mass accuracy. All these developments have been instrumental in driving forward instrumental and biologically based advances in metabolomics.

The initial definitions of both the metabolome and metabolomics were described in the late 1990s,[10,11] and further defined in 2002.[12] However, the initial development of metabolomics occurred 20 years before and were mass spectrometry driven. In the late 1960s and early 1970s, the Horning group employed the rapidly developing technique of GC-MS to collect the first reported multicomponent metabolic profiles of mammalian biofluids, and suggested that profiles could be used to define normal and pathological states.[13] During the late 1970s, Sauter and colleagues at BASF developed strategies for obtaining metabolic profiles of plant extracts for the classification of herbicidal modes of action.[14]

Developments in instrumentation, together with the rapid growth of computational power and the affordability of these technologies during the 1990s resulted in a 'second dawn'. Three publications in 2000 and 2001 initiated an increase in developments and applications. These employed gas chromatography–mass spectrometry (GC–MS) and chromatographic deconvolution software.[15–17] This strategy increased the number of metabolite peaks detectable, from initially over 300 to subsequently more than 1000.[18,19] Comprehensive GCxGC-ToF-MS has increased this number to even higher values,[20,21] and the requirement for instrument optimisation has been demonstrated.[19,20] The complementary technique of HPLC-MS, and the recent introduction of systems operating with sub-2 µm chromatographic particles capable of withstanding pressures of up to 15 000 psi [[Ultra Performance Liquid Chromatography (UPLC), or Ultra High Performance Liquid Chromatography (UHPLC)], have reported the detection of thousands of metabolite features.[22] These advances have provided a much wider scope for the applications of mass spectrometry in metabolomics, including metabolic profiling, targeted analyses and the imaging of cells or tissues,[23] with a substantial increase in the volume of publications relating to metabolomics being observed from the beginning of the twenty-first century.[23,24] However, there are current 'bottlenecks' and limitations associated with the applications of this technique, including inherent instability in large-scale studies (which are discussed and solutions provided in Section 6.5), and the ability to identify, in an automated manner, all detected metabolites in holistic metabolic profiling studies. These have been discussed in detail previously[23]). Novel solutions for metabolite identification in metabolic profiling studies have been reported recently.[25–29]

Three generalised experimental strategies which employ mass spectrometry platforms are applied in metabolomics.[23] Metabolic profiling (or metabolite profiling or untargeted analysis) focuses on the holistic study of a metabolome, where relative quantification data are collected for 100 s–1000 s of metabolites.[30–33] No calibration curves are constructed to provide absolute quantification, and sample preparation is minimal in order to eliminate the loss of specific metabolite classes in sample preparation processes and hence minimise chemical bias. However, analytical bias is introduced, since no single analytical platform is able to detect all metabolites in a metabolome, and the application of a combination of analytical platforms (for example, GC-MS, UPLC-MS and Nuclear Magnetic Resonance (NMR) spectroscopy are the three most frequently applied platforms) is recommended in order to maximise coverage of the detected metabolome. Mass spectrometry provides the ability to differentiate between the multitude of metabolites based on accurate masses and gas-phase fragmentation experiments, and therefore 100 s–1000 s of metabolite features are detected; notwithstanding, the chemical identification of these features is a current major limitation of the strategy. Typical platforms applied include direct infusion (or injection) mass spectrometry (DIMS), GC-MS, HPLC-MS and associated platforms (for example, UHPLC or UPLC), and capillary electrophoresis-mass spectrometry (CE-MS).

Targeted analysis provides absolute quantification data related to a small number of metabolites (typically fewer than ten) of known biological interest.[34] The sample preparation required is, however, extensive, but necessary to separate analytes of interest from the sample matrix, and mass spectrometers which provide high specificity and sensitivity are applied, the most common being the triple quadrupole mass spectrometer (QqQ or QQQ). Calibration curves are constructed with authentic chemical standard solutions of different concentrations analysed on the same analytical platform in the same analytical batch. These are then used to calculate the absolute concentration of metabolites. The inclusion of internal standards (for example, isotopic analogues such as $^{13}C_6$ glucose to quantify $^{12}C_6$ glucose) are recommended to compensate for analytical or systematic variation.

A third strategy, which the authors define as semi-targeted analysis, combines the advantages observed in metabolic profiling and targeted analysis.[35,36] Here, relative or absolute quantification is performed for 10–300 metabolites, applying minimal sample preparation so as not to bias metabolites detected, and using QQQ mass spectrometers to provide specificity and sensitivity to the analytical method. The metabolites to be detected are pre-defined (based on authentic chemical standard availability and biological knowledge, where available) and the time required to convert analytical data to 'biological knowledge' is low for this reason: no time-consuming metabolite identification process is required such as that for metabolic profiling.

Metabolic profiling and semi-targeted analysis are inductive or hypothesis-generating strategies.[37] They are applied to define new discoveries related to identifying novel biomarkers or new molecular pathophysiological mechanisms. Indeed, the metabolites or areas of metabolism of biological importance are not known prior to the collection of analytical data. Instead, a robust experiment is designed to collect analytical data from which the metabolites of biological interest are defined subsequent to data acquisition. A hypothesis is constructed from these data; such hypotheses can be tested or validated applying targeted analysis methods.

In mammalian systems, mass spectrometry is applied for the profiling of metabolomes to define molecular pathophysiological mechanisms of disease (for example, refs 38,39) or drug efficacy or toxicity (for example, ref. 40), to discover prognostic or diagnostic biomarkers of disease (for example, refs 31,41) or to discover biomarkers applied in drug toxicity or efficacy studies (for example, refs 42,43). A range of these applications are described in further chapters of this volume. Further studies focus on the spatial distribution of metabolites, which is typically destroyed during sample preparation for metabolic profiling or targeted analysis. Here, techniques including Secondary Ion Mass Spectrometry (SIMS),[44,45] Matrix-Assisted Laser Desorption Ionisation (MALDI)[46] or Direct Electrospray Ionisation (DESI)[47,48] are applied for imaging.[49] These studies provide complementary data on the location and potential mechanistic roles of metabolites.

For a more detailed description of the principles of mass spectrometry and its applications in metabolomic studies, there are a number of relevant references[23,24,50–54] Of these, the two books (one written by Downard and one by Gross) are highly recommended, although there are many other suitable textbooks available. Detailed descriptions are beyond the scope of the current chapter. The objectives of this contribution are (i) to introduce the concepts and instrumentation applied in metabolomics studies and (ii) to discuss the role mass spectrometry will have in future studies. The second objective will specifically discuss improvements in experimental design which are required to allow the traverse from small-scale (n = 10 s) to large-scale (n = 100 s–1000 s) studies, and also include the use of Quality Control (QC) samples and the application of robust Quality Assurance (QA) procedures.

6.2 Mass Spectrometry Instrumentation

Mass spectrometers have changed little in their general principles of operation (ionisation, mass analysis based on *m/z* and detection) since the first mass spectrometer was developed more than 100 years ago. However, the

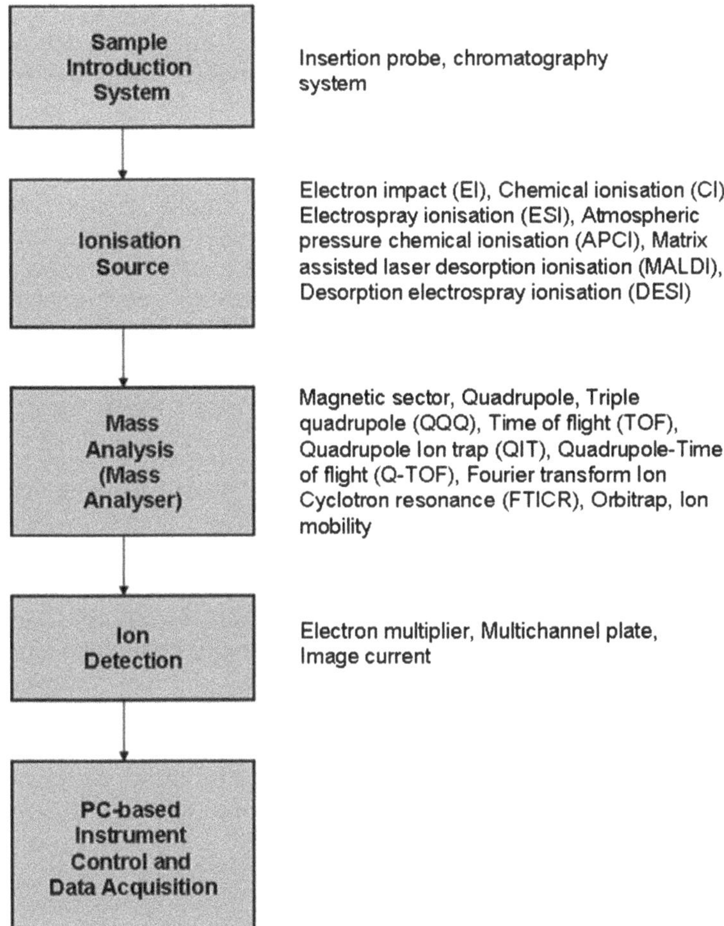

Figure 6.1 The five simplified components of a mass spectrometer and examples of the different types of sample introduction, ionisation sources, mass analysers and detection systems commercially available.

variety of different mechanisms of ion formation, mass analysis and detection have provided a range of different mass spectrometer components and configurations. The five basic components of a mass spectrometer, and examples of the different types of sample introduction, ionisation sources, mass analysers and ion detection systems available, are shown in Figure 6.1.

6.2.1 Sample Introduction

Gas, liquid or solid samples are introduced using a range of different strategies, including chromatography and direct infusion/direct insertion into the ion source. The operating pressure of the ionisation source has to be considered during sample introduction. For example, if the ionisation source is operating at a vacuum pressure (typically 10^{-6} atmospheres), then a high gas flow-rate would result in a significant increase in the vacuum pressure, which would be detrimental to instrument operation and the provision of robust analytical data. Therefore, GC-MS gas flow-rates of 0.5–2.0 ml min^{-1} are typically applied and allow the vacuum pressure to be maintained at an appropriate operating level. However, the introduction and vaporisation of typical liquid chromatography mobile phase flow-rates would overwhelm vacuum pressures typically applied. The invention of atmospheric pressure ionisation (API) sources was essential to allow routine on-line coupling of liquid chromatographs with mass spectrometers. The most frequently applied systems are discussed in Section 6.3.

6.2.2 Ion Formation

Positively or negatively charged ions are created from a gas, liquid or solid sample. The formation of ions is essential, since the subsequent control of ion trajectories and mechanisms of m/z separation and analysis are based on the influence of electrical or electromagnetic properties on charged species. Ion formation occurs in an ion source, traditionally operated under vacuum pressure but more recently API alternatives have become available, including electrospray ionisation (ESI) and atmospheric pressure chemical ionisation (APCI) sources. All ionisation techniques create a molecular (or quasimolecular) ion (M), which can be analysed directly or after gas-phase fragmentation. Some ionisation processes induce fragmentation of molecular (or quasimolecular) ions (for example, electron impact ionisation), whereas alternative fragmentations have to be induced (for example, collision-induced dissociation, CID). Brief descriptions of the range of ionisation sources applied in metabolomics are provided in Table 6.2.

6.2.3 Mass Ion Separation According to Mass-to-charge Ratio

The third stage of the mass spectrometry experiment is the separation of ions according to their mass-to-charge ratio (m/z). For Time-of-Flight (TOF) mass analysers, ions are separated in time so that they reach a single detector at different times, the time from source to detector representing a measure of the m/z ratio. Quadrupole mass analysers allow ion packets of different m/z values to traverse the mass analyser sequentially in order to construct a mass spectrum.

Alternatively, some instruments are based on ion storage in a single space with the separation/detection of ion packets of different m/z ratios. These include Quadrupole Ion Trap (QIT), Linear Ion Trap (LIT) and Fourier-Transform Ion Cyclotron Resonance (FTICR) instruments, where ions are separated spatially. In QIT and LIT mass analysers, the orbits of ions are sequentially destabilised to move ions to the external detector in m/z 'packets'. In FTICR instruments (including the Orbitrap mass analyser), the image current on a detector plate is detected; this is produced by the cyclic

Table 6.2 Description of the commonly applied ionisation sources in metabolomics.

Electron Impact (EI)	Frequently applied technique in GC-MS where gaseous molecules are bombarded with energetic electrons (at an energy of 70 eV) to form positively charged radical ions. These ions are highly energetic and fragment *via* covalent bond dissociation in a highly reproducible pattern characteristic of the metabolite structure. The fragmentation pattern allows identification of metabolites by comparison with mass spectral libraries or mass spectrum interpretation.
Chemical Ionisation (CI)	Less commonly applied technique in GC-MS where ionisation is performed in an EI source into which a chemical reagent gas (for example, methane or ammonia) is infused. The reagent gas, present at a significantly higher concentration than metabolites, is ionised and metabolite ionisation occurs by ion or charge transfer. Ionisation does not result in significant fragmentation of the molecular ion producing a less complex mass spectrum than EI. Fragmentation can be performed by subsequent gas-phase fragmentation.
Electrospray Ionisation (ESI)	Most common ionisation technique available to interface liquid chromatography and mass spectrometry instruments. The source is an electrochemical cell where metabolites present in a liquid solution are pumped through a metal needle held at high kV voltages in which ions form by addition or removal of protons or adduct ions. Ions are subsequently transferred to the gas phase by liquid nebulisation and droplet desolvation. The source operates at atmospheric pressure and minimal fragmentation is normally detected. Fragmentation of molecular ions can be performed in-source or by gas-phase fragmentation. Two mechanisms are proposed for ionisation, Ion Evaporation Model (IEM) and Charged Residue Model (CRM).
Atmospheric Pressure Chemical Ionisation (APCI)	Another technique employed to interface liquid chromatography and mass spectrometry. This is a form of chemical ionisation operating with a nebulised and desolvated liquid using an atmospheric pressure corona discharge. Ionisation of solvent molecules occurs followed by subsequent ion or charge transfer to metabolite molecules present at lower concentrations in relation to solvent. Minimal fragmentation of molecular ions is observed, though degradation by heat (up to 500 °C) may occur.
Atmospheric Pressure Photo Ionisation (APPI)	A complementary tool to ESI and APCI for ionisation of low polarity metabolites at atmospheric pressure. Photons from discharge UV lamps cause

Table 6.2　(*Continued*)

	electron ejection from metabolite molecules. Minimal ion fragmentation is observed.
Matrix Assisted Laser Desorption Ionisation (MALDI)	Applied most frequently to the analysis of large biomolecules compared to metabolites and is a soft ionisation technique producing minimal fragmentation. The sample is diluted in a low molecular weight matrix and dried. Absorption of UV laser energy by the sample-matrix composite results in desorption, ionisation of matrix components (present at a higher concentration than sample molecules) and subsequent charge transfer to sample molecules. Applied for profiling and imaging.
Secondary Ion Mass Spectrometry (SIMS)	SIMS is applied for imaging of metabolomic samples and operates by sputtering the surface of a sample with a primary source of ions. This provides ionisation (and fragmentation) and the secondary beam of ions from the sample are passed for mass analysis.
Desorption Electrospray Ionisation (DESI)	An ambient ionisation source which provides ionisation of gases, liquids and solids with minimal sample preparation. An electrically charged mist of solvent molecules (*e.g.* water/ methanol) is directed at the sample to provide ionisation. Ions are transferred from atmospheric pressure to the mass spectrometer. Ionisation of metabolites operates by charge transfer with minimal fragmentation observed.
Direct Analysis in Real Time (DART)	Ambient ionisation method which can operate with no or minimal sample preparation. The ionisation mechanism is different from DESI and operates through energy transfer from an excited gas molecule to the metabolite to cause electron ejection and formation of a radical cation.

precessional orbits of ions of different m/z ion packets (a composite of the frequencies of orbital precession of all m/z packets). Brief descriptions of the range of mass analysers applied in metabolomics are provided in Table 6.3.

6.2.4　Ion Detection and Data Acquisition

A range of ion detectors are commercially available, and of these the electron multiplier (EM) and microchannel plate (MCP) detect ions from either a relatively focused single ion beam or defocused ion beams. These are applied for quadrupole, QIT, LIT and TOF instruments. Ion image currents are detected for FTICR and Orbitrap instruments, and detected ion or image currents are generally amplified prior to their final conversion to a mass spectrum, in order to provide higher sensitivity.

Table 6.3 Description of the commonly applied mass analysers in metabolomics.

Linear Quadrupole (Q)	Four precisely spaced parallel rods construct the mass analyser, opposite pairs are electrically connected. DC and RF potentials are applied to the rods which act as a mass filter, normally by increasing the RF and DC voltages but keeping their ratio constant. At a given RF/DC ratio ions of a specific m/z are transmitted to the detector and ions of lower and higher mass are lost by collisions with the quadrupoles. The mass analyser is cheap and simple to operate and nominal mass resolution is normally achievable.
Quadrupole Ion Trap (QIT)	This mass analyser is composed of a ring and two outer cap electrodes. With the application of DC and RF voltages and a helium gas bath, ions are trapped in stable oscillatory orbits within the trap, dependent on the m/z. Destabilisation and ejection of ions to a detector enables ion detection, normally with an increasing RF/DC field to perform a mass scan. The mass analyser can be employed for multi-stage mass spectrometry experiments by selection of a precursor ion, ejection of all other m/z ions and gas-phase precursor ion excitation in a collision-induced dissociation mechanism.
Linear Ion Trap (LIT)	Linear Ion Traps operate in a similar manner to QIT but ion trapping is performed in a two-dimensional quadrupole field, imposed by sets of linear quadrupoles, instead of a three-dimensional field. These mass analysers are larger in construction and can therefore trap larger ion populations which result in an improved sensitivity.
Triple Quadrupole (QQQ)	A system used for tandem mass spectrometry experiments where two quadrupoles (Q1 and Q3) are separated by a collision cell (Q2) operating at a higher pressure and containing a collision gas. Ions transferred through Q1 are accelerated into Q2 to produce collision induced dissociation (CID) followed by mass analysis in Q3. A number of experiments can be performed; product ion scanning, precursor ion scanning, neutral loss scanning, single and multiple reaction monitoring (SRM and MRM).
Time of Flight (TOF)	A simple mass analyser which applies a constant accelerating voltage to all ions and the resulting flight time to reach the detector is dependent on m/z: lower m/z ions reach the detector before higher m/z ions. The use of one or more ion mirrors (reflectron) focuses ions of the same m/z but different kinetic energies and increases mass resolution. Higher specification instruments provide mass resolutions of 4000–40 000 FWHM.

Table 6.3   (*Continued*)

Quadrupole-Time of Flight (Q-TOF)	A hybrid instrument combining quadrupole and TOF mass analysers, with a higher pressure collision cell separating the two mass analysers, to provide tandem mass spectrometry experiments. The added benefit is that product ion mass spectra are detected at high mass resolution with accurate mass measurements, unlike the case for triple quadrupole instruments.
Fourier Transform Ion Cyclotron Resonance (FTICR)	Measurement of the image current of RF excited ions orbiting in a high strength magnetic field ($>$7 Tesla) at low pressures (10^{-10} atmospheres). The orbital frequency is dependent on m/z and the detected time-dependent image current can be Fourier Transformed to the frequency domain which is related to m/z. This technique offers the highest achievable mass resolution (R $>$ 100 000) and mass accuracies (ppb) currently achievable.
Orbitrap	A recent addition to the toolbox which employs a mass analyser composed of an outer barrel-like electrode and a coaxial inner spindle-like electrode. An electric field is applied between the electrodes and when ions are introduced, the combination of electrical attraction and centrifugal forces traps ions in an orbiting arrangement. Ions oscillate around and along the central electrode and the frequency of oscillations along the central electrode is dependent on the m/z. This is independent of ion energy. A Fourier Transform is employed to convert to the frequency domain. This is another high mass resolution (up to 450 000) and mass accuracy ($<$1 ppm) instrument. In reality this is a hybrid instrument with a linear ion trap (LTQ) used to store ions from an API source before periodic introduction to the Orbitrap *via* a C-trap which acts to collisionally dampen and focus ions.

6.2.5   Instrument Control and Data Processing

Today's mass spectrometers are complex instruments with many operating parameters that require continual alteration and detection for feedback control. The quantity of data produced in high-throughput and/or large studies is immense, and requires data acquisition, storage and analysis. These processes are integrated with a PC-based instrument control and data acquisition/processing software. The vast volumes of data produced require computational power for acquisition and storage processes. For example, a 60-minute comprehensive GCxGC-MS analysis with an acquisition rate of 125 Hz produces $>$500 MB of data per sample.

Raw data processing is dependent on the mass spectrometer configuration applied, the raw data format and analytical strategy employed. Data formats can be specific for an instrument manufacturer, and may require conversion to a standard format (for example, NetCDF) prior to further data processing on open source software packages. The use of mzML will be a requirement in Systems Biology research in the future. This extensible mark-up language (XML) allows standardisation of mass spectrometry raw data formats, and also annotation of sample information and instrument operating parameters. There is a major requirement in Systems Biology studies to facilitate integration and the searching of multiple XML-based databases using workflows, or Web Services including Taverna.[55]

Targeted and semi-targeted analysis data, in which the metabolites detected are already known, require specific software packages to quantify peak areas for samples and standard solutions, and therefore calculate metabolite concentrations in each sample. These are widely available from instrument manufacturers as a component of instrument purchase.

Metabolic profiling data, where the metabolites detected are not known *a priori*, are complex and require non-biased peak deconvolution or peak integration software to define and quantify peak areas or heights prior to data analysis (for example, see refs 56–58). Chromatographic separation of complex metabolomic samples is not comprehensive, and the co-elution or minimal chromatographic separation of metabolites requires a mathematical isolation approach to separate these metabolites with similar retention times *in silico* on the basis of differences in their mass spectra, retention time and peak shape. All deconvolution software assume a consistent pattern of these measured variables of a metabolite feature across a chromatographic peak, and use this variable to define the peak. The variable for GC-MS is the electron impact mass spectrum and generally for LC-MS is the nominal or accurate mass of a molecular feature. A number of freely available software packages are available. These include XCMS,[57] MZMine,[58] MSFACTS,[59] COMSPARI,[60] MET-IDEA,[61] MathDAMP[62] and MetAlign.[63] A range of instrument vendors and associated suppliers also provide appropriate software packages including MarkerLynx (Waters), MarkerView (AB Sciex), ChromaToF (Leco), SIEVE (ThermoScientific), MassHunter (Agilent) and AnalyserPro (SpectralWorks). A number of other approaches to interrogate raw datasets are also available, mainly based on feature selection strategies.[64–71]

6.2.6 Other Considerations

Mass spectrometers operate under high vacuum pressures, typically pressures of 10^{-6} to 10^{-10} atmospheres in the mass analyser and detector regions, and also potentially in the ionisation source. This is essential to eliminate the majority of ion-molecule collisions present at higher pressures, resulting in ion scattering and a loss of ion path cohesion and sensitivity. A vacuum is maintained by a combination of rotary pumps, which maintain pressures to 10^{-3} atmospheres, and turbomolecular pumps, which maintain pressures at 10^{-3}–10^{-10} atmospheres. Generally, rotary pumps obtain a relatively low vacuum pressure before turbomolecular pumps start to operate in order to increase turbomolecular pump lifetimes and efficiencies. Some older instruments (for example, the Sciex API III) apply cryopumps operating at temperatures of 20 K to cryogenically cool and remove all gases (nitrogen gas freezes at 63 K at one atmosphere pressure). However, these require daily maintenance by thawing and the removal of vapours, and therefore instrument down-time.

6.3 Sample Introduction Systems

As shown in Figure 6.1, there are a range of sample introduction systems coupled with mass spectrometry instruments in metabolomic studies. In discovery-phase studies, five are applied with high frequencies and these are discussed below.

6.3.1 Direct Infusion Mass Spectrometry (DIMS)

Direct infusion (or injection) mass spectrometry (DIMS) introduces a liquid sample directly to an ESI, or less commonly an APCI instrument. This allows rapid profiling (sometimes described as metabolic or metabolite fingerprinting) for high-throughput investigations used for sample classification studies or disease diagnostics. Sample introduction can be automated using autosamplers and pumps, flow injection technologies, infusion pumps or nanoelectropray sources (*e.g.* Triversa Nanomate systems). High mass resolution and mass accuracy instruments (for example, FTICR-MS[72]) are required in order to provide the mass discrimination of metabolites of the same nominal mass but different monoisotopic mass. Other considerations specific to FTICR instruments have also been investigated and optimised for the purpose of maximising mass accuracy and sensitivity.[72]

The associated problem of ionisation suppression, occurring in the electrospray source in highly complex samples, is commonly observed.[73,74] Such samples contain many hundreds of metabolites and other ionic species such as inorganic salts which compete for ion formation in the liquid phase or ion release from liquid to the gas phase. Relative quantification studies are normally performed, and these assume that the sample matrix will not significantly differ between different samples (however, such assumptions are not necessarily satisfied). Therefore, changes in measured ion responses are assumed to be biological, and not a consequence of modifications in sample matrices. It has been shown that for similar sample matrices this assumption holds true.[75]

The identification of metabolites is also difficult without chromatographic separation, since multiple metabolites with the same chemical formula or accurate mass are detected together without prior separation. Methods for data preprocessing, analysis and metabolite identification for DIMS–acquired data have been reported.[27–29,76] Structural isomers have identical empirical formulae and accurate masses, and gas-phase fragmentation is required to potentially differentiate between such isomers. The application of tandem mass spectrometry with QQQ instruments improves sensitivity and selectivity, and is employed efficiently in neonatal screening for metabolic disorders (collectively referred to as 'inborn errors of metabolism'), where there is the requirement for many thousands of samples to be analysed, and therefore a high-throughput strategy is required.[77]

6.3.2 Gas Chromatography-Mass Spectrometry

Capillary gas chromatography operates with a silica capillary, a solid or liquid stationary phase coated on the inner surface of the capillary and a gaseous mobile phase (carrier gas), which flows through the capillary column (typically helium or nitrogen at a flow-rate of 0.5–2.0 ml min^{-1}). The capillary column is typically of 5–60 metres in length and has an internal diameter of 0.18–0.30 mm. The stationary phase film thickness is typically 0.10–0.50 µm.

The chromatography column operates in a temperature-controlled oven at temperatures greater than room temperature. Small volumes (0.5–2 µl) of liquid samples are introduced with a low volume syringe into a heated injection inlet containing an injection liner. The injection inlet operates at high temperatures (typically >200 °C) and is swept with the carrier gas. Rapid vaporisation of the liquid sample occurs on introduction to the injection inlet, and the vapour is transferred to the top of the column by the carrier gas. Two types of injection are routinely performed, split and splitless. Splitless injections introduce the complete vaporised sample onto the top of the column. Split injections separate the vaporised sample with a fraction passing onto the column, and the other fraction being passed to waste through a split vent. The split ratio is controlled, and can vary from 1 : 10 to greater than 1 : 100, and allows low sample volumes to be introduced onto the column, which would not be reproducibly achievable with a splitless injection. Each injection method has differing advantages and limitations, although splitless injections are more common in metabolomics for a number of reasons, including the potential for discrimination and reductions in the detected concentrations of high-boiling-point metabolites.

The majority of metabolites in mammalian metabolomes do not have sufficiently low boiling points to be directly analysed by GC-MS. Hence, chemical derivatisation is commonly applied to reduce the boiling points of metabolites. Oximation followed by trimethylsilylation (TMS) is the most frequently applied process since it provides derivatisation of a wide range of chemical functionalities, and therefore allows profiling of a wide range of metabolite classes including amino acids, organic and fatty acids, sugars and thiols. Carbonyl functional groups are converted to oximes prior to TMS derivatisation, since the direct TMS derivatisation of carbonyl groups is slow and does not proceed to completion. Following oximation, TMS derivatisation is applied to replace active hydrogens which contribute to inter- and intra-molecular hydrogen bonding (the elimination of hydrogen bonding reduces the boiling point of metabolites). A range of derivatisation conditions have been applied.[78-80] Other methods of derivatisation are employed, although less frequently, and are typically more specific. For example, methyl chloroformate derivatisation is specific for amino and non-amino organic acids, phosphorylated organic acids and fatty acid intermediates.[81] Automated systems for chemical derivatisation prior to injection or, alternatively, within the injection liner have been described.[82]

Although liquid samples are the most frequently analysed samples, gas samples can also be analysed (*e.g.* human breath). Here, gas samples are collected onto a sorbent trap (for example, in a clinic) and transferred to the instrument where the metabolites are thermally desorbed and introduced onto the GC column.[83]

The separation of different metabolites is dependent on their distribution between the stationary and gaseous phases. This distribution is controlled by the stationary phase composition, the temperature of the column, the internal diameter of the column, the stationary phase thickness and the carrier gas flow-rate. The most commonly applied stationary phase compositions are 5/95% methyl/phenyl and 50/50% methyl/phenyl, although other stationary phase compositions are available.[34] To optimise chromatographic separations of complex samples containing metabolites with a range of boiling points (including those observed in metabolomics), a temperature ramp is applied, which, in combination with the gas-phase velocity and column dimensions, serves to optimise the separation of metabolites according to their boiling points and chemical selectivities for the stationary phase. This allows appropriate, but not necessarily perfect, chromatographic separations of complex mixtures of metabolites.

Electron impact ionisation is the traditional method employed to ionise gas-phase metabolites introduced from a gas chromatograph in a range of metabolomic applications. The source is composed of a filament which is a thin metallic ribbon or wire (for example, tungsten or rhenium) through which a voltage is applied to create resistive heating. Electrons are thermally emitted from the filament and then accelerated at a constant velocity through a vacuum region, through which the gaseous sample traverses. The electron beam is monitored in a 'closed-loop' approach in order to ensure a consistent electron current. The electrons interact with gas-phase molecules in a quantum-mechanical manner to generate the ejection of a single electron which creates a positively charged radical ion for mass analysis. Only positive ions are created; the process of electron addition can occur, but reaction rates are negligible.

$$M + e^- \rightarrow M^+ + 2e^-$$

An electron energy of 70 eV is universally employed. The ionisation mechanism incorporates an excess of energy to the molecular ion which, since this ion is present in a vacuum, cannot be lost by ion-molecule collisions, and therefore energy loss results from the fission of covalent bonds and fragmentation of molecular ions in a pattern characteristic of the structure of the metabolite (this allows reproducible fragmentation patterns to be employed for identification purposes). Fragmentation can be characteristic of the derivatisation process. For example, trimethylsilylation derivatives infrequently produce a detectable molecular ion, but more commonly lose a methyl group, and an ion at *m/z* 15 lower than the mass of the molecular ion (M-15 ion) is detected. All these processes occur in the ion source which operates at temperatures up to 300 °C, and ions are extracted from the source to the mass analyser.

Electron impact ionisation is referred to as a 'hard' ionisation technique which provides excess energy to the molecular ion and results in fragmentation prior to mass analysis. Other ionisation techniques discussed are referred to as 'soft' ionisation techniques, where fragmentation is minimal since excess energy is lost by ion-molecule collisions rather than covalent bond fission. These include chemical ionisation (CI).

The source design for CI is similar to that which is used for electron impact ionisation, with the exception that a reagent gas is infused to maintain a pressure of approximately 10^{-4} atmospheres. The reagent gas (methane, ethane or ammonia) is in excess to other sample molecules and is ionised by the electron impact ionisation mechanism, and subsequent charge or ion transfer occurs between reagent gas ion and neutral sample molecules *via* ion molecule collisions. When ionised, methane creates a positively charged radical ion from which the transfer of a proton to a separate methane molecule provides a CH_5^+ ion as an efficient proton donor available for the protonation of metabolites; the resulting spectra are less complex with minimal fragmentation. Ion sources are usually more 'closed' to avoid overloading the mass analyser and detector regions of the mass spectrometer.

$$CH_4 + e^- \rightarrow CH_4^+ + 2e^-$$

$$CH_4^+ + CH_4 \rightarrow CH_3 + CH_5^+$$

$$CH_5^+ + M \rightarrow CH_4 + [M+H]^+$$

GC-MS offers a number of advantages. It is a sensitive technique which allows the detection of metabolites at physiological concentrations (mM to µM, generally), and separation is observed with high chromatographic resolution in order to maximise the biological information content of any given sample. Only in a few minor cases is any reduction in ionisation or detector efficiency observed, such as when a minor component is present in a large excess of a major component, unlike the problems of ionisation suppression observed in electrospray ionisation.[34,84] However, the technique does provide bias in the metabolites which are detected (with or without chemical derivatisation). Central metabolism (including the glycolysis, citric acid cycle and amino acid metabolic pathways) is well represented in detectable metabolites, although higher MW lipids cannot be detected. A range of mass analysers can be applied including TOF, fast quadrupole and FT instruments. With high chromatographic resolution now available, these mass analysers should allow multiple scans to be collected per second, or at acquisition rates greater than 1 Hz. Application of EI ionisation sources provides a near universal method of ion formation, independent of physicochemical properties. Both polar (for example, amino acid) and non-polar (for example, alkane) metabolites can be ionised with the application of EI ionisation sources (alkanes are not easily ionised with API ionisation sources).

6.3.3 Comprehensive GCxGC-MS

Although GC-MS plays an extensive role in metabolomics, issues related to the complexity of the samples and the chromatographic resolution achievable frequently arise. In high-throughput applications with complex samples and short analysis times, chromatographic resolution is highly compromised and, even when applying longer analysis times, the perfect chromatographic resolution and separation of all components is not achievable. Chromatographic deconvolution software can provide *in silico* separation of co-eluting species, but this process is not perfect for all applications.

A method which provides an increased chromatographic resolution is preferential. Comprehensive GCxGC-MS is one option being applied, and two columns, each with differing but complementary stationary phases, are used to increase chromatographic resolution. This is generally performed in a continuous approach (comprehensive GCxGC-MS), though collection and analysis of specific fractions can also be performed ('heart-cutting').

Column 1 is of similar characteristics to that employed for GC-MS, whereas the second column is normally less than 2 metres in length and has a thinner stationary phase film thickness. Column 1 provides analysis times similar to GC-MS whereas, depending on the length of column 2, retention times of less than 10 s can be achieved. Samples are collected from column 1 at periodic stages in a modulator using cryogenic modulation with a liquid nitrogen jet cryofocusing 4–10 s fractions of eluant from column 1, and then releasing the effluent fraction as peaks of width <0.2 s onto column 2 *via* the introduction of a hot nitrogen jet. This can be performed twice in a series in order to increase peak focusing. Other types of modulators may become commercially available in the future, which will eliminate the requirement for liquid N_2 and a further gas supply, and also reduce operational costs.[85]

The different stationary phases applied in columns 1 and 2 create orthogonal and complementary separations in dimensions 1 and 2. Column 1 is usually a non-polar column which separates according to boiling point, whilst the second column has a greater polar fraction composition and separates according to polarity. Therefore, metabolites of similar volatility but differing polarities can be resolved. Careful consideration has to be made regarding the 'on-column' volume in view of the shorter length and thinner stationary phase applied for column 2. Overloading of column 2 is achievable where that on column 1 is not observed. Columns 1 and 2 are normally operated in two separate ovens, with column 2 mirroring the temperature of column 1 with a small positive temperature offset. Both GC-MS and comprehensive GCxGC-MS techniques require high computing power, including chromatographic deconvolution software. For reproducible results, a high acquisition rate is required (generally >10 and >100 Hz for GC and comprehensive GCxGC, respectively). However, narrow peak widths can be observed in capillary GC applications, and the requirement for multiple mass spectra (>15 data points) across each chromatographic peak requires a fast acquisition rate mass analyser for efficient peak definition and chromatographic deconvolution. At present, TOF instruments are the only mass analysers available that can provide these acquisition rates whilst concomitantly maintaining sensitivity.

Comprehensive GCxGC-ToF-MS offers many advantages. These include improved chromatographic resolution, higher sensitivity, increased numbers of detected metabolites and improved mass spectral quality, advantages which arise from the chromatographic separation of background chemicals from metabolites of interest. Limitations revolve around the greater complexity of method development and longer analysis times employed when compared to those of GC–MS. Currently there are limited published applications, although a growth in applications of this technique is anticipated.[21,68,86–88]

6.3.4 High Performance Liquid Chromatography-Mass Spectrometry

The requirements for high performance liquid chromatography (HPLC) are different from those for gas chromatography. The chromatographic column is composed of stainless steel tubes containing a stationary phase, through which a liquid mobile phase is pumped under pressure (up to 6000 psi on HPLC systems). The column is of wider internal diameters (typically, 1.0–4.6 mm), but shorter lengths (typically, 2–25 cm) than those of GC columns. Columns are packed with small spherical silica particles of diameter 1.7–5 µm (or less commonly as a monolithic polymer structure) on which the stationary phase is coated or chemically bound to the surface. Capillary LC using monolithic silica columns has also been reported in metabolomics, but at significantly lower frequencies of application. When coupled to a nano-LC system interfaced to a mass spectrometer, limits of detection should be lower than that of conventional HPLC.[89,90]

The components of a liquid chromatography system include the mobile phase reservoirs, pumps, injector, column, detector and a column oven. Mobile phases are delivered from the reservoir to the column by high-performance pumps operating against high back pressures at flow-rates of 0.1–2 ml min^{-1}. Binary pumping systems allow the use of two solvents, whereas quaternary pumping systems allow the use of four solvents and a greater degree of solvent mixtures for more extensive method development.

Most methods operate with a gradient elution system in which the relative proportions of two (or more) solvents are adjusted in a linear, curved or stepped gradient. In reversed-phase applications, solvent A is generally an aqueous solution and solvent B is an organic solvent, commonly methanol or acetonitrile. Mobile phases can be supplemented with modifiers to improve chromatographic performance (for example, ion-pairing reagents such as tetra-*t*-butyl ammonium bromide) or ionisation (for example, formic acid or ammonium acetate). Liquid samples of volumes 5–20 µl are typically introduced onto the column.

Chromatographic separation is achieved through the appropriate choice of column, stationary- and mobile-phases. The column length and internal diameter can influence the number of theoretical plates available. In general, higher internal diameters may result in a lower sensitivity caused by peak broadening. Both stationary- and mobile-phases influence the retention time of sample components, and for complex samples a gradient elution program is often utilised to provide optimal chromatographic separation for metabolome samples containing 100 s–1000 s of metabolites. Elevated temperatures can also be used to lower the back pressure by reducing solvent viscosity, and this process has been shown to improve chromatographic separation.[91]

A wide range of stationary phases are commercially available, and provide for the separation and detection of a wide variety of metabolite classes. HPLC-MS provides some bias in the classes of metabolites detectable; this depends on the ionisation source applied. However, metabolites with a wider diversity of molecular weights and boiling points can be detected when compared to that achievable by GC-MS, and include high-molecular-mass lipids (for example, glycerophosphatidylcholine metabolites).

C_{18} or C_8 stationary phases are commonly employed in reversed-phase applications, with chromatographic separations based on hydrophobicity/hydrophilicity. Here, a gradient elution is applied to increase the % composition of an organic solvent expressed relative to an aqueous solvent (for example, begin with a 95/5% ratio of water/methanol, and increase the percentage of methanol during the analysis to 5/95% water/methanol). Polar metabolites elute quickly, whereas apolar ones (including many lipids) are retained on the column and elute at later retention times. Many applications of these columns have been reported.[30,92,93]

In comparison, normal-phase chromatography achieves separations on silica particles generally with no stationary phase attached. Here, the surface is composed of polar silanol groups, and non-polar mobile phases are employed, including hexane and iso-propanol. These solvents are not easily introduced into an electrospray ionisation source. Hydrophilic Interaction Liquid Chromatography (HILIC) is a derivation of normal phase chromatography, although it operates with relatively polar mobile phases and polar surfaces (typically, silica or silica-bound with polar groups such as amino, amide, cationic or anionic groups). The gradient elution commences with a high percentage of organic solvent and proceeds through an increasing aqueous gradient. A limited number of applications have been reported, although they offer an alternative to the GC-MS analysis of polar metabolites.[94–96]

The introduction of liquid chromatography systems with sub-2 µm particle columns operating at higher than normal linear velocities, and coupled with pumps capable of withstanding back-pressures of up to 15 000 psi have had a major impact on metabolomic studies. Indeed, improvements in sensitivity, chromatographic resolution and analytical throughput have all been observed. The first such system was introduced by Waters, and termed Ultra Performance Liquid Chromatography (UPLC™), and since then other systems have been commercially introduced (defined as Ultra High Performance Liquid Chromatography UHPLC), and a wide range of applications for this technique have been described.[30,31,92,97–101]

2D-HPLC-MS approaches might also be considered when the complexity of the sample is such that it cannot be resolved through a 'normal' HPLC-MS run (two orthogonal separations enables improved separations, as is the case for comprehensive GCxGC-MS). Traditionally this type of analysis has been reserved for 'shotgun' style proteomics experiments; although the approach is expected to gain popularity in metabolomics,[102,103] even though it is more technically demanding than comprehensive GCxGC-MS and requires a customised HPLC system.

Electrospray ionisation (ESI) is the most frequently applied ionisation source for coupling HPLC/UPLC and MS. ESI is a soft ionisation process in which ions created in solution are transferred to the gas phase prior to mass analysis.[104] The analyte of interest is dissolved in an appropriate solvent and delivered to the mass spectrometer *via* a metal capillary. Volatile additives such as acids (for example, formic acid) or bases (for example, ammonium acetate) are often included in the mobile phase in order to aid ionisation. The analyte of interest may, in solution, form an ion *via* the addition of a proton or a cation (such as Na^+), or exist in an anionic form through the loss of a proton (the formation of ions is dependent on the acidity or basicity of the metabolite). For example, organic acids are acidic and therefore are typically detected as deprotonated ions. However, unexpected metabolite features may also be detected. For example, the metabolite feature detected with highest intensity during the analysis of glucose is not the deprotonated ion in negative ion mode (as might be expected), but is the sodium adduct ion detected in positive ion mode.

The application of a high voltage to the capillary, at a positive or negative potential, leads to the formation of an aerosol *via* the production of a 'Taylor cone'. This term is used to describe the effect of exposing a conductive liquid to an electric field, and was first described in 1964.[105] N_2 gas is used to nebulise the liquid flow and provide a heat source for droplet desolvation, and the droplets are dispersed as the solvent evaporates and the charged analyte molecules repel each other. However, opinion remains divided regarding the final stages of ion production, with two competing theories: the Charged Residue Model (CRM)[106] and the Ion Evaporation Model (IEM).[107] The former suggests that there are a series of evaporation and droplet fission cycles, until the last solvent molecules are removed, whilst the latter purports that when a certain droplet radius is reached, an increase in field strength at the surface leads to single ion expulsion.

Nanoelectrospray (nanoESI) operates at flow-rates of 200–1000 nL min^{-1}, produces smaller droplet diameters and results in an improved ionisation efficiency. Nanoelectrospray requires an LC system capable of delivering the reduced flow-rates, and is generally currently used for studying macromolecules such as peptides.

6.3.5 Capillary Electrophoresis-Mass Spectrometry

Capillary Electrophoresis (CE) is the collective name for a group of separation techniques, of which the most common is Capillary Zone Electrophoresis (CZE). In CE, narrow-bore fused-silica capillaries are employed to separate samples in high electric fields, on the basis of charge, size and hydrophobicity; highly polar and charged analytes are particularly suited to this application. The capillary tubes have a high surface-to-volume ratio, leading to heat dissipation, and also enabling higher voltages to be applied than has routinely been the case for traditional electrophoresic techniques.

Capillary Electrophoresis-Mass Spectrometry (CE-MS) is a hyphenated technique, which offers several advantages. This technique is highly sensitive, and suitable for the analysis of a large array of compounds from low-molecular-weight metabolites to proteins. Another major advantage is that sample and solvent requirements are low, typically nanolitre sample volumes (compared to microlitres for standard LC applications), and its use enables the analysis of *e.g.* biological fluids where the sample volumes available are low and sometimes difficult to obtain (*e.g.* cerebrospinal or gingival crevicular fluids).

However, whilst desirable, the interfacing of CE with MS is not a trivial process. Compatibility of buffers can be an issue, and those that are commonly used for CE include phosphate and borate buffers. These non-volatile buffers are not well suited to electrospray ionisation, since they can accumulate in the ESI source leading to ionisation suppression effects, and requirements for frequent maintenance of the source. Furthermore, high voltages required for the capillary in CE present difficulties for maintaining a stable current for CE-MS experiments. Nilsson *et al.*, however, showed that the APPI process, compared to ESI, is less affected by non-volatile salts in the CE buffers, and thus a range of CE buffers can be applied when APPI is the ionisation method of choice.[108]

There is, however, a lower frequency of reported applications of CE-MS applications compared to those offered by GC-MS and HPLC-MS studies.[109–114]

6.4 Moving from Small-scale to Large-scale Metabolomic Studies

Mammalian systems are investigated in two different types of environments.[23] The first is the laboratory environment where many environmental (for example, growth media and animal food/water availability) and genetic factors can be well controlled. Here, mammalian cell or tissue culture systems,[115,116] or animal models,[117–119] are studied. In this well-controlled environment, 'intra-class' variance is low and the 'treatment' which (we may hope) separates the classes can cause large perturbations in the measured metabolome. To define statistically significant differences, a low number of biological samples in each investigated class are required, typically 6–20 biological replicates. The second environment is the 'real-world', where the system being studied is most frequently humans. Here, the control of genetic and environmental factors is limited, intra-class variability is much higher and, depending on the study and perturbations, inter-class components of variance may be low. The application of longitudinal studies can reduce the impact of intra-subject variability by the application of each subject as their own control.[120]

Many currently published metabolic profiling studies applied in either of the environments discussed above have involved relatively small numbers of samples, typically <200. When applying mass spectrometry as the analytical platform, these samples are typically analysed as a single analytical batch with minimal effects on the quality of data. However, the analysis of larger numbers of samples (100 s–1000 s) gives rise to a reduction in the quality and reproducibility of data acquired. This includes a significant reduction in response, and can include changes in chromatographic properties (including retention time). Previous studies have shown the effect of large numbers of sample injections on analytical data quality.[101,121] Indeed, the combination of sample matrix components and metabolites contaminate the analytical platform, and in metabolic profiling studies (where sample preparation is minimal), the lack of removal of high concentration matrix components or metabolites (which represent major contaminants) further hinders the analysis. In 'targeted' analysis, a more complex sample preparation process should remove the majority or all of these components. A greater degree of sample fractionation, applying solid-phase extraction, has been shown to be beneficial in the metabolic profiling of blood samples.[98] Instrument components, including the GC injector, GC and HPLC columns, and ion sources, are contaminated, and this contamination gives rise to changes in chromatographic and mass spectrometric measured properties, including retention time and MS response. Clearly, when required, routine maintenance and cleaning is necessary. However, it has been observed that the changes in response and retention time are metabolite-dependent, *i.e.* all metabolites do not show the same trend over a defined time period. The number of injections in an analytical batch required to diminish the analytical quality of data decreases, and the maintenance and cleaning required is instrument-dependent; this should be experimentally determined for all mass spectrometry platforms applied.[101] However, there is an increasing requirement to study 100 s–1000 s of subjects in epidemiological-style studies of the human population, to investigate either health in general, or a multitude of diseases. Until recently this has not been easily achievable and, as noted above, this cannot be performed in a single analytical batch. Instead, multiple datasets acquired across multiple analytical batches have to be collected and integrated into a single dataset subsequent to the data acquisition process. This strategy ensures that the quality of data within each analytical batch is high, and also within acceptable tolerance limits. Specific data preprocessing tools are then required to integrate the data from multiple analytical batches into a single dataset. This is a complex process, since responses measured across multiple analytical batches for analysis of the same sample will possibly be subject to modification in view of instrumental contamination and maintenance.

Experimental methods have been designed to overcome these problems:[30] these methods require the application of Quality Control (QC) samples and their intermittent analysis within all analytical batches. QC samples are biological samples of identical or near-identical qualitative and quantitative composition in relation to the test sample matrix and metabolites. Two types of QC samples can be applied, the most appropriate being that which is constructed *via* a pooling of small aliquots from all samples to be analysed together into a single pooled QC sample. This represents the sample (both qualitatively and quantitatively) most identical in composition to those of the biological test samples to be analysed. However, in large studies of 1000 s of samples, the capacity to pool aliquots from all samples is limited (either with required resources, or because sample analysis is required to commence prior to the collection of all samples required for the investigation). Here, a sub-set of samples can be applied to construct a pooled QC sample, or alternatively a sample not lying as close to the compositional content of biological samples can be acquired. For example, for urine (which can be acquired non-invasively), samples can be collected from a subset of individuals and applied or, for plasma and serum, a sample can be acquired commercially (for example, the purchase of serum or plasma from the Sigma-Aldrich Chemical Company or alternative suppliers). The important principle related to the analysis of QC samples is that the replicate analyses of these samples should produce identical results for each sample analysed. In reality, there will be the introduction of small levels of variation from sample preparation, data acquisition and data preprocessing steps. However, in general this level of variation should be lower than that of the related biological variation, and therefore should lie within pre-defined tolerance limits.

QC samples are applied for multiple reasons. The samples are used to equilibrate the analytical platform prior to the analysis of a subject's samples. For urine, it has been shown that the first five injections on a UPLC-MS system show a poor reproducibility and therefore the injection of non-precious samples provides an appropriate level of contamination of the system which leads to the acquisition of data of appropriate reproducibility after these pre-conditioning injections.[122] The number of injections required to provide appropriate pre-conditioning is both sample-type- and instrument-dependent. For example, five and ten injections are suitable for the GC-MS and UPLC-MS analysis of blood serum, respectively.[78,99] After the equilibration period, QC samples are analysed intermittently, generally every fourth to seventh injection. As many of 33% of all injections in an analytical batch can be QC sample-based, and QC samples allow a number of post-data acquisition processes to be applied with data representing the QC samples.[30]

The first application is to perform the correction of small levels of drift in response observed in a single analytical batch, and allow signal correction across multiple analytical batches in order to aid data integration. This provides a reduced level of variation within an analytical batch, and an increased quality of data. One univariate method applies Quality Control-based Loess Signal Correction (QC-RLSC) where the drift in response in bracketing QC samples is applied to correct for that observed in biological samples for GC-MS acquired data.[30]

The second application is to quantitatively assess the reproducibility of data acquired. Here, data for the set of 'pre-conditioning' QC samples are removed and then the relative standard deviations (RSDs) for each metabolite feature across all QC samples are calculated to define the reproducibility across a single analytical batch for each feature. Those features with an RSD greater than a pre-determined tolerance limit should be removed since they show a poor reproducibility. An RSD of 20% has been recommended by several groups for UPLC-MS acquired data for urine and blood serum/plasma,[101,123] and the author and colleagues have recommended a level of 30% for GC-MS acquired data for blood serum (and plasma).[78] This is a Quality Assurance (QA) process; the increased number of sample preparation stages (including chemical derivatisation) and lower reproducibility of sample volume injections result in a higher level of variation in GC-MS data.

The above processes ensure that reproducible data are acquired in single analytical batches, whether the biological study requires data from single or multiple analytical batches to be collected. Indeed, the authors recommend their application in all metabolic profiling projects. In large-scale studies, there is a final step required to integrate data from multiple analytical batches into a single dataset available for further data analysis processes. If the same QC sample is applied in all analytical batches, then responses for each metabolite feature are scaled to the same level in the QC-RLSC process and require no further normalisation. However, the matching of metabolite features by the application of accurate mass and retention time similarities may be required to construct a single data matrix, depending on the data preprocessing software being applied. Software which applies an identical reference table for each analytical batch does not require this final matching process. However, software applying no unique reference table for each separate analytical batch does require this matching process.

Although these methods have not been applied routinely in metabolic profiling studies, the methods have been developed and applied in multiple analytical/biological experiments performed in Manchester, UK. These include the discovery of prognostic biomarkers of pre-eclampsia,[31] and in the study of 'healthy' humans in the HUSERMET project.[30,124] These will enable an increase in size of biological studies using multiple analytical experiments. Appropriate experimental design is required to ensure randomisation of samples across the multiple analytical batches, and also to ensure that no bias is introduced in these processes (*e.g.* to ensure all class 1 samples are not analysed in batches 1–2, and all class 2 samples are not analysed in batches 3–4). Although the use of QC samples is recommended in targeted analysis for quality assurance, there is no requirement for signal correction in these studies since calibration curves are constructed within each analytical batch and are applied for signal correction or calibration.

6.5 Concluding Remarks

Developments in mass spectrometry platforms have led to their increased applications in biological research, including metabolomics experiments. The application of different mass spectrometry platforms provides a diverse field of applications. Each mass spectrometer provides a range of advantages, including sensitivity, specificity, the ability to detect and classify 100 s–1000 s of metabolites in a single sample and a number of experimental methods to provide the structural characterisation of metabolites. Although still relatively under-developed in comparison to proteomics, further analytical method and technological developments will undoubtedly lead to new biological discoveries relating to human health, disease and therapeutic interventions. Some of these will be discussed in subsequent chapters in this book.

Acknowledgements

The authors would like to thank the BBSRC for financial support of The Manchester Centre for Integrative Systems Biology (BB/C008219). This work was supported by the NIHR Manchester Biomedical Research Centre.

References

1. I. W. Griffiths, J. J. Thomson – The centenary of his discovery of the electron and of his invention of mass spectrometry, *Rapid Comm. Mass Spectrom.*, 1997, **11**(1), 3–16.
2. L. P. Lindeman and J. L. Annis, Use of a conventional mass spectrometer as a detector for gas chromatography, *Anal. Chem.*, 1960, **32**(13), 1742–1749.

3. W. M. A. Niessen, Advances in instrumentation in liquid chromatography mass spectrometry and related liquid-introduction techniques, *J. Chrom. A*, 1998, **794**(1–2), 407–435.
4. J. Van der greef, W. M. A. Niessen and U. R. Tjaden, Liquid-Chromatography Mass-Spectrometry – the need for a multidimensional approach, *J. Chrom.*, 1989, **474**(1), 5–19.
5. M. Dole, H. L. Cox and J. Gieniec, Electrospray Mass-Spectroscopy, *Adv. Chem.*, 1973, **125**, 73–84.
6. M. Yamashita and J. B. Fenn, Electrospray ion-source – another variation on the free-jet theme, *J. Phys. Chem.*, 1984, **88**(20), 4451–4459.
7. R. S. Gohlke, Time-of-flight mass spectrometry – application to capillary column gas chromatography, *Anal. Chem.*, 1962, **34**(10), 1332.
8. W. H. McFadden, R. Terabishi, D. R. Black and J. C. Day, Use of capillary gas chromatography with a time-of-flight mass spectrometer, *J. Food Sci.*, 1963, **28**(3), 316–319.
9. Q. Hu, R. J. Noll, H. Li, A. Makarov, M. Hardman and R. G. Cooks, The Orbitrap: a new mass spectrometer, *J. Mass Spectrom.*, 2005, **40**(4), 430–443.
10. S. G. Oliver, M. K. Winson, D. B. Kell and F. Baganz, Systematic functional analysis of the yeast genome, *Trends Biotechnol.*, 1998, **16**(9), 373–378.
11. H. Tweeddale, L. Notley-McRobb and T. Ferenci, Effect of slow growth on metabolism of Escherichia coli, as revealed by global metabolite pool ("Metabolome") analysis, *J. Bacteriol.*, 1998, **180**(19), 5109–5116.
12. O. Fiehn, Metabolomics – The link between genotypes and phenotypes, *Plant Mol. Biol.*, 2002, **48**(1–2), 155–171.
13. E. C. Horning and M. G. Horning, Metabolic profiles: gas-phase methods for analysis of metabolites, *Clin Chem.*, 1971, **17**(8), 802–809.
14. H. Sauter, M. Lauer and H. Fritsch, Metabolic profiling of plants – a new diagnostic-technique, *ACS Symposium Series*, 1991, **443**, 288–299.
15. O. Fiehn, *et al.*, Metabolite profiling for plant functional genomics, *Nat. Biotechnol.*, 2000, **18**(11), 1157–1161.
16. U. Roessner, A. Luedemann, D. Brust, O. Fiehn, T. Linke, L. Willmitzer and A. Fernie, Metabolic profiling allows comprehensive phenotyping of genetically or environmentally modified plant systems, *Plant Cell*, 2001, **13**(1), 11–29.
17. U. Roessner, C. Wagner, J. Kopka, R. N. Trethewey and L. Willmitzer, Simultaneous analysis of metabolites in potato tuber by gas chromatography-mass spectrometry, *Plant J.*, 2000, **23**(1), 131–142.
18. W. Weckwerth, M. E. Loureiro, K. Wenzel and O. Fiehn, Differential metabolic networks unravel the effects of silent plant phenotypes, *Proc. Natl Acad. Sci. USA*, 2004, **101**(20), 7809–7814.
19. S. O'Hagan, W. B. Dunn, M. Brown, J. D. Knowles and D. B. Kell, Closed-loop, multiobjective optimization of analytical instrumentation: Gas chromatography/time-of-flight mass spectrometry of the

metabolomes of human serum and of yeast fermentations, *Anal. Chem.*, 2005, **77**(1), 290–303.

20. S. O'Hagan, W. B. Dunn, J. D. Knowles, D. Broadhurst, R. Williams, J. J. Ashworth, M. Cameron and D. B. Kell, Closed-loop, multiobjective optimization of two-dimensional gas chromatography/mass spectrometry for serum metabolomics, *Anal. Chem.*, 2007, **79**(2), 464–476.
21. W. Welthagen, R. A. Shellie, J. Spranger, M. Ristow, R. Zimmermann and O. Fiehn, Comprehensive two-dimensional gas chromatography-time-of-flight mass spectrometry (GC x GC-TOF) for high resolution metabolomics: biomarker discovery on spleen tissue extracts of obese NZO compared to lean C57BL/6 mice, *Metabolomics*, 2005, **1**(1), 65–73.
22. I. D. Wilson, J. K. Nicholson, J. Castro-Perez, J. H. Granger, K. A. Johnson, B. W. Smith and R. S. Plumb, High resolution "Ultra performance" liquid chromatography coupled to oa-TOF mass spectrometry as a tool for differential metabolic pathway profiling in functional genomic studies, *J. Proteome Res.*, 2005, **4**(2), 591–598.
23. W. B. Dunn, D. I. Broadhurst, H. J. Atherton, R. Goodacre and J. L. Griffin, Systems level studies of mammalian metabolomes: the roles of mass spectrometry and nuclear magnetic resonance spectroscopy, *Chem. Soc. Rev.*, 2011, **40**(1), 387–426.
24. K. Dettmer, P. A. Aronov and B. D. Hammock, Mass spectrometry-based metabolomics, *Mass Spectrom. Rev.*, 2007, **26**(1), 51–78.
25. M. Brown, D. C. Wedge, R. Goodacre, D. B. Kell, P. N. Baker, L. C. Kenny, M. A. Mamas, L. Neyses and W. B. Dunn, Automated workflows for accurate mass-based putative metabolite identification in LC/MS-derived metabolomic datasets, *Bioinformatics*, 2011, **27**(8), 1108–1112.
26. R. J. M. Weber and M. R. Viant, MI-Pack: Increased confidence of metabolite identification in mass spectra by integrating accurate masses and metabolic pathways, *Chemometr. Intell. Lab. Syst.*, 2010, **104**(1), 75–82.
27. M. Beckmann, D. Parker, D. P. Enot, E. Duval and J. Draper, High-throughput, nontargeted metabolite fingerprinting using nominal mass flow injection electrospray mass spectrometry, *Nat. Protocol.*, 2008, **3**(3), 486–504.
28. D. P. Overy, D. P. Enot, K. Tailliart, H. Jenkins, D. Parker, M. Beckmann and J. Draper, Explanatory signal interpretation and metabolite identification strategies for nominal mass FIE-MS metabolite fingerprints, *Nat. Protocol.*, 2008, **3**(3), 471–485.
29. J. Draper, D. P. Enot, D. Parker, M. Beckmann, S. Snowdon, W. Lin and H. Zubair, Metabolite signal identification in accurate mass metabolomics data with MZedDB, an interactive m/z annotation tool utilising predicted ionisation behaviour 'rules', *BMC Bioinformatics*, 2009, **10**, 227.
30. W. B. Dunn, D. Broadhurst, P. Begley, E. Zelena, S. Francis-McIntyre, N. Anderson, M. Brown, J. D. Knowles, A. Halsall, J. N. Haselden,

A. W. Nicholls, I. D. Wilson, D. B. Kell and R. Goodacre, and The Human Serum Metabolome (HUSERMET) Consortium, Procedures for large-scale metabolic profiling of serum and plasma using gas chromatography and liquid chromatography coupled to mass spectrometry, *Nat. Protocol.*, 2011, **6**(7), 1060–1083.

31. L. C. Kenny, D. I. Broadhurst, W. Dunn, M. Brown, R. A. North, L. McCowan, C. Roberts, G. J. S. Cooper, D. B. Kell and P. N. Baker, on behalf of the Screening for Pregnancy Endpoints Consortium, Robust early pregnancy prediction of later preeclampsia using metabolomic biomarkers, *Hypertension*, 2010, **56**(4), 741–749.

32. H. G. Gika, G. A. Theodoridis and I. D. Wilson, Liquid chromatography and ultra-performance liquid chromatography-mass spectrometry fingerprinting of human urine. Sample stability under different handling and storage conditions for metabonomics studies, *J. Chrom. A*, 2008, **1189**(1–2), 314–322.

33. H. H. Draisma, T. H. Reijmers, I. Bobeldijk-Pastorova, J. J. Meulman, G. F. Estourgie-Van Burk, M. Bartels, R. Ramaker, J. van der Greef, D. I. Boomsma and T. Hankemeier, Similarities and differences in lipidomics profiles among healthy monozygotic twin pairs, *OMICS: A Journal of Integrative Biology*, 2008, **12**(1), 17–31.

34. K. Robards, P. R. Haddad and P. E. Jackson, *Principles and Practice of Modern Chromatographic Methods*, Academic Press, London, 1997.

35. G. D. Lewis, R. Wei, E. Liu, E. Yang, X. Shi, M. Martinovic, L. Farrell, A. Asnani, M. Cyrille, A. Ramanathan, O. Shaham, G. Berriz, P. A. Lowry, I. F. Palacios, M. Taşan, F. P. Roth, J. Min, C. Baumgartner, H. Keshishian, T. Addona, V. K. Mootha, A. Rosenzweig, S. A. Carr, M. A. Fifer, M. S. Sabatine and R. E. Gerszten, Metabolite profiling of blood from individuals undergoing planned myocardial infarction reveals early markers of myocardial injury, *J. Clin. Invest.*, 2008, **118**(10), 3503–3512.

36. M. S. Sabatine, E. Liu, D. A. Morrow, E. Heller, R. McCarroll, R. Wiegand, G. F. Berriz, F. P. Roth and R. E. Gerszten, Metabolomic identification of novel biomarkers of myocardial ischemia, *Circulation*, 2005, **112**(25), 3868–3875.

37. D. B. Kell and S. G. Oliver, Here is the evidence, now what is the hypothesis? The complementary roles of inductive and hypothesis-driven science in the post-genomic era, *Bioessays*, 2004, **26**(1), 99–105.

38. A. Sreekumar, L. M. Poisson, T. M. Rajendiran, A. P. Khan, Q. Cao, J. Yu, B. Laxman, R. Mehra, R. J. Lonigro, Y. Li, M. K. Nyati, A. Ahsan, S. Kalyana-Sundaram, B. Han, X. Cao, J. Byun, G. S. Omenn, D. Ghosh, S. Pennathur, D. C. Alexander, A. Berger, J. R. Shuster, J. T. Wei, S. Varambally, C. Beecher and A. M. Chinnaiyan, Metabolomic profiles delineate potential role for sarcosine in prostate cancer progression, *Nature*, 2009, **457**(7231), 910–914.

39. M. Oresic, S. Simell, M. Sysi-Aho, K. Näntö-Salonen, T. Seppänen-Laakso, V. Parikka, M. Katajamaa, A. Hekkala, I. Mattila, P. Keskinen, L. Yetukuri, A. Reinikainen, J. Lähde, T. Suortti, J. Hakalax, T. Simell,

H. Hyöty, R. Veijola, J. Ilonen, R. Lahesmaa, M. Knip and O. Simell, Dysregulation of lipid and amino acid metabolism precedes islet autoimmunity in children who later progress to type 1 diabetes, *J. Exp. Med.*, 2008, **205**(13), 2975–2984.

40. C. Chen, K. W. Krausz, Y. M. Shah, J. R. Idle and F. J. Gonzalez, Serum metabolomics reveals irreversible inhibition of fatty acid beta-oxidation through the suppression of PPAR alpha activation as a contributing mechanism of acetaminophen-induced hepatotoxicity, *Chem. Res. Toxicol.*, 2009, **22**(4), 699–707.

41. T. J. Wang, M. G. Larson, R. S. Vasan, S. Cheng, E. P. Rhee, E. McCabe, G. D. Lewis, C. S. Fox, P. F. Jacques, C. Fernandez, C. J. O'Donnell, S. A. Carr, V. K. Mootha, J. C. Florez, A. Souza, O. Melander, C. B. Clish and R. E. Gerszten, Metabolite profiles and the risk of developing diabetes, *Nat. Med.*, 2011, **17**(4), 448–453.

42. L. K. Schnackenberg and R. D. Beger, The role of metabolic biomarkers in drug toxicity studies, *Toxicol. Mech. Meth.*, 2008, **18**(4), 301–311.

43. H. C. Keun, Metabonomic modeling of drug toxicity, *Pharmacol. Therapeut.*, 2006, **109**(1–2), 92–106.

44. J. S. Fletcher, Cellular imaging with secondary ion mass spectrometry, *Analyst*, 2009, **134**(11), 2204–2215.

45. S. Mas, R. Perez, R. Martinez-Pinna, J. Egido and F. Vivanco, Cluster TOF-SIMS imaging: A new light for *in situ* metabolomics?, *Proteomics*, 2008, **8**(18), 3735–3745.

46. D. Touboul, *et al.*, MALDI-TOF and cluster-TOF-SIMS imaging of Fabry disease biomarkers, *Int. J. Mass Spectrom.*, 2007, **260**(2–3), 158–165.

47. Z. Z. Pan, H. Gu, N. Talaty, H. Chen, N. Shanaiah, B. E. Hainline, R. G. Cooks and D. Raftery, Principal component analysis of urine metabolites detected by NMR and DESI-MS in patients with inborn errors of metabolism, *Anal. Bioanal. Chem.*, 2007, **387**(2), 539–549.

48. Z. Takats, J. M. Wiseman and R. G. Cooks, Ambient mass spectrometry using desorption electrospray ionization (DESI): instrumentation, mechanisms and applications in forensics, chemistry, and biology, *J. Mass Spectrom.*, 2005, **40**(10), 1261–1275.

49. E. R. A. van Hove, D. F. Smith and R. M. A. Heeren, A concise review of mass spectrometry imaging, *J. Chrom. A*, 2010, **1217**(25), 3946–3954.

50. K. Downard, *Mass Spectrometry: A Foundation Course*, Royal Society of Chemistry, Cambridge, 2004.

51. J. H. Gross, *Mass Spectrometry, A Textbook*, Springer-Verlag, Berlin, 1st edn, 2004, p. 518.

52. S. G. Villas-Bôas, S. Mas, M. Akesson, J. Smedsgaard and J. Nielsen, Mass spectrometry in metabolome analysis, *Mass Spectrom. Rev.*, 2005, **24**(5), 613–646.

53. S. G. Villas-Boas, U. Roessner, M. A. E. Hansen, J. Smedsgaard and J. Nielsen, *Metabolome Analysis: An Introduction*, John Wiley and Sons, Inc., New York, 2007.

54. W. B. Dunn, Current trends and future requirements for the mass spectrometric investigation of microbial, mammalian and plant metabolomes, *Phys. Biol.*, 2008, **5**(1), 011001.

55. T. Oinn, M. Addis, J. Ferris, D. Marvin, M. Senger, M. Greenwood, T. Carver, K. Glover, M. R. Pocock, A. Wipat and P. Li, Taverna: a tool for the composition and enactment of bioinformatics workflows, *Bioinformatics*, 2004, **20**(17), 3045–3054.

56. J. M. Halket, A. Przyborowska, S. E. Stein, W. G. Mallard, S. Down and R. A. Chalmers, Deconvolution gas chromatography mass spectrometry of urinary organic acids – Potential for pattern recognition and automated identification of metabolic disorders, *Rapid Comm. Mass Spectrom.*, 1999, **13**(4), 279–284.

57. C. A. Smith, E. J. Want, G. O'Maille, R. Abagyan and G. Siuzdak, XCMS: Processing mass spectrometry data for metabolite profiling using nonlinear peak alignment, matching, and identification, *Anal. Chem.*, 2006, **78**(3), 779–787.

58. M. Katajamaa, J. Miettinen and M. Oresic, MZmine: toolbox for processing and visualization of mass spectrometry based molecular profile data, *Bioinformatics*, 2006, **22**(5), 634–636.

59. A. L. Duran, J. Yang, L. Wang and L. W. Sumner, Metabolomics spectral formatting, alignment and conversion tools (MSFACT), *Bioinformatics*, 2003, **19**(17), 2283–2293.

60. J. E. Katz, D. S. Dumlao, S. Clarke and J. Hau, A new technique (COMSPARI) to facilitate the identification of minor compounds in complex mixtures by GC/MS and LC/MS: Tools for the visualization of matched datasets, *J. Am. Soc. Mass Spectrom.*, 2004, **15**(4), 580–584.

61. C. D. Broeckling, I. R. Reddy, A. L. Duran, X. Zhao and L. W. Sumner, MET-IDEA: Data extraction tool for mass spectrometry-based metabolomics, *Anal. Chem.*, 2006, **78**(13), 4334–4341.

62. R. Baran, H. Kochi, N. Saito, M. Suematsu, T. Soga, T. Nishioka, M. Robert and M. Tomita, MathDAMP: a package for differential analysis of metabolite profiles, *BMC Bioinformatics*, 2006, **7**, 530.

63. A. Lommen, MetAlign: Interface-driven, versatile metabolomics tool for hyphenated full-scan mass spectrometry data preprocessing, *Anal. Chem.*, 2009, **81**(8), 3079–3086.

64. J. C. Hoggard and R. E. Synovec, Parallel factor analysis (PARAFAC) of target analytes in GC x GC-TOFMS data: Automated selection of a model with an appropriate number of factors, *Anal. Chem.*, 2007, **79**(4), 1611–1619.

65. J. J. Jansen, H. C. Hoefsloot, H. F. Boelens, J. van der Greef and A. K. Smilde, Analysis of longitudinal metabolomics data, *Bioinformatics*, 2004, **20**(15), 2438–2446.

66. P. Jonsson, J. Gullberg, A. Nordström, M. Kusano, M. Kowalczyk, M. Sjöström and T. Moritz, A strategy for identifying differences in large series of metabolomic samples analyzed by GC/MS, *Anal. Chem.*, 2004, **76**(6), 1738–1745.

67. P. Jonsson, S. J. Bruce, T. Moritz, J. Trygg, M. Sjöström, R. Plumb, J. Granger, E. Maibaum, J. K. Nicholson, E. Holmes and H. Antti, Extraction, interpretation and validation of information for comparing samples in metabolic LC/MS data sets, *Analyst*, 2005, **130**(5), 701–707.

68. R. E. Mohler, K. M. Dombek, J. C. Hoggard, E. T. Young and R. E. Synovec, Comprehensive two-dimensional gas chromatography time-of-flight mass spectrometry analysis of metabolites in fermenting and respiring yeast cells, *Anal. Chem.*, 2006, **78**(8), 2700–2709.

69. K. M. Pierce, J. C. Hoggard, J. L. Hope, P. M. Rainey, A. N. Hoofnagle, R. M. Jack, B. W. Wright and R. E. Synovec, Fisher ratio method applied to third-order separation data to identify significant chemical components of metabolite extracts, *Anal. Chem.*, 2006, **78**(14), 5068–5075.

70. P. Shen, Y. F. Kang and Y. Cheng, Pattern feature discovery for metabonomics of breast cancer and HPLC/MS/MS analysis of characteristic metabolites, *Chemical Journal of Chinese Universities – Chinese*, 2005, **26**(10), 1798–1802.

71. A. K. Smilde, J. J. Jansen, H. C. Hoefsloot, R. J. Lamers, J. van der Greef and M. E. Timmerman, ANOVA-simultaneous component analysis (ASCA): a new tool for analyzing designed metabolomics data, *Bioinformatics*, 2005, **21**(13), 3043–3048.

72. A. D. Southam, T. G. Payne, H. J. Cooper, T. N. Arvanitis and M. R. Viant, Dynamic range and mass accuracy of wide-scan direct infusion nanoelectrospray Fourier transform ion cyclotron resonance mass spectrometry-based metabolomics increased by the spectral stitching method, *Anal. Chem.*, 2007, **79**(12), 4595–4602.

73. R. King, R. Bonfiglio, C. Fernandez-Metzler, C. Miller-Stein and T. Olah, Mechanistic investigation of ionization suppression in electrospray ionization, *J. Am. Soc. Mass Spectrom.*, 2000, **11**(11), 942–950.

74. C. Böttcher, E. V. Roepenack-Lahaye, E. Willscher, D. Scheel and S. Clemens, Evaluation of matrix effects in metabolite profiling based on capillary liquid chromatography electrospray ionization quadrupole time-of-flight mass spectrometry, *Anal. Chem.*, 2007, **79**(4), 1507–1513.

75. W. B. Dunn, S. Overy and W. P. Quick, Evaluation of automated electrospray-TOF mass spectrometry for metabolic fingerprinting of the plant metabolome, *Metabolomics*, 2005, **1**(2), 137–148.

76. D. P. Enot, W. Lin, M. Beckmann, D. Parker, D. P. Overy and J. Draper, Preprocessing, classification modeling and feature selection using flow injection electrospray mass spectrometry metabolite fingerprint data, *Nat. Protocol.*, 2008, **3**(3), 446–470.

77. M. S. Rashed, Clinical applications of tandem mass spectrometry: ten years of diagnosis and screening for inherited metabolic diseases, *J. Chrom. B*, 2001, **758**(1), 27–48.

78. P. Begley, S. Francis-McIntyre, W. B. Dunn, D. I. Broadhurst, A. Halsall, A. Tseng, J. Knowles, HUSERMET Consortium, R. Goodacre and D. B. Kell, Development and performance of a gas chromatography-time-of-flight mass spectrometry analysis for large-scale nontargeted

metabolomic studies of human serum, *Anal. Chem.*, 2009, **81**(16), 7038–7046.

79. W. B. Dunn, D. Broadhurst, D. I. Ellis, M. Brown, A. Halsall, S. O'Hagan, I. Spasic, A. Tseng and D. B. Kell, A GC-TOF-MS study of the stability of serum and urine metabolomes during the UK Biobank sample collection and preparation protocols, *Int. J. Epidemiol.*, 2008, **37**, 23–30.

80. K. K. Pasikanti, P. C. Ho and E. C. Y. Chan, Development and validation of a gas chromatography/mass spectrometry metabonomic platform for the global profiling of urinary metabolites, *Rapid Comm. Mass Spectrom.*, 2008, **22**(19), 2984–2992.

81. K. F. Smart, R. B. Aggio, J. R. Van Houtte and S. G. Villas-Bôas, Analytical platform for metabolome analysis of microbial cells using methyl chloroformate derivatization followed by gas chromatography-mass spectrometry, *Nat. Protocol.*, 2010, **5**(10), 1709–1729.

82. M. M. Koek, F. Bakels, W. Engel, A. van den Maagdenberg, M. D. Ferrari, L. Coulier and T. Hankemeier, Metabolic profiling of ultrasmall sample volumes with GC/MS: from microliter to nanoliter samples, *Anal. Chem.*, 2010, **82**(1), 156–162.

83. M. Basanta, R. M. Jarvis, Y. Xu, G. Blackburn, R. Tal-Singer, A. Woodcock, D. Singh, R. Goodacre, C. L. Thomas and S. J. Fowler, Non-invasive metabolomic analysis of breath using differential mobility spectrometry in patients with chronic obstructive pulmonary disease and healthy smokers, *Analyst*, 2010, **135**(2), 315–320.

84. R. L. Grob and E. F. Barry, *Modern Practice of Gas Chromatography*, John Wiley and Sons, Inc., New York, 2004.

85. J. V. Seeley, N. J. Micyus, S. V. Bandurski, S. K. Seeley and J. D. McCurry, Microfluidic Deans switch for comprehensive two-dimensional gas chromatography, *Anal. Chem.*, 2007, **79**(5), 1840–1847.

86. R. Mayadunne, T. T. Nguyen and P. J. Marriott, Amino acid analysis by using comprehensive two-dimensional gas chromatography, *Anal. Bioanal. Chem.*, 2005, **382**(3), 836–847.

87. J. Beens and U. A. T. Brinkman, Comprehensive two-dimensional gas chromatography – a powerful and versatile technique, *Analyst*, 2005, **130**(2), 123–127.

88. K. M. Pierce, J. L. Hope, J. C. Hoggard and R. E. Synovec, Principal component analysis based method to discover chemical differences in comprehensive two-dimensional gas chromatography with time-of-flight mass spectrometry (GC x GC-TOFMS) separations of metabolites in plant samples, *Talanta*, 2006, **70**(4), 797–804.

89. V. V. Tolstikov, A. Lommen, K. Nakanishi, N. Tanaka and O. Fiehn, Monolithic silica-based capillary reversed-phase liquid chromatography/electrospray mass spectrometry for plant metabolomics, *Anal. Chem.*, 2003, **75**(23), 6737–6740.

90. A. Maruska and O. Kornysova, Application of monolithic (continuous bed) chromatographic columns in phytochemical analysis, *J. Chrom. A*, 2006, **1112**, 319–330.

91. H. G. Gika, G. Theodoridis, J. Extance, A. M. Edge and I. D. Wilson, High temperature-ultra performance liquid chromatography-mass spectrometry for the metabonomic analysis of Zucker rat urine, *J. Chrom. B*, 2008, **871**(2), 279–287.

92. R. S. Plumb, P. D. Rainville, W. B. Potts, K. A. Johnson, E. Gika and I. D. Wilson, Application of ultra performance liquid chromatography-mass spectrometry to profiling rat and dog bile, *J. Proteome Res.*, 2009, **8**(5), 2495–2500.

93. H. G. Gika, G. A. Theodoridis and I. D. Wilson, Liquid chromatography and ultra performance liquid chromatography-mass spectrometry fingerprinting of human urine – Sample stability under different handling and storage conditions for metabonomics studies, *J. Chrom. A*, 2008, **1189**(1–2), 314–322.

94. J. J. Pesek, M. T. Matyska, J. A. Loo, S. M. Fischer and T. R. Sana, Analysis of hydrophilic metabolites in physiological fluids by HPLC-MS using a silica hydride-based stationary phase, *J. Separ. Sci.*, 2009, **32**(13), 2200–2208.

95. D. L. Callahan, D. De Souza, A. Bacic and U. Roessner, Profiling of polar metabolites in biological extracts using diamond hydride-based aqueous normal phase chromatography, *J. Separ. Sci.*, 2009, **32**(13), 2273–2280.

96. H. G. Gika, G. A. Theodoridis and I. D. Wilson, Hydrophilic interaction and reversed-phase ultra-performance liquid chromatography TOF-MS for metabonomic analysis of Zucker rat urine, *J. Separ. Sci*, 2008, **31**(9), 1598–1608.

97. E. J. Want, I. D. Wilson, H. Gika, G. Theodoridis, R. S. Plumb, J. Shockcor, E. Holmes and J. K. Nicholson, Global metabolic profiling procedures for urine using UPLC-MS, *Nat. Protocol.*, 2010, **5**(6), 1005–1018.

98. F. Michopoulos, L. Lai, H. Gika, G. Theodoridis and I. D. Wilson, UPLC-MS-based analysis of human plasma for metabonomics using solvent precipitation or solid phase extraction, *J. Proteome Res.*, 2009, **8**(4), 2114–2121.

99. W. B. Dunn, D. Broadhurst, M. Brown, P. N. Baker, C. W. Redman, L. C. Kenny and D. B. Kell, Metabolic profiling of serum using Ultra Performance Liquid Chromatography and the LTQ-Orbitrap mass spectrometry system, *J. Chrom. B*, 2008, **871**(2), 288–298.

100. P. D. Rainville, C. L. Stumpf, J. P. Shockcor, R. S. Plumb and J. K. Nicholson, Novel application of reversed-phase UPLC-oaTOF-MS for lipid analysis in complex biological mixtures: A new tool for lipidomics, *J. Proteome Res.*, 2007, **6**(2), 552–558.

101. E. Zelena, W. B. Dunn, D. Broadhurst, S. Francis-McIntyre, K. M. Carroll, P. Begley, S. O'Hagan, J. D. Knowles, A. Halsall, I. D. Wilson and D. B. Kell, Development of a robust and repeatable UPLC-MS method for the long-term metabolomic study of human serum, *Anal. Chem.*, 2009, **81**(4), 1357–1364.

102. S. E. Porter, D. R. Stoll, S. C. Rutan, P. W. Carr and J. D. Cohen, Analysis of four-way two-dimensional liquid chromatography-diode array data: application to metabolomics, *Anal. Chem.*, 2006, **78**(15), 5559–5569.

103. M. Gaspari, K. C. Verhoeckx, E. R. Verheij and J. van der Greef, Integration of two-dimensional LC-MS with multivariate statistics for comparative analysis of proteomic samples, *Anal. Chem.*, 2006, **78**(7), 2286–2296.

104. S. J. Gaskell, Electrospray: Principles and Practices, *J. Mass Spectrom.*, 1997, **32**, 677–688.

105. G. Taylor, Disintegration of Water Droplets in an Electric Field, *Proc. Royal Soc. A*, 1964, **280**(1382), 383.

106. M. Dole, L. L. Mack and R. L. Hines, Molecular beams of macroions, *J. Chem. Phys.*, 1968, **49**(5), 2240–&.

107. J. V. Iribarne and B. A. Thomson, Evaporation of small ions from charged droplets, *J. Chem. Phys.*, 1976, **64**(6), 2287–2294.

108. S. L. Nilsson, C. Andersson, P. J. Sjöberg, D. Bylund, P. Petersson, M. Jörntén-Karlsson and K. E. Markides, Phosphate buffers in capillary electrophoresis/mass spectrometry using atmospheric pressure photoionization and electrospray ionization, *Rapid Comm. Mass Spectrom.*, 2003, **17**(20), 2267–2272.

109. T. Soga, Y. Ohashi, Y. Ueno, H. Naraoka, M. Tomita and T. Nishioka, Quantitative metabolome analysis using capillary electrophoresis mass spectrometry, *J. Proteome Res.*, 2003, **2**(5), 488–494.

110. A. C. Servais, J. Crommen and M. Fillet, Capillary electrophoresis-mass spectrometry, an attractive tool for drug bioanalysis and biomarker discovery, *Electrophoresis*, 2006, **27**(13), 2616–2629.

111. W. C. Yang, F. E. Regnier and J. Adamec, Comparative metabolite profiling of carboxylic acids in rat urine by CE-ESI MS/MS through positively pre-charged and H-2-coded derivatization, *Electrophoresis*, 2008, **29**(22), 4549–4560.

112. R. Ramautar, G. W. Somsen and G. J. de Jong, CE-MS in metabolomics, *Electrophoresis*, 2009, **30**(1), 276–291.

113. E. Nevedomskaya, R. Ramautar, R. Derks, R. I. Westbroek, G. Zondag, I. van der Pluijm, A. M. Deelder and O. A. Mayboroda, CE-MS for metabolic profiling of volume-limited urine samples: application to accelerated aging TTD mice, *J. Proteome Res.*, 2010, **9**(9), 4869–4874.

114. R. Ramautar, O. A. Mayboroda, A. M. Deelder, G. W. Somsen and G. J. de Jong, Metabolic analysis of body fluids by capillary electrophoresis using noncovalently coated capillaries, *J. Chrom. B*, 2008, **871**(2), 370–374.

115. C. A. Sellick, R. Hansen, A. R. Maqsood, W. B. Dunn, G. M. Stephens, R. Goodacre and A. J. Dickson, Effective quenching processes for physiologically valid metabolite profiling of suspension cultured mammalian cells, *Anal. Chem.*, 2009, **81**(1), 174–183.

116. J. B. Ritter, Y. Genzel and U. Reichl, Simultaneous extraction of several metabolites of energy metabolism and related substances in mammalian cells: Optimization using experimental design, *Anal. Biochem.*, 2008, **373**(2), 349–369.

117. H. M. Lin, S. I. Edmunds, N. A. Helsby, L. R. Ferguson and D. D. Rowan, Nontargeted urinary metabolite profiling of a mouse model of Crohn's disease, *J. Proteome Res.*, 2009, **8**(4), 2045–2057.

118. J. Shearer, G. Duggan, A. Weljie, D. S. Hittel, D. H. Wasserman and H. J. Vogel, Metabolomic profiling of dietary-induced insulin resistance in the high fat-fed C57BL/6J mouse, *Diabetes Obes. Metabol.*, 2008, **10**(10), 950–958.

119. H. J. Atherton, N. J. Bailey, W. Zhang, J. Taylor, H. Major, J. Shockcor, K. Clarke and J. L. Griffin, A combined 1H-NMR spectroscopy- and mass spectrometry-based metabolomic study of the PPAR-alpha null mutant mouse defines profound systemic changes in metabolism linked to the metabolic syndrome, *Physiol. Genomics*, 2006, **27**(2), 178–186.

120. E. J. van Velzen, J. A. Westerhuis, J. P. van Duynhoven, F. A. van Dorsten, C. H. Grün, D. M. Jacobs, G. S. Duchateau, D. J. Vis and A. K. Smilde, Phenotyping tea consumers by nutrikinetic analysis of polyphenolic end-metabolites, *J. Proteome Res.*, 2009, **8**(7), 3317–3330.

121. P. A. Guy, I. Tavazzi, S. J. Bruce, Z. Ramadan and S. Kochhar, Global metabolic profiling analysis on human urine by UPLC-TOFMS: Issues and method validation in nutritional metabolomics, *J. Chrom. B*, 2008, **871**(2), 253–260.

122. H. G. Gika, E. Macpherson, G. A. Theodoridis and I. D. Wilson, Evaluation of the repeatability of ultra-performance liquid chromatography-TOF-MS for global metabolic profiling of human urine samples, *J. Chrom. B*, 2008, **871**(2), 299–305.

123. T. Sangster, H. Major, R. Plumb, A. J. Wilson and I. D. Wilson, A pragmatic and readily implemented quality control strategy for HPLC-MS and GC-MS-based metabonomic analysis, *Analyst*, 2006, **131**(10), 1075–1078.

124. http://www.husermet.org/.

125. T. Kind and O. Fiehn, Metabolomic database annotations via query of elemental compositions: Mass accuracy is insufficient even at less than 1 ppm, *BMC Bioinformatics*, 2006, **7**, 234.

126. T. Kind and O. Fiehn, Seven golden rules for heuristic filtering of molecular formulas obtained by accurate mass spectrometry, *BMC Bioinformatics*, 2007, **8**, 105.

127. M. Brown, W. B. Dunn, P. Dobson, Y. Patel, C. L. Winder, S. Francis-McIntyre, P. Begley, K. Carroll, D. Broadhurst, A. Tseng, N. Swainston, I. Spasic, R. Goodacre and D. B. Kell, Mass spectrometry tools and metabolite-specific databases for molecular identification in metabolomics, *Analyst*, 2009, **134**(7), 1322–1332.

Recent Advances in the Multivariate Chemometric Analysis of Cancer Metabolic Profiling

KENICHI YOSHIDA*[a] AND MARTIN GROOTVELD*[b]

[a] Department of Life Sciences, Meiji University, 1-1-1 Higashimita, Tama-ku, Kawasaki, Kanagawa 214-8571, Japan; [b] Leicester School of Pharmacy, De Montfort University, The Gateway, Leicester LE1 9BH, United Kingdom
*Email: yoshida@isc.meiji.ac.jp; mgrootveld@dmu.ac.uk

7.1 Introduction

Accurate and global descriptions of the molecular composition of human tissues, particularly during pathogenesis, are of much current interest and concern. Amongst diseases, cancer is extremely heterogeneous; therefore, classical histopathological examinations of surgical biopsies are often incapable of providing a precise diagnosis, which is crucial for accurate therapy and better patient care. However, we now have an era for taking advantage of the evolving technologies for 'omics' that have come to be recognised as powerful tools for dissecting the molecular etiologies underlying human tumourigenesis. Metabolomics is defined as 'the quantitative measurement of the multiparametric metabolic responses of living systems to pathophysiological stimuli or genetic modification',[1] and this area of

Issues in Toxicology No. 21
Metabolic Profiling: Disease and Xenobiotics
Edited by Martin Grootveld
© The Royal Society of Chemistry 2015
Published by the Royal Society of Chemistry, www.rsc.org

study is being used to investigate a wide range of biomedical research topics since molecular metabolites have been shown to be useful for the diagnosis, and also assessments of the treatment and prognosis of human tumours.

To apply metabolomics to routine clinical diagnosis, efficient, robust, reproducible and cheap methods for obtaining metabolic profiles in biological specimens are highly desirable. Spectral imaging, reproduction and colour appearance modelling for images have been developed for biomedical purposes, and also combined with instrumentation for metabolomics. For example, high-resolution spectroscopic images of biological samples have been processed using multivariate (MV) analysis methodologies such as principal component analysis (PCA), a powerful technique for investigating the metabolic profiles of biological events and certain diseases.[2,3] Indeed, PCA is widely used as a statistical technique for the MV analysis of high-throughput data in order to extract information from complex multi-dimensional datasets, to discover associations between and amongst variables, and also to stabilise estimates.[4] In addition to conventional PCA, multidimensional techniques can be performed for the purpose of statistical examinations. As examples of popular classifiers,[5] linear discriminant analysis (LDA) and quadratic discriminant analysis (QDA),[6] logistic regression (LR), K-nearest neighbour (KNN) analysis,[7] the Wilcoxon signed rank test,[8] artificial neural networks (ANN)[9] and support vector machines (SVM)[10] are routinely used. These MV data analysis methods can be coupled not only with infrared spectroscopy, but also with nuclear magnetic resonance (NMR) spectroscopy and mass spectrometry (MS) techniques to facilitate data processing and reconstitution. Metabolomics data can be statistically analysed using MetaboAnalyst 2.0, which includes PCA, hierarchical clustering and partial least squares discriminant analysis (PLS-DA) (see http:// www.metaboanalyst.ca/).[11] Indeed, the number of metabolomics databases available is growing rapidly.[12–15]

For the analysis of tissue sections, the chemical composition of such specimens can be measured using an infrared spectral approach.[16] Fourier transform infrared (FTIR) spectroscopy is a powerful tool that can probe the vibrational properties of amino acids and co-factors.[17] To apply FTIR imaging to cancer diagnosis, an integrative approach with optimal data acquisition, classification and validation has been established.[18,19] On the other hand, mid-infrared (MIR) spectroscopy has been shown to detect DNA, RNA, proteins, carbohydrates and lipids, and to be applicable to monitoring changes in the molecular composition or structures of cells and tissues, and also the biological fluids of cancer patients. MV data analysis techniques using MIR spectroscopy are superior to simple (univariate) data analysis approaches in view of their elimination of overlapping signals originating from the complexity of such biological samples; therefore, this modality has been demonstrated to be a powerful tool for cancer detection and analysis.[20]

Currently, metabolic profiles for the *in vivo* and *ex vivo* diagnosis of human diseases are predominantly obtained using chemometric strategies, and also an improved MV statistical data analyses of high-resolution spectra obtained using NMR spectroscopy and MS.[21] These approaches are essential for the identification of current biomarkers used for cancer diagnosis, prognosis and the evaluation of therapeutic strategies.[22–24] To apply NMR spectroscopy to clinical metabolic profiling for tumour diagnosis, accuracy and sensitivity are required to distinguish healthy tissues from tumour samples, and also to discriminate between tumour stages and cell-type specificities. Towards this goal, several experiments analysing tumour specimens with the help of chemometric and MV data analysis strategies have been reported.[25]

Further important considerations are that a simple blood or urine test providing highly specific metabolomics cancer biomarkers is much more cost-effective than genome sequencing, or indeed a complete proteomic analysis. Such an approach may also serve to provide critical diagnostic information concerning the early detection of cancer, and perhaps aid us in choosing an optimal therapeutic strategy for patients.

Just recently, there has been a recent alteration in precisely how cancers are viewed, assessed and subsequently treated. Specifically, tumours are defined and stratified by their molecular characteristics in addition to their location (*e.g.*, colon, breast, and brain). Indeed, the detection of mutations which affect hormone receptors and oncogenes (*e.g.*, the HER-2 receptors in breast, and K-RAS in colorectal tumours), now plays a significant role when treatment plans are actuated.[26,27] Notwithstanding, currently such markers are lacking (or await validation and confirmation) for the majority of cancers. However, it is of much importance to note that in at least some cases, patients who are provided with the same diagnosis can differentially respond to exactly the same or very similar therapies, and therefore have contrasting prognostic outcomes. Intriguingly, targeted therapies have recently been developed in order to address the problem of tumour heterogeneities, and such agents have the capacity to focus on cancer sub-classes and hence represent a more effective type of therapeutic option with higher specificities and efficacies, together with less adverse side-effects.[28]

Hence, metabolomics investigations, targeted or untargeted, can in principle offer a more customised form of cancer diagnosis *via* the identification of sub-classes of patients that may benefit from a particular drug or therapeutic regimen, together with those who may develop a resistance to such treatments, or, for that matter, experience deleterious toxic side-effects arising therefrom. In conjunction with metabolic set enrichment and topological pathway analysis, multianalytical technique-linked metabolomics studies can seek and find new or novel biotargets for selectively-designed drugs, can provide valuable information regarding the mechanisms of action of therapeutic agents or regimens, and can provide new applications for already available ones. Indeed, the application of metabolomics techniques to assessments of the relative efficacies of chemotherapy and radiation treatments has been recently reported.[29,30]

Below, the current progress on cancer metabolomics using infrared and NMR spectroscopies and MS coupled with MV chemometric analysis is reviewed.

7.2 MV Chemometric Analysis of Cancer

7.2.1 Infrared Spectroscopy

Conventional infrared vibrational spectroscopy has a low throughput for data acquisition and lacks spatial resolution. Consequently, intensive numerical computations of the resulting data are required to maximise the extraction of information. To represent the spatial heterogeneity of tissues accurately, FTIR spectroscopy has been coupled with a data processing algorithm.[31] As an optical molecular spectroscopic technique, FTIR spectroscopy has been developed to detect the spatial distributions of vibrational spectroscopic 'signatures' of biopsies, and this modality has been shown to offer advantages over classical histological examinations. FTIR spectroscopy is a non-destructive optical technique that can rapidly provide a valuable representation of a sample's biochemistry, and also identify variations between healthy and diseased tissues.[32] To date, this analytical technique has been used to discriminate between benign and malignant prostate cells embedded as paraffin sections.[33] FTIR spectroscopy can provide differential spectral 'signatures' for differing primary tumours. Therefore, micrometastases can be characterised using this method. Indeed, the combination of FTIR spectroscopy with histological staining has increased the molecular specificity, and has been used to discriminate between the biochemical characteristics of metastatic prostate cancer cells in bone marrow.[34]

Chemometric and statistical MV analyses of spectral data have enabled the intrinsic chemical compositions of tissues to be confirmed based on images of the reconstituted spectrum. Such analyses can be applied directly to paraffin-embedded sections without previous dewaxing. Moreover, the tumour heterogeneity of the tested samples can be determined in a manner which is independent of the visual morphology. Indeed, the acquisition of spectral data from tissues embedded in paraffin using FTIR spectroscopy coupled with MV analysis has enabled the detection of breast cancer micrometastases in lymph nodes.[35,36] This approach has also been effectively used to characterise xenografts of human colon carcinomas.[37] In addition to paraffin-embedded sections, the ANN analysis of FTIR spectra has been used to distinguish adenomatous polyp and malignant cells in biopsy specimens.[38]

Harvey *et al.* discriminated between different types of prostate cells using FTIR spectra and an MV chemometric analysis. They showed that the culture medium and the nucleus-to-cytoplasm ratio had no effect on this form of analysis.[39] Moreover, FTIR spectroscopy coupled with chemometric and MV spectral processing techniques has been shown to be an effective method for the diagnosis of skin carcinomas and cervical cancer.[40–42] FTIR spectroscopic imaging has been coupled with tissue micro-arrays, and the statistical pattern recognition of spectra has led to the identification of characteristics of the endogenous molecular composition, which are representative of either benign or malignant prostate epithelium.[43]

The FTIR spectra of biopsies from malignant colon and healthy tissues were analysed using attenuated total reflectance-Fourier transform infrared (ATR-FTIR) spectroscopy; the prediction accuracy was then evaluated using chemometric methods including PCA and soft independent modelling of class analogy (SIMCA). According to the results acquired, SIMCA showed a relatively high accuracy, specificity and sensitivity for the discrimination of malignant colon tissues from healthy ones.[44] Compared with PCA and cluster analysis, the SIMCA chemometric technique showed a superior accuracy when coupled with ATR-FTIR spectroscopy for the discrimination of blood samples obtained from healthy volunteers and patients with basal cell carcinoma.[45] ATR-FTIR spectroscopy and linear discriminant analysis (LDA) have also been reportedly used to classify cancerous blood samples from non-cancerous ones with a high degree of accuracy.[46]

Stone and Matousek[47] have demonstrated that near-infrared (NIR) transmission Raman spectroscopy combined with high spectral variance PCA could be applied to the diagnosis of *in vivo* human breast cancer, and NIR Raman spectroscopy combined with an MV statistical analysis method has also been shown to be a promising diagnostic tool for colon and ovarian cancer.[48–50] This modality can identify the molecular composition and distributions of lipids, proteins, mucus and collagens in normal and malignant tissues. In view of the ability of Raman microspectroscopy to probe subcellular compartments of the cell, spectral information for the nucleus-to-cytoplasm ratio can be extracted using MV statistical methods, and this approach can be applied to discriminatory applications at the single cell level.[51]

7.2.2 Nuclear Magnetic Resonance Spectroscopy

The *in vivo* application of magnetic resonance spectroscopy (MRS) imaging, particularly with proton (^{1}H) spectra, has been developed for human tissues such as muscle.[52,53] As part of a cancer metabolomics approach, ^{1}H NMR spectroscopy has been used to identify the metabolic signatures of tumours, and compared with those of normal, healthy tissues, for the development of diagnostic biomarkers.[54] An improvement in spectral resolution has been achieved both *in vivo* and *in vitro* by recording the MR spectra in more than one dimension.[55] In the field of cancer diagnosis, ^{1}H MRS imaging has been extensively developed for the diagnosis of child brain tumours,[56–58] and this rapidly advancing technique has become a reliable, non-invasive method for monitoring patients with brain tumours.[59] MR imaging using a KNN algorithm has been used for the quantitative analysis of the MV images of brain tumour patients.[60] Moreover, MRS combined with PCA and PLS-DA has been applied to the characterisation of brain metastases originating from different primary cancers, providing useful information for clinical treatment.[61] Other than brain tumours, *in vivo* ^{1}H MRS has been used to reveal the presence of a triglyceride bulk-chain fatty acid (-CH$_2$-)$_n$ resonance, which putatively serves as a marker of clinical cervical cancers.[62]

As an analytical tool for examining the multicomponent composition of small molecules in biofluids such as urine and blood serum, ^{1}H NMR spectroscopy has commonly been employed. In addition, ^{1}H spectra obtained on patients plasma and analysed using statistical pattern recognition techniques has been frequently applied in the biomedical field.[63,64] ^{1}H NMR analysis was combined with MV statistical analyses including unsupervised PCA, supervised SIMCA and receiver operating characteristics (ROC)[65] has been employed for the pattern recognition of biochemical data and the interpretation of serum samples from patients with epithelial ovarian cancer; these methods showed an appropriate sensitivity and specificity for the discrimination of epithelial ovarian cancer from healthy control tissue.[66,67] Tiziani *et al.* characterised the metabolic 'signature' of blood plasma samples obtained from oral cancer patients.[68] Similarly, the ^{1}H MRS spectra of plasma derived from oral squamous cell carcinoma patients and healthy controls were processed using both unsupervised PCA and supervised PLS-DA,[69] and these researchers found that such an approach can be applicable to the diagnosis of early oral cancer. Again, ^{1}H NMR spectroscopy coupled with MV data analysis of human serum samples has also been used for the metabolic profiling of renal cell carcinoma (RCC).[70] Also, ^{1}H MRS combined with chemometric techniques has been shown to be an effective screening tool for the diagnosis of colorectal cancer.[71] Faecal extracts of colorectal cancer patients have been used for the metabolic profiling of short-chain fatty acids. As well-known markers associated with colorectal cancer, acetate and *n*-butyrate were down regulated in faecal extracts of colorectal cancer patients when compared with samples from healthy controls.[72]

High-resolution magic angle spinning MRS (HR-MAS MRS) allows a significant reduction in the line widths of resonances present in the ^{1}H spectra of tissues, which are of a high quality, a major advantage over conventional MRS when applied *in vivo*; however, *in vivo* MRS must be further improved in order to achieve better localisation pulse sequences, better coil design and higher magnetic fields.[64] Recently, *ex vivo* HR-MAS MRS has been used for the direct monitoring of human clinical tumour specimens. For example, reductions in lipid levels, and corresponding increases in those of lactate and choline, have been identified in gastric cancer *ex vivo*.[73] In addition, elevations in lipid levels both *in vivo* and *ex vivo* were detected in the biopsy specimens of pre-invasive and invasive cervical cancers.[74] A recent report has indicated that the intensity/intensities of the above noted bulk-chain (-CH$_2$-)$_n$ lipid resonance(s) in cervical cancer should be further examined in order to determine whether it serves as a reliable 'biomarker'.[75] Calabrese *et al.* have examined various clinical settings of human gastric carcinoma, including normal, autoimmune atrophic gastritis, *Helicobacter pylori* infection and adenocarcinoma using *ex vivo* HR-MAS MRS to determine whether any diagnostic biomarkers are related to the clinical stages of gastric pathogenesis. Their results showed that mono-dimensional and bi-dimensional HR-MAS MRS successfully detected glycine, alanine, free choline and triacylglycerols as possible biomarkers associated with the differentiation process of gastric mucosa into neoplasm.[76] Similarly, Righi *et al.* have found that *ex vivo* HR-MAS MRS combined with an MV data analysis successfully discriminated between the metabolic profiles of human colorectal tumours and those of normal tissue.

To determine further biochemical differences between healthy and colorectal cancer tissues, PCA and PLS-DA have been utilised for NMR spectral data processing. These efforts have led to the identification of increased levels of taurine, acetate, lactate and lipids, and decreased levels of polyols and sugars in tumour samples. In the same experiments, the authors clearly demonstrated that neighbouring regions adjacent to adenocarcinoma that were histologically-verified as being non-cancerous showed an identical metabolic profile to that of the cancerous tissue.[77] The spectra obtained using HR-MAS MRS to examine biopsy specimens obtained from breast cancer patients were analysed using PLS-DA, probabilistic neural networks (PNNs), and Bayesian belief networks (BBNs), and PLS-DA was found to be superior for the prediction of estrogen and progesterone receptor statuses, whilst BBNs were superior for the prediction of the lymph node status; all these parameters are known as important prognostic factors in breast cancer.[78] HR-^{31}P MRS combined with LDA has been used to reveal the phospholipid profiles of biopsy specimens of various human brain tumours;[79] HR-MAS MR spectra were examined using PCA and PLS-DA and were found to be predictors of the brain metastasis classification.[80] Moreover HR-MAS ^{1}H NMR spectroscopy combined with PCA has shown to be a potential option for the diagnosis and prognosis of liver tumours.[81] Indeed, metabolic changes in normal cerebellum compared with tumour progression and recurrence after the resection of a posterior fossa cerebellar tumour and chemotherapy have been characterised using ^{1}H MRS and PLS-DA, together with orthogonal signal correction (OSC) spectral filtering.[82] Barba *et al.* have demonstrated that HR-MAS ^{1}H NMR spectroscopy combined with a pattern-recognition method such as PCA, or PLS-DA, successfully discriminated follicular lymphoma from diffuse large B cell lymphoma. The metabolic changes observed using this approach revealed a relative increase in taurine and alanine contents in follicular lymphoma and diffuse large B cell lymphoma, respectively.[83] Together, these findings suggest that HR-MAS MRS coupled with selected MV data analysis serves as a promising modality for the early diagnosis of various classes of tumours.

More recently, NMR-linked metabolomics approaches have been successfully applied to explore a wide range of cancers, including bladder,[84] breast,[85] colorectal,[86,87] esophageal,[88] lung,[89] oral,[90] ovarian,[91] pancreatic,[92] and prostate malignancies.[93] From these detailed investigations, it has been demonstrated that cancer patients have metabolic patterns that are clearly distinct from those of healthy age-matched controls and benign disease patients. Notably, the tumour location,[86] site,[94] and prognostic development[95] have all been shown to additionally influence the human metabolome. These observations are, of course, not unexpected, since is well known that cancer cells express a significantly modified metabolism, and this phenomenon is frequently interpreted in terms of the well-known Warburg effect.[96] Intriguingly, the characteristic elevation in the level of glycolysis observed is often of a highly complex nature, and this has recently been explored through the application of an extensive range of metabolomics strategies.[97,98]

7.2.3 Mass Spectrometry

For the identification of relatively large numbers of proteins in a relatively short time period, liquid chromatography (LC) combined with MS has been widely used in the area of proteomics; currently, its value is expanding to metabolomics.[99,100] Recently, Scherer *et al.* validated the usefulness of LC-tandem MS (LC-MS/MS) for the quantitative detection of lysophosphatidic acid (LPA), sphingosine 1-phosphate (S1P) and sphinganine 1-phosphate (SA1P) as potential biomarkers for various diseases including cancer.[101] In addition, the plasma lysophospholipid levels in control subjects and patients with benign or malignant breast tumours were quantified using LC-MS/MS.[102] These bioactive lipids are considered to be novel drug targets for cancer treatment.[103] MS coupled with soft ionisation techniques such as electrospray ionisation (ESI) and matrix-assisted laser desorption/ionisation (MALDI) for proteome analysis have already been well established.[104–106] MALDI and its variant surface-enhanced laser desorption/ionisation time-of-flight MS (SELDI-TOF-MS) have been successfully applied for the early diagnosis of breast cancer using samples derived from blood serum and plasma, tissue, nipple fluid and ductal lavage.[107,108] SELDI-TOF-MS, together with ANN analysis, has been applied to discriminate the serum and tear fluids of breast cancer patients from those of healthy controls.[109] Hammad *et al.* performed MS and chemometric statistical analysis using PCA, analysis of variance (ANOVA) and ROC for characterisation of the serum phospholipid content of breast cancer patients.[110] In addition, MALDI-TOF spectrometry of serum samples has obviously shown some potential for the differentiation of such samples between early breast cancer patients and healthy controls.[111]

Using hepatocellular carcinoma (HCC) as a model case, proteomics and glycoproteomics based on SELDI-TOF-MS have been rapidly developed to identify biomarkers related to the diagnosis of liver tumourigenesis, especially since conventional serological diagnostic tests of the serum alpha-fetoprotein (AFP) level lacks both the sensitivity and the specificity required for use as a screening tool for HCC.[112–119] Additionally, rapid resolution LC, reversed-phased (RP) LC and hydrophilic interaction chromatography (HILIC), coupled with quadrupole TOF-MS (Q-TOF-MS), were combined with chemometric analyses in an attempt to profile the serum samples of patients with HCC or liver cirrhosis. Accordingly, glycocholic acid, glycochenodeoxycholic acid, taurocholic acid and taurochenodesoxycholic acid were associated with liver cirrhosis, whereas dihydrosphingosine and phytosphingosine were associated with HCC.[120] Recently, Nanni *et al.* used label-free MALDI-TOF-MS and a chemometric analysis to investigate the metabolite patterning of serum samples from patients with Crohn's disease.[121] Collectively, TOF-MS coupled with chemometric and MV data analysis is a powerful tool for the metabolic profiling of cancer patients, including the early stages of such malignant conditions.

Finally, inductively coupled plasma (ICP)-optical emission spectrometry (ICP-OES), and inductively coupled plasma-mass spectrometry (ICP-MS) have been used for tumour and adjacent non-tumour paired colon biopsies, and different components were detected for these tissues when differing chemometric tools were employed for the evaluation of data.[122]

7.2.4 Other Methods

Chemometric methods have been used to evaluate the voltametric responses of biological samples. Indeed, high performance LC (HPLC) using a diode array detection (DAD) system and a chemometric method revealed that the voltametric response of the cytoplasm of human breast cancer MCF-7 cells and human prostate cancer PC-3 cells were correlated with modifications in the levels of xanthine and guanine.[123,124] The monitoring of *in vivo* cancer behaviour based on voltametric responses may be possible, but the tedious steps involved would probably limit the application of this method for routine clinical study.

The application of auto-fluorescence techniques for the biomedical discrimination of pathogenesis has also been reported.[125,126] Kamath *et al.* reported that laser-induced fluorescence (LIF) using 325-nm pulsed laser excitation combined with a PCA-based non-parametric KNN analysis showed a high specificity, sensitivity and accuracy for the discrimination of normal, benign and malignant ovarian tissues, as well as normal and malignant colonic tissues.[127,128] PCA has also been used to evaluate the diagnostic potential of the extracted intrinsic fluorescence spectra. This suggests that an intrinsic fluorescence analysis could be used to improve the diagnostic specificity of fluorescence spectroscopy and imaging.[129]

Electrical impedance tomography (EIT) has been developed as an imaging and diagnostic tool.[130,131] Based on the differing impedance values of breast cancer and normal breast tissues, EIT represents a potentially useful diagnostic tool for this disease.[132] However, the major problem associated with this technique is that the resolution of EIT is too low for the accurate reconstruction of an impedance image of the tissue, but these disadvantages could be solved using standard classification tools such as SVM. Indeed, EIT has already been performed in conjunction with SVM.[133] Electrical spectroscopy measurements for malignant and benign breast cancer tissues have been performed using SVM as a classifier, and the database created by a mathematical simulation model.[134] Electrical impedance spectroscopy combined with multifeature-based ANN and the 'leave-one-case-out' CV method, together with ROC analysis, has enabled application of the tissue resonance frequency of the breast for the purpose of diagnosing breast biopsy abnormalities with a high specificity.[135]

7.2.5 Further Considerations

Although the multicomponent analysis of cultured cells and their culture media can often provide valuable insights, biofluid and tissue biopsy samples collected during patient-based metabolomics investigations are, of course, more relevant to chemopathological processes developed or developing in the human body. Notwithstanding, highly progressed tumours rarely comprise greater than 1% of the total body mass, and therefore it is improbable that all metabolic disturbances implicated from the MV analysis of such samples are actually ascribable to the cancer process; indeed, there may also be significant contributions arising from the immune response. For example, Hodgkin's lymphoma is viewed as an uncontrolled inflammatory disease in which cancerous B lymphocytes secrete an extensive range of cytokines that serve to chemotactically-attract a series of normal leukocytes, which predominantly comprise the tumour mass.[136,137] Cultured cell lines are less susceptible to disturbances exerted by external factors, and hence should primarily be employed for drug testing episodes focused on the personalisation of cancer treatment. However, the results acquired from such studies should, of course, be validated in fully-approved clinical trials.

The human metabolome is a paradigm which serves to represent ongoing processes in the human body, for example, the maintenance of homeostasis, including energy metabolism and a variety of dynamic fluxes which may involve responses to disease processes. As expected, factors such as gender, diet, body mass index and above all, age and body composition exert highly significant effects on it,[138] together with drug intake. In view of its high level of sensitivity and rapid response to pathological changes in selected biological environments, the metabolome can frequently accurately indicate the phenotype, unlike alternative '-omes', particularly the genome and the proteome.[139] However, such investigations necessarily require both reliable and highly reproducible experimental probes, in addition to meaningful experimental designs and the generation of correctly-validated biomarker molecules.

7.3 Summary

MV chemometric analyses have been applied to the non-invasive assessment of both soft and hard tissues, and human biofluids. Data classification algorithms, including those unsupervised such as cluster analysis or PCA, and those supervised algorithms such as LDA, SIMCA, ANN, SVM machines, Bayesian classification and PLS-DA are highly useful for the characterisation of metabolites and for discriminating between healthy and cancerous tissues. Ongoing developing technologies can also be applied to various classes of tumours to identify biomarkers; however, metabolites should be further examined in multiple experiments.

References

1. J. C. Lindon, E. Holmes and J. K. Nicholson, Metabonomics in pharmaceutical R&D, *FEBS J.*, 2007, **274**, 1140–1151.
2. J. K. Nicholson, J. Connelly, J. C. Lindon and E. Holmes, Metabonomics: A platform for studying drug toxicity and gene function, *Nat. Rev. Drug Discov.*, 2002, **1**, 153–161.
3. F. Fava, J. A. Lovegrove, R. Gitau, K. G. Jackson and K. M. Tuohy, The gut microbiota and lipid metabolism: Implications for human health and coronary heart disease, *Curr. Med. Chem.*, 2006, **13**, 3005–3021.
4. J. J. Song, Y. Ren and F. Yan, Classification for high-throughput data with an optimal subset of principal components, *Comput. Biol. Chem.*, 2009, **33**, 408–413.
5. C. Krafft, G. Steiner, C. Beleites and R. Salzer, Disease recognition by infrared and Raman spectroscopy, *J. Biophotonics*, 2009, **2**, 13–28.
6. D. J. Hand, Statistical methods in diagnosis, *Stat. Methods Med. Res.*, 1992, **1**, 49–67.
7. H. A. Fayed and A. F. Atiya, A novel template reduction approach for the K-nearest neighbor method, *IEEE Trans. Neural Network.*, 2009, **20**, 890–896.
8. G. Tudor and G. G. Koch, Review of nonparametric methods for the analysis of crossover studies, *Stat. Methods Med. Res.*, 1994, **3**, 345–381.
9. J. L. Patel and R. K. Goyal, Applications of artificial neural networks in medical science, *Curr. Clin. Pharmacol.*, 2007, **2**, 217–226.
10. A. Barla, G. Jurman, S. Riccadonna, S. Merler, M. Chierici and C. Furlanello, Machine learning methods for predictive proteomics, *Brief Bioinform.*, 2008, **9**, 119–128.
11. J. Xia, R. Mandal, I. V. Sinelnikov, D. Broadhurst and D. S. Wishart, MetaboAnalyst 2.0—a comprehensive server for metabolomic data analysis, *Nucleic Acids Res.*, 2012, **40**, W127–W133.
12. D. S. Wishart, D. Tzur, C. Knox, R. Eisner, A. C. Guo, N. Young, D. Cheng, K. Jewell, D. Arndt, S. Sawhney, C. Fung, L. Nikolai, M. Lewis, M. A. Coutouly, I. Forsythe, P. Tang, S. Shrivastava, K. Jeroncic, P. Stothard, G. Amegbey, D. Block, D. D. Hau, J. Wagner, J. Miniaci, M. Clements, M. Gebremedhin, N. Guo, Y. Zhang, G. E. Duggan, G. D. Macinnis, A. M. Weljie, R. Dowlatabadi, F. Bamforth, D. Clive, R. Greiner, L. Li, T. Marrie, B. D. Sykes, H. J. Vogel and L. Querengesser, HMDB: The human metabolome database, *Nucleic Acids Res.*, 2007, **35**, D521–526.
13. Q. Cui, I. A. Lewis, A. D. Hegeman, M. E. Anderson, J. Li, C. F. Schulte, W. M. Westler, H. R. Eghbalnia, M. R. Sussman and J. L. Markley,

Metabolite identification via the madison metabolomics consortium database, *Nat. Biotechnol.*, 2008, **26**, 162–164.

14. C. A. Smith, G. O'Maille, E. J. Want, C. Qin, S. A. Trauger, T. R. Brandon, D. E. Custodio, R. Abagyan and G. Siuzdak, Metlin: A metabolite mass spectral database, *Ther. Drug Monit.*, 2005, **27**, 747–751.

15. J. Kopka, N. Schauer, S. Krueger, C. Birkemeyer, B. Usadel, E. Bergmüller, P. Dörmann, W. Weckwerth, Y. Gibon, M. Stitt, L. Willmitzer, A. R. Fernie and D. Steinhauser, GMD@CSB.DB: The golm metabolome database, *Bioinformatics*, 2005, **21**, 1635–1638.

16. P. Lasch, L. Chiriboga, H. Yee and M. Diem, Infrared spectroscopy of human cells and tissue: Detection of disease, *Technol. Cancer Res. Treat.*, 2002, **1**, 1–7.

17. C. Berthomieu and R. Hienerwadel, Fourier transform infrared (FTIR) spectroscopy, *Photosynth. Res.*, 2009, **101**, 157–170.

18. R. Bhargava, Towards a practical Fourier transform infrared chemical imaging protocol for cancer histopathology, *Anal. Bioanal. Chem.*, 2007, **389**, 1155–1169.

19. C. H. Petter, N. Heigl, M. Rainer, R. Bakry, J. Pallua, G. K. Bonn and C. W. Huck, Development and application of Fourier-transform infra-red chemical imaging of tumour in human tissue, *Curr. Med. Chem.*, 2009, **16**, 318–326.

20. L. Wang and B. Mizaikoff, Application of MV data-analysis techniques to biomedical diagnostics based on mid-infrared spectroscopy, *Anal. Bioanal. Chem.*, 2008, **391**, 1641–1654.

21. G. A. Gowda, S. Zhang, H. Gu, V. Asiago, N. Shanaiah and D. Raftery, Metabolomics-based methods for early disease diagnostics, *Expert Rev. Mol. Diagn.*, 2008, **8**, 617–633.

22. J. L. Spratlin, N. J. Serkova and S. G. Eckhardt, Clinical applications of metabolomics in oncology: A review, *Clin. Cancer Res.*, 2009, **15**, 431–440.

23. N. J. Serkova and K. Glunde, Metabolomics of cancer, *Methods Mol. Biol.*, 2009, **520**, 273–295.

24. H. J. Issaq, Q. N. Van, T. J. Waybright, G. M. Muschik and T. D. Veenstra, Analytical and statistical approaches to metabolomics research, *J. Sep. Sci.*, 2009, **32**, 2183–2199.

25. N. J. Serkova and C. U. Niemann, Pattern recognition and biomarker validation using quantitative 1H-NMR-based metabolomics, *Expert Rev. Mol. Diagn.*, 2006, **6**, 717–731.

26. G. Orphanos and P. Kountourakis, Targeting the HER2 receptor in metastatic breast cancer, *Hematol. Oncol. Stem Cell Ther.*, 2012, **5**, 127–137.

27. M. Aiello, N. Vella, C. Cannavo, A. Scalisi, D. A. Spandidos, G. Toffoli, A. Buonadonna, M. Libra and F. Stivala, Role of genetic polymorphisms and mutations in colorectal cancer therapy, *Mol. Med. Rep.*, 2011, **4**, 203–208.

28. National Cancer Institute. Available online: http://m.cancer.gov/topics/factsheets/targeted/ (accessed on 21 January 2013).

29. H. Lyng, B. Sitter, T. F. Bathen, L. R. Jensen, K. Sundfor, G. B. Kristensen and I. S. Gribbestad, Metabolic mapping by use of high-resolution magic angle spinning 1H NMR spectroscopy for assessment of apoptosis in cervical carcinomas, *BMC Cancer*, 2007, 7, 11.
30. F. G. Blankenberg, P. D. Katsikis, R. W. Storrs, C. Beaulieu, D. Spielman, J. Y. Chen, L. Naumovski and J. F. Tait, Quantitative analysis of apoptotic cell death using proton nuclear magnetic resonance spectroscopy, *Blood*, 1997, **89**, 3778–3786.
31. R. Bhargava, D. C. Fernandez, S. M. Hewitt and I. W. Levin, High throughput assessment of cells and tissues: Bayesian classification of spectral metrics from infrared vibrational spectroscopic imaging data, *Biochim. Biophys. Acta*, 2006, **1758**, 830–845.
32. I. W. Levin and R. Bhargava, Fourier transform infrared vibrational spectroscopic imaging: Integrating microscopy and molecular recognition, *Annu. Rev. Phys. Chem.*, 2005, **56**, 429–474.
33. E. Gazi, J. Dwyer, P. Gardner, A. Ghanbari-Siahkali, A. P. Wade, J. Miyan, N. P. Lockyer, J. C. Vickerman, N. W. Clarke, J. H. Shanks, L. J. Scott, C. A. Hart and M. Brown, Applications of Fourier transform infrared microspectroscopy in studies of benign prostate and prostate cancer. A pilot study, *J. Pathol.*, 2003, **201**, 99–108.
34. E. Gazi, J. Dwyer, N. P. Lockyer, P. Gardner, J. H. Shanks, J. Roulson, C. A. Hart, N. W. Clarke and M. D. Brown, Biomolecular profiling of metastatic prostate cancer cells in bone marrow tissue using FTIR microspectroscopy: A pilot study, *Anal. Bioanal. Chem.*, 2007, **387**, 1621–1631.
35. B. Bird, M. Romeo, N. Laver and M. Diem, Spectral detection of micro-metastases in lymph node histo-pathology, *J. Biophotonics*, 2009, **2**, 37–46.
36. B. Bird, K. Bedrossian, N. Laver, M. Miljković, M. J. Romeo and M. Diem, Detection of breast micro-metastases in axillary lymph nodes by infrared micro-spectral imaging, *Analyst*, 2009, **134**, 1067–1076.
37. R. Wolthuis, A. Travo, C. Nicolet, A. Neuville, M. P. Gaub, D. Guenot, E. Ly, M. Manfait, P. Jeannesson and O. Piot, IR spectral imaging for histopathological characterization of xenografted human colon carcinomas, *Anal. Chem.*, 2008, **80**, 8461–8469.
38. S. Argov, J. Ramesh, A. Salman, I. Sinelnikov, J. Goldstein, H. Guterman and S. Mordechai, Diagnostic potential of Fourier-transform infrared microspectroscopy and advanced computational methods in colon cancer patients, *J. Biomed. Opt.*, 2002, 7, 248–254.
39. T. J. Harvey, E. Gazi, A. Henderson, R. D. Snook, N. W. Clarke, M. Brown and P. Gardner, Factors influencing the discrimination and classification of prostate cancer cell lines by FTIR microspectroscopy, *Analyst*, 2009, **134**, 1083–1091.
40. E. Ly, O. Piot, A. Durlach, P. Bernard and M. Manfait, Differential diagnosis of cutaneous carcinomas by infrared spectral micro-imaging combined with pattern recognition, *Analyst*, 2009, **134**, 1208–1214.

41. B. R. Wood, L. Chiriboga, H. Yee, M. A. Quinn, D. McNaughton and M. Diem, Fourier transform infrared (FTIR) spectral mapping of the cervical transformation zone, and dysplastic squamous epithelium, *Gynecol. Oncol.*, 2004, **93**, 59–68.

42. W. Steller, J. Einenkel, L. C. Horn, U. D. Braumann, H. Binder, R. Salzer and C. Krafft, Delimitation of squamous cell cervical carcinoma using infrared microspectroscopic imaging, *Anal. Bioanal. Chem.*, 2006, **384**, 145–154.

43. D. C. Fernandez, R. Bhargava, S. M. Hewitt and I. W. Levin, Infrared spectroscopic imaging for histopathologic recognition, *Nat. Biotechnol.*, 2005, **23**, 469–474.

44. M. Khanmohammadi, A. B. Garmarudi, K. Ghasemi, H. K. Jaliseh and A. Kaviani, Diagnosis of colon cancer by attenuated total reflectance-Fourier transform infrared microspectroscopy and soft independent modeling of class analogy, *Med. Oncol.*, 2009, **26**, 292–297.

45. M. Khanmohammadi, R. Nasiri, K. Ghasemi, S. Samani and A. Bagheri Garmarudi, Diagnosis of basal cell carcinoma by infrared spectroscopy of whole blood samples applying soft independent modeling class analogy, *J. Cancer. Res. Clin. Oncol.*, 2007, **133**, 1001–1010.

46. M. Khanmohammadi, M. A. Ansari, A. B. Garmarudi, G. Hassanzadeh and G. Garoosi, Cancer diagnosis by discrimination between normal and malignant human blood samples using attenuated total reflectance-Fourier transform infrared spectroscopy, *Cancer Invest.*, 2007, **25**, 397–404.

47. N. Stone and P. Matousek, Advanced transmission Raman spectroscopy: A promising tool for breast disease diagnosis, *Cancer Res.*, 2008, **68**, 4424–4430.

48. A. Beljebbar, O. Bouché, M. D. Diébold, P. J. Guillou, J. P. Palot, D. Eudes and M. Manfait, Identification of Raman spectroscopic markers for the characterization of normal and adenocarcinomatous colonic tissues, *Crit. Rev. Oncol. Hematol.*, 2009, **72**, 255–264.

49. K. Maheedhar, R. A. Bhat, R. Malini, N. B. Prathima, P. Keerthi, P. Kushtagi and C. M. Krishna, Diagnosis of ovarian cancer by Raman spectroscopy: A pilot study, *Photomed. Laser Surg.*, 2008, **26**, 83–90.

50. M. V. Chowdary, K. K. Kumar, K. Thakur, A. Anand, J. Kurien, C. M. Krishna and S. Mathew, Discrimination of normal and malignant mucosal tissues of the colon by Raman spectroscopy, *Photomed. Laser Surg.*, 2007, **25**, 269–274.

51. F. Draux, P. Jeannesson, A. Beljebbar, A. Tfayli, N. Fourre, M. Manfait, J. Sulé-Suso and G. D. Sockalingum, Raman spectral imaging of single living cancer cells: A preliminary study, *Analyst*, 2009, **134**, 542–548.

52. C. Boesch and R. Kreis, Dipolar coupling and ordering effects observed in magnetic resonance spectra of skeletal muscle, *NMR Biomed.*, 2001, **14**, 140–148.

53. C. Boesch and R. Kreis, Observation of intramyocellular lipids by 1H-magnetic resonance spectroscopy, *Ann. NY Acad. Sci.*, 2000, **904**, 25–31.

54. C. Mountford, S. Ramadan, P. Stanwell and P. Malycha, Proton MRS of the breast in the clinical setting, *NMR Biomed.*, 2009, **22**, 54–64.
55. M. A. Thomas, S. Lipnick, S. S. Velan, X. Liu, S. Banakar, N. Binesh, S. Ramadan, A. Ambrosio, R. R. Raylman, J. Sayre, N. DeBruhl and L. Bassett, Investigation of breast cancer using two-dimensional MRS, *NMR Biomed.*, 2009, **22**, 77–91.
56. J. H. Hwang, G. F. Egnaczyk, E. Ballard, R. S. Dunn, S. K. Holland and W. S. Ball Jr, Proton MR spectroscopic characteristics of pediatric pilocytic astrocytomas, *AJNR Am. J. Neuroradiol.*, 1998, **19**, 535–540.
57. S. E. Byrd, T. Tomita, P. S. Palka, C. F. Darling, J. P. Norfray and J. Fan, Magnetic resonance spectroscopy (MRS) in the evaluation of pediatric brain tumours, part I: Introduction to MRS, *J. Natl Med. Assoc.*, 1996, **88**, 649–654.
58. A. Broniscer, A. Gajjar, R. Bhargava, J. W. Langston, R. Heideman, D. Jones, L. E. Kun and J. Taylor, Brain stem involvement in children with neurofibromatosis type 1: Role of magnetic resonance imaging and spectroscopy in the distinction from diffuse pontine glioma, *Neurosurgery*, 1997, **40**, 331–337.
59. T. R. McKnight, Proton magnetic resonance spectroscopic evaluation of brain tumour metabolism, *Semin. Oncol.*, 2004, **31**, 605–617.
60. M. C. Lee and S. J. Nelson, Supervised pattern recognition for the prediction of contrast-enhancement appearance in brain tumours from MV magnetic resonance imaging and spectroscopy, *Artif. Intell. Med.*, 2008, **43**, 61–74.
61. T. E. Sjøbakk, R. Johansen, T. F. Bathen, U. Sonnewald, K. A. Kvistad, S. Lundgren and I. S. Gribbestad, Metabolic profiling of human brain metastases using in vivo proton MR spectroscopy at 3T, *BMC Cancer*, 2007, **7**, 141.
62. M. M. Mahon, A. D. Williams, W. P. Soutter, I. J. Cox, G. A. McIndoe, G. A. Coutts, R. Dina and N. M. deSouza, 1H magnetic resonance spectroscopy of invasive cervical cancer: An in vivo study with ex vivo corroboration, *NMR Biomed.*, 2004, **17**, 1–9.
63. M. Ala-Korpela, Critical evaluation of 1H NMR metabonomics of serum as a methodology for disease risk assessment and diagnostics, *Clin. Chem. Lab. Med.*, 2008, **46**, 27–42.
64. J. L. Griffin, Metabonomics: NMR spectroscopy and pattern recognition analysis of body fluids and tissues for characterisation of xenobiotic toxicity and disease diagnosis, *Curr. Opin. Chem. Biol.*, 2003, **7**, 648–654.
65. M. H. Zweig and G. Campbell, Receiver-operating characteristic (ROC) plots: A fundamental evaluation tool in clinical medicine, *Clin. Chem.*, 1993, **39**, 561–577.
66. K. Odunsi, R. M. Wollman, C. B. Ambrosone, A. Hutson, S. E. McCann, J. Tammela, J. P. Geisler, G. Miller, T. Sellers, W. Cliby, F. Qian, B. Keitz, M. Intengan, S. Lele and J. L. Alderfer, Detection of epithelial ovarian cancer using 1H-NMR-based metabonomics, *Int. J. Cancer*, 2005, **113**, 782–788.

67. K. Odunsi, Cancer diagnostics using 1H-NMR-based metabonomics, *Ernst Schering Found Symp. Proc.*, 2007, **4**, 205–226.
68. S. Tiziani, V. Lopes and U. L. Günther, Early stage diagnosis of oral cancer using 1H NMR-based metabolomics, *Neoplasia*, 2009, **11**, 269–276.
69. J. Zhou, B. Xu, J. Huang, X. Jia, J. Xue, X. Shi, L. Xiao and W. Li, 1H NMR-based metabonomic and pattern recognition analysis for detection of oral squamous cell carcinoma, *Clin. Chim. Acta*, 2009, **401**, 8–13.
70. H. Gao, B. Dong, X. Liu, H. Xuan, Y. Huang and D. Lin, Metabonomic profiling of renal cell carcinoma: High-resolution proton nuclear magnetic resonance spectroscopy of human serum with MV data analysis, *Anal. Chim. Acta*, 2008, **624**, 269–277.
71. T. Bezabeh, R. Somorjai, B. Dolenko, N. Bryskina, B. Levin, C. N. Bernstein, E. Jeyarajah, A. H. Steinhart, D. T. Rubin and I. C. Smith, Detecting colorectal cancer by 1H magnetic resonance spectroscopy of fecal extracts, *NMR Biomed.*, 2009, **22**, 593–600.
72. D. Monleón, J. M. Morales, A. Barrasa, J. A. López, C. Vázquez and B. Celda, Metabolite profiling of fecal water extracts from human colorectal cancer, *NMR Biomed.*, 2009, **22**, 342–348.
73. C. W. Mun, J. Y. Cho, W. J. Shin, K. S. Choi, C. K. Eun, S. S. Cha, J. Lee, Y. I. Yang, S. H. Nam, J. Kim and S. Y. Lee, Ex vivo proton MR spectroscopy (1H-MRS) for evaluation of human gastric carcinoma, *Magn. Reson. Imaging*, 2004, **22**, 861–870.
74. M. M. Mahon, I. J. Cox, R. Dina, W. P. Soutter, G. A. McIndoe, A. D. Williams and N. M. deSouza, 1H magnetic resonance spectroscopy of preinvasive and invasive cervical cancer: In vivo-ex vivo profiles and effect of tumour load, *J. Magn. Reson. Imaging*, 2004, **19**, 356–364.
75. S. S. De Silva, G. S. Payne, V. A. Morgan, T. E. Ind, J. H. Shepherd, D. P. Barton and N. M. deSouza, Epithelial and stromal metabolite changes in the transition from cervical intraepithelial neoplasia to cervical cancer: An in vivo 1H magnetic resonance spectroscopic imaging study with ex vivo correlation, *Eur. Radiol.*, 2009, **19**, 2041–2048.
76. C. Calabrese, A. Pisi, G. Di Febo, G. Liguori, G. Filippini, M. Cervellera, V. Righi, P. Lucchi, A. Mucci, L. Schenetti, V. Tonini, M. R. Tosi and V. Tugnoli, Biochemical alterations from normal mucosa to gastric cancer by ex vivo magnetic resonance spectroscopy, *Cancer Epidemiol. Biomarkers Prev.*, 2008, **17**, 1386–1395.
77. V. Righi, C. Durante, M. Cocchi, C. Calabrese, G. Di Febo, F. Lecce, A. Pisi, V. Tugnoli, A. Mucci and L. Schenetti, Discrimination of healthy and neoplastic human colon tissues by ex vivo HR-MAS NMR spectroscopy and chemometric analyses, *J. Proteome Res.*, 2009, **8**, 1859–1869.
78. G. F. Giskeødegård, M. T. Grinde, B. Sitter, D. E. Axelson, S. Lundgren, H. E. Fjøsne, S. Dahl, I. S. Gribbestad and T. F. Bathen, MV modeling and prediction of breast cancer prognostic factors using MR metabolomics, *J. Proteome Res.*, 2010, **9**, 972–979.
79. J. Solivera, S. Cerdán, J. M. Pascual, L. Barrios and J. M. Roda, Assessment of 31P-NMR analysis of phospholipid profiles for potential

differential diagnosis of human cerebral tumours, *NMR Biomed.*, 2009, **22**, 663–674.

80. T. E. Sjøbakk, R. Johansen, T. F. Bathen, U. Sonnewald, R. Juul, S. H. Torp, S. Lundgren and I. S. Gribbestad, Characterization of brain metastases using high-resolution magic angle spinning MRS, *NMR Biomed.*, 2008, **21**, 175–185.

81. Y. Yang, C. Li, X. Nie, X. Feng, W. Chen, Y. Yue, H. Tang and F. Deng, Metabonomic studies of human hepatocellular carcinoma using high-resolution magic-angle spinning 1H NMR spectroscopy in conjunction with MV data analysis, *J. Proteome Res.*, 2007, **6**, 2605–2614.

82. L. Boguszewicz, S. Blamek and M. Sokół, Pattern recognition methods in 1H MRS monitoring in vivo of normal appearing cerebellar tissue after treatment of posterior fossa tumours, *Acta Neurochir. Suppl.*, 2010, **106**, 171–175.

83. I. Barba, C. Sanz, A. Barbera, G. Tapia, J. L. Mate, D. Garcia-Dorado, J. M. Ribera and A. Oriol, Metabolic fingerprinting of fresh lymphoma samples used to discriminate between follicular and diffuse large b-cell lymphomas, *Exp. Hematol.*, 2009, **37**, 1259–1265.

84. M. Cao, L. Zhao, H. Chen, W. Xue and D. Lin, NMR-based metabolomic analysis of human bladder cancer, *Anal. Sci.*, 2012, **28**, 451–456.

85. A. M. Weljie, A. Bondareva, P. Zang and F. R. Jirik, 1H NMR metabolomics identification of markers of hypoxia-induced metabolic shifts in a breast cancer model system, *J. Biomol. NMR*, 2011, **49**, 185–193.

86. E. Chun, Y. Chan, P. Koon Koh, M. Mal, P. Yean Cheah, K. Weng Eu, A. Backshall, R. Cavill, J. K. Nicholson and H. C. Keun, Metabolic profiling of human colorectal cancer using high-resolution Magic angle spinning nuclear magnetic resonance (HR-MAS NMR) spectroscopy and gas chromatography mass spectrometry (GC/MS), *J. Proteome Res.*, 2009, **8**, 352–361.

87. F. Farshidfar, A. M. Weljie, K. Kopciuk, W. D. Buie, A. Maclean, E. Dixon, F. R. Sutherland, A. Molckovsky, H. J. Vogel and O. F. Bathe, Serum metabolomic profile as a means to distinguish stage of colorectal cancer, *Genome Med.*, 2012, **4**, 42.

88. A. Hasim, H. Ma, B. Mamtimin, A. Abudula, M. Niyaz, L. W. Zhang, J. Anwer and I. Sheyhidin, Revealing the metabonomic variation of EC using 1H-NMR spectroscopy and its association with the clinicopathological characteristics, *Mol. Biol. Rep.*, 2012, **39**, 8955–8964.

89. J. Carrola, C. M. Rocha, A. S. Barros, A. M. Gil, B. K. Goodfellow, I. M. Carreira, J. Bernardo, A. Gomes, S. Sousa and L. Carvalho, *et al.*, Metabolic signatures of lung cancer in biofluids: NMR-based metabonomics of urine, *J. Proteome Res.*, 2011, **10**, 221–230.

90. S. Tiziani, L. Lopes and U. L. Gunther, Early stage diagnosis of oral cancer using ¹H NMR–based metabolomics, *Neoplasia*, 2009, **11**, 269–276.

91. D. Ben Sellem, K. Elbayed, A. Neuville, F. M. Moussallieh, G. Lang-Averous, M. Piotto, J. P. Bellocq and I. J. Namer, Metabolomic

characterization of ovarian epithelial carcinomas by hrmas-NMR spectroscopy, *J. Oncol.*, 2011, **10**, 174019.

92. O. F. Bathe, R. Shaykhutdinov, K. Kopciuk, A. M. Weljie, A. McKay, F. R. Sutherland, E. Dixon, N. Dunse, D. Sotiropoulos and H. J. Vogel, Feasibility of identifying pancreatic cancer based on serum metabolomics, *Cancer Epidemiol. Biomar. Prev.*, 2011, **20**, 140–147.

93. O. Teahan, C. L. Bevan, J. Waxman and H. C. Keun, Metabolic signatures of malignant progression in prostate epithelial cells, *Int. J. Biochem. Cell. Biol.*, 2011, **43**, 1002–1009.

94. C. M. Slupsky, H. Steed, T. H. Wells, K. Dabbs, A. Schepansky, V. Capstick, W. Faught and M. B. Sawyer, Urine metabolite analysis offers potential early diagnosis of ovarian and breast cancers, *Clin. Cancer Res.*, 2010, **16**, 5835–5841.

95. M. Y. Fong, J. McDunn and S. S. Kakar, Identification of metabolites in the normal ovary and their transformation in primary and metastatic ovarian cancer, *PloS One*, 2011, **6**, e19963.

96. O. Warburg, On the origin of cancer cells, *Science*, 1956, **123**, 309–314.

97. J. W. Locasale, A. R. Grassian, T. Melman, C. A. Lyssiotis, K. R. Mattaini, A. J. Bass, G. Heffron, C. M. Metallo, T. Muranen and H. Sharfi, *et al.*, Phosphoglycerate dehydrogenase diverts glycolytic flux and contributes to oncogenesis, *Nat. Genet.*, 2011, **43**, 869–874.

98. M. G. Vander Heiden, J. W. Locasale, K. D. Swanson, H. Sharfi, G. J. Heffron, D. Amador-Noguez, H. R. Christofk, G. Wagner, J. D. Rabinowitz and J. M. Asara, *et al.*, Evidence for an alternative glycolytic pathway in rapidly proliferating cells, *Science*, 2010, **329**, 1492–1499.

99. J. W. Wong, M. J. Sullivan and G. Cagney, Computational methods for the comparative quantification of proteins in label-free LCN-MS experiments, *Brief Bioinform.*, 2008, **9**, 156–165.

100. K. Morgenthal, S. Wienkoop, F. Wolschin and W. Weckwerth, Integrative profiling of metabolites and proteins: Improving pattern recognition and biomarker selection for systems level approaches, *Methods Mol. Biol.*, 2007, **358**, 57–75.

101. M. Scherer, G. Schmitz and G. Liebisch, High-throughput analysis of sphingosine 1-phosphate, sphinganine 1-phosphate, and lysophosphatidic acid in plasma samples by liquid chromatography-tandem mass spectrometry, *Clin. Chem.*, 2009, **55**, 1218–1222.

102. M. Murph, T. Tanaka, J. Pang, E. Felix, S. Liu, R. Trost, A. K. Godwin, R. Newman and G. Mills, Liquid chromatography mass spectrometry for quantifying plasma lysophospholipids: Potential biomarkers for cancer diagnosis, *Methods Enzymol.*, 2007, **433**, 1–25.

103. O. Peyruchaud, Novel implications for lysophospholipids, lysophosphatidic acid and sphingosine 1-phosphate, as drug targets in cancer, *Anticancer Agents Med. Chem.*, 2009, **9**, 381–391.

104. I. C. Guerrera and O. Kleiner, Application of mass spectrometry in proteomics, *Biosci. Rep.*, 2005, **25**, 71–93.

105. X. Han, A. Aslanian and J. R. Yates 3rd, Mass spectrometry for proteomics, *Curr. Opin. Chem. Biol.*, 2008, **12**, 483–490.
106. P. L. Ferguson and R. D. Smith, Proteome analysis by mass spectrometry, *Annu. Rev. Biophys. Biomol. Struct.*, 2003, **32**, 399–424.
107. C. Laronga and R. R. Drake, Proteomic approach to breast cancer, *Cancer Control*, 2007, **14**, 360–368.
108. M. C. Gast, J. H. Schellens and J. H. Beijnen, Clinical proteomics in breast cancer: A review, *Breast Cancer Res. Treat.*, 2009, **116**, 17–29.
109. A. Lebrecht, D. Boehm, M. Schmidt, H. Koelbl and F. H. Grus, Surface-enhanced laser desorption/ionisation time-of-flight mass spectrometry to detect breast cancer markers in tears and serum, *Cancer Genomics Proteomics*, 2009, **6**, 75–83.
110. L. A. Hammad, G. Wu, M. M. Saleh, I. Klouckova, L. E. Dobrolecki, R. J. Hickey, L. Schnaper, M. V. Novotny and Y. Mechref, Elevated levels of hydroxylated phosphocholine lipids in the blood serum of breast cancer patients, *Rapid Commun. Mass Spectrom.*, 2009, **23**, 863–876.
111. M. Pietrowska, L. Marczak, J. Polanska, K. Behrendt, E. Nowicka, A. Walaszczyk, A. Chmura, R. Deja, M. Stobiecki, A. Polanski, R. Tarnawski and P. Widlak, Mass spectrometry-based serum proteome pattern analysis in molecular diagnostics of early stage breast cancer, *J. Transl. Med.*, 2009, **7**, 60.
112. Z. Dai, J. Zhou, S. J. Qiu, Y. K. Liu and J. Fan, Lectin-based glycoproteomics to explore and analyze hepatocellular carcinoma-related glycoprotein markers, *Electrophoresis*, 2009, **30**, 2957–2966.
113. A. El-Aneed and J. Banoub, Proteomics in the diagnosis of hepatocellular carcinoma: Focus on high risk hepatitis B and C patients, *Anticancer Res.*, 2006, **26**, 3293–3300.
114. D. G. Ward, Y. Cheng, G. N'Kontchou, T. T. Thar, N. Barget, W. Wei, L. J. Billingham, A. Martin, M. Beaugrand and P. J. Johnson, Changes in the serum proteome associated with the development of hepatocellular carcinoma in hepatitis C-related cirrhosis, *Br. J. Cancer*, 2006, **94**, 287–292.
115. C. Wu, Z. Wang, L. Liu, P. Zhao, W. Wang, D. Yao, B. Shi, J. Lu, P. Liao, Y. Yang and L. Zhu, Surface enhanced laser desorption/ionization profiling: New diagnostic method of HBV-related hepatocellular carcinoma, *J. Gastroenterol. Hepatol.*, 2009, **24**, 55–62.
116. X. Geng, F. Wang, Y. G. Li, G. P. Zhu and W. M. Zhang, SELDI-TOF MS proteinchip technology for screening of serum markers of HBV-induced hepatocellular carcinoma, *J. Exp. Clin. Cancer Res.*, 2007, **26**, 505–508.
117. J. F. Cui, Y. K. Liu, H. J. Zhou, X. N. Kang, C. Huang, Y. F. He, Z. Y. Tang and T. Uemura, Screening serum hepatocellular carcinoma-associated proteins by SELDI-based protein spectrum analysis, *World J. Gastroenterol.*, 2008, **14**, 1257–1262.
118. J. Cui, X. Kang, Z. Dai, C. Huang, H. Zhou, K. Guo, Y. Li, Y. Zhang, R. Sun, J. Chen, Y. Li, Z. Tang, T. Uemura and Y. Liu, Prediction of chronic hepatitis B, liver cirrhosis and hepatocellular carcinoma by

SELDI-based serum decision tree classification, *J. Cancer Res. Clin. Oncol.*, 2007, **133**, 825–834.

119. F. X. Wu, Q. Wang, Z. M. Zhang, S. Huang, W. P. Yuan, J. Y. Liu, K. C. Ban and Y. N. Zhao, Identifying serological biomarkers of hepatocellular carcinoma using surface-enhanced laser desorption/ionization-time-of-flight mass spectroscopy, *Cancer Lett.*, 2009, **279**, 163–170.

120. P. Yin, D. Wan, C. Zhao, J. Chen, X. Zhao, W. Wang, X. Lu, S. Yang, J. Gu and G. Xu, A metabonomic study of hepatitis B-induced liver cirrhosis and hepatocellular carcinoma by using RP-LC and HILIC coupled with mass spectrometry, *Mol. Biosyst.*, 2009, **5**, 868–876.

121. P. Nanni, F. Levander, G. Roda, A. Caponi, P. James and A. Roda, A label-free nano-liquid chromatography-mass spectrometry approach for quantitative serum peptidomics in Crohn's disease patients, *J. Chromatogr. B Analyt. Technol. Biomed. Life Sci.*, 2009, **877**, 3127–3136.

122. I. Lavilla, M. Costas, P. S. Miguel, J. Millos and C. Bendicho, Elemental fingerprinting of tumourous and adjacent non-tumourous tissues from patients with colorectal cancer using ICP-MS, ICP-OES and chemometric analysis, *Biometals*, 2009, **22**, 863–875.

123. J. T. Wang, X. E. Li, J. G. Liu, Y. Zhang, Z. Y. Zhang, T. Zhang, S. G. Jiang, D. M. Wu and Y. G. Zu, Voltammetric behavior of the MCF-7 cell cytoplasm and the effect of taxol on voltammetric response, *Anal. Biochem.*, 2009, **394**, 229–236.

124. D. M. Wu, G. L. Fu, H. Z. Fang, L. Hu, J. L. Li, X. Yuan and Z. Y. Zhang, Studies on the origin of the voltammetric response of the PC-3 cell suspension, *Talanta*, 2009, **78**, 602–607.

125. G. S. Nayak, S. Kamath, K. M. Pai, A. Sarkar, S. Ray, J. Kurien, L. D'Almeida, B. R. Krishnanand, C. Santhosh, V. B. Kartha and K. K. Mahato, Principal component analysis and artificial neural network analysis of oral tissue fluorescence spectra: Classification of normal premalignant and malignant pathological conditions, *Biopolymers*, 2006, **82**, 152–166.

126. S. D. Kamath and K. K. Mahato, Optical pathology using oral tissue fluorescence spectra: Classification by principal component analysis and k-means nearest neighbor analysis, *J. Biomed. Opt.*, 2007, **12**, 014028.

127. S. D. Kamath, R. A. Bhat, S. Ray and K. K. Mahato, Autofluorescence of normal, benign, and malignant ovarian tissues: A pilot study, *Photomed. Laser Surg.*, 2009, **27**, 325–335.

128. S. D. Kamath and K. K. Mahato, Principal component analysis (PCA)-based k-nearest neighbor (k-NN) analysis of colonic mucosal tissue fluorescence spectra, *Photomed. Laser Surg.*, 2009, **27**, 659–668.

129. Y. Fawzy and H. Zeng, Intrinsic fluorescence spectroscopy for endoscopic detection and localization of the endobronchial cancerous lesions, *J. Biomed. Opt.*, 2008, **13**, 064022.

130. M. H. Choi, T. J. Kao, D. Isaacson, G. J. Saulnier and J. C. Newell, A reconstruction algorithm for breast cancer imaging with electrical

impedance tomography in mammography geometry, *IEEE Trans. Biomed. Eng.*, 2007, **54**, 700–710.

131. R. Halter, A. Hartov and K. D. Paulsen, Design and implementation of a high frequency electrical impedance tomography system, *Physiol. Meas.*, 2004, **25**, 379–390.

132. J. Jossinet and M. Schmitt, A review of parameters for the bioelectrical characterization of breast tissue, *Ann. NY Acad. Sci.*, 1999, **873**, 30–41.

133. Y. Wu, L. Guo, G. Dong, Q. Wu, X. Shen, G. Xu and W. Yan, Tissue conductivity estimation in two-dimension head model based on support vector machine, *Conf. Proc. IEEE Eng. Med. Biol. Soc.*, 2006, **1**, 1130–1133.

134. S. Laufer and B. Rubinsky, Tissue characterization with an electrical spectroscopy SVM classifier, *IEEE Trans. Biomed. Eng.*, 2009, **56**, 525–528.

135. B. Zheng, M. L. Zuley, J. H. Sumkin, V. J. Catullo, G. S. Abrams, G. Y. Rathfon, D. M. Chough, M. Z. Gruss and D. Gur, Detection of breast abnormalities using a prototype resonance electrical impedance spectroscopy system: A preliminary study, *Med. Phys.*, 2008, **35**, 3041–3048.

136. S. M. Varnum, B-JM Webb-Robertson, N. A. Hessol, R. D. Smith and R. C. Zangar, Plasma biomarkers for detecting Hodgkin's Lymphoma in HIV patients, *PLoS ONE*, 2011, **6**(12), e29263.

137. G. Khan, Epstein-Barr virus, cytokines, and inflammation: a cocktail for the pathogenesis of Hodgkin s lymphoma?, *Exp. Hematol.*, 2006, **34**, 399–406.

138. C. Jourdan, A. K. Petersen, C. Gieger, A. Doring, T. Illig, R. Wang-Sattler, C. Meisinger, A. Peters, J. Adamski and C. Prehn, *et al.*, Body fat free mass is associated with the serum metabolite profile in a population-based study, *PLoS ONE*, 2012, **7**, e40009.

139. S. P. Putri, Y. Nakayama, F. Matsuda, T. Uchikata, S. Kobayashi, A. Matsubara and E. Fukusaki, Current metabolomics: Practical applications, *J. Biosci. Bioeng.*, 2013, 579–589.

Group-specific Internal Standard Technology (GSIST) for Mass Spectrometry-based Metabolite Profiling

JIRI ADAMEC[†]

Department of Biochemistry and Redox Biology Center, University of
Nebraska – Lincoln, Lincoln, NE 68588, USA
Email: jadamec2@unl.edu

8.1 Introduction

Whilst metabolomics, a new 'omics' platform technology for system biology,
is parallel to genomics, transcriptomics and proteomics in concept, its study
tools, such as sampling, sample preparation, instrumental analysis, data
processing and data interpretation, represent unique challenges. These
challenges arise from the large population of metabolites, their wide variations in chemical and physical properties, as well as a broad dynamical
range of concentration distributions, phenomena giving rise to difficulties
regarding a fully comprehensive metabolome analysis of small molecules.
Alternatively, the identification and quantification of a selected number of
pre-defined metabolites, generally related to one or more specific metabolic

---

[†]Present address: University of Nebraska – Lincoln, Department of Biochemistry, BEAD N151,
  1901 Vine Street, Lincoln, NE 68588, USA.

Issues in Toxicology No. 21
Metabolic Profiling: Disease and Xenobiotics
Edited by Martin Grootveld

Published by the Royal Society of Chemistry, www.rsc.org

pathway(s), specifically metabolite-profiling, currently serves as a feasible approach.[1]

Over the years, many techniques have been developed to provide reliable and reproducible means of quantification of metabolites in complex samples. Although enzyme-based assays for individually determining certain metabolites have been available for some time, these assays are usually time-consuming, and limited to small numbers of metabolites, and also critically depend on the availability of the enzymes, and the simultaneous quantification of multiple metabolites is highly preferred for the purpose of achieving comprehensive studies of cellular metabolism.[2–4] Originally, liquid- or gas-chromatography coupled with UV/VIS, pulsed amperometric,[5–7] potentiometric,[8] conductimetric[5,9–12] and other detection methods were predominantly employed for these studies. Although these techniques work well for simple mixtures, they have major limitations in the 'global' analysis of complex biological samples in view of the inadequate separation of individual molecules.

Currently, the most effective manner of analysing metabolites is through their separation by various chromatographic techniques, followed by identification and quantification through mass-spectrometry (MS),[13,14] which offers an additional separation dimension since it differentiates molecules based on their molecular mass. Unfortunately, MS-based methods remain imperfect in quantification, since the signal intensity of an analyte in MS depends on both its concentration and ionisation efficiency, which not only varies 'between-analytes', but can also depend on other components in the matrix investigated, particularly in the case of electrospray ionisation used in LC-MS analysis. Some researchers have used standard addition methods to overcome this problem,[15–18] however, the MS response can change over time in view of modifications in the MS instrument[19] and hence compromise results. A widely employed strategy is comparative quantification using *in vivo* uniform isotopic labelling with ^{2}H, ^{13}C or ^{15}N, but it may exclude the case of humans, a phenomenon ascribable to its inherent feature. Another option is to quantify components through the use of stable isotope-coded internal standards that co-elute with the analytes, and have an ionisation environment identical to the analyte. When added in known concentrations to a sample, they serve to compensate for inter-analysis variations in sample manipulation. A limitation of this method is that in the analysis of large numbers of metabolites, isotopically coded standards must be synthesised for each molecule, and the requisite number of syntheses can become prohibitive. Unfortunately, a comprehensive collection of internal standard metabolites is generally not available. To address the above issues, an enhanced post-biosynthetic (*in vitro*) stable isotope encoding strategy, defined as Group-specific Internal Standard Technology (GSIST), has been developed.[20–25]

8.2 Basic Principles of GSIST

It has been shown in proteomics that one strategy that can be adopted to deal with this problem is to label large numbers of peptides through the derivatisation of primary amines with stable isotope-coded labelling agents. Peptides in one sample are globally coded with a single isotope of the coding agent (light form), whilst those of a second sample are globally coded with a second isotope (heavy form).[26] After the two samples are mixed, the relative amounts of isotopically coded peptides can be determined in a single analysis.[27] A similar approach is used in GSIST, in which a group of 'targeted' metabolites is derivatised with well-designed reagents, including usually permanent positive charge, length-adjustable alkyl chain and stable isotope atoms, such as ^{2}H and/or ^{13}C.

A permanent positive charge provides an enhanced ionisation efficiency; therefore, the ionisation of non-targeted co-existing molecules can be suppressed in view of the absence of this charge. Recently, the effect of charge on ionisation efficacy of derivatised metabolites was examined by Yang *et al.*[28] In this study, the authors compared the ionisation of nicotinoate- and benzoate-derivatised glutamate (Glu), tryptophan (Try) and lysine (Lys). As predicted, the enhancement of sensitivity was observed for the amino acids derivatised with permanently charged derivatising reagents. Interestingly, MS response was also substantially affected by the length of the alkyl chain in *N*-alkylnicotinic acid *N*-hydroxysuccinimide ester derivatives.[28] Compared to nicotinoate-derivatised tryptophan (NA-Trp), C_4-NA-Trp gave a 12.5 times greater MS signal intensity.[28] This phenomenon has been explained by the fact that lengthening the alkyl chain in C_{1-4}-NA-AA causes the derivatised amino acid to have more surfactant-like properties. This in turn increases AA concentration at droplet surfaces in ESI-MS, and furthermore enhances ionisation. The net effect is that, in comparison to BA-$^{13}C_6$ derivatives, C_4-NA derivatives of Glu, Lys and Trp showed 30-, 6- and 83-times greater ESI-MS signal intensity, respectively.[28]

The length-adjustable alkyl chain in derivatisation reagents is also valuable in terms of modifications of the physicochemical properties of analytes for separation purposes. For example, by selecting various lengths of alkyl chains, hydrophilic and hydrophobic molecules can be analysed together *via* a single chromatographic platform such as Reversed Phase C-18 chromatography.

Finally, the incorporation of stable isotope atoms into the derivatisation reagent structures enables the coding of analytes according to their origin as well as their accurate, matrix complexity and composition-independent MS quantification.

GSIST can be employed for both relative and absolute quantification. In relative quantification applications, metabolites from control and experimental samples are derivatised with two different labelling agents that are chemically identical, but isotopically distinct. After mixing these derivatised samples, each molecule from the control sample serves as an internal standard for determining the relative concentration of the chemically identical component in the experimental sample. Absolute quantification of target metabolites is achieved by the addition of isotope-labelled derivatives as internal standards.

8.3 Application of GSIST

8.3.1 Absolute Quantification Targeting Specific Functional Groups: Determination of Estrogens

Estrogens generally are present in tissues and biological fluids at low concentrations. When combined with the complexity of the matrix within which they are found, the analysis of estrogens in biological samples is challenging. The recent review by Giese[29] shows that immunoassays, high-performance liquid chromatography (HPLC) coupled with electrochemical or mass-spectrometric (MS) detection[30–37] and gas-chromatography (GC) with electron-capture detection (ECD) or MS detection[38–45] are amongst the most widely used methods for estrogen analysis. However, major complications with accurate determination of estrogens include their structural diversity and possible conjugations with other species, such as DNA, sulfate and glucuronide.[29] In addition, immunoassays such as radioimmunoassay[46] fail to measure all the forms of estrogen in a single assay. Another issue that must be addressed is the large number of samples that have to be examined for a meaningful epidemiological analysis and sensitivity.

The challenges noted above were addressed through the development of a comparative HPLC-ESI-MS method centring on a stable isotope-coding agent that facilitates ionisation and quantification, in addition to accelerating HPLC separation. Indeed, a stable isotope-based internal standard quantification of 16 unique estrogen metabolites was achieved in less than 7 min.[24]

8.3.1.1 Derivatisation Strategy for Estrogen Metabolites

Endogenous estrogen metabolites[29,35] possess the aromatic steroid core structure with phenolic and alcoholic hydroxyl groups. These hydroxyl groups are very weakly acidic with ionisation of the phenol hydroxyl group being greater than that of alcohols. At the pH of the HPLC mobile phase, ionisation of ethinylestradiol is theoretically less than 0.001%.[30] Since ionisation in the gas phase is directly related to ionisation in solution,[47] the limited ionisation of estrogens in solution implies poor ionisation efficacy in ESI-MS. One way to enhance ESI sensitivity is to directly introduce a quaternary amine (permanent positive charge) into analytes (Figure 8.1), using

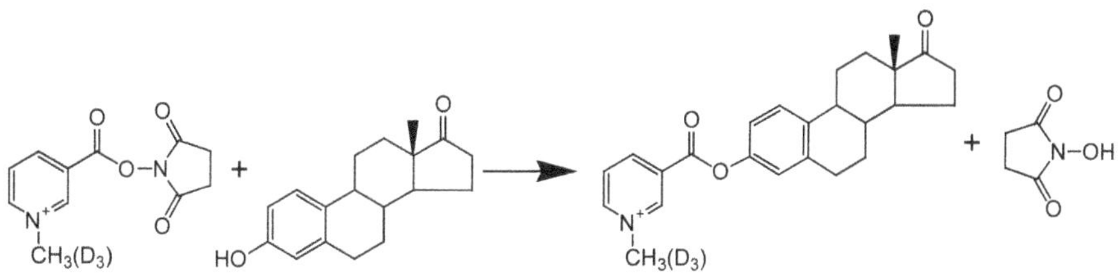

Figure 8.1 Representative derivatisation reactions of estrogens (adapted from ref. 24).

N-methyl-nicotinic acid *N*-hydroxysuccinimide ester (C1-NA-NHS).[24] This strategy has been applied to amino acids and peptides,[28,48,49] alcohols, phenols and thiols,[50,51] sugars[52,53] and carboxylic acids.[21] In terms of reactivity, the phenolic hydroxyl group in estrogens is similar to the amine group in amino acids, suggesting that the amino acid derivatisation procedure can be used, with a few modifications, for estrogen labelling as well. Problems with the solubility of analytes were addressed by changing to a non-aqueous reaction system in which the estrogens are soluble [small amounts of water (<10% in volume) in the reaction did not adversely affect the reaction]. The second modification was implemented in order to stabilise the derivatives by neutralising the reaction with formic acid after reaction completion. Although the reaction is initiated at basic pH, the derivatives generated are unstable under these conditions (similar problems have been observed in the case of O-acylation of tyrosine with C1-NA-NHS[28]). In neutral solution, the derivatives are stable for at least one week at 4 °C. It has been shown that under optimised conditions, C1-NA-NHS only reacted with phenolic hydroxyl groups, and not with alcohol hydroxyls. This property is significant, since it increases the analytical selectivity of estrogens in complex biological matrices.

8.3.1.2 LC-MS Analysis of Estrogen Derivatives

Estrogens have very similar structures, sharing an aromatic steroid backbone, hydroxyl or methoxyl groups at similar positions and several pairs of positional isomers or enantiomers. Detailed analysis of extracted ion chromatograms of the derivatised estrogens subjected to LC-MS (Figure 8.2A, Table 8.1) shows that, except for 2-hydroxyestrone and 7β-estradiol, all the other estrogens are well resolved within retention times of 7 min.[24] The mass difference between 2-hydroxyestrone and 7β-estradiol is, however, 14 amu and mass spectrometry easily differentiates between the two. Compared to the separation of underivatised estrogens with an aqueous ammonium acetate mobile phase, these results suggest that derivatisation increased the chromatographic selectivity, and the presence of the quaternary amine derivatising 'tag' on the aromatic ring of estrogens diminishes interaction of that portion of the molecule with the C18 stationary phase.[24]

A unique characteristic of C1-NA-estrogen spectra is the presence of a single intense quaternary amine molecular ion, with no fragmentation or formation of adduct ions (Figure 8.2B). Consequently, maximum intensity was gained in selected ion monitoring, and hence quantification was enhanced.[24] This is advantageous when collision-induced dissociation of the parent ion is employed to increase detection selectivity. Derivatisation decreased the detection limit by 1–2 orders of magnitude when compared with the published detection limits for underivatised estrogens using ESI-MS (*e.g.* 10 ng ml^{-1} *versus* 0.44 ng ml^{-1} for 17β-estradiol, and 5 ng ml^{-1} *versus* 0.36 ng ml^{-1} for estrone: Table 8.1).[24]

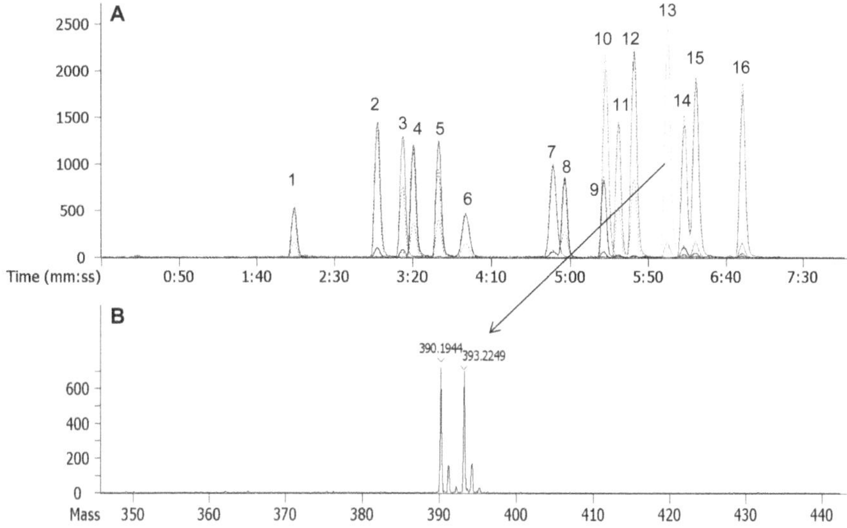

Figure 8.2 **A** – Extracted ion chromatograms of the C1-NA-NHS (solid lines) and C1-NA-NHS-d3 (dashed lines) derivatives of estrogens; **B** – corresponding spectra of the C1-NA-NHS and C1-NA-NHS-d3 derivatives of estrone (adapted from ref. 24). Peak identification is given in Table 8.1.

8.3.1.3 *Comparative Quantification through Isotopic Labelling*

A deuterated version of C1-NA-NHS (C1-d3-NA-NHS) was synthesised and used for the derivatisation of a known amount of estrogen, resulting in a set of heavy-labelled estrogen internal standards. It is desirable for isotope-coded internal standards to co-elute with the non-deuterated analyte derivatives in order to minimise quantification errors. A 1 : 1 mixture of C1-NA-estrogens and C1-d3-NA-estrogens was examined to determine the degree of analytical similarity.[24] Extracted ion chromatograms of all the derivatives in the mixture, together with a representative MS spectrum for a single chromatographic peak are shown in Figure 8.2A and 8.2B, respectively (the two sets of peaks completely overlapped chromatographically). Doublet clusters of ions separated by 3 atomic mass units were observed in mass spectra of nearly the same peak intensity.[24] When used in comparative quantification, the stable isotope-coded internal standard method was found to be linear throughout concentration ratios ranging from 1 : 1 to 1 : 40 for all the estrogens (Table 8.1) *via* the measurement of concentration ratio points at 1 : 1, 1 : 5, 1 : 10, 1 : 15, 1 : 20, 1 : 2 0, 1 : 25, 1 : 30, 1 : 35, 1 : 40, 1 : 50, 1 : 100.[24]

8.3.1.4 *Method Validation in Complex Sample*

Recovery was explored in order to examine the extraction and labelling efficacies of estrogens from blood samples. Aliquots of pooled normal serum was extracted, derivatised and analysed by procedures described by

Table 8.1 Linearity and dynamical range for comparative quantifications and their detection limits (reprinted from ref. 24).

Peak number	Derivatised estrogens	m/z d_3 labelled	m/z Non-d_3 labelled	Linearity[a]	Avg LOD [ng ml^{-1}, (pg/injection), n = 5]
1	16-epiestriol	411.2911	408.2675	$y = 0.9901x + 0.032$; $r^2 = 0.9925$	2.34 (11.71)
2	16-keto-17b-estradiol	409.2611	406.2359	$y = 0.9243x + 0.013$; $r^2 = 0.9953$	0.39 (1.99)
3	16a-hydroxyestrone	409.2714	406.2373	$y = 1.1043x + 0.010$; $r^2 = 0.9989$	0.62 (3.12)
4	estriol	411.2944	408.2596	$y = 0.8445x + 0.052$; $r^2 = 0.9969$	0.69 (3.47)
5	17-epiestriol	411.2924	408.2642	$y = 0.8582x + 0.044$; $r^2 = 0.9957$	0.46 (2.34)
6	4-hydroxyestradiol	411.2786	408.255	$y = 0.8376x + 0.047$; $r^2 = 0.9952$	0.75 (3.75)
7	4-hydroxyestrone	409.2586	406.2433	$y = 0.8411x + 0.060$; $r^2 = 0.9885$	0.56 (2.84)
8	2-hydroxyestradiol	411.2882	408.258	$y = 1.1139x + 0.019$; $r^2 = 0.9966$	0.69 (3.47)
9	2-hydroxyestrone	409.2812	406.2363	$y = 0.8512x + 0.054$; $r^2 = 0.9989$	0.75 (3.75)
10	17β-estradiol	395.2286	392.1844	$y = 0.8421x + 0.016$; $r^2 = 0.9988$	0.44 (2.23)
11	4-methoxyestradiol	425.3775	422.3347	$y = 0.8921x + 0.033$; $r^2 = 0.9987$	0.48 (2.40)
12	2-methoxyestradiol	425.3782	422.3412	$y = 0.8656x + 0.041$; $r^2 = 0.9988$	0.37 (1.87)
13	estrone	393.2098	390.1731	$y = 0.8797x + 0.039$; $r^2 = 0.9928$	0.36 (1.80)
14	4-methoxyestrone	423.3415	420.3153	$y = 0.8856x + 0.027$; $r^2 = 0.9986$	0.62 (3.12)
15	2-methoxyestrone	423.3394	420.3125	$y = 0.8934x + 0.021$; $r^2 = 0.9980$	0.46 (2.34)
16	2-hydroxyestrone-3-methyl ether	423.3481	420.315	$y = 0.9089x + 0.029$; $r^2 = 0.9989$	0.62 (3.12)

[a] x = Concentration ratio of non-d3 labelled to d3-labelled analytes; y = average extracted ion chromatographic peak intensity ratio of this pair of analytes. Linearity range for all is $1:1 \sim 1:40$.

Yang *et al.*[24] No chromatographic peaks corresponding to ion masses of the 16 estrogen standards were found (using extracted ion chromatograms), indicating that the concentrations of estrogens in the pooled serum matrix are present at levels lower than the detection limits and therefore serum can be used as a blank for this study. To determine recovery, known amounts of estrogen standards were added to the serum, and the sample carried through the extraction, drying and C1-d3-NA-NHS (heavy) derivatisation process.[24] Another aliquot of the standards was directly derivatised with the C1-NA-NHS reagent (light form) and added to 'heavy'-labelled serum. Recovery was determined by comparing the extracted chromatographic peak area of the deuterated (heavy) and non-deuterated (light) derivatives, and these data showed that except for 16α-hydroxyestrone [(42.0 ± 0.53% (mean ± SD, n = 3)] and 17-epiestriol [(73.0 ± 0.46% (mean ± SD, n = 3)], extraction efficiency was >95%;[24] no substances in human serum that interfered with the analysis were found. Understanding the lower recovery for 16α-hydroxyestrone and 17-epiestriol is, however, not straightforward. Based on the current recovery study protocol, the extraction and drying steps, rather than derivatisation reaction, affect the recovery measurement. The use of the same mixed standard solution for 'spiking' serum after derivatisation with C1-d3-NA-NHS, and for direct derivatisation with C1-NA-NHS, is a possible reason for the additional sample error. Indeed, the authors speculated that these two estrogens are less stable or less extractable than the others.[24]

8.3.1.5 Determination of Estrogens in Blood Serum Collected from Breast Cancer Patients

The above method was also used to determine estrogen levels in blood serum collected from breast cancer patients, where estrogen concentrations are known to be increased. A heavy-labelled standard solution was mixed with an equivalent volume of light-labelled human cancer serum sample solution, and the mixture was analysed by the LC-ESI-MS technique (Figure 8.3). Except for 16-epiestriol, the other 15 estrogens listed in Table 8.1 were identified.[24] The concentrations of estrogens were estimated (based on the comparative quantitation protocol). For most estrogens, the concentrations were in the range of 80–530 pg ml^{-1}. An exception is estrone, which, with a concentration of 1.2 ng ml^{-1}, was much higher than all the others. This is not unexpected since it is the major form of estrogen.

8.3.2 Absolute Quantification Targeting Multiple Functional Groups: Analysis of Central Carbon and Energy Metabolism

Most of the GSIST methods have been primarily focused on the development and application of derivatising agents targeting single functional

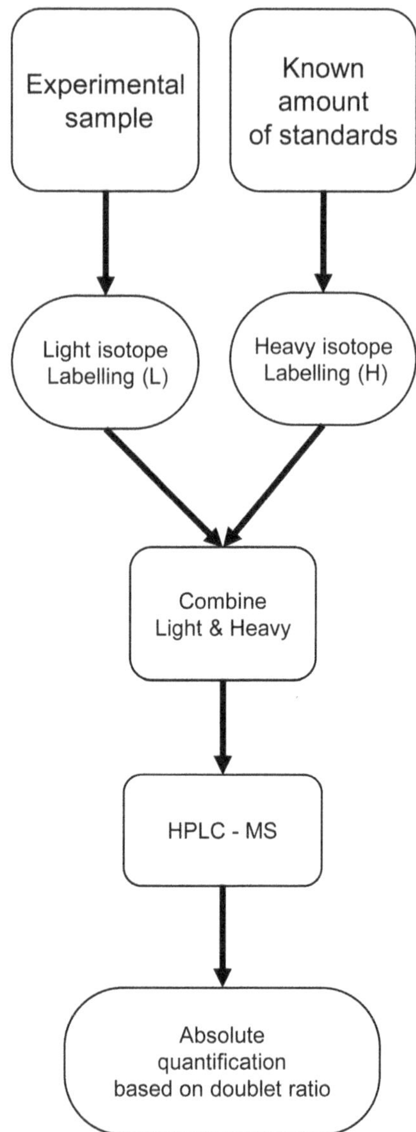

Figure 8.3 Work-flow for estrogen quantification by GSIST (adapted from ref. 24).

groups.[20–24,28] Although these coding agents work well for specific classes of molecules, they have some limitations in global- and pathway-targeted approaches, since not all molecules contain the same functional groups. For this reason, a new derivatisation reagent targeting multiple functional groups, which is more suitable for the quantification of specific metabolic pathways as well as differential global metabolomics, has been introduced.[25]

Amongst the whole cellular metabolic network, central carbon metabolism composed of glycolysis, the pentose-phosphate pathway and the tricarboxylic acid cycle (TCA) play a key function in substrate degradation, energy and co-factor regeneration and biosynthetical precursor supply. The compounds directly involved in central carbon metabolism contain carbonyl, phosphate and carboxyl groups. Therefore, a relatively 'globalised' labelling approach, which could introduce ^{13}C-coded hydrophobic moieties into all metabolites involved in these pathways, and that would allow the determination of all these compounds in a single RPLC-MS run, had to be developed. Reductive amination with an amino group-containing reagent is a common method of labelling carbonyl groups.[54] Amino group-containing reagents have also been reported to label carboxyl and phosphate groups *via* nucleophilic addition reactions which utilise a water-soluble carbodiimide such as EDC.[55,56] Based on these observations, isoforms of aniline (including aniline-^{13}C$_6$) were selected and employed for the derivatisation reactions summarised in Figure 8.4.[25]

8.3.2.1 Derivatisation and Analytical Conditions

Phosphomonoesters are typically labelled with aniline at 20 °C and pH values in the 4.5–5.5 range for 1.00 hr using EDC catalysis.[56] These conditions are similar to those in primary amino group-labelling of carboxyl groups using EDC.[55] However, carbonyl labelling with primary amines is often achieved in a non-aqueous solvent such as methanol with ~30% acetic acid at *ca.* 50 °C. Since the intermediate Schiff base adduct formed in carbonyl-labelling is unstable under acidic conditions, it was found to be valuable to reduce the adducts arising with NaCNBH$_3$ to form stable secondary amines. Carbonyl groups in glucose, xylose and phosphosugars were also derivatised under these conditions, even without the addition of acetic acid.[25] The inclusion of acid in the reaction possibly converts ketoses to aldoses, but this process was found not to bring an advantage here since the Schiff base intermediate is unstable at a pH value of 4.5. Indeed, adjusting the pH to 10.0 *via* the addition of 2.0 µl of TEA on completion of the reaction appreciably increased stability. No significant degradation was observed

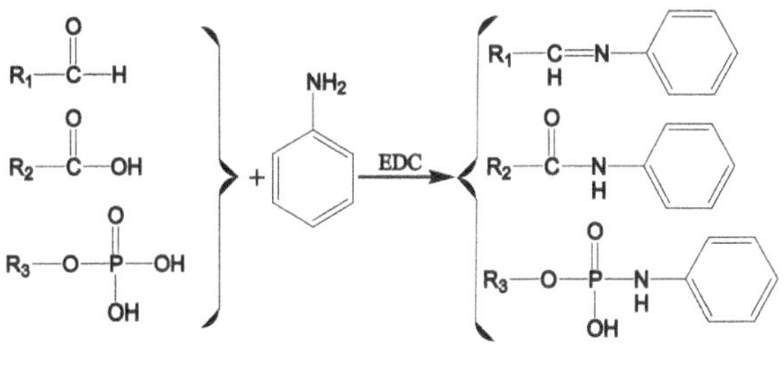

Figure 8.4 General labelling schemes for carbonyl, phosphoryl and carboxyl functional groups with aniline (adapted from ref. 25).

throughout a 3-day period when a labelled sample was placed in the auto-sampler at 10 °C. The addition of $NaCNBH_3$ was also investigated, and the overall LC separation deteriorated with the use of this reagent.[25] Therefore, this reduction step was omitted.

Aniline concentrations ranging from 0.3 to 6.0 M were used to further optimise the primary labelling conditions above using extracted ion-chromatographic peak intensities.[25] With increasing concentrations, the labelling yield increased, especially for carbonyl-containing analytes and some carboxylic acids, such as succinic and furmaric acids. Labelling time was examined from 10 to 150 min., and the yield was approximately 70% in 10 min., which slowly increased to nearly quantitative derivatisation at a time period of 105 min., where most unlabelled analytes were present at levels below 0.1%. Raising the labelling temperature from ambient to 50 °C decreased the labelling efficacy, a phenomenon presumably attributable to the acceleration of EDC hydrolysis. The final optimised protocol was set to label at ambient temperature for 2 hr with at least a 300-fold excess of aniline at a pH value of 4.5.[25]

Labelled standards were analysed using ion-paired RPLC followed by ESI-MS in the negative ion mode of ionisation. TBA was adapted as an ion-pairing agent in view of its promising performance in the separation of unlabelled central carbon metabolites.[17] Optimisation of the separation focused on mobile phase pH and TBA concentration, and a pH value of 5.0 was found to be the optimum for all the analytes. To shorten analysis time, 5.00 mM TBA was used.[25]

8.3.2.2 *Method Evaluation and Validation*

Most of the metabolites studied here have more than one functional group which could be labelled. In order to validate the labelling reaction, standard metabolites were individually labelled with aniline and aniline-$^{13}C_6$. A mixture of equivalent amounts of the two labelled solutions was analysed. Labelling patterns can be easily recognised by examining the spectrum (Figure 8.5).[25] Chromatographic peaks should contain a doublet set of ions, *e.g.* two major ions of similar peak intensity and a mass difference of $6 \times n$ (where $n = 1$, 2 or 3), where n is the number of functional groups in the molecule that were labelled. Moreover, the *m/z* value of the first ion in the doublet should be the molecular weight of the metabolite plus 75 amu$\times$n ($n = 1$, 2, 3). Retention time, labelling pattern and MS species identified are summarised in Table 8.2.[25]

However, some special cases need to be addressed. Indeed, since simple labelled sugars are neutral, their chloride adducts were detectable by the negative ESI-MS technique. Instead of aniline-labelling, the dehydrated glycerol 3-P ion was found. This ion probably arose *via* inter-molecular addition of a hydroxyl group at C-1 that attached to the EDC-activated C3-phosphate. Phosphoaldoses such as glucose 6-phosphate and ribose 5-phosphate were *bis*-labelled, whilst phosphoketoses such as fructose

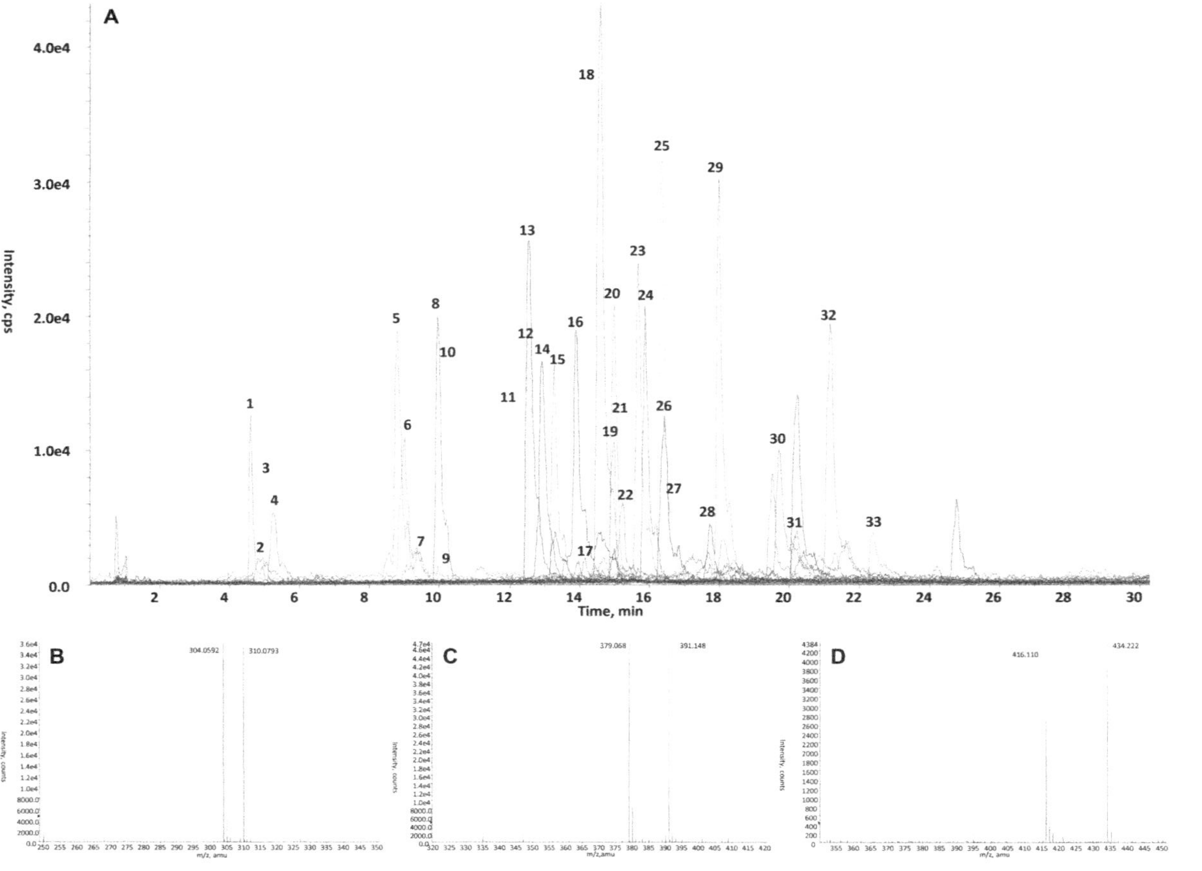

Figure 8.5 A – Overlapped extracted ion chromatograms of 33 metabolites in a 14.3 µM agent-containing standard mixture (adapted from ref. 25). Peak identification is given in Table 8.2. Representative MS doublets from aniline-labelled and aniline-^{13}C$_6$-labelled metabolites mixed at a 1 : 1 ratio; B – D-ribulose 5-phosphate-momo-aniline; C – D-ribose 5-phosphate-bis-aniline; D – citrate tri-aniline.

Table 8.2 Labelling and identification of standard metabolites by the RPLC-MS technique (reprinted from ref. 25).

Peak No.	Compound	Retention time (min.)	m/z value			Labelling pattern	MS species
			^{12}C labelling	^{13}C labelling	Non-labelling		
1	Glycerol 3-phosphate	4.60			152.99	Non-	[M-H$_2$O-H]$^-$
2	Xylose	4.96	260.07	266.09		Mono-	[M + Cl]$^-$
3	NAD	5.01			698.08	Non-	[M + Cl]$^-$
4	Glucose	5.27	290.08	296.10		Mono-	[M + Cl-H]$^-$
5	Fructose 6-phosphate	8.81	334.07	340.09		Mono-	[M-H]$^-$
6	Lactic acid	8.93	164.07	170.09		Mono-	[M-H]$^-$
7	D-Ribulose 5-phosphate	9.37	304.06	310.08		Mono-	[M-H]$^-$
8	AMP	9.97	421.10	421.12		Mono-	[M-H]$^-$
9	Dihydroxyacetone 1-P	10.07	244.04	250.06		Mono-	[M-H]$^-$
10	NADP	10.31			724.06	Non-	[M-H]$^-$
11	D-(−)Glycerate 3-P	11.94	242.03	248.06		Mono-	[M-H]$^-$
12	FAD	12.56			784.15	Non-	[M-H]$^-$
13	ADP	12.56	501.07	507.03		Mono-	[M-H]$^-$
14	Fructose 1,6-bisphosphate	12.93	396.03	402.05		Mono-	[M-H]$^-$
15	Gluconate 6-phosphate	13.29	425.11	437.15		Bi-	[M-H]$^-$
16	Glucose 6-phosphate	13.90	409.12	421.16		Bi-	[M-H]$^-$
17	NADH	14.16	633.11	639.13		Mono-	[M-nicotinamide + H$_2$O-H]$^-$
18	Ketoglutarate	14.61	295.01	307.14		Bi-	[M-H]$^-$
19	DL-Glyceraldehyde 3-P	14.99	319.09	331.13		Bi-	[M-H]$^-$
20	Malate	15.01	283.11	295.15		Bi-	[M-H]$^-$
21	ATP	15.14	581.03	587.05		Mono-	[M-H]$^-$
22	D-Ribose 5-phosphate	15.25	379.11	391.15		Bi-	[M-H]$^-$
23	Acetyl Co A	15.69			790.11	Non-	[M-H]$^-$
24	D-Erythrose 4-phosphate	15.88	349.09	361.13		Bi-	[M-H]$^-$
25	Phospho(enol)pyruvate	16.38	317.07	329.11		Bi-	[M-H]$^-$
26	Succinate	16.43	267.12	279.16		Bi-	[M-H]$^-$
27	NADPH	16.47	695.07	701.09		Mono-	[M-nicotinamide-H]$^-$
28	Fumarate	17.72	265.09	277.13		Bi-	[M-H]$^-$
29	Glycerate 1,3-bisphosphate	17.99	490.09	508.15		Tri-	[M-H]$^-$
30	Oxalacetate	19.70	280.98	293.02		Bi-	[M-H]$^-$
31	Isocitrate	20.14	398.15	416.21		Tri-	[M-H]$^-$
32	Citrate	21.19	416.16	434.22		Tri-	[M-H]$^-$
33	*Cis*-aconitate	22.35	380.98	399.04		Tri-	[M-H$_2$O-H]$^-$

6-phosphate and ribulose 5-phosphate were *mono*-labelled. This illustrates that ketoses do not convert to aldoses in the weakly acid medium employed. However, this labelling pattern benefits the separation of these isomers. They are easily separated without the requirement to 'fine-tune' the separation.[25] NADH and NADPH were labelled, but with loss of the nicotinamide moiety. It is unclear whether this release is ascribable to 'in-source' fragmentation, or the labelling process itself.[25] On the other hand, oxidised forms of the coenzymes NAD, FAD and NADP were not labelled and quantification had to be performed *via* standard addition. It is presumed that this is attributable to the formation of an intra-molecular salt between the quaternary amine on the pyridine ring and the negatively charged phosphate group.[25] In cases where multiple phosphates exist within a molecule, only one phosphate group was labelled, such as with fructose 1,6-*bis*phosphate, ADP and ATP.[25]

Figure 8.5A shows overlapped extracted ion chromatograms from an equimolar mixture of 33 metabolites in which individual components were present at a concentration of 14.3 μM each.Whilst some of the components co-elute, they differ in *m/z* values and are easily differentiated. Some representative MS spectra with doublet ions are shown in Figure 8.5B–D, giving further confirmation of labelling patterns. Clearly, subsequent to labelling, RPLC-MS discriminated between 33 central carbon intermediates within a period of 30 min.[25]

The method was validated by determination of its limit of detection (LOD), limit of quantification (LOQ), linearity range correlation coefficient and within-assay precision for the 33 analytes by analysing aniline-labelled standard mixtures of variable concentrations 'spiked' with a fixed amount of the same aniline-$^{13}C_6$-labelled standards. For most compounds, the LOD (S/N = 3) and LOQ (S/N = 10) were established below 1.0 and 2.5 μM, respectively.[25] The 3-phosphoenol pyruvate LOD was one of the lowest (at 0.09 μM) with a 20 μl injection volume. Calibration curves for each compound were computed by plotting the peak intensity ratios between variable (light) and constant (heavy) amounts of standard *versus* the additive nominal concentrations. Linearity was calculated using a non-weighted least squares linear regression method which generally spanned 2 to 3 orders of magnitude with correlation coefficients greater than 0.995. A linear regression of all compounds showed a unit slope with an interception point close to zero. Analytical precision was calculated from 'within-assay' variability by measuring the peak intensity ratios of analytes to its $^{13}C_6$-reference at a concentration ratio of 1:1, and expressed as the percentage relative standard deviation (coefficient of variation). This variation was generally below 5%.[25] This approach reflects variations involved in the overall process, including sample preparation, labelling and LC-MS analysis.

The method was also validated by determination of energy metabolites in a yeast cell extract.[25] In view of the unavailability of standards or structural analogues, 5 out of 35 metabolites could not be quantified. The remainder of the metabolites were divided into three categories based on the quantification approach. Typically, GSIST quantification includes derivatisation of sample and standard solutions with aniline and aniline-^{13}C$_6$, respectively. After labelling, mixtures of the sample and standards were mixed at a specific ratio and analysed by LC-MS. The concentrations of individual metabolites were then determined from the ratio between the intensity of corresponding light (experimental sample) and heavy (standard) peaks in the doublet sets of ions (*approach A- labelled metabolites*). Most of the intermediates in the cell extract were quantified in this manner but, as noted above, some of the metabolites were not labelled or standards were unavailable. In the *labelled but no standard available approach (B)*, quantification of D-6-phospho-glucono-δ-lactone was based on the ^{13}C-labelled standard of its contiguous peak, fructose 6-phosphate, which served as a structural analogue. With the *unlabelled metabolites approach (C)*, underivatised analytes were quantified *via* the standard addition technique as described by Huck *et al.*[16] Quantification results arising from yeast central carbon and energy metabolism are summarised in Figure 8.6. Generally, the RSD was below 10.0%.[25]

8.3.3 Relative Quantification of Unknown Metabolites in Complex Samples: Determination of Triterpenoid Metabolomic Fingerprints

The GSIST approach is also a viable technique for the *in situ* characterisation of the 'molecular fingerprints' of complex mixtures (relative amount of bioavailable metabolites in tissue can be determined without a corresponding purified standard). The relative concentration of the individual compounds in the tissue is calculated as a ratio between the light form (from experimental tissue) and the heavy form (from a natural source of investigated compounds that serves as an internal standard). The method also allows for an improved identification of unknown compounds sharing common active functional groups. Whilst the detection of a co-eluting peak doublet in the LC-MS chromatogram indicates the presence of a derivatisation 'tag' (targeted group), the mass difference between the light and heavy forms specifies the number of active groups in the molecule. A typical example of GSIST application for molecular fingerprinting is the characterisation of a precious Asian mushroom *Ganoderma lucidum* that was recognised in China, Korea and Japan more than 4000 years ago.[57] Two major biologically active classes of compounds isolated from *G. lucidum* have been identified as polysaccharides (mainly glucans and glycoproteins) and lanostane-type triterpenes (ganoderic acids, ganoderic alcohols and their derivatives).[58] Since the first purification of two triterpenes, ganoderic acids A and B, from *G. lucidum* in 1982,[59] more than 130 triterpenes have been isolated from the fruiting bodies, spores, mycelia and culture media of this species.[60] These bioactive molecules are predominantly oxygenated lanostane-type triterpenes that can be divided according to the number of carbons (C24, C27 and C30 compounds) and their functional groups.[61] Some

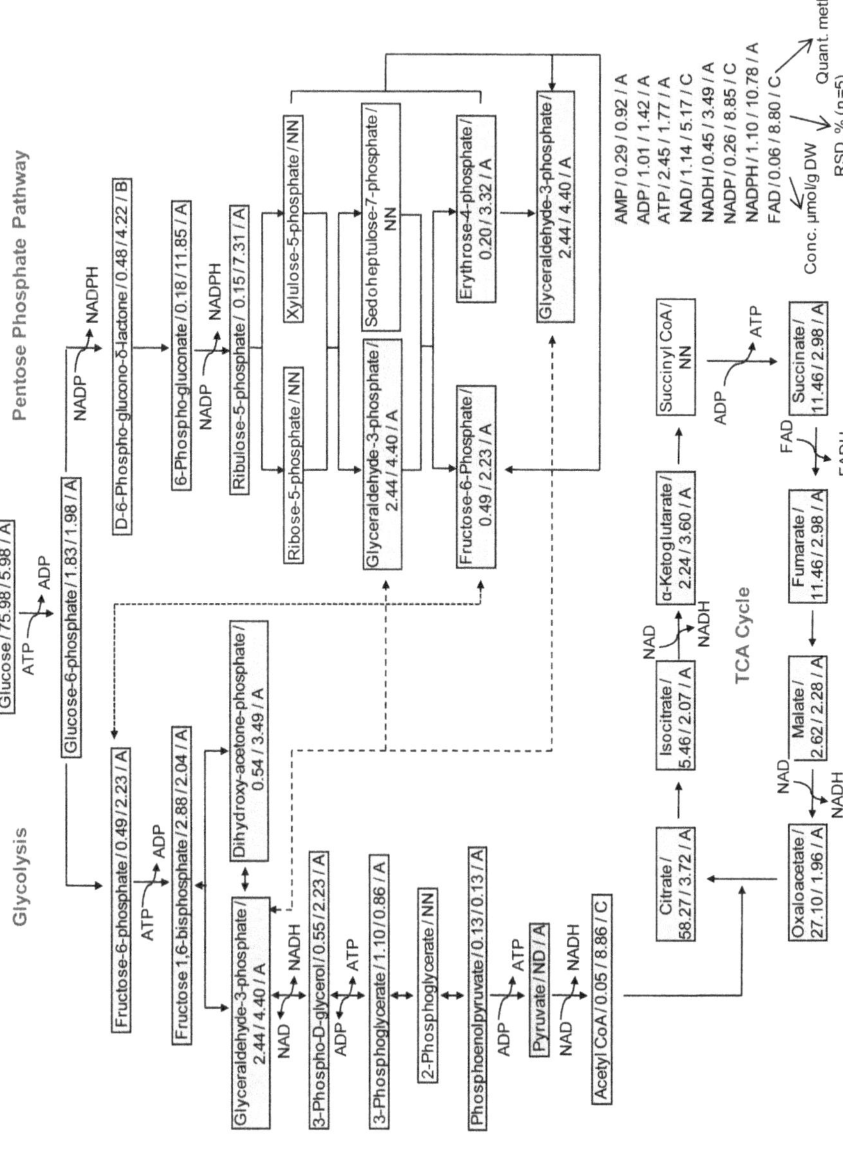

Figure 8.6 Central carbon metabolism map and determined metabolite concentrations in yeast (adapted from ref. 25). The quantification approaches using: A – GILISA, B – structural analogue and C – Standard addition are shown. Further abbreviations: NN – no standard available; ND – not determined.

of the *G. lucidum* triterpene-containing extracts, or isolated triterpenes, modulate specific signalling pathways in cancer cells and hence they exert anti-cancer activity in cell culture and animal models.[62–65] The bioactivity of these natural products is, however, often through the synergism of multiple metabolites acting pleiotropically. This 'cocktail' effect renders the isolation, identification and characterisation of active components extremely complex and challenging. Using GSIST, we were able to compare the ganoderic acid contents between two commercially available extracts, tentatively identify and relatively quantify some of the metabolites in pharmacokinetic studies.

8.3.3.1 Evaluation of Derivatisation Step

The linearity of the response and concentration ratios of the two isoforms is an important aspect in stable isotope-based quantification. If the GSIST technique is used for quantification, then two issues must be addressed.[27] Firstly, the coded isoforms must be chromatographically co-eluted in order to minimise differential suppression from the ESI ionisation process. Secondly, the MS peak height (area) ratio of this pair of isotopomers should be proportional to their concentration ratio within a specified concentration range in both the sample and the control. These issues were examined with typical representatives of ganoderic acids, specifically ganoderic acid A, F and H standard solutions, and CMP and CMP-d$_3$ derivatising agents (Figure 8.7). The results acquired indicated that CMP and CMP-d$_3$ labelled standards are completely co-eluted, and the method exhibits linearity for concentration ratios ranging from 0.1:1 to 1:10, with correlation co-efficients greater than 0.995.[20] Moreover, the derivatisation reaction was >99% efficient, and no chromatographic isotope effect was found.[20]

8.3.3.2 Tentative Identification of Ganoderic Acids in Mushroom Extracts

ReishiMax and GLT extracts were used to examine the ability of the GSIST technique for the tentative identification of ganoderic acids. The structure of most ganoderic acids contains the carboxylic acid functional group. In order to identify the number of carboxylic acid group-containing compounds in mushroom extract samples, they were divided into two equivalent volumes (Figure 8.8A) and each aliquot was individually derivatised with the CMP (light) and CMP-d3 (heavy) forms of the labelling reagent.[20] Subsequent to derivatisation, the aliquots were combined and analysed using LC-MS. Visual examination of the chromatograms showed that most of the high-intensity peaks were potentially derivatised ganoderic acids. For detailed analysis, we then used a rule in which two co-eluting ions with a mass difference of **n**×3 amu must be found in the chromatogram corresponding to CMP (light) and CMP-d3 (heavy) derivatives (in which **n** represents the number of derivatising 'tags' per labelled molecule, and 3 amu is the

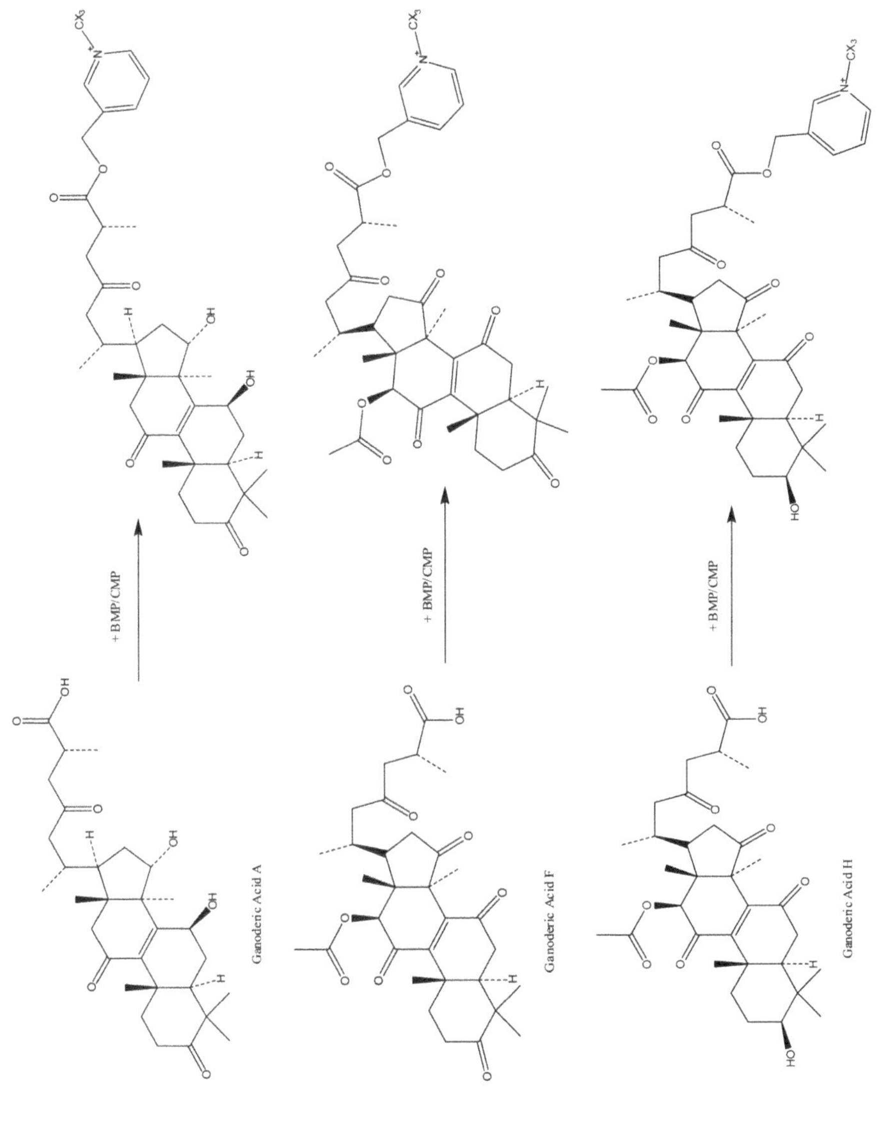

Figure 8.7 Structures of ganoderic acid A, F and H, and representative derivatisation reactions (adapted from ref. 20).

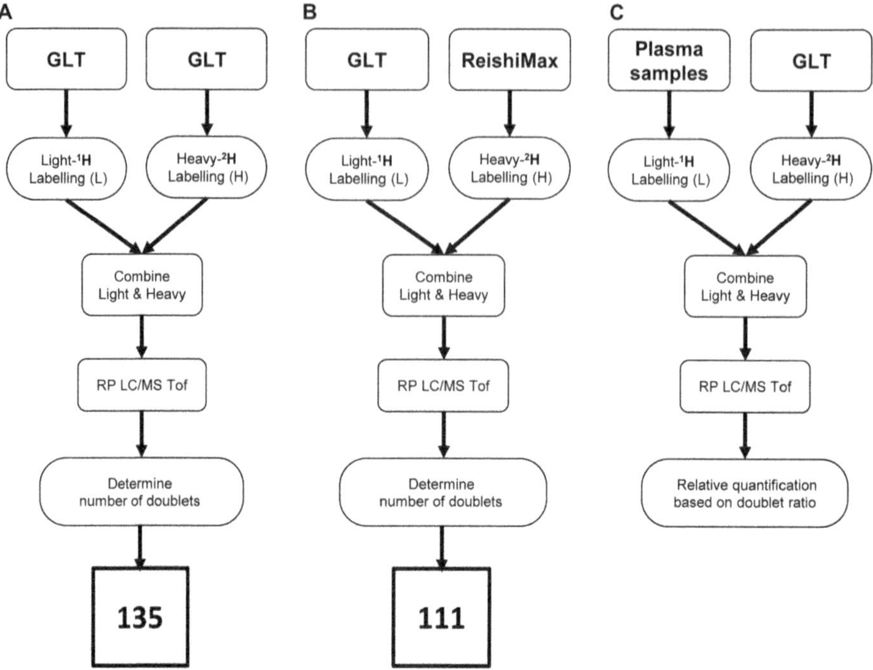

Figure 8.8 Labelling strategy for the determination of **A** – carboxylic acid functional group-containing metabolites in GLT extracts; **B** – common carboxylic acid group-containing metabolites between ReishiMax and GLT extracts; and **C** – labelling schema for the relative quantification of ganoderic acids in plasma samples (adapted from ref. 20).

molecular weight difference between CMP and CMP-d3 tags).[20] Ganoderic acids were then identified by assuming that the ratio of ion intensities of corresponding doublets must be in the range of 0.80–1.25. The molecular mass of the original (underivatised) compound was calculated from the molecular mass of the ion found in the MS scan by subtracting the molecular mass of the derivatising 'tag'. The calculated molecular masses were then employed to search our 'in house' carboxyl group-containing ganoderic acid database derived from the manuscript by Feng and Shen.[66] Multiple-carboxyl group-containing compounds and doublets outside the defined range of ratios were excluded from the list obtained.

Overall, 135 doublets were found in the GLT extract. Based on the criteria specified above, 57 doublets were identified as potential ganoderic acids, and 11 of those matched with the database (Table 8.3). In addition, ganoderic acids A, F and H were confirmed with standards.[20]

According to the manufacturer, the GLT extract represents the triterpene-enriched fraction of ReshiMax. To determine the similarity between GLT and ReishiMax, extracts were labelled with the CMP and CMP-d3 'tag', respectively, equivalent volumes mixed, and then analysed in a single LC-MS run (Figure 8.8B). We found 111 doublets, which suggested a high degree of

Table 8.3 Ganoderic acid peaks found in GLT-extract: identification results (reprinted from ref. 20).

Peak #	m/z	RT (min.)	Ratio[a]	Identification[b]	Compound
1	552.35	14.76	0.96	T	Lucidenic Acid H
2	558.37	18.52	1.03	T	Ganoderic Acid S1
3	562.30	14.11	1.06	T	Lucidenic Acid D1
4	564.32	13.01	0.96	T	Lucidenic Acid D
5	566.32	4.39	1.21	T	Lucidenic Acid B
					Lucidenic Acid E1
					Lucidenic Acid L
6	566.33	7.22	0.93	T	Methyl Lucidenate G
7	622.36	13.33	1.25	S	Ganoderic Acid A
8	632.31	15.15	1.00	T	Ganosporeric Acid
9	676.33	15.13	1.03	S	Ganoderic Acid F
10	678.35	13.81	1.09	S	Ganoderic Acid H
11	708.32	15.27	1.16	T	Ganoderic Acid Mh

[a]Ratio of ion intensities of corresponding light and heavy forms of doublets.
[b]T – Tentatively identified compounds; S – compounds confirmed by standards.

similarity between the two extracts. In general, the intensities of peaks corresponding to the GLT extract were two-fold higher than that observed in ReshiMax.[20]

The above results illustrate the strengths of the GSIST technique for the analysis of carboxylic acid-containing molecules such as ganoderic acids. The high selectivity and sensitivity obtained through the derivatisation process utilised greatly facilitates the profiling of carboxylic acids in 'real' biological samples.

8.3.3.3 Bioavailability Study

Current techniques available for determinations of the bioavailabilities of specific compounds require purified standards. If standards are not available, samples are usually analysed in separate LC-MS runs, and individual chromatograms are then deconvoluted, aligned and normalised. Peaks of interest are then statistically analysed based on their intensities. When substantially different matrices are compared (such as those of mushroom and plasma), the results are affected not only by ionisation suppression effects, but also by small 'matrix-dependent' shifts in the elution time of corresponding peaks, leading to the misalignment and identification of a large number of false-positive/-negative peaks. This can be eliminated using *in vitro* labelling strategies, and the original source of the compounds of interest as an internal standard. To evaluate the GSIST technique for the profiling of potentially interesting anti-cancer compounds from complex natural products, Sprague-Dawley female rats received a dose of GLT extract (500 mg kg^{-1} of body weight) *via* gastric gavage, and aliquots of plasma were collected at 0, 15, 30, 45, 60, 90 and 120 min. periods thereafter.[20] Plasma samples and GLT extracts were labelled with the light and heavy forms of the derivatisation 'tag', mixed and then analysed by LC-MS. Resulting chromatograms were further analysed by extracting the specific heavy form ions identified in the GLT extract (Table 8.3), and then seeking the corresponding light forms that originated from the plasma samples.[20] The concentrations of the individual compounds in the plasma were calculated as a ratio between the light (from plasma) and the heavy forms of GLT internal standard (Figure 8.8C). The time-dependent profiles of compounds with $m/z = 622.35$, 676.33 and 678.35 (corresponding to ganoderic acids A, F and H) are shown in Figure 8.9. All three ganoderic acids reached a maximum concentration at 90min., and then rapidly declined within the next 30 min.[20] Our data are in partial agreement with a recent study that demonstrated the rapid absorption of *G. lucidum* triterpenes into this biofluid from the gastrointestinal tract after an oral administration of *G. lucidum* extract. These authors detected maximum plasma concentrations by HPLC in 16.79 min. (ganoderic acid C_2), 6.26 min. (ganoderic acid B), 32.10 min. (ganoderic acid K) and 24.88 min. (ganoderic acid H).[67] One potential reason for the difference in pharmocokinetics between our study and that of Wang *et al.* could be the amount of biologically active triterpenes available in the samples suggested to analysis. In our study, we used an oral administration of 500 mg kg^{-1} of GLT, which contains 3.88 mg g^{-1} of ganoderic acid A, 0.95 mg g^{-1} of ganoderic acid F and 1.74 mg g^{-1} of ganoderic acid H. In contrast, Wang *et al.* used 1.2 g kg^{-1} of *G. lucidum* extract, which contains 0.0553 mg g^{-1} of acid C_2, 0.258 mg g^{-1} of ganoderic acid B, 0.0758 mg g^{-1} of ganoderic acid K and 0.155 mg g^{-1} of ganoderic acid H.[67] In addition to the identified triterpenes, these extracts may contain additional molecules that can affect pharmacokinetic parameters. A further investigation of an oral administration of purified ganoderic acid A (25 mg kg^{-1}) resulted in its serum detection by an enzyme immunoassay at 20 min.[68] In summary, three different methods have shown that specific ganoderic acids can be detected in plasma within a short period of time after oral administration, observations confirming the bioavailability of these biologically active molecules *in vivo*.

8.3.4   Discovery of Novel Metabolites

Most of the metabolites in 'real' biological samples are present at very low concentrations, and their detection requires the use of highly sensitive methods. The process for discovering a novel metabolite using MS is more difficult than that required for detecting or confirming a pre-conceived one. In complex samples, large numbers of ions with a wide dynamic range of concentrations generates many chromatographic peaks which vary in peak intensities, and the use of high accuracy/resolution instruments such as TOF-MS, Orbitrap-MS or FT-ICR-MS is often necessary to identify peaks that may potentially correspond to novel metabolites. Structural determination and confirmation is then achieved by Q-TOF, Q-Trap, ion trap or other instruments capable of MS^n analysis. Even with these mass spectrometers, it may be still difficult to provide an unambiguous identification of

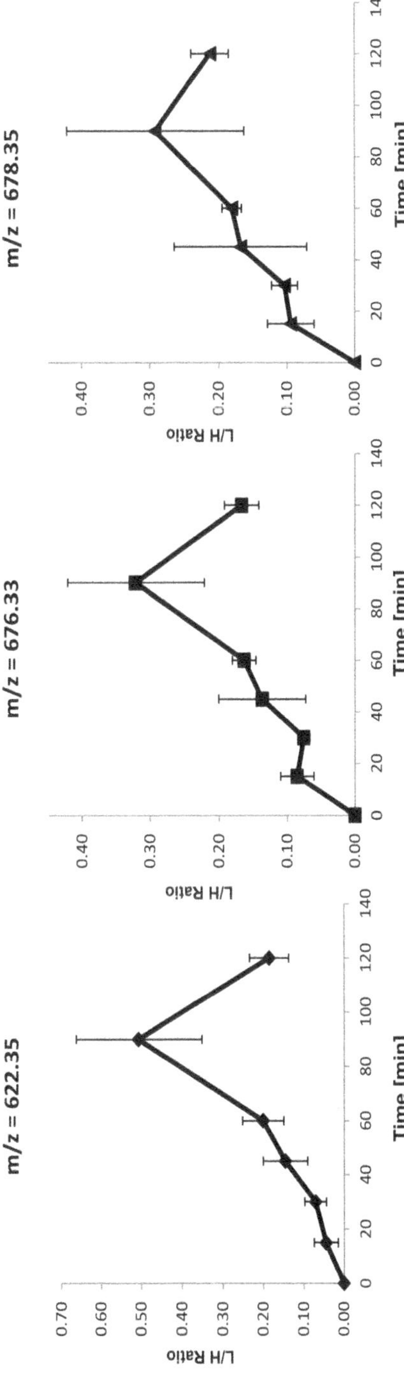

Figure 8.9 Plasma/time profiles of peaks with *m/z* values of 622.35, 676.33 and 678.35 following the administration of GLT extract (adapted from ref. 20). Peaks were identified and confirmed as ganoderic acids A, F and H (see Table 8.3). The relative quantities of individual compounds is expressed as a ratio between their peak intensities detected in plasma (light form), and the corresponding peak intensity found in GLT extract (heavy form). A 'spike' of the GLT extract served as an internal standard.

endogenous metabolites. Ultimately, the best approach would be the synthesis of possible metabolites, and then direct comparison of the LC-MS data achieved for the experimentally detected metabolites and synthesised compounds. This approach is, however, expensive and time-consuming, and also requires much effort to conduct the synthesis of the desired structures.

Tocopherol represents the most biologically relevant form of vitamin E in many diets, and this agent (vitamin E) provides antioxidant activities through the electron-trapping of lipophilic electrophiles and reactive nitrogen and oxygen species.[69] Plasma γ-tocopherol has been shown to be inversely associated with the risk of prostate cancer and coronary heart diseases.[69] The first metabolite of γ-tocopherol, 7,8-dimethyl-2-(β-carboxy-ethyl)-6-hydroxychromanol (γ-CEHC), was isolated from human urine in 1996.[70,71] The structure of γ-CEHC suggests that it is metabolised *via* phytyl chain oxidation of γ-tocopherol, without oxidative modification of the chroman ring. A cytochrome P450 ω-hydroxylase pathway of γ-tocopherol catabolism was elucidated and confirmed in 2002.[72,73] This pathway involves cytochrome P450-mediated ω-hydroxylation of the tocopherol phytyl side chain, followed by a stepwise removal of two- or three-carbon moieties, ultimately yielding γ-CEHC as shown in Figure 8.10. Recently, several novel sulfate conjugates of γ-tocopherol metabolites were reported,[74] and some intermediate metabolites have been shown to be potent inhibitors for cyclooxygenases.[75] The same mechanism of γ-tocopherol side-chain degradation has also been reported in other cells, such as HepG2 cells.[76] However, except for γ-CEHC, all other intermediates have been readily available for use in metabolic studies, and therefore they have not been further confirmed with radioactive or stable isotope-labelled forms of γ-tocopherol or, alternatively, *via* direct comparison with authentic compounds.

Tocopherol metabolites contain carboxylic acid anion groups (Figure 8.10), suggesting that maximum sensitivity would be achieved *via* ESI in negative ion mode. For reversed-phase HPLC separation, a small amount of an organic acid is typically added to the mobile phase in order to maintain separation reproducibility, and also to extend the lifetime of a column. This additive may suppress the ionisation efficacy of the acidic analytes, and significantly reduce the detection sensitivity.[28] To circumvent this obstacle, organic acid metabolites can be derivatised with a quaternised derivatising reagent, C1-NANHS, as illustrated in Figure 8.11A. This derivatisation reaction introduces a permanent positive charge to the analytes, and hence ESI efficacies and detection sensitivities are greatly increased.[21] Moreover, the added 'tag' decreases the hydrophobicities of derivatised compounds and consequently reduces their retention times and the overall time required for the analysis. It is, however, important to note that the parent molecule and its metabolites must carry the same functional group, such as phenolic moiety in the case γ-tocopherol catabolism. Metabolites with modified functional groups cannot be derivatised and identified in this manner. In addition, products of the derivatisation reaction involved must be stable in order to provide consistent results during LC-MS analysis.

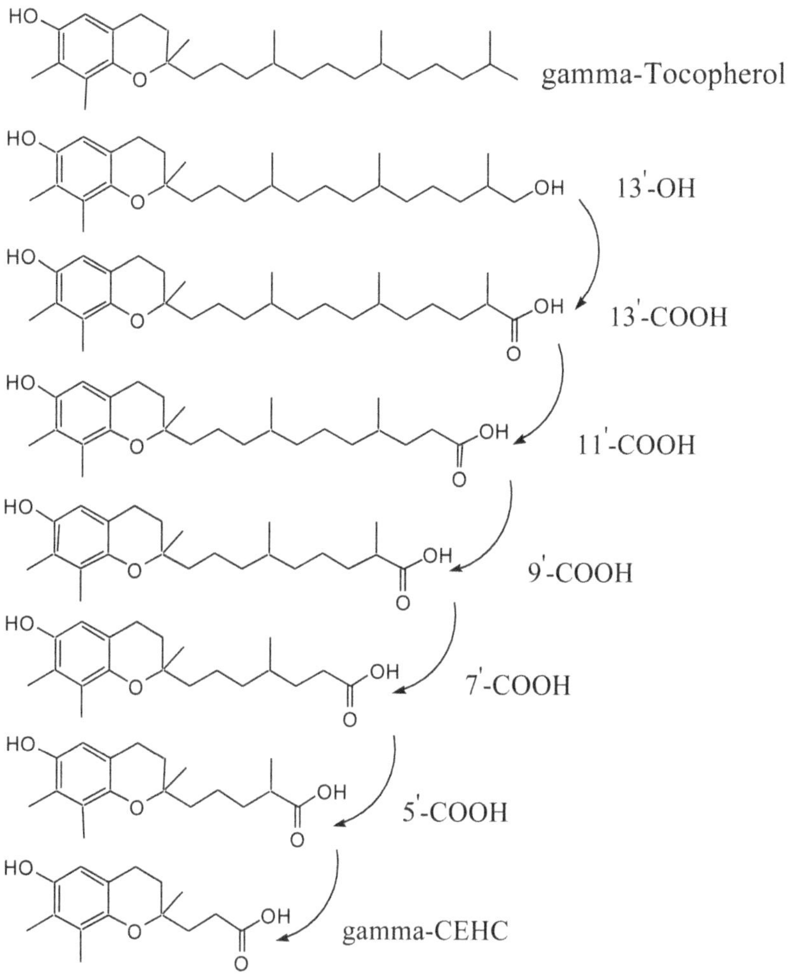

Figure 8.10 Metabolic pathway of γ-tocopherol (adapted from ref. 23).

Although derivatisation increases the ionisation efficiencies of tagged molecules, a major challenge presented by complex samples is the method required to differentiate between peaks that correspond to derivatised and non-derivatised metabolites.

8.3.4.1 Identification of γ-Tocopherol Metabolites

The method illustrated in Figure 8.11B was used to verify various metabolites derived from γ-tocopherol, including long-chain carboxychromanols generated in A549 cells.[75] The experimental sample was split into two portions, and these aliquots were simultaneously derivatised with C1-NA-NHS (light form) or C1-d3-NA-NHS (heavy form).[23] The resulting deuterated and

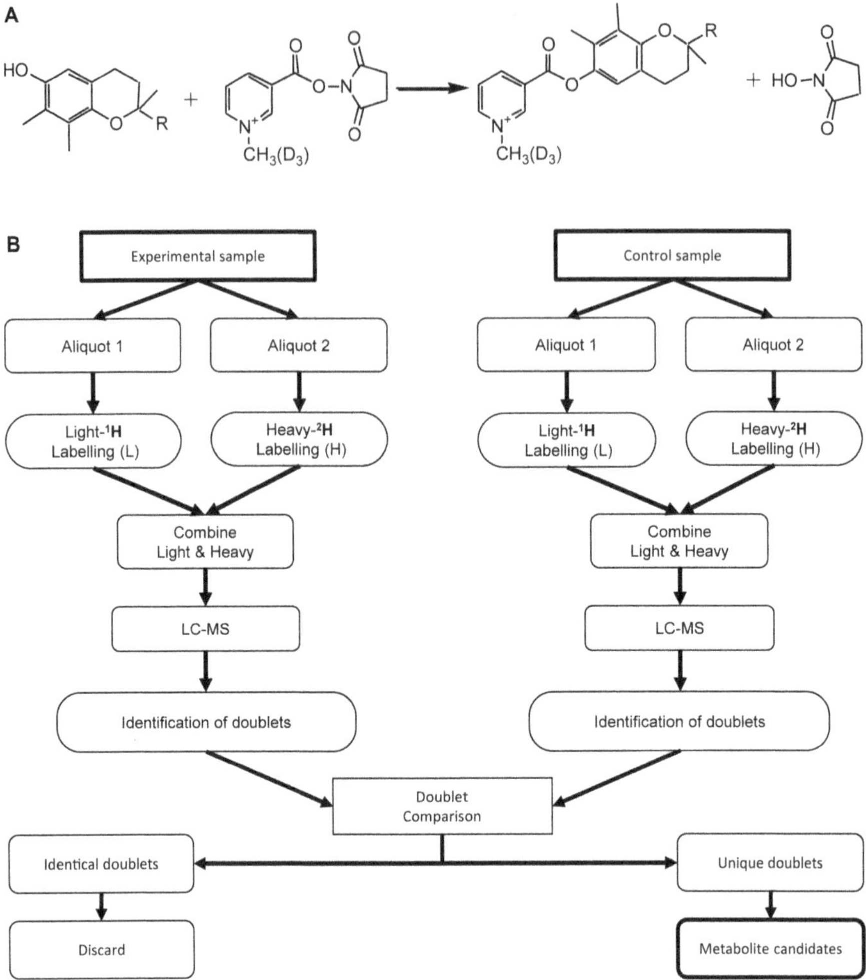

Figure 8.11 **A** – Derivatisation reaction of γ-tocopherol and its metabolites; **B** – Strategy for metabolite discovery (adapted from ref. 23).

non-deuterated experimental samples were then mixed and analysed by LC-MS. Chromatographic peaks were detected by LECO Unique MS software, and then were examined manually for the presence of ion doublets.[23] Since all metabolites of γ-tocopherol contain a single phenol group, and equivalent volumes of samples were derivatised by the reagents with a mass difference of 3 amu, only doublets with similar intensities and a mass difference of 3 were selected for further analysis (the control sample was treated and analysed in the same manner). The selected ion doublets from both experimental and control samples were then compared, and those detected in both samples disregarded:[23] the doublets exclusively found in the experimental sample were identified as a potential metabolite of γ-tocopherol catabolism. This conclusion was based on two facts: (1) cells for experimental and control samples were treated identically, except that the experimental cells were cultivated in the presence of γ-tocopherol and (2) γ-tocopherol and its metabolites contain a single derivatisable functional group. This procedure assured the exclusion of false-positive/-negative candidates *via* the use of these three additional selection criteria for identification, specifically the derivatisation of targeted molecules, the mass difference and a 1 : 1 ratio in peak intensities between the light and heavy form of the same compound. Using this screening process, six ions with m/z values of 496.30 (ion **1**), 593.32 (ion **2**), 621.33 (ion **3**), 566.38 (ion **4**), 663.40 (ion **5**) and 552.40 (ion **6**) for non-deuterated derivatisation were uniquely found in the experimental sample (Figure 8.12A). None of these ions were detectable in the analytical ion chromatogram of the control, non-γ-tocopherol-supplemented cell culture media (Figure 8.12B).[23]

8.3.4.2 Structural Determination of Identified Ions

The above evidence strongly suggests that the six ion candidates arose from γ-tocopherol catabolism (Figure 8.12A). After subtracting the mass of the derivatising 'tag' (mass is 120.13), the ion **1** (m/z 376.17) matches the metabolite 9′-COOH, ion **4** (m/z 446.25) matches the 13′-COOH one and ion **6** (m/z 432.27) matches the 13′-OH one.[23] No match, however, was found for ions **2**, **3** and **5**. Further examination revealed that the m/z values of ions **2**, **3** and **5** correspond to the derivatised 9′-COOH, 11′-COOH and 13′-COOH metabolites plus the mass addition of 97.02, respectively. This number represents the mass of N-hydroxysuccinimide (NHS) (Mr 115.09) minus H_2O (Mr 18.02). NHS is a by-product of the derivatisation process, and also a reactant required for synthesis of the derivatising reagent. Therefore, we suspected that these ions might be generated from the secondary derivatisation through the carboxylate groups of the metabolites, and NHS produced in the derivatisation of the phenolic group.[23] In order to verify this hypothesis, a standard metabolite (γ-CEHC) was derivatised with C1-NANHS and C1-d3-NANHS and the 1 : 1 mixture was analysed. The expected derivatised ions arising from the phenolic group (Peak 1, $m/z = 384.14$ and 387.17) and also a major ion with an extra mass of 97.02 (Peak 2, $m/z = 481.16$ and 484.18) appeared. In both cases, similar intensities between the light and heavy forms of the corresponding ions indicated that they are direct products of this derivatisation reaction.[23] These results are consistent with those obtained for ions **2**, **3** and **5**, and suggest the esterification of samples. Reaction between the carboxylic acid groups and N-hydroxysuccinamide (NHS) requires activation of the former through a coupling reagent, such as dicyclohexylcarbodiimide. In this case, however, no coupling reagent is present, and the precise mechanism of the reaction process remains unclear.

Molecular structures of derivatised γ-tocopherol, and secondarily derivatised 13′-COOH (ion **5**) were also confirmed by MS/MS analysis. Indeed, corresponding fractions were collected during LC separation and directly

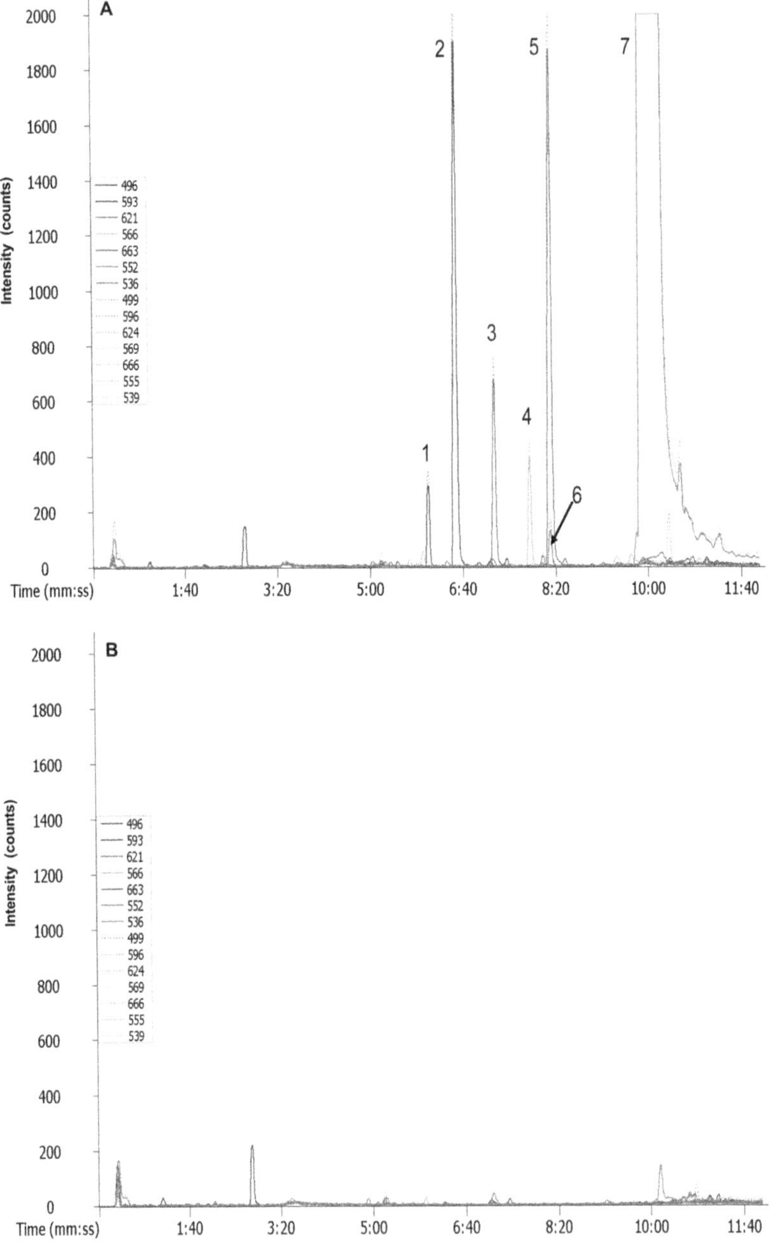

Figure 8.12 Extracted ion chromatograms (EIC) suggesting the presence of intermediates of γ-tocopherol catabolism in cell culture media (adapted from ref. 23). **A:** cells incubated in the media with γ-tocopherol; **B:** cells incubated in the media without γ-tocopherol. The numbered peaks represent the ions which were uniquely found in cell culture media (except for γ-tocopherol itself). Dotted and solid lines represent deuterated and non-deuterated derivatives, respectively.

Table 8.4 Ions corresponding to metabolites of
γ-tocopherol in cell culture media
(reprinted from ref. 23).

Observed m/z value of ions	Corresponding metabolite
496.30	9′-COOH (mono-derivatised)
593.32	9′-COOH (bis-derivatised)
621.33	11′-COOH (bis-derivatised)
566.38	13′-COOH (mono-derivatised)
663.40	13′-COOH (bis-derivatised)
552.40	13′-OH

analysed using tandem mass spectrometry. The MS/MS fragments originating in precursor molecules can be easily recognised, since they contain characteristic deuterated and non-deuterated ion doublet patterns.[23] Moreover, the fragment similar to ion *m/z* 270.1 has been previously reported in the EI spectra of γ-tocopherol metabolites.[73] The MS/MS spectrum of ion **5** confirmed the hypothesis of the secondary derivatisation process, and therefore the six unique ions found in the experimental sample were identified as four metabolites of γ-tocopherol (Table 8.4).

8.4 Conclusion

GSIST is a new, highly sensitive LC-MS method that enables the analysis of metabolites at the levels required in life science research. Novel derivatisation reagents, and also the derivatisation methods, benefit the LC-ESI-MS analysis of metabolites in several manners including (1) an increase in the detection sensitivity of ESI-MS by one to two orders of magnitude (when expressed relative to that observed with underivatised metabolites); (2) adjustment of the hydrophobicity/hydrophilicity of analysed compounds, retention times and band spreading in chromatography, a process which provides an enhanced analysis speed and a higher level of peak resolution than those of existing methods; and (3) an increased efficacy of both comparative quantitation and recovery studies, *via* an allowance of multiple sample (or standard) analyses to be performed in a single process (by a combination of the use of isotopic versions of derivatisation reagents).

Acknowledgements

This work was supported by grants: AG13319 (NIH), 5R33DK070290-03 (NIH), R01AT001821 (NIH), DBI-0421102 (NSF) and the US Department of Energy Biomass Program. The text and figures are, in part, reprinted from: 1) W.-C. Yang, F. Regnier and J. Adamec, Stable isotope-coded quaternization for comparative quantification of estrogen metabolites by high-performance liquid chromatography-electrospray ionization mass spectrometry, *J. Chrom. B*, 2008, **870**(2), 233–240, Copyright (2008), with permission from Elsevier; 2)

W.-C. Yang, M. Sedlak, F. Regnier, N. Mosier, N. Ho and J. Adamec, Simultaneous quantification of metabolites involved in central carbon and energy metabolism using reverse phase liquid chromatography-mass spectrometry and *in vitro* ^{13}C labeling, *Anal. Chem.*, 2008, **80**(24), 9508–9516, Copyright (2008), with permission from American Chemical Society; 3) J. Adamec, A. Jannasch, S. Dudhgaonkar, A. Jedinak, M. Sedlak and D. Sliva, Development of a new method for improved identification and relative quantification of unknown metabolites in complex samples: Determination of a triterpenoid metabolic fingerprint for the *in situ* characterization of Ganoderma bioactive compounds, *J. Sep. Sci.*, 2009, **32**(23–24), 4052–4058, Copyright Wiley-VCH Verlag GmbH & Co. KGaA, reproduced with permission; and 4) W.C. Yang, F. E. Regnier, Q. Jiang and J. Adamec, *In vitro* stable isotope labeling for discovery of novel metabolites by LC-MS: Confirmation of γ-tocopherol metabolism in human A549 cell, *J. Chrom. A*, 2010, **1217**(5), 667–675, Copyright 2010, with permission from Elsevier.

References

1. O. Fiehn, J. Kopka, R. N. Trethewey and L. Willmitzer, Identification of uncommon plant metabolites based on calculation of elemental compositions using gas chromatography and quadrupole mass spectrometry, *Anal. Chem.*, 2000, **72**, 3573–3580.
2. H. Hajjaj, P. J. Blanc, G. Goma and J. Francois, Sampling techniques and comparative extraction procedures for quantitative determination of intra-and extracellular metabolites in filamentous fungi, *FEMS Microbiol. Lett.*, 1998, **164**, 195–200.
3. G. J. G. Ruijter and J. Visser, Determination of intermediary metabolites in Aspergillus niger., *J. Microbiol. Methods*, 1996, **25**, 295–302.
4. U. Theobald, W. Mailinger, M. Bates, M. Rizzi and M. Reuss, *In vivo* analysis of metabolic dynamics in Saccharomyces Cerevisiae: I. Experimental observations, *Biotechnol. Bioeng.*, 1997, **55**, 305–316.
5. E. Groussac, M. Ortiz and J. Francois, Improved protocols for quantitative determination of metabolites from biological samples using high performance ionic-exchange chromatography with conductimetric and pulsed amperometric detection, *Enzyme Microb. Technol.*, 2000, **26**, 715–723.
6. N. B. Jensen, K. V. Jokumsen and J. Villadsen, Determination of the phosphorylated sugars of the Embden-Meyerhoff-Parnas pathway in Lactococcus lactis using a fast sampling technique and solid phase extraction, *Biotechnol. Bioeng.*, 1999, **63**, 356–362.
7. H. P. Smits, A. Cohen, T. Buttler, J. Nielsen and L. Olsson, Cleanup and analysis of sugar phosphates in biological extracts by using solid-phase extraction and anion-exchange chromatography with pulsed amperometric detection, *Anal. Biochem.*, 1998, **261**, 36–42.
8. S. Picioreanu, I. Poels, J. Frank, J. C. van Dam, G. W. van Dedem and L. J. Nagels, Potentiometric detection of carboxylic acids, phosphate

esters, and nucleotides in liquid chromatography using anion-selective coated-wire electrodes, *Anal. Chem.*, 2000, **72**, 2029–2034.

9. M. Bhattacharya, L. Fuhrman, A. Ingram, K. W. Nickerson and T. Conway, Single-run separation and detection of multiple metabolic intermediates by anion-exchange high-performance liquid chromatography and application to cell pool extracts prepared from *Escherichia coli*, *Anal. Biochem.*, 1995, **232**, 98–106.

10. S. R. Hull and R. Montgomery, Separation and analysis of 4′-epimeric UDP-sugars, nucleotides, and sugar phosphates by anion-exchange high-performance liquid chromatography with conductimetric detection, *Anal. Biochem.*, 1994, **222**, 49–54.

11. J. B. Ritter, Y. Genzel and U. Reichl, High-performance anion-exchange chromatography using on-line electrolytic eluent generation for the determination of more than 25 intermediates from energy metabolism of mammalian cells in culture, *J. Chrom. B: Analyt. Technol. Biomed. Life Sci.*, 2006, **843**, 216–226.

12. A. M. Vogt, C. Ackermann, T. Noe, D. Jensen and W. Kubler, Simultaneous detection of high energy phosphates and metabolites of glycolysis and the Krebs cycle by HPLC, *Biochem. Biophys. Res. Commun.*, 1998, **248**, 527–532.

13. G. Stephanopoulos, H. Alper and J. Moxley, Exploiting biological complexity for strain improvement through systems biology, *Nat. Biotechnol.*, 2004, **22**, 1261–1267.

14. M. M. Wamelink, E. A. Struys, J. H. Huck, B. Roos, M. S. van der Knaap, C. Jakobs and N. M. Verhoeven, Quantification of sugar phosphate intermediates of the pentose phosphate pathway by LC-MS/MS: application to two new inherited defects of metabolism, *J. Chrom. B: Analyt. Technol. Biomed. Life Sci.*, 2005, **823**, 18–25.

15. A. Buchholz, R. Takors and C. Wandrey, Quantification of intracellular metabolites in Escherichia coli K12 using liquid chromatographic-electrospray ionization tandem mass spectrometric techniques, *Anal. Biochem.*, 2001, **295**, 129–137.

16. J. H. Huck, E. A. Struys, N. M. Verhoeven, C. Jakobs and M. S. van der Knaap, Profiling of pentose phosphate pathway intermediates in blood spots by tandem mass spectrometry: application to transaldolase deficiency, *Clin. Chem.*, 2003, **49**, 1375–1380.

17. B. Luo, K. Groenke, R. Takors, C. Wandrey and M. Oldiges, Simultaneous determination of multiple intracellular metabolites in glycolysis, pentose phosphate pathway and tricarboxylic acid cycle by liquid chromatography-mass spectrometry, *J. Chrom. A*, 2007, **1147**, 153–164.

18. J. C. van Dam, M. R. Eman, J. Frank, H. C. Lange, G. W. K. van Dedem and S. J. Heijnen, Analysis of glycolytic intermediates in Saccharomyces cerevisiae using anion exchange chromatography and electrospray ionization with tandem mass spectrometric detection, *Anal. Chem. Acta*, 2002, **460**, 209–218.

19. L. Coulier, R. Bas, S. Jaspersen, E. Verheij, M. J. van der Werf and
    T. Hankemeier, Simulatenous Quantitative Analysis of Metabolites
    Using Ion-Pair Liquid Chromatography-Electrospray Ionization Mass
    Spectrometry, *Anal. Chem.*, 2006, **78**, 6573–6582.
20. J. Adamec, A. Jannasch, S. Dudhgaonkar, A. Jedinak, M. Sedlak and
    D. Sliva, Development of a new method for improved identification and
    relative quantification of unknown metabolites in complex samples:
    determination of a triterpenoid metabolic fingerprint for the in situ
    characterization of Ganoderma bioactive compounds, *J. Sep. Sci.*, 2009,
    32, 4052–4058.
21. W. C. Yang, J. Adamec and F. E. Regnier, Enhancement of the LC/MS
    analysis of fatty acids through derivatization and stable isotope coding,
    Anal. Chem., 2007, **79**, 5150–5157.
22. W. C. Yang, F. E. Regnier and J. Adamec, Comparative metabolite pro-
    filing of carboxylic acids in rat urine by CE-ESI MS/MS through positively
    pre-charged and (2)H-coded derivatization, *Electrophoresis*, 2008, **29**,
    4549–4560.
23. W. C. Yang, F. E. Regnier, Q. Jiang and J. Adamec, *In vitro* stable isotope
    labeling for discovery of novel metabolites by liquid chromatography-
    mass spectrometry: Confirmation of gamma-tocopherol metabolism in
    human A549 cell, *J. Chrom. A*, 2010, **1217**, 667–675.
24. W. C. Yang, F. E. Regnier, D. Sliva and J. Adamec, Stable isotope-coded
    quaternization for comparative quantification of estrogen metabolites
    by high-performance liquid chromatography-electrospray ionization
    mass spectrometry, *J. Chrom. B: Analyt. Technol. Biomed. Life Sci.*, 2008,
    870, 233–240.
25. W. C. Yang, M. Sedlak, F. E. Regnier, N. Mosier, N. Ho and J. Adamec,
    Simultaneous quantification of metabolites involved in central carbon
    and energy metabolism using reversed-phase liquid chromatography-
    mass spectrometry and *in vitro* 13C labeling, *Anal. Chem.*, 2008, **80**,
    9508–9516.
26. A. Chakraborty and F. E. Regnier, Global internal standard technology
    for comparative proteomics, *J. Chrom. A*, 2002, **949**, 173–184.
27. R. Zhang, C. S. Sioma, R. A. Thompson, L. Xiong and F. E. Regnier,
    Controlling deuterium isotope effects in comparative proteomics, *Anal.
    Chem.*, 2002, **74**, 3662–3669.
28. W. C. Yang, H. Mirzaei, X. Liu and F. E. Regnier, Enhancement of amino
    acid detection and quantification by electrospray ionization mass
    spectrometry, *Anal. Chem.*, 2006, **78**, 4702–4708.
29. R. W. Giese, Measurement of endogenous estrogens: analytical chal-
    lenges and recent advances, *J. Chrom. A*, 2003, **1000**, 401–412.
30. M. R. Anari, R. Bakhtiar, B. Zhu, S. Huskey, R. B. Franklin and
    D. C. Evans, Derivatization of ethinylestradiol with dansyl chloride to
    enhance electrospray ionization: application in trace analysis of
    ethinylestradiol in rhesus monkey plasma, *Anal. Chem.*, 2002, **74**,
    4136–4144.

31. V. F. Fredline, P. J. Taylor, H. M. Dodds and A. G. Johnson, A reference method for the analysis of aldosterone in blood by high-performance liquid chromatography-atmospheric pressure chemical ionization-tandem mass spectrometry, *Anal. Biochem.*, 1997, **252**, 308–313.

32. S. S. Tai and M. J. Welch, Development and evaluation of a reference measurement procedure for the determination of estradiol-17beta in human serum using isotope-dilution liquid chromatography-tandem mass spectrometry, *Anal. Chem.*, 2005, **77**, 6359–6363.

33. X. Xu, L. K. Keefer, D. J. Waterhouse, J. E. Saavedra, T. D. Veenstra and R. G. Ziegler, Measuring seven endogenous ketolic estrogens simultaneously in human urine by high-performance liquid chromatography-mass spectrometry, *Anal. Chem.*, 2004, **76**, 5829–5836.

34. X. Xu, J. M. Roman, T. D. Veenstra, J. Van Anda, R. G. Ziegler and H. J. Issaq, Analysis of fifteen estrogen metabolites using packed column supercritical fluid chromatography-mass spectrometry, *Anal. Chem.*, 2006, **78**, 1553–1558.

35. X. Xu, T. D. Veenstra, S. D. Fox, J. M. Roman, H. J. Issaq, R. Falk, J. E. Saavedra, L. K. Keefer and R. G. Ziegler, Measuring fifteen endogenous estrogens simultaneously in human urine by high-performance liquid chromatography-mass spectrometry, *Anal. Chem.*, 2005, **77**, 6646–6654.

36. X. Xu, R. G. Ziegler, D. J. Waterhouse, J. E. Saavedra and L. K. Keefer, Stable isotope dilution high-performance liquid chromatography-electrospray ionization mass spectrometry method for endogenous 2- and 4-hydroxyestrones in human urine, *J. Chrom. B: Analyt. Technol. Biomed. Life Sci.*, 2002, **780**, 315–330.

37. H. Yamada, K. Yoshizawa and T. Hayase, Sensitive determination method of estradiol in plasma using high-performance liquid chromatography with electrochemical detection, *J. Chrom. B: Analyt. Technol. Biomed. Life Sci.*, 2002, **775**, 209–213.

38. H. Adlercreutz, S. L. Gorbach, B. R. Goldin, M. N. Woods, J. T. Dwyer and E. Hamalainen, Estrogen metabolism and excretion in Oriental and Caucasian women *J. Natl. Cancer Inst.*, 1994, **86**, 1076–1082.

39. H. Adlercreutz, P. Kiuru, S. Rasku, K. Wahala and T. Fotsis, An isotope dilution gas chromatographic-mass spectrometric method for the simultaneous assay of estrogens and phytoestrogens in urine, *J. Steroid Biochem. Mol. Biol.*, 2004, **92**, 399–411.

40. L. A. Castagnetta, O. M. Granata, F. P. Arcuri, L. M. Polito, F. Rosati and G. P. Cartoni, Gas chromatography/mass spectrometry of catechol estrogens, *Steroids*, 1992, **57**, 437–443.

41. L. A. Dehennin and R. Scholler, Preparation and physico-chemical properties of some steroid heptafluorobutyrates and 3-enol heptafluorobutyrates, *Steroids*, 1969, **13**, 739–761.

42. T. Fotsis, The multicomponent analysis of estrogens in urine by ion exchange chromatography and GC-MS–II. Fractionation and quantitation of the main groups of estrogen conjugates, *J. Steroid Biochem.*, 1987, **28**, 215–226.

43. K. D. Pinnella, B. K. Cranmer, J. D. Tessari, G. N. Cosma and D. N. Veeramachaneni, Gas chromatographic determination of catecholestrogens following isolation by solid-phase extraction, *J. Chrom. B: Biomed. Sci. Appl.*, 2001, **758**, 145–152.

44. X. Xiao and D. McCalley, Quantitative analysis of estrogens in human urine using gas chromatography/negative chemical ionisation mass spectrometry, *Rapid Comm. Mass Spectrom.*, 2000, **14**, 1991–2001.

45. X. Xu, A. M. Duncan, B. E. Merz-Demlow, W. R. Phipps and M. S. Kurzer, Menstrual cycle effects on urinary estrogen metabolites, *J. Clin. Endocri. Metab.*, 1999, **84**, 3914–3918.

46. T. Chard, *An Introduction to Radioimmunoassay and Related Techniques*, Elsevier Biomedical Press, Amsterdam, 1982.

47. P. Kebarle, A brief overview of the present status of the mechanisms involved in electrospray mass spectrometry, *J. Mass Spectrom.*, 2000, **35**, 804–817.

48. D. Y. Ren, S. Julka, H. D. Inerowicz and F. E. Regnier, Enrichment of cysteine-containing peptides from tryptic digests using a quaternary amine tag, *Anal. Chem.*, 2004, **76**, 4522–4530.

49. N. A. Stewart, V. T. Pham, C. T. Choma and H. Kaplan, Improved peptide detection with matrix-assisted laser desorption/ionization mass spectrometry by trimethylation of amino groups, *Rapid Commun. Mass Spectrom.*, 2002, **16**, 1448–1453.

50. J. M. E. Quirke, C. L. Adams and G. J. Vanberkel, Chemical Derivatization for Electrospray-Ionization Mass-Spectrometry 1. Alkyl-Halides, Alcohols, Phenols, Thiols, and Amines, *Anal. Chem.*, 1994, **66**, 1302–1315.

51. J. M. E. Quirke and G. J. Van Berkel, Electrospray tandem mass spectrometric study of alkyl 1-methylpyridinium ether derivatives of alcohols, *J. Mass Spectrom.*, 2001, **36**, 1294–1300.

52. S. Broberg, A. Broberg and J. O. Duus, Matrix-assisted laser desorption/ionization time-of-flight mass spectrometry of oligosaccharides derivatized by reductive amination and N,N-dimethylation, *Rapid Commun. Mass Spectrom.*, 2000, **14**, 1801–1805.

53. J. Hsu, S. J. Chang and A. H. Franz, MALDI-TOF and ESI-MS analysis of oligosaccharides labeled with a new multifunctional oligosaccharide tag, *J. Am. Soc. Mass Spectrom.*, 2006, **17**, 194–204.

54. E. W. Baxter and A. B. Reitz, Reductive aminations of carbonyl compounds with borohydride and borane reducing agents, *Org. React.*, 2002, **59**, 1–714.

55. Q. L. Ford, J. M. Burns and J. L. Ferry, Aqueous *in situ* derivatization of carboxylic acids by an ionic carbodiimide and 2,2,2-trifluoroethylamine for electron-capture detection, *J. Chrom. A*, 2007, **1145**, 241–245.

56. M. G. Ivanovskaya, M. B. Gottikh and Z. A. Shabarova, Modification of oligo(poly)nucleotide phosphomonoester groups in aqueous solutions, *Nucleos. Necleot.*, 1987, **6**, 913–934.

57. S. P. Wasser, *Reishi (Ganoderma lucidum)*, CRC Press, Boca Raton, FL, 2005.

58. Y. Gao, S. Zhou, W. Jiang, M. Huang and X. Dai, Effects of ganopoly (a Ganoderma lucidum polysaccharide extract) on the immune functions in advanced-stage cancer patients, *Immunol. Invest.*, 2003, **32**, 201–215.

59. T. Kubota, Y. Asaka, I. Miura and H. Mori, Structures of Ganoderic Acid-a and Acid-B, 2 New Lanostane Type Bitter Triterpenes from Ganoderma-Lucidum (Fr) Karst, *Helv. Chim. Acta*, 1982, **65**, 611–619.

60. C. W. Huie and X. Di, Chromatographic and electrophoretic methods for Lingzhi pharmacologically active components, *J. Chrom. B: Analyt. Technol. Biomed. Life Sci.*, 2004, **812**, 241–257.

61. J. L. Gao, Z. L. Yu, S. P. Li and Y. T. Wang, *Edible Fungi China*, 2005, **24**, 6–11.

62. J. Jiang, B. Grieb, A. Thyagarajan and D. Sliva, Ganoderic acids suppress growth and invasive behavior of breast cancer cells by modulating AP-1 and NF-kappaB signaling, *Int. J. Mol. Med.*, 2008, **21**, 577–584.

63. Y. Kimura, M. Taniguchi and K. Baba, Antitumor and antimetastatic effects on liver of triterpenoid fractions of Ganoderma lucidum: mechanism of action and isolation of an active substance, *Anticancer Res.*, 2002, **22**, 3309–3318.

64. D. Sliva, Cellular and physiological effects of Ganoderma lucidum (Reishi), *Mini-Rev. Med. Chem.*, 2004, **4**, 873–879.

65. W. Tang, J. W. Liu, W. M. Zhao, D. Z. Wei and J. J. Zhong, Ganoderic acid T from Ganoderma lucidum mycelia induces mitochondria mediated apoptosis in lung cancer cells, *Life Sci.*, 2006, **80**, 205–211.

66. M. Feng and J. Shen, *Chemical Composition of Lingzhi*, Science Press, Beijing, 2005.

67. X. Wang, R. Liu, J. Sun, S. Guan, M. Yang, K. Bi and D. Guo, HPLC method for the determination and pharmacokinetic studies of four triterpenoids in rat plasma after oral administration of Ganoderma lucidum extract, *Biomed. Chrom.*, 2007, **21**, 389–396.

68. M. Hattori, International Symposium on Ganoderma Science, Auckland, 2001.

69. Q. Jiang, S. Christen, M. K. Shigenaga and B. N. Ames, gamma-tocopherol, the major form of vitamin E in the US diet, deserves more attention, *Am. J. Clin. Nutr.*, 2001, **74**, 714–722.

70. E. D. Murray, Jr., D. Kantoci, S. A. DeWind, A. E. Bigornia, D. C. D'Amico, J. G. King, Jr., T. Pham, B. H. Levine, M. E. Jung and W. J. Wechter, Endogenous natriuretic factors 3: isolation and characterization of human natriuretic factors LLU-alpha, LLU-beta 1, and LLU-gamma, *Life Sci.*, 1995, **57**, 2145–2161.

71. W. J. Wechter, D. Kantoci, E. D. Murray, Jr. , D. C. D'Amico, M. E. Jung and W. H. Wang, A new endogenous natriuretic factor: LLU-alpha, *Proc. Natl. Acad. Sci. U.S.A.*, 1996, **93**, 6002–6007.

72. R. S. Parker, T. J. Sontag, J. E. Swanson and C. C. McCormick, Discovery, characterization, and significance of the cytochrome P450 omega-hydroxylase pathway of vitamin E catabolism, *Ann. N. Y. Acad. Sci.*, 2004, **1031**, 13–21.

73. T. J. Sontag and R. S. Parker, Cytochrome P450 omega-hydroxylase pathway of tocopherol catabolism. Novel mechanism of regulation of vitamin E status, *J. Biol. Chem.*, 2002, **277**, 25290–25296.

74. Q. Jiang, H. Freiser, K. V. Wood and X. Yin, Identification and quantitation of novel vitamin E metabolites, sulfated long-chain carboxychromanols, in human A549 cells and in rats, *J. Lipid Res.*, 2007, **48**, 1221–1230.

75. Q. Jiang, X. Yin, M. A. Lill, M. L. Danielson, H. Freiser and J. Huang, Long-chain carboxychromanols, metabolites of vitamin E, are potent inhibitors of cyclooxygenases, *Proc. Natl. Acad. Sci. U.S.A.*, 2008, **105**, 20464–20469.

76. M. Birringer, P. Pfluger, D. Kluth, N. Landes and R. Brigelius-Flohe, Identities and differences in the metabolism of tocotrienols and tocopherols in HepG2 cells, *J. Nutr.*, 2002, **132**, 3113–3118.

^{18}O-assisted ^{31}P NMR and Mass Spectrometry for Phosphometabolomic Fingerprinting and Metabolic Monitoring

EMIRHAN NEMUTLU,*[a,b,†] SONG ZHANG,[a] ANDRE TERZIC[a] AND PETRAS DZEJA*[a,‡]

[a] Division of Cardiovascular Diseases, Departments of Medicine, Molecular Pharmacology and Experimental Therapeutics, Mayo Clinic, Rochester, MN, USA; [b] Department of Analytical Chemistry, Faculty of Pharmacy, University of Hacettepe, Ankara, Turkey
*Email: enemutlu@hacettepe.edu.tr; dzeja.petras@mayo.edu

9.1 Introduction

Metabolomic analyses require comprehensive and simultaneous systematic fingerprinting of multiple metabolites. These are to be identified and quantified along with their cellular and systemic variations in response to diseases, drugs, toxins and human lifestyle, as well as in the context of

---

†Present address: Department of Analytical Chemistry, Faculty of Pharmacy, University of Hacettepe, 06100 Ankara, Turkey.
‡Present address: Mayo Clinic, 200 First Street SW, Stabile 5, Rochester, MN 55905, USA.

Issues in Toxicology No. 21
Metabolic Profiling: Disease and Xenobiotics
Edited by Martin Grootveld

genetic or environmental challenges.[1–8] Analytical platforms developed for metabolomics studies allow screening of hundreds of metabolites from complex biological samples with analytical precision, comprehensiveness and sample throughput.[6,9–12] The physicochemical diversity of metabolites, from ionic inorganic species to hydrophilic carbohydrates, volatile alcohols and ketones, amino- and non-amino-organic acids, hydrophobic lipids and complex natural products, necessitates application of different complementary analytical techniques.[2,3,9] Currently, no single platform fulfils all requirements for an ideal global metabolite profiling tool. Application of advanced and information-rich spectroscopic techniques is typically essential for the generation of metabolic profiles required for metabolomic studies.[13] The main spectroscopic techniques employed for metabolomic studies are based on NMR spectroscopy (^{1}H, ^{31}P, ^{13}C and ^{17}O, amongst others) and mass spectrometry (direct infusion or combined with GC, LC or CE). Both techniques can give extensive structural and conformational information on multiple chemical classes in a single analytical procedure; however, they have differing analytical strengths and weaknesses.[1,11,13]

Characterisation of a metabolic phenotype requires knowledge not only of metabolite levels, but also of their turnover rates from which metabolic fluxes and, therefore, the dynamic state of a metabolic system can be determined (Figure 9.1).[14–16] Since many metabolites are present at low concentrations and associated with high flux/turnover rates through the metabolite pools, significant changes in metabolic flux could occur without changes in metabolite concentrations.[17] Therefore, dynamic metabolomic profiling and flux measurements are essential for a complete understanding of metabolic phenotypes.[2,16–20]

Stable isotope tracer-based metabolomic technologies allow simultaneous determinations of metabolite levels and their turnover rates with the subsequent evaluation of metabolic network dynamics.[14,15,21,22] ^{13}C labelling is widely used to track turnover of the carbon backbone of metabolites and label propagation through metabolic networks.[23–25] This technique alone, however, does not allow acquisition of a full 'picture' of metabolic dynamics and of the status of the cell energetic system. ^{18}O isotopes are suitable for the following of cellular phosphorus turnover and metabolic dynamics of phosphoryls in energetically and signal transduction-important biomolecules, as well as label distribution through phosphotransfer networks.[15,22,26–31] ^{18}O is a natural, stable and non-radioactive isotope of oxygen. When tissues or cells are exposed to media containing water with a known percentage of ^{18}O, $H_2{}^{18}$O rapidly equilibrates with cellular water, and then ^{18}O from water is incorporated into cellular phosphate metabolites to an extent proportional to the rate of enzymatic reactions involved.[30] The percentage of ^{18}O incorporation into phosphate metabolites of interest can be determined by ^{31}P NMR or mass spectrometry.[15,32,33] Incorporation of ^{18}O into phosphoryls as a consequence of cellular metabolic activity induces an isotope shift in the ^{31}P NMR spectrum ascribable to differences in the shielding effects of ^{16}O *versus* ^{18}O on the ^{31}P nucleus, in addition to a shift in the mass spectrum of

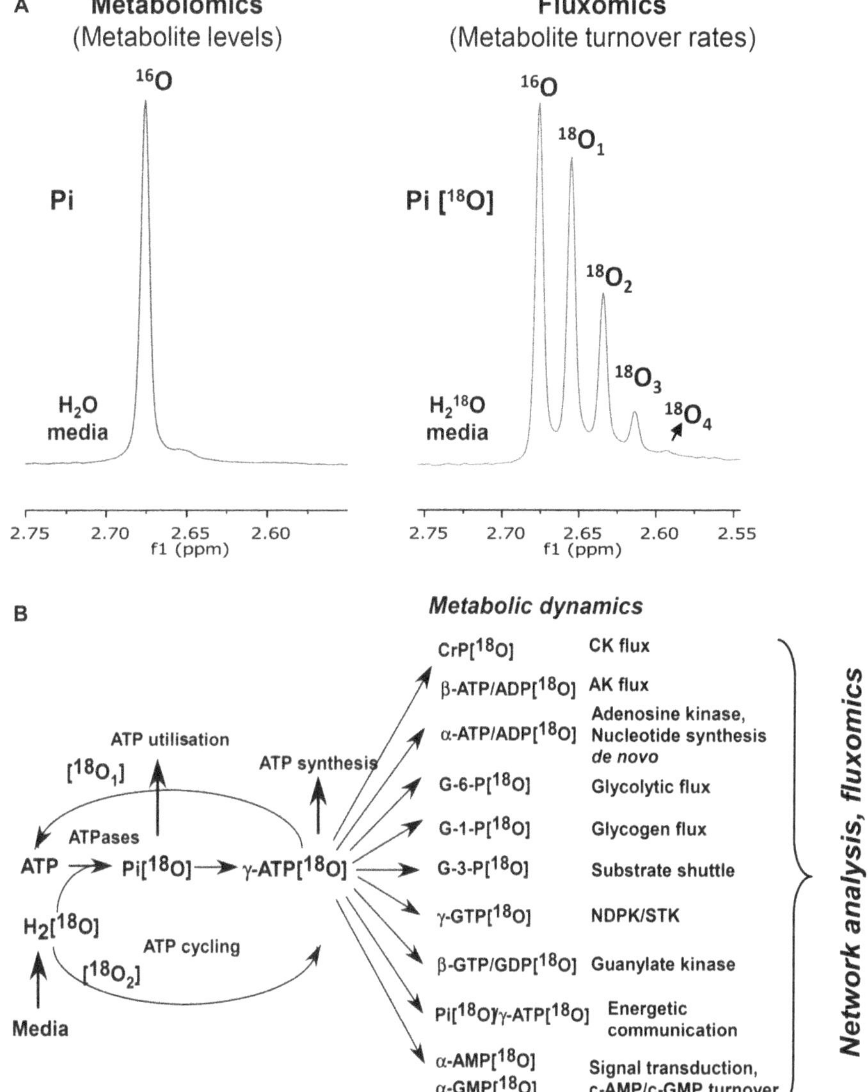

Figure 9.1 Principles of ^{18}O-labelling methodology for dynamic metabolomic profiling. **A** – Analytical differences between metabolomic (metabolite levels) and fluxomic (metabolite turnover rates) analyses using metabolic flux-dependent ^{18}O-labelling resulting in induced shift in ^{31}P NMR spectra of phosphoryl containing metabolites. **B** – Schematic representation of ^{18}O-labelling procedure for comprehensive characterisation of cellular energetic system and distribution of multiple phosphotransfer fluxes.

phosphoryl-containing metabolite species.[15,31,34] Calculation of the percentage of ^{18}O incorporation into phosphate metabolites from the induced isotope shift in ^{31}P NMR spectra acquired can be employed to determine (1) turnover rates and (2) phosphotransfer fluxes through specific energetic circuits (Figure 9.1A).

The ^{18}O labelling procedure is based on the incorporation of one ^{18}O atom, provided from $H_2^{18}O$, into P_i, with each act of ATP hydrolysis and the subsequent distribution of ^{18}O-labelled phosphoryls amongst other phosphate-carrying molecules (Figure 9.1B). In conjunction with ^{18}O-assisted ^{31}P NMR spectroscopy and mass spectrometry, the ^{18}O labelling procedure provides a versatile methodology for the simultaneous measurement of metabolite levels and metabolic fluxes through phosphotransfer systems, allowing a characterisation of different energetic pathways[15,16,22,27–29,33,35,36] (Figure 9.1A). This includes simultaneous recordings of ATP synthesis and utilisation, phosphotransfer fluxes through adenylate kinase, creatine kinase and glycolytic pathways, as well as mitochondrial Krebs cycle activity, glycogen turnover and intra-cellular energetic communication (Figure 9.1B). Another advantage of this ^{18}O methodology is that it can measure almost every phosphotransfer reaction taking place in the cell, including important signalling molecules such as cAMP, cGMP and AMP turnovers and their metabolically active pool sizes.[22,30,37,38] The ^{18}O-phosphoryl labelling procedure detects only newly generated molecules containing ^{18}O-labelled phosphoryls, reflecting their turnover rates and net fluxes through individual metabolic pathways.[15,35,39] Theoretically, up to one-third of all metabolites containing phosphorus,[40] and their turnover rates, can be quantified using high-resolution ^{31}P NMR spectroscopy and mass spectrometry. Thus, these combined technologies permit determination of phosphometabolites and multiple phosphotransfer fluxes within metabolic networks.

All metabolomic studies result in complex multivariate datasets that require visualisation software and chemometric methods for interpretation. The aim of these procedures is to produce biochemically based fingerprints that are of diagnostic or other classification value, and to identify potentially complex sets of biomarkers supporting the diagnosis or classification.[1,41–44] Here, multivariate datasets obtained from different analytical techniques and ^{18}O-labelling ratios were combined and interpreted using principal component analysis (PCA) and partial least squares discriminant analysis (PLS-DA) chemometric techniques to extract latent metabolic information, and hence enable sample classification and biomarker discovery.

In this chapter, we describe the principles and methodology of metabolic profiling and analysis of phosphometabolite turnover rates using stable isotope ^{18}O-assisted ^{31}P NMR analysis and mass spectrometry. This advanced phosphometabolomic platform is a valuable tool in studies of intact muscle energetics and phosphotransfer networks, and unique for measurements of intra-cellular energetic communication and metabolic signal dynamics. Basic concepts of the ^{18}O-labelling technique are explained and illustrated with several examples. Special focus is placed on sample preparation, the calculation of labelling rates and multivariate data analyses.

9.2 Methodology

9.2.1 Phosphometabolomic Platforms

Phosphorous is an essential element indispensable to life activity, such as genetic inheritance, signal transduction, metabolism and energy conversion.[45] Phosphate is the most common fragment *via* the frequency of occurrence in the metabolome of living organisms.[40] In the Human Metabolome Database (http://www.hmdb.ca/), there are 744 compounds containing 'phospho' and 419 with 'phosphate' in their structures from 8536 metabolites. Origins of comprehensive analysis of phosphorus-containing metabolites can be traced to Besman's phosphate analyser where ^{32}P-labelling coupled with chromatographic separation and quantification of phosphometabolites was performed.[46] Most phosphorus-containing metabolites are highly polar and their separation and analysis represent a major challenge. Phosphometabolites can be measured simultaneously by several analytical techniques, including ^{31}P NMR, LC/MS, GC/MS, CE/MS and HPLC analyses.[45,47,48] Although these methods are generally successful in determining the concentration of a range of metabolites, it is not possible to measure all phosphometabolites using one technique in view of their stabilities, concentrations or the dynamic range of instruments. For example, sugar phosphates are best separated using GC/MS,[12] whilst phospholipids are best investigated by ^{1}H and ^{31}P-NMR,[49] and nucleotides by LC.[50]

We established a dynamic phosphometabolomic platform (Figure 9.2) that includes ^{18}O-assisted GC/MS, ^{18}O-assisted ^{31}P NMR, together with ^{1}H NMR and HPLC. We are also developing an LC/MS method for the quantification of ^{18}O-labelling of mono- or oligo-phosphometabolites. ^{18}O-assisted GC/MS technology, which originally was developed in Nelson Goldberg's laboratory,[27,32,35,37] allows separation and quantitation of ^{18}O/^{16}O isotope ratios in phosphoryl metabolites with a molecular mass <500 Da. Higher molecular weight phosphates and oligo-phosphates, such as ATP or GTP, can be analysed after enzymatic transfer of corresponding phosphoryls to glycerol.[27,36] The ^{18}O-assisted ^{31}P NMR technique is dependent on the magnitude of an ^{18}O-induced shift in ^{31}P NMR spectra in order to determine the percentage of ^{18}O-labelling of phosphoryl metabolites.[15,31]

This technology, which has been used for enzymatic mechanism analyses *in vitro*,[31,34] is adapted and developed for tracking phosphoryl metabolic dynamics in intact tissues.[15,22] The critical advantage of the ^{18}O-assisted ^{31}P NMR technique is that it does not require prior metabolite separation and derivatisation; it is stable, and quantitative, and allows simultaneous single-run recordings of multiple metabolite phosphoryls, and those of separate phosphoryls within one molecule such as the α-, β- and γ-phosphoryls of

Phosphometabolomic platform

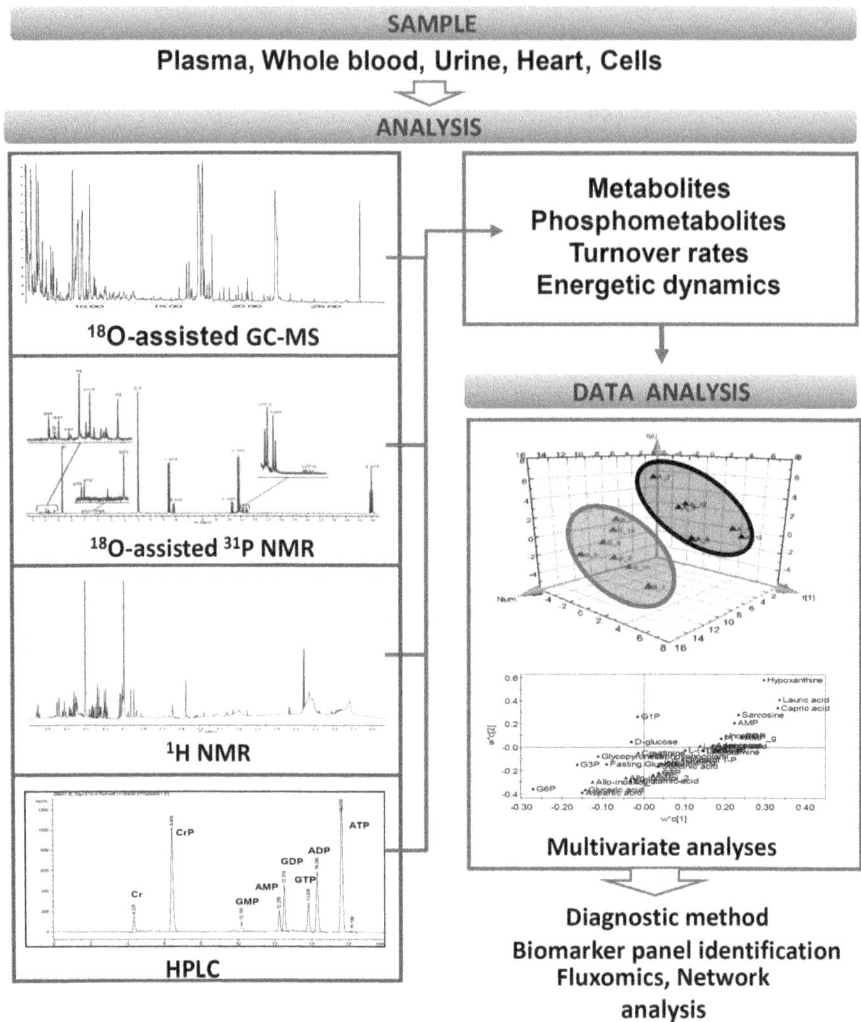

Figure 9.2 Stable isotope-based analytical platform for phosphometabolite analysis and phosphometabolomic fingerprinting of metabophenotypes. Combination of ¹⁸O-assisted GC/MS and ³¹P NMR with ¹H NMR and HPLC provides a powerful platform for dynamic phosphometabolomic profiling of energetic and signalling processes and network analysis cellular bioenergetics system.

ATP.[15,22] However, compared to GC/MS, ¹⁸O-assisted ³¹P NMR is less sensitive, and requires a larger amount of sample and a longer analysis time. In our studies, ¹H NMR analysis is employed as a complementary technology for the quantification of phosphometabolite levels in tissue extracts and biological fluids.[22] HPLC using ion-exchange, reversed-phase, hydrophobic and hydrophilic interaction chromatography is a versatile technique for the separation and quantification of major phosphometabolite classes.[15,27,36] The use of triethylammonium bicarbonate (TEAB) buffer, introduced by Khorana,[51] is preferential since its volatility facilitates sample recovery after HPLC chromatographic separation, and renders it suitable for the mass spectrometric analysis of phosphometabolites.

9.2.2 ^{18}O Metabolic Labelling Procedure

^{18}O is a natural, stable and non-radioactive isotope of oxygen. When tissue or cells are exposed to media containing a known percentage (20–30%) of ^{18}O, $H_2^{18}O$ rapidly equilibrates with cellular water, and then water-containing ^{18}O from water is transferred to cellular phosphate metabolites proportionally to the rate of enzymatic reactions involved. The rates of sequential enzymatic reactions between P_i, γ-ATP and CrP are high (Figure 9.3A) and upon ^{18}O labelling display exponential kinetics with saturation occurring within 2 min.[22,29] (Figure 9.3B). Therefore the labelling of metabolites should be performed within the initial linear phase (0–1 min.) of the ^{18}O labelling curve, whilst for β-ADP and β-ATP, which have lower turnover rates, labelling can be performed within a 5 min. time window. After the desired time of exposure with $H_2^{18}O$, cellular metabolism is instantaneously quenched by immersing cells or tissue into liquid N_2.

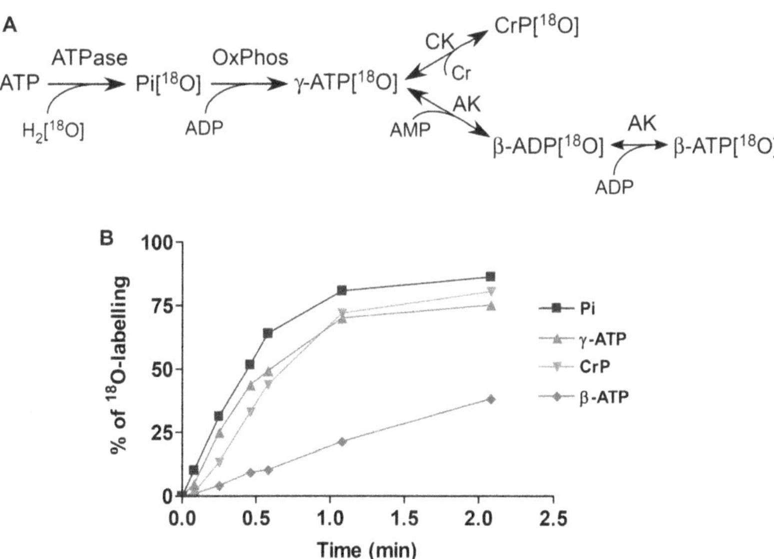

Figure 9.3 Analysis of heart phosphotransfer dynamics. **A)** Schematic representation of ^{18}O-labelling reaction sequence and **B)** kinetics of ^{18}O-labelling of major heart phosphometabolites.

Heart perfusion and ^{18}O phosphoryl labelling

Hearts from heparinised (50 U *ip*) and anaesthetised (75 mg kg^{-1} pento-barbital sodium *ip*) wild-type or transgenic mice are excised and retrogradely perfused with a 95% O$_2$–5% CO$_2$-saturated Krebs–Henseleit (K–H) solution (in mM: 118 NaCl, 5.3 KCl, 2.0 CaCl$_2$, 19 NaHCO$_3$, 1.2 MgSO$_4$, 11.0 glucose, 0.5 EDTA; 37 °C) at a perfusion pressure of 70 mmHg. Hearts were paced at 400 beats min^{-1}, and then perfused for 30 min. and subjected to labelling with ^{18}O, which was introduced for 30–60 s with the K–H buffer supplemented with 20–30% of ^{18}O-labelled H$_2$O (Isotec). Then hearts were freeze-clamped, pulverised under liquid N$_2$ and extracted in a solution containing 0.6 M HClO$_4$ and 1.00 mM EDTA. Extracts were neutralised with 2 M KHCO$_3$ and used to determine ^{18}O incorporation into metabolite phosphoryls.[28,33]

^{18}O-labelling of cultured cells or isolated cardiomyocytes

Cells were washed with PBS and pre-incubated with ADS or an alternative medium.[52,53] After 15 min., the medium was removed and replaced with a 2.00 ml volume of this matrix (for a 35 mm dish), enriched with a 20–30% solution of H$_2$$^{18}O$ and incubated for 2 min. at 37 °C. The incubation was terminated by rapid removal of H$_2$$^{18}O$-enriched ADS medium and the immediate addition of ice-cold 0.60 M perchloric acid containing 1.00 mM EDTA. Whilst on ice, the cells were scraped from the surface and transferred along with the HClO$_4$ to a test tube. Subsequently, acid extracts were neutralised with 2.00 M KHCO$_3$. The final extracts obtained from cell or heart tissue were then analysed using ^{18}O-assisted GC-MS or ^{31}P-NMR analysis in order to determine ^{18}O labelling ratios in phosphate metabolites of interest, and also calculate phosphotransfer rates. The tissue levels of metabolites were analysed using GC-MS, HPLC, ^{1}H NMR and ^{31}P NMR spectroscopy for metabolomic 'fingerprinting'.[15,16,22,28,29]

9.2.3 GC/MS Analysis of ^{18}O-labelling of Metabolite Phosphoryls

^{18}O labelling ratios of monophosphates (such as G3P, G6P and G1P) were evaluated using GC-MS following their purification with HPLC in view of their low concentrations in samples evaluated in this manner. Although P$_i$ has high concentration in the sample, it must be separated from other phosphate-containing metabolites, since some are very unstable during GC-MS analysis, and metabolites such as CrP and GA3P are easily degraded to liberate P$_i$, which, of course, interferes with the free P$_i$ level in the sample. Therefore, samples were fractionated and concentrated using HPLC. Consequently, the labelling ratio can be precisely determined.

Cellular phophometabolites are purified and quantified with HPLC (Figure 9.4A) using a Mono Q HR 5/5 ion-exchange column (Pharmacia Biotech) with triethylammonium bicarbonate buffer (pH 8.8) at a 1.00 ml min^{-1}

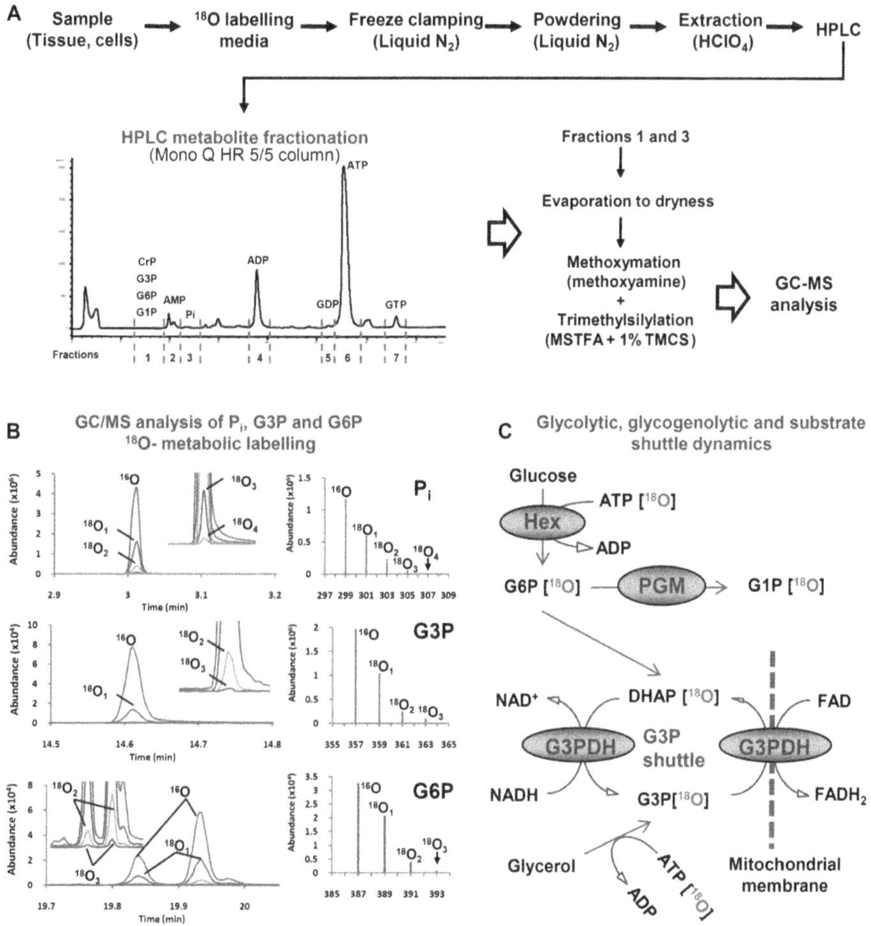

Figure 9.4 Phosphometabolite ^{18}O-labelling analysis using GC-MS. **A)** Sample preparation and HPLC fractionation for subsequent GC-MS analysis, **B)** analysis of ^{18}O-metabolic labelling of P$_i$, G3P and G6P using GC-MS and **C)** ^{18}O-labelling based fluxomic analysis of glycolytic and glycogenolytic phosphotransfers and mitochondrial substrate shuttle activity.

flow-rate.[33,36,52] From each sample, seven fractions were collected. The first fraction contained G6P, G3P, G1P and CrP, and fractions from the second to the seventh contained AMP, P$_i$, ADP, GDP, ATP and GTP, respectively (Figure 9.4A). Fractions were dried using vacuum centrifugation (SpeedVac, Savant), and reconstituted with water. Monophosphates were then transferred to GC-MS vials for silylation, whilst oligo-phosphates were subjected to enzymatic reactions in Eppendorf tubes to transfer each phosphoryl group to glycerol. The γ-phosphoryl of ATP or GTP was transferred to glycerol by glycerokinase, and β-phosphoryls of ATP and ADP were transferred to glycerol by a combined catalytic action of adenylate kinase and glycerokinase. The β-phosphoryls of GTP and GDP were transferred to glycerol by the combined catalytic action of guanylate kinase and glycerokinase. The phosphoryl group of CrP was transferred to γ-ATP by creatine kinase, and then to glycerol with glycerokinase. Samples containing the γ-ATP, γ-GTP, β-ATP, β-ADP, β-GTP/GDP phosphoryl groups as G3P, P_i, G6P, G1P, G3P and CrP were converted to their respective trimethylsilyl derivatives with Tri-Sil/BSA (Pierce) as a derivatisation agent.[22,33] The ^{18}O-enrichments of phosphoryls were determined with GC-MS operated in the select ion-monitoring mode. GC-MS analysis of ^{18}O-labelling in P_i, G3P and G6P labelling is presented in Figure 9.4B. The left panel represents GC-MS chromatograms of the metabolites, whilst in the right panel the isotope abundance of oxygen is shown. Another phosphometabolite, G1P, can be analysed in this HPLC fraction too (data not shown). Using this approach in a single run, the metabolic dynamics of glycolysis and glycogenolysis, and mitochondrial substrate shuttle activity can be monitored (Figure 9.4C). Our data indicate that G-3-P metabolic dynamics is altered in transgenic animal models, an observation indicating defects in substrate shuttle and the supply of reducing equivalents to mitochondria. This is of much importance, since G-3-P turnover abnormalities and metabolic arrest are linked to human diseases such as 'sudden-death' syndrome. Mass ions (m/z) of selected metabolites monitored as trimethylsilyl derivatives are given in the Table 9.1. Monophosphates can be labelled with up to three oxygens, whilst P_i and PP_i can be labelled with up to four and seven oxygens, respectively. Mass ions (m/z) of monophosphates corresponding to phosphoryl species of ^{16}O, [18]O$_1$, [18]O$_2$ and [18]O$_3$ are monitored as the parent ion (containing ^{16}O) + 2, +4 and +6, respectively.[33,35]

Table 9.1 Mass ions (m/z) of selected phosphometabolites that correspond to ^{18}O-labelled phosphoryl species monitored as trimethylsilyl derivatives.

	[16]O	[18]O$_1$	[18]O$_2$	[18]O$_3$	[18]O$_4$
P_i	299	301	303	305	307
G3P	357	359	361	363	
G6P	387	389	391	393	
G1P	217	219	221	223	
CrP(G3P)[a]	357	359	361	363	
PEP	369	371	373	375	
PP_i[b]	451	453	455	457	459
3-PG	357	359	361	363	
IMP	315	317	319	321	
AMP	315	317	319	321	
R5P	315	317	319	321	
γATP(G3P)[a]	357	359	361	363	
βATP(G3P)[a]	357	359	361	363	
αATP(G3P)[a]	357	359	361	363	
βADP(G3P)[a]	357	359	361	363	
αADP(G3P)[a]	357	359	361	363	

[a]Phosphate labelling was determined as G3P after enzymatically transferring it to glycerol.
[b]PP_i can be labelled up to seven oxygen atoms. For simplicity, four of them were given in the table.

9.2.4 ^{31}P NMR Analysis of ^{18}O Incorporation into Phosphoryl Metabolites

Samples were pre-cleansed for 1.0 h with Chelex-100 resin (Sigma) supplemented with the ^{31}P NMR spectroscopy internal standard methylene diphosphonate, and concentrated *via* vacuum centrifugation (Savant) to a volume of 0.30 ml. Concentrated extracts were then filtered (centrifuge filter; 0.22 μm, Milipore) and supplemented with 0.10 ml of D_2O (Isotec) and 0.10 ml of a 1.00 mM EDTA solution. Samples were additionally cleansed with the Chelex resin by rotation at 4 °C for 12 hr. To maximise resolution of ^{18}O-induced shifts in ^{31}P NMR spectra, and also to increase sample stability, HC104-extracted tissue samples were subjected to extensive chelation in order to remove divalent cations.[15,22,28,29]

^{31}P NMR data acquisition was performed at 202.5 MHz using a Bruker 11 T (Avance) spectrometer in high-quality 5-mm diameter NMR tubes (535-PP-7 Wilmad Glass) at ambient temperature and a sample spinning rate of 20 Hz. 9000 scans were acquired without relaxation delay (acquisition time 1.61 s) using a pulse width of 10 μs (53° angle) with proton decoupling during data acquisition (WALTZ-16 with 90° angle, pulse width of 506 μs for ^{1}H). Prior to Fourier-transformation, FIDs were zero-filled to 32 K, and multiplied by an exponential window function with 0.30 Hz line-broadening (Figure 9.5A). Peak areas were integrated using the Bruker software after automatic corrections of phase and baseline. Typical line-widths at half height of various cellular phosphates in ^{31}P NMR spectra were *ca.* 0.0080 ppm (1.5 Hz on 202.5 MHz), a value significantly less than the ^{18}O-induced shift ranging between 0.0210 and 0.0280 ppm. The internal standard was used to reference chemical shift values to 16.00 ppm, and also to determine metabolite levels. The metabolite levels normalised to the internal standard were corrected for NOE (by factors determined in a typical sample recorded both with and without decoupling), and incomplete relaxation (by factors calculated from T_1 times in a typical sample, measured by the inversion-recovery technique) as previously described.[28,33]

A typical ^{31}P NMR spectrum of heart extract is shown in Figure 9.5A. Incorporation of ^{18}O resulting from cellular metabolic activity induces an isotope shift in the ^{31}P NMR spectrum of phosphoryl containing metabolites.[31] Although the ^{18}O-induced isotope shift is rather small (around 0.020 ppm), it can be visualised and quantified using high-resolution NMR spectroscopy (Figure 9.5B). Incorporation of each ^{18}O isotope induces shifts of between 0.0210 and 0.0250 ppm in the ^{31}P NMR spectrum of P$_i$, CrP, γ-ATP, β-ATP, α-ATP, β-ADP, α-ADP, AMP, PC, G6P and G3P. It should also be noted that the isotope shift in the spectrum of β ATP was different for bridging and non-bridging ^{18}O oxygens, specifically 0.0170 and 0.0287 ppm, respectively. Moreover, since G6P exists in equatorial and axial forms, the ^{16}O and ^{18}O species of G6P were represented as two peaks corresponding to each of the two forms (Figure 9.5B). During the integration procedure, the bridging and non-bridging forms of β-ATP, as well as the equatorial and axial

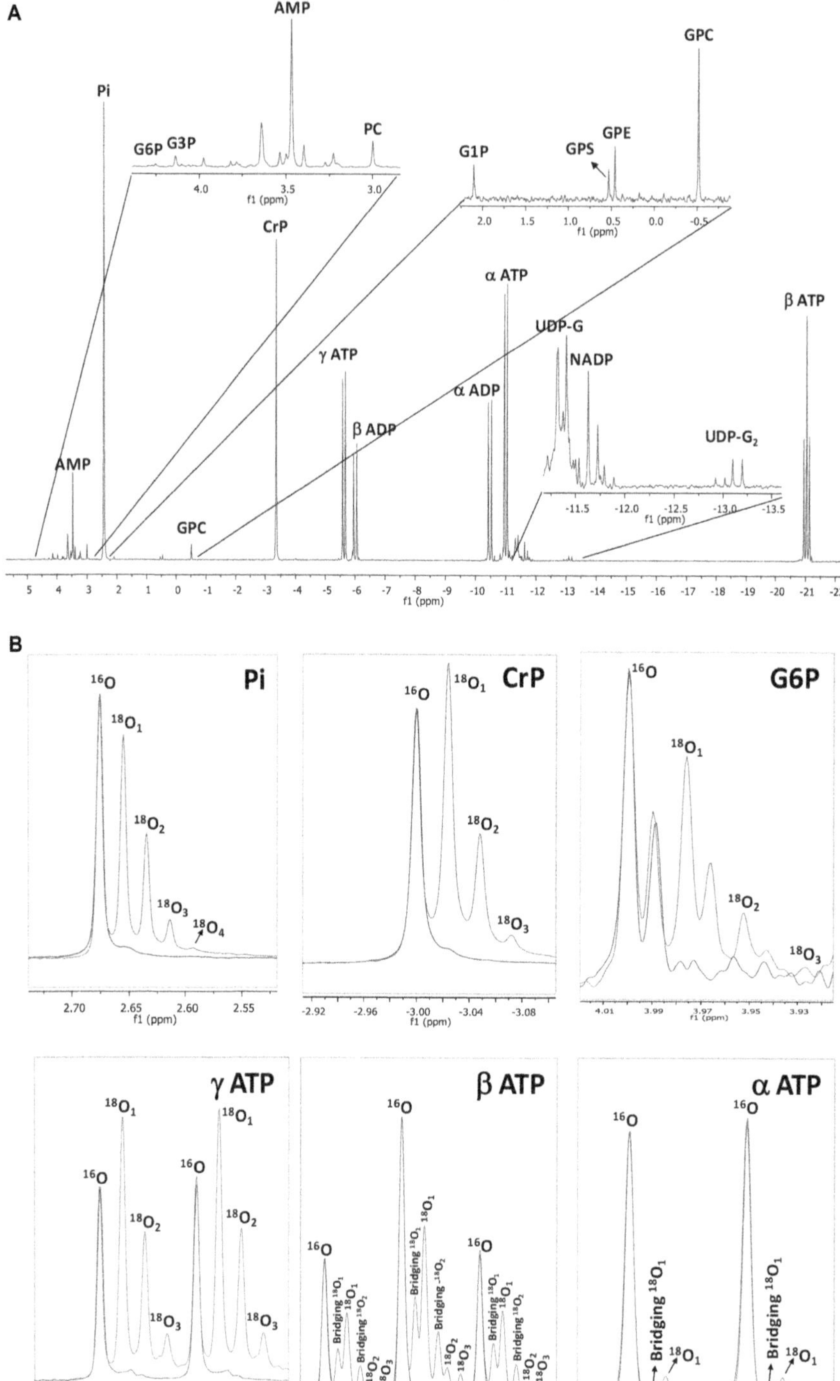

forms of G6P for particular ^{16}O or ^{18}O species, were integrated as single peaks.

9.2.5 Phosphometabolite Analysis by ^{1}H-NMR

^{1}H NMR provides a robust and precise method for metabolite quantification including the number of phosphometabolites. ^{1}H NMR data acquisition was performed at 500 MHz using a Bruker 11 T (Avance) spectrometer at ambient temperature and sample spinning at a rate of 20 Hz. 128 scans were accumulated under fully relaxed conditions (12.8 s relaxation delay), with a pulse width of 9 μs (90° angle). FIDs were zero-filled to 32 K, and Fourier-transformed without filtering. Phase and baselines were manually adjusted before integration and deconvolution. Chemical shifts were assigned relative to that of the trimethylsilyl propionate (TSP) signal at 0.00 ppm. Metabolite levels such as those of AMP, ATP, ADP, IMP, CrP, glycolytic intermediates and phospholipids were calculated by the expression of their resonance areas relative to that of TSP used as an internal standard. The identity of metabolites was conducted using Chenomx NMR software suite, which provides a pattern recognition technique, an efficient method for identifying metabolites in biofluids; these identities were confirmed by standard additions.

9.2.6 Data Analysis and Calculations of Phosphoryl Turnover and Phosphotransfer Fluxes

Introduction of ^{18}O-labelled water in tissues of interest leads to ^{18}O incorporation into cellular phosphates according to the rate of involved phosphotransfer reactions (see Figure 9.1).[15,27,30,36] Such a property allows the tracking of high-energy phosphoryl transfer routes, and the quantification of respective enzymatic fluxes at different levels of cellular activity.[15,22,27–30,33,36] Up to three ^{18}O atoms can be incorporated in monophosphate (G3P, G6P, G1P and CrP) and phosphate at different positions in oligo-phosphate (γ-, β- and α- for triphosphates, and β- and α- for diphosphates), and up to four and seven for P$_i$ and PP$_i$, respectively. The percentages of ^{16}O, ^{18}O$_1$, ^{18}O$_2$, ^{18}O$_3$ and ^{18}O$_n$ are proportional to the integrals of their respective resonances in the ^{31}P NMR spectrum, or in the GC-MS chromatograms[15,22,28,29] (see Figures 9.4 and 9.5). The cumulative percentage of phosphoryl oxygens replaced by ^{18}O in the metabolites is calculated as $[\%^{18}O_1 + 2(\%^{18}O_2) + 3(\%^{18}O_3) + \cdots \cdot n(\%^{18}O_n)]/[n(\%^{18}O\ \text{in}\ H_2O)]$.[15,22]

Figure 9.5 Non-destructive phosphometabolite and ^{18}O-labelling analysis using ^{31}P NMR spectroscopy. **A)** A typical ^{31}P NMR spectrum of major phosphometabolites in heart extract; **B)** ^{18}O assisted ^{31}P NMR spectra of ^{18}O-labelled P$_i$, CrP, G6P, γ-ATP, β-ATP and α-ATP in rat heart extract. Incorporation of ^{18}O induces an isotopic shift in ^{31}P NMR spectra of phosphoryl contained metabolites. ^{16}O, ^{18}O$_1$, ^{18}O$_2$, ^{18}O$_3$ and ^{18}O$_4$ designate phosphoryls containing 0, 1, 2, 3 and 4 atoms of ^{18}O.

The total cellular ATP turnover can be estimated from the total number of ^{18}O atoms that appeared in the phosphoryl-containing metabolites and orthophosphate.[22,33,36] The kinetics of ^{18}O-labelled phosphoryl appearance in γ-ATP reflects the cellular ATP synthesis rate, whilst the kinetics of P_i ^{18}O-labelling indicates cellular ATPase activity.[33] The P_i/γ-ATP ^{18}O-labelling ratio, an index of intra-cellular energetic communication,[54] is calculated using the amount or percentage of ^{18}O-incorporated into P_i and γ-ATP. ^{18}O-induced shifts in ^{31}P NMR spectra and the kinetics of ^{18}O-labelling of P_i and γ-ATP are presented in Figure 9.6. Indeed, the incorporation of ^{18}O into P_i and γ-ATP induces very robust multiple shifts in ^{31}P NMR spectra depending on the number of oxygens replaced (Figure 9.6A). From each shift, the labelling ratio can be calculated at different cycle levels (Figure 9.6B), or total labelling from the sum of these different cycles. Labelling reaches saturation within

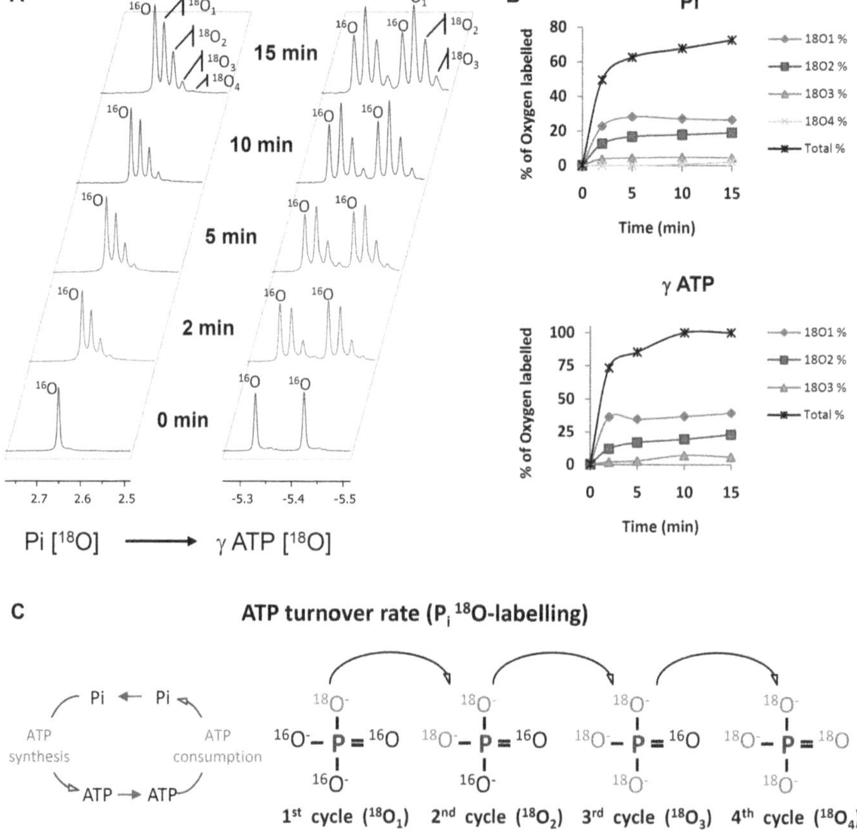

Figure 9.6 Dynamics of heart ATP utilisation and synthesis processes. A) ^{31}P NMR spectra of unlabelled and ^{18}O-labelled P_i and γ-ATP at different time points; incorporation of ^{18}O induces an isotopic shift in ^{31}P NMR spectra of P_i and γ-ATP; B) kinetics of ^{18}O-labelling of P_i and γ-ATP; C) schematic representation of $P_i \leftrightarrow$ ATP cycling and sequential ^{18}O incorporation into P_i during cell energetic cycle.

2–5 min., from which the metabolically active pool size can be determined. At saturation, almost 100% of γ-ATP, and *ca.* 80% of P_i are metabolically active (^{18}O labelled) (Figure 9.6B). Incorporation of one, two, three and four atoms of ^{18}O into phosphoryl groups reflects $P_i \leftrightarrow$ ATP cycling between ATP consumption and ATP production sites (Figure 9.6C).

Adenylate kinase phosphotransfer fluxes can be determined from the rate of appearance of ^{18}O-containing β-phosphoryls in ADP and ATP using a computer model based on Stella software[22,35] or CWave,[55] FiatFlux,[56] Flux-Simulator[57] or other available software. To obtain AK velocity, the total number of ^{18}O-labelled phosphoryls in β-ADP and β-ATP produced by the AK catalysis is counted. The pool of metabolically active ADP, obtained from labelling studies, is usually larger than 'free' ADP calculated from the CK equilibrium,[32,58] and is in dynamic equilibrium between the free and bound states.[59,60] The best fits to experimental data are obtainable using a metabolically active (^{18}O-labelled) pool size of 90% for β-ATP, and 30% for β-ADP.[32] Total AMP turnover (AK- and non-AK-mediated) is estimated from the kinetics of AMP α-phosphoryl (non-AK-mediated) and β-ATP/β-ADP phosphoryl (mediated by AK) ^{18}O-labelling. The metabolically active AMP or other phosphometabolite pool size is determined after prolonged (20–30 min.) ^{18}O-labelling performed in order to establish isotopic equilibrium.[32] At saturation, almost 100% of γ-ATP and CrP, and about 80% of P_i are labelled and metabolically active. The calculation of α-AMP turnover time is conducted using the formula: $SA_t = 1 - (2^{-N})$, where SA_t is specific activity of α-AMP ^{18}O-labelling at a given time t, and N is equal to the number of turnover cycles observed during the incubation period.[61,62] Thus, AK independent turnover time of the AMP pool can be calculated from the expression $T = t/N$, where T is the turnover time in s. AK-dependent AMP turnover can be calculated using the formula:

$$dN/dt = \rho(P^*/P - N^*/N) \tag{2.1}$$

where N^*/N is the specific ^{18}O-labelling of adenine nucleotide β-phosphoryls, P^*/P the specific ^{18}O-labelling of precursor adenine nucleotide γ-phosphoryls and ρ the rate of ^{18}O-labelling in the nucleotide pool per time unit.[61,62]

The creatine kinase phosphotransfer rate is determined from the rate of appearance of CrP species containing ^{18}O-labelled phosphoryls, and can be modelled using Stella[22,35] and other available software.[55–57] The glycolytic flux and glycerol phosphate shuttle is determined from the rate of appearance of ^{18}O-labelled G6P and G3P, respectively,[16,22] whereas glycogen flux is determined from the rate of appearance of ^{18}O-labelled G1P. The activity of NDPK/Succinyl-CoA synthase is determined from γ-GTP ^{18}O-labelling, whilst β-GTP/GDP ^{18}O-labelling indicates guanylate kinase activity.

9.2.7 Multivariate Statistical Analysis

Multivariate datasets obtained from different analytical techniques and labelling ratios were combined and interpreted using principal component analysis (PCA) and partial least squares discriminant analysis (PLS-DA) methods. Initially, data are examined with PCA scatter plots of the first two score vectors (t1–t2) in order to reveal the homogeneity of the data, together with any groupings, outliers and trends. Then PLS-DA is applied to acquire additional information, increase the class separation and simplify interpretation, and detect potential biomarkers.[63,64] The additional information (significant metabolites in group classification) may assist with VIP (variable importance in the projection), loading and regression coefficients plots. The VIP (variable importance in the projection) values,[63,65,66] a weighted sum of squares of the PLS weight which indicates the importance of the variable to the whole model, are calculated to identify the most important molecular variables for the 'clustering' of specific groups, whilst the regression co-efficient plots of metabolic variables in the PLS-DA model show the effect of variables on the groups' larger coefficient values (positive or negative) have a stronger correlation with group metabolic profile classification. Examination of the corresponding loading plot indicated those metabolites responsible for the clustering of groups. Metabolites located in the centre of the plot do not contribute to the clustering of the patient groups, whereas those in the same geographical region of a sample group in the corresponding scores plot are responsible for the separation. Attention must be given to PLS-DA analysis, since it is a supervised method. Even if the two groups are not different from each other; the method is forced to separate them.[67] Therefore, the PLS-DA model must be validated. For validation, R^2 (the fraction of variance explained by a component) and Q^2 (the fraction of the total variation predicted by a component) values are considered as measures of goodness-of-model and the model robustness, respectively. The value of Q^2 ranges from 0 to 1, and typically a Q^2 value of >0.4 is considered a good model, and those with Q^2 values over 0.5 are viewed as robust.[63,68] Additionally, the validation of the PLS-DA model can be performed by comparison to the classification statistics of models generated after random permutations of the class matrix. If the model R^2 and Q^2 values are higher than those obtained in random permuted models across all iterations, the method is valid. Calculation of the PCA and PLS-DA model parameters was carried out using SIMCA-P+ (v12.0, Umetrics AB, Umea, Malmö, Sweden) and the MetaboAnalyst web browser.[66]

9.3 Results

9.3.1 Phosphometabolomic Profiling of Transgenic Animal Models

9.3.1.1 *Adenylate Kinase AK1 Knockout Hearts*

Maintenance of optimal cardiac function requires precise control of cellular nucleotide ratios and high-energy phosphoryl fluxes. Within the cellular energetic infrastructure, adenylate kinase has been recognised as an important phosphotransfer enzyme that catalyses adenine nucleotide exchange (ATP + AMP ⇋ 2ADP) and facilitates transfer of both β- and γ-phosphoryls in ATP. In this manner, adenylate kinase doubles the energetic potential of ATP as a high-energy-phosphoryl carrying molecule, and provides an additional energy source under conditions of increased demand and/or compromised metabolic state. By regulating adenine nucleotide processing, adenylate kinase has been implicated in metabolic signal transduction. Indeed, phosphoryl flux through adenylate kinase has been shown to correlate with functional recovery in the metabolically compromised heart, and facilitates intra-cellular energetic communication.[15,20–22,28,29,32,33,35,36,54,69] Deletion of the major adenylate kinase AK1 isoform, which catalyses adenine nucleotide exchange, disrupts cellular energetic economy and compromises metabolic signal transduction and ischemia-reperfusion response.[16,28,29,69,70] Here, we compare the metabolomic phenotypes, phosphometabolite and phosphotransfer dynamics in the hearts of wild-type and AK1 knockout mice at baseline. Male homozygous AK1 knockout (AK1−/−) mice were compared with age- and sex-matched wild-type controls.[16,29]

In hearts with a null mutation of the AK1 gene, which encodes the major adenylate kinase isoform, the total adenylate kinase activity and ATP/ADP β-phosphoryl transfer was reduced by 94% and 36%, respectively. Knockout of the major adenylate kinase isoform, AK1, disrupted the synchrony between inorganic phosphate P_i turnover at ATP-consuming sites, and γ-ATP exchange at ATP synthesis sites, as revealed by ^{18}O-assisted ^{31}P NMR analysis.[70] This reduced energetic signal communication in the post-ischemic heart.[29] Moreover, AK1 gene deletion 'blunted' vascular adenylate kinase phosphotransfer, compromised the contractility-coronary flow relationship and precipitated inadequate coronary reflow following ischemia-reperfusion.[70] This was associated with up-regulation of phosphoryl flux through the remaining minor adenylate kinase isoforms, and the glycolytic phosphotransfer enzyme 3-phosphoglycerate kinase.[28]

Data acquired from ^{18}O labelling rate, together with those from ^{31}P and ^{1}H NMR analysis, are transformed into meaningful data through multivariate analysis of global profiling by unsupervised PCA and supervised PLS-DA. Initially, data were examined with a PCA score plot of the first two score vectors (t1–t2) in order to reveal the homogeneity of the data, plus any groupings, outliers and trends. As seen in Figure 9.7A, there is clear separation between the groups without any outliers and trends. To improve the visualisation, these profiles were displayed as hierarchical cluster analysis (Figure 9.7B). The heat map represents the unsupervised hierarchical clustering of the data grouped by sample type (rows), which also enabled visualisation of the up- or down-regulation of each metabolite (columns). Hierarchical clustering was performed with Spearman's rank correlation for similarity measurement, and Ward's linkage for clustering using MetaboAnalyst web server.[66] As noted in Figure 9.7, a very clear clustering is visible between two groups. Subsequently, PLS-DA was applied to gain

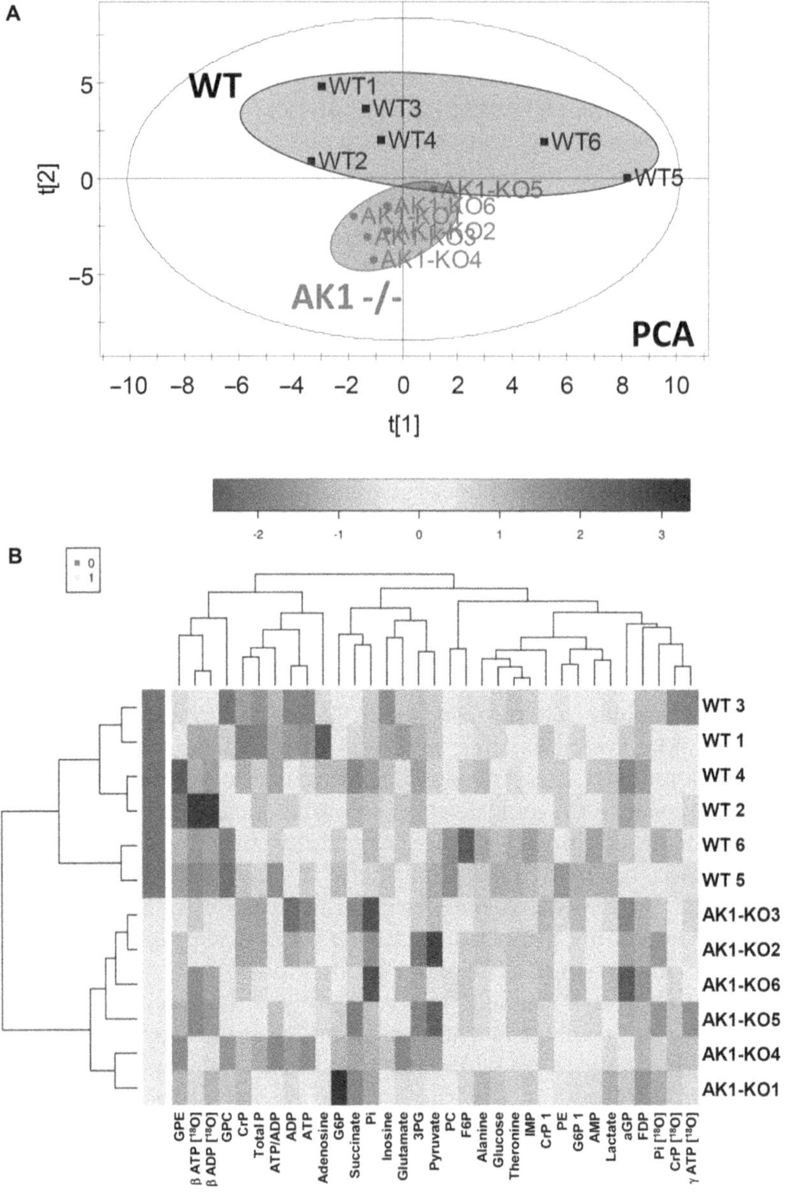

Figure 9.7   Metabolomic profiling of wild-type and AK1–/– mutant hearts. **A**) PCA score plot shows clear separation between metabolomic profiles of WT and AK1–/– mice; **B**) hierarchical clustering and heat map representation of the metabolomic dataset. The dendrogram on the left of the figure shows the WT and AK1–/– mice, while observed metabolic differences are indicated by colour changes.

additional information, increase the class separation, and simplify interpretation, and also to discover potential biomarkers.[64]

Genetic deletion of AK1 removed all but 6% of total myocardial adenylate kinase activity, yet the intra-cellular adenylate kinase phosphotransfer flux was only halved in AK1 knockout hearts. The reduced adenylate kinase-catalysed phosphotransfer-induced rearrangements in adenine nucleotide and glycolytic metabolism shifted cellular energetics into an apparently new steady state. These changes produced a differential metabolomic profile of the WT and AK1−/− KO mice heart as noted in the PCA and PLS-DA scores plots (Figure 9.8A). In order to determine significant metabolites in the group differentiation, VIP, loading and regression coefficient plots were used (Figure 9.8B–D). From these plots, it can be concluded that glycolytic and nucleotide metabolism, and adenylate kinase flux, has been altered significantly. Adenylate kinase fluxomic (β-ATP[^{18}O] and β-ADP[^{18}O] turnover), alanine, glucose, threonine, CrP, GPE and nucleotide levels (ADP, AMP and IMP) were decreased in AK1−/− mice, whilst those of 3-PG, pyruvate, P_i, G3P, G6P, γ-ATP[^{18}O] and CrP[^{18}O] turnover, glutamate, succinate and F6P all were increased. Alterations in 3-PG, G3P, G6P and F6P metabolites indicate adaptations in glycolytic and substrate shuttle activities, whilst changes in glutamate and succinate levels point to altered mitochondrial Krebs cycle activity. Taken together, these changes indicate a system-wide response of cellular energy metabolism to the deletion of one significant node in the network. With PLS-DA analysis performed to model the metabolic changes associated with gene deletion, a robust predictive model was produced ($R^2(X) = 0.68$; $R^2(Y) = 0.98$; $Q^2 = 0.89$ for the three components) (Figure 9.8E). This model passed cross-validation according to a random 100 permutations of the class matrix. The model R^2 and Q^2 values on the right were higher than those obtained in random permuted models across all 100 iterations, which indicates validity of the method. Thus, phosphometabolomic profiling of adenylate kinase-deficient hearts revealed rearrangements and adaptations in its energetic system, with an induced shift in glycolytic and creatine kinase phosphotransfer pathways and substrate utilisation networks.

9.3.1.2 Creatine Kinase M-CK Knockout Hearts

Creatine kinase (CK)-catalysed phosphotransfer is the major component of energy transfer and distribution network in the heart, and compromised CK function represents a 'hallmark' of abnormal bioenergetics in diseased hearts.[39,71–77] Studies of transgenic animal models have demonstrated an inherent plasticity of the cellular energetic system, and the development of cytoarchitectural and metabolic compensatory mechanisms in striated muscles.[16,20,28,59,78–83] These studies have led to the concept that the interchangeability and rearrangement of phosphotransfer networks provide an intra-cellular energetic continuum which couples discrete mitochondrial energetic units with ATP utilisation sites.[39,84–86]

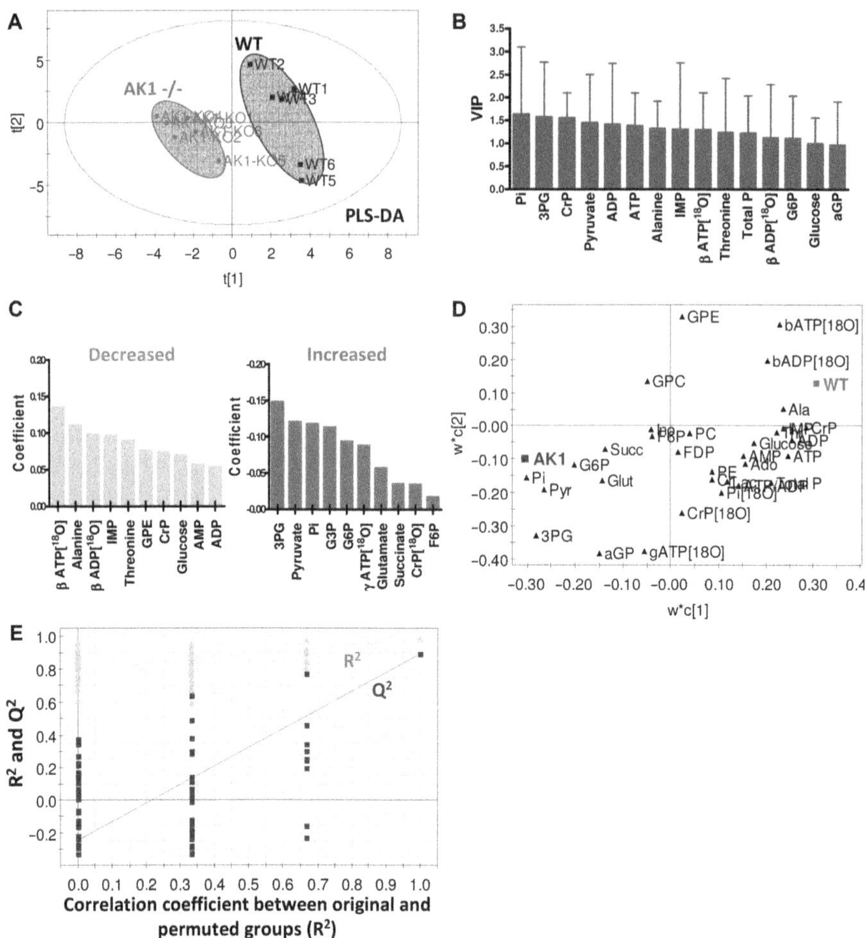

Figure 9.8 Phosphometabolomic profiling of wild-type and AK1−/− mutant hearts. **A**) PLS-DA score plot of metabolomic profiles shows clear separation between groups; **B**) VIP plot of the PLS-DA method represents importance of metabolites in discriminating between metabolomic profiles of the groups; **C**) regression coefficient plots of metabolic variables in the PLS-DA model; larger coefficient values (positive or negative) indicate a stronger correlation with group metabolic profile classification; **D**) loading plot of the PLS-DA plot. Dots correspond to the mean position of WT and AK1−/− group in the plot; **E**) validation of the PLS-DA model by comparison to the classification statistics of models generated after 100 random permutations of the class matrix.

Although hearts deficient in the major CK isoforms have no gross basal functional abnormalities, under increased load they cannot sustain normal global ATP/ADP ratios, a phenomenon indicating a compromised communication between ATP-consuming and ATP-generating cellular sites.[58,81,87–89] This renders contractions to be more energetically costly, forcing the heart to operate under less efficient cardiac bioenergetics.[58,89] Such energetic abnormalities reduce the ability of the myocardium to respond to β-adrenergic stimulation,[90] and CK-deficient hearts are more vulnerable to ischemia-reperfusion injury.[91] In addition, CK-deficient hearts cannot maintain adequate sub-sarcolemmal nucleotide exchange, and also have increased electrical instability under metabolic stress.[92] It is likely that CK-deficient hearts develop cytoarchitectural and metabolic adaptations that modulate energetic disturbances.[82,93–95] However, the adaptive metabolomic phenotype, and rearrangements in the bioenergetic system in CK-deficient hearts, are still poorly understood.

Here, adult wild-type mice (strain C57/BL6) and transgenic mice lacking cytosolic CK isoform (M-CK−/−), were employed.[78,96] Male homozygous M-CK−/− mice were compared with age- and sex-matched wild-type controls. Hearts were perfused and labelled with ^{18}O as outlined in Section 2.2.

^{18}O labelling procedure. Metabolic signatures for M-CK knockout hearts were revealed using PLS-DA analysis. As demonstrated in the PLS-DA scores plot (Figure 9.9A), a good separation was obtained between wild-type and M-CK knockout hearts based on metabolite levels and their turnover/^{18}O-labelling rates, and substrate metabolism. In order to determine significant metabolites in group discrimination, VIP, plus loading and regression co-efficient plots, were used (Figure 9.9B–D). With PLS-DA analysis conducted to model the metabolic changes associated with gene deletion, a robust predictive model was produced ($R^2(X) = 0.59$; $R^2(Y) = 0.99$; $Q^2 = 0.86$ for the three components) (Figure 9.9E).

The CK activity of M-CK−/− hearts was reduced by 71%, leading to decreases in CK flux assessed by a rate of appearance of ^{18}O-labelled phosphoryls in PCr of 23%. However, the overall ATP synthesis rate measured as the rate of appearance of ^{18}O-labelled phosphoryls in γ-ATP did not differ amongst wild-type and M-CK deficient hearts, an observation suggesting a robustness of cellular energetic system. The trend of an increased γ-ATP ^{18}O-labelling and a smaller pool size of metabolically active P_i, together with the decreased P_i/γ-ATP ^{18}O-labelling ratio (an indicator of intra-cellular energetic communication), observed here for M-CK deficient hearts, indicate less efficient phosphotransfer energetics. The VIP results show the importance of parameters of glycolytic metabolism (G6P ^{18}O-labelling), AK phosphotransfer (β-ATP/β-ADP ^{18}O-labelling), P_i/ATPase rate (P_i ^{18}O-labelling, P_i, TP) and adenine nucleotide metabolism and ATP turnover (γ-ATP ^{18}O-labelling, ADP and AMP levels) in group classification (Figure 9.9B). Glycolysis, in addition to its traditional role in ATP production, also catalyses rapid phosphoryl exchange, and has been implicated in intra-cellular energy transfer and distribution.[20,85] Here, changes in glycolytic phosphotransfer in wild-type and M-CK knockout hearts were assessed by monitoring the appearance of ^{18}O-labelled phosphoryls in G6P as a result of a reaction catalysed by hexokinase, the entry point into glycolysis. In wild-type hearts, ^{18}O-labelling of G6P was 8.1 ± 0.5%, which was more than 10% of γ-ATP turnover. Deletion of M-CK resulted in an increase of G6P ^{18}O-labelling to

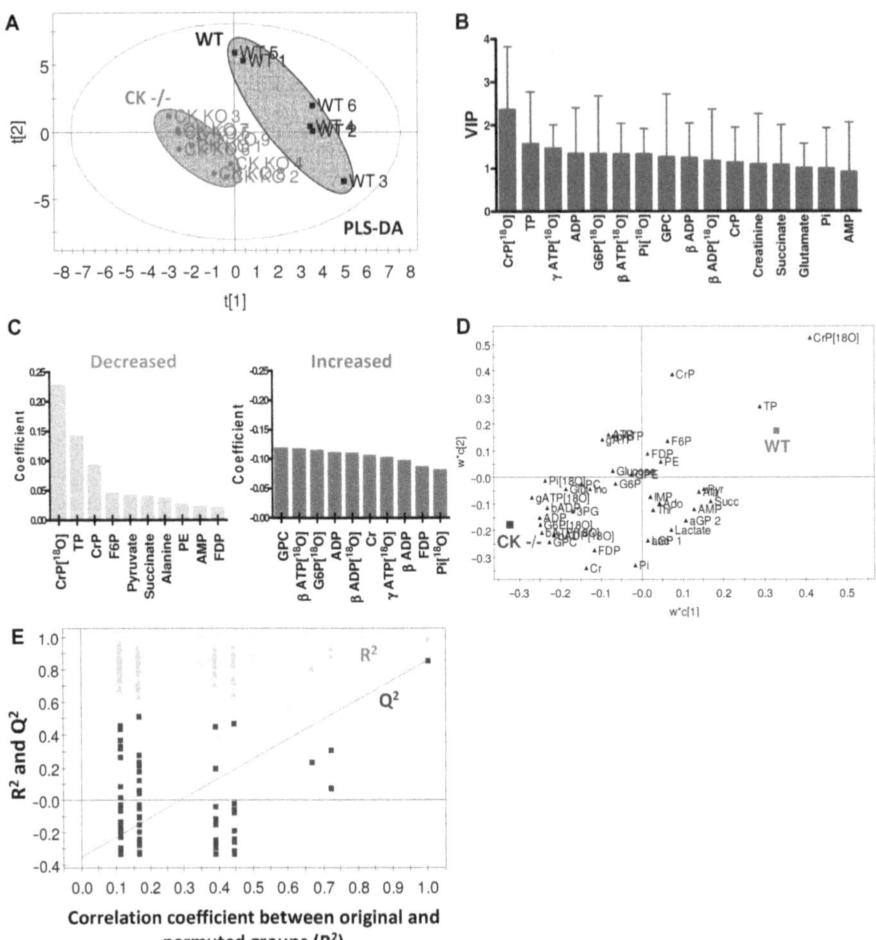

Figure 9.9   Phosphometabolomic profiling of wild-type and M-CK–/– mutant hearts. **A)** PLS-DA score plot shows clear separation between groups; **B)** VIP plot of the PLS-DA method represents importance of metabolites in discriminating between metabolomic profiles of the groups; **C)** regression coefficient plots of metabolic variables in the PLS-DA model; larger coefficient values (positive or negative) indicate a stronger correlation with group metabolic profile classification; **D)** loading plot of the PLS-DA plot. Dots correspond to the mean position of WT and M-CK–/– group in the plot; **E)** validation of the PLS-DA model by comparison to the classification statistics of models generated after 100 random permutations of the class matrix.

$13.3 \pm 0.8\%$, which corresponded to 27% of γ-ATP turnover. Therefore, glycolytic phosphotransfer is accelerated in M-CK knockout hearts and may represent an important compensation factor which alleviates myocardial energetic disturbances.

These results are consistent with studies of CK-deficient hearts performed by other researchers. Increased activities of glycolytic enzymes such as pyruvate kinase and GAPDH were also found in the hearts of CK knockout animals.[94] M-CK deficient cardiomyocytes display a higher sensitivity to glycolytic inhibition manifested in premature opening of ATP-sensitive potassium channels, and a shortening of action potential when compared to the wild-type mice,[92] suggesting a greater reliance on glycolytic metabolism. To this end, compensation provided by adenylate kinase and glycolytic phosphotransfers in CK-deficient muscles indicate their integral role in facilitating intra-cellular high-energy phosphoryl exchange, especially under conditions of genetic or metabolic stress. Thus, metabolomic profiling and flux analysis reveal plasticity and restructuring of the cellular bioenergetic system in response to genetic deficiency.

9.4 Conclusions

The ^{18}O-assisted ^{31}P NMR and mass spectrometric analysis techniques provide a versatile methodology, allowing simultaneous recordings of multiple parameters of cellular bioenergetics, and also the characterisation of metabolic fluxes through different energetic pathways. This includes the simultaneous recordings of ATP synthesis and utilisation, phosphotransfer fluxes through adenylate kinase, creatine kinase and glycolytic pathways, as well as mitochondrial Krebs cycle-associated nucleotide turnover and glycogen metabolism. This methodology has also a unique capability to measure intra-cellular energetic communication *via* comparisons of the kinetics of P_i ^{18}O-labelling (in the ATPase compartment) to that of γ-ATP (in the ATP synthesis compartment). Integrated kinetic data obtained using ^{18}O-labelling technology provides a basis for a novel cardiac system bioenergetics concept where major ATP-consuming and ATP-generating processes are inter-connected by phosphotransfer network composed by adenylate kinase and creatine kinase circuits, together with glycolytic/glycogenolytic network nodes. Metabolomic and fluxomic profiling of phosphotransfer enzyme-deficient transgenic animals (AK1−/− and M-CK−/−) using GC/MS, plus ^{1}H and ^{18}O-assisted ^{31}P NMR analyses, indicate metabolic perturbations and adaptations in the whole energetic system.

In summary, the ^{18}O-labelling technique has the capacity to monitor phosphotransfer reactions and energetic dynamics in all systems of interest in living tissues. Our studies demonstrate that this approach is valuable for metabolomic and fluxomic profiling of pre-conditioned and failing hearts, as well as transgenic animal models simulating human diseases, and also the diagnosis of mitochondrial energetic deficiency.[15,20,22,28,29] Hence, metabolomic analyses in conjunction with system and network approaches provide new avenues for an increased level of understanding of cellular energetic systems in health and diseases.

Abbreviations

3-PG	3-Phosphoglyceric acid
6-PG	6-Phosphogluconate
ADP	Adenosine diphosphate
AMP	Adenosine monophosphate
ATP	Adenosine triphosphate
cAMP	Cyclic adenosine monophosphate
CE	Capillary electrophoresis
Cr	Creatine
CrP	Creatine phosphate
DHAP	Dihydroxyacetone phosphate
F6P	Fructose 1,6-bisphosphate
FAD	Flavin adenine dinucleotide
FADH	Flavin adenine dinucleotide reduced
FDP	Fructose 1,6-bisphosphate
G1P	Glucose 1-phosphate
G3P	Glycerol 3-phosphate
G6P	Glucose 6-phosphate
GA3P	Glyceraldehyde 3-phosphate
GC	Gas chromatography
GDP	Guanosine diphosphate
GMP	Guanosine monophosphate
GPC	Glycerophosphocholine
GPE	Glycerophosphoethanolamine
GPS	Glycerol 3-phosphoserine
GTP	Guanosine triphosphate
IMP	Inosine monophosphate
LAC	Lactate
LC	Liquid chromatography
NADP	Nicotinamide adenine dinucleotide phosphate
NADPH	Nicotinamide adenine dinucleotide phosphate reduced
NMR	Nuclear magnetic resonance
PC	Phosphocholine
PCA	Principal component analysis
PEP	Phospho(enol)pyruvic acid
P_i	Inorganic phosphate
PLS DA	Partial least squares discriminant analysis
PP_i	Pyrophosphate
R5P	Ribose 5-phosphate
TP	Total phosphate

Acknowledgments

Supported by National Institutes of Health, Marriott Heart Disease Research Program, Marriott Foundation and The Mayo Clinic.

References

1. O. Beckonert, H. C. Keun, T. M. D. Ebbels, J. G. Bundy, E. Holmes, J. C. Lindon and J. K. Nicholson, Metabolic profiling, metabolomic and metabonomic procedures for NMR spectroscopy of urine, plasma, serum and tissue extracts, *Nat. Protocol.*, 2007, **2**, 2692–2703.
2. W. Weckwerth, *Metabolomics: Methods and Protocols*, Humana PR Inc., 2007, p. 312.
3. M. Brown, W. B. Dunn, P. Dobson, Y. Patel, C. L. Winder, S. Francis-McIntyre, P. Begley, K. Carroll, D. Broadhurst, A. Tseng, N. Swainston, I. Spasic, R. Goodacre and D. B. Kell, Mass spectrometry tools and metabolite-specific databases for molecular identification in metabolomics, *Analyst*, 2009, **134**, 1322–1332.
4. G. D. Lewis, R. Wei, E. Liu, E. Yang, X. Shi, M. Martinovic, L. Farrell, A. Asnani, M. Cyrille, A. Ramanathan, O. Shaham, G. Berriz, P. A. Lowry, I. F. Palacios, M. Tasan, F. P. Roth, J. Y. Min, C. Baumgartner, H. Keshishian, T. Addona, V. K. Mootha, A. Rosenzweig, S. A. Carr, M. A. Fifer, M. S. Sabatine and R. E. Gerszten, Metabolite profiling of blood from individuals undergoing planned myocardial infarction reveals early markers of myocardial injury, *J. Clin. Invest.*, 2008, **118**, 3503–3512.
5. G. D. Lewis, A. Asnani and R. E. Gerszten, Application of metabolomics to cardiovascular biomarker and pathway discovery, *J. Am. Coll. Cardiol.*, 2008, **52**, 117–123.
6. D. S. Wishart, C. Knox, A. C. Guo, R. Eisner, N. Young, B. Gautam, D. D. Hau, N. Psychogios, E. Dong, S. Bouatra, R. Mandal, I. Sinelnikov, J. G. Xia, L. Jia, J. A. Cruz, E. Lim, C. A. Sobsey, S. Shrivastava, P. Huang, P. Liu, L. Fang, J. Peng, R. Fradette, D. Cheng, D. Tzur, M. Clements, A. Lewis, A. De Souza, A. Zuniga, M. Dawe, Y. P. Xiong, D. Clive, R. Greiner, A. Nazyrova, R. Shaykhutdinov, L. Li, H. J. Vogel and I. Forsythe, HMDB: a knowledgebase for the human metabolome, *Nucleic Acids Res.*, 2009, **37**, D603–D610.
7. R. Kaddurah-Daouk, Metabolomic mapping of schizophrenic patients treated with atypical antipsychotics discloses drug-specific differences in global lipid changes, *Int. J. Neuropsychopharmacol.*, 2008, **11**, 31–32.
8. R. Kaddurah-Daouk, B. S. Kristal and R. M. Weinshilboum, Metabolomics: A global biochemical approach to drug response and disease, *Ann. Rev. Pharmacol. Toxicol.*, 2008, **48**, 653–683.
9. S. G. Villas-Boas and P. Bruheim, The potential of metabolomics tools in Bioremediation studies, *OMICS: A Journal of Integrative Biology*, 2007, **11**, 305–313.
10. I. R. Lanza, S. C. Zhang, L. E. Ward, H. Karakelides, D. Raftery and K. S. Nair, Quantitative Metabolomics by H-1-NMR and LC-MS/MS Confirms Altered Metabolic Pathways in Diabetes, *PLOS One*, 2010, **5**, e10538.
11. J. C. Lindon and J. K. Nicholson, Spectroscopic and statistical techniques for information recovery in metabonomics and metabolomics, *Annu. Rev. Anal. Chem. (Palo Alto Calif.)*, 2008, **1**, 45–69.

12. W. B. Dunn, Current trends and future requirements for the mass spectrometric investigation of microbial, mammalian and plant metabolomes, *Phys. Biol.*, 2008, **5**, 011001.

13. E. M. Lenz and I. D. Wilson, Analytical strategies in metabonomics, *J. Proteome Res.*, 2007, **6**, 443–458.

14. J. L. Griffin and C. Des Rosiers, Applications of metabolomics and proteomics to the mdx mouse model of Duchenne muscular dystrophy: lessons from downstream of the transcriptome, *Genome Med.*, 2009, **1**, 32.

15. D. Pucar, P. P. Dzeja, P. Bast, N. Juranic, S. Macura and A. Terzic, Cellular energetics in the preconditioned state: protective role for phosphotransfer reactions captured by 18O-assisted 31P NMR, *J. Biol. Chem.*, 2001, **276**, 44812–44819.

16. E. Janssen, P. P. Dzeja, F. Oerlemans, A. W. Simonetti, A. Heerschap, A. de Haan, P. S. Rush, R. R. Terjung, B. Wieringa and A. Terzic, Adenylate kinase 1 gene deletion disrupts muscle energetic economy despite metabolic rearrangement, *EMBO J.*, 2000, **19**, 6371–6381.

17. N. J. Kruger and R. G. Ratcliffe, Insights into plant metabolic networks from steady-state metabolic flux analysis, *Biochimie*, 2009, **91**, 697–702.

18. R. G. Ratcliffe and Y. Shachar-Hill, Revealing metabolic phenotypes in plants: inputs from NMR analysis, *Biol. Rev.*, 2005, **80**, 27–43.

19. A. Cornish-Bowden and M. L. Cardenas, From genome to cellular phenotype - a role for metabolic flux analysis?, *Nat. Biotechnol.*, 2000, **18**, 267–268.

20. P. P. Dzeja, A. Terzic and B. Wieringa, Phosphotransfer dynamics in skeletal muscle from creatine kinase gene-deleted mice, *Mol. Cell. Biochem.*, 2004, **256**, 13–27.

21. P. P. Dzeja and A. Terzic, Phosphotransfer networks and cellular energetics, *J. Exp. Biol.*, 2003, **206**, 2039–2047.

22. D. Pucar, P. P. Dzeja, P. Bast, R. J. Gumina, C. Drahl, L. Lim, N. Juranic, S. Macura and A. Terzic, Mapping hypoxia-induced bioenergetic rearrangements and metabolic signaling by O-18-assisted P-31 NMR and H-1 NMR spectroscopy, *Mol. Cell. Biochem.*, 2004, **256**, 281–289.

23. E. Fischer, N. Zamboni and U. Sauer, High-throughput metabolic flux analysis based on gas chromatography-mass spectrometry derived C-13 constraints, *Anal. Biochem.*, 2004, **325**, 308–316.

24. K. Noh, K. Gronke, B. Luo, R. Takors, M. Oldiges and W. Wiechert, Metabolic flux analysis at ultra short time scale: Isotopically non-stationary C-13 labeling experiments, *J. Biotechnol.*, 2007, **129**, 249–267.

25. N. Zamboni, ^{13}C metabolic flux analysis in complex systems, *Curr. Opin. Biotechnol.*, 2011, **22**, 103–108.

26. K. E. Stempel and P. D. Boyer, Refinement in oxygen-18 methodology for the study of phosphorylation mechanisms, *Meth. Enzymol.*, 1986, **126**, 618–639.

27. R. J. Zeleznikar and N. D. Goldberg, Kinetics and compartmentation of energy metabolism in intact skeletal muscle determined from ^{18}O labeling of metabolite phosphoryls, *J. Biol. Chem.*, 1991, **266**, 15110–15119.

28. D. Pucar, E. Janssen, P. P. Dzeja, N. Juranic, S. Macura, B. Wieringa and A. Terzic, Compromised energetics in the adenylate kinase AK1 gene knockout heart under metabolic stress, *J. Biol. Chem.*, 2000, **275**, 41424–41429.

29. D. Pucar, P. Bast, R. J. Gumina, L. Lim, C. Drahl, N. Juranic, S. Macura, E. Janssen, B. Wieringa, A. Terzic and P. P. Dzeja, Adenylate kinase AK1 knockout heart: energetics and functional performance under ischemia-reperfusion, *Am. J. Physiol. Heart Circ. Physiol.*, 2002, **283**, H776–H782.

30. S. M. Dawis, T. F. Walseth, M. A. Deeg, R. A. Heyman, R. M. Graeff and N. D. Goldberg, Adenosine triphosphate utilization rates and metabolic pool sizes in intact cells measured by transfer of ^{18}O from water, *Biophys. J.*, 1989, **55**, 79–99.

31. M. Cohn and A. Hu, Isotopic (^{18}O) shift in ^{31}P nuclear magnetic resonance applied to a study of enzyme-catalyzed phosphate–phosphate exchange and phosphate (oxygen)–water exchange reactions, *Proc. Natl Acad. Sci. USA*, 1978, **75**, 200–203.

32. R. J. Zeleznikar, R. A. Heyman, R. M. Graeff, T. F. Walseth, S. M. Dawis, E. A. Butz and N. D. Goldberg, Evidence for compartmentalized adenylate kinase catalysis serving a high energy phosphoryl transfer function in rat skeletal muscle, *J. Biol. Chem.*, 1990, **265**, 300–311.

33. P. P. Dzeja, K. T. Vitkevicius, M. M. Redfield, J. C. Burnett and A. Terzic, Adenylate kinase-catalyzed phosphotransfer in the myocardium – Increased contribution in heart failure, *Circ. Res.*, 1999, **84**, 1137–1143.

34. D. D. Hackney, G. Rosen and P. D. Boyer, Subunit interaction during catalysis: alternating site cooperativity in photophosphorylation shown by substrate modulation of [^{18}O]ATP species formation, *Proc. Natl Acad. Sci. USA*, 1979, **76**, 3646–3650.

35. R. J. Zeleznikar, P. P. Dzeja and N. D. Goldberg, Adenylate kinase-catalyzed phosphoryl transfer couples ATP utilization with its generation by glycolysis in intact muscle, *J. Biol. Chem.*, 1995, **270**, 7311–7319.

36. P. P. Dzeja, R. J. Zeleznikar and N. D. Goldberg, Suppression of creatine kinase-catalyzed phosphotransfer results in increased phosphoryl transfer by adenylate kinase in intact skeletal muscle, *J. Biol. Chem.*, 1996, **271**, 12847–12851.

37. S. M. Dawis, R. M. Graeff, R. A. Heyman, T. F. Walseth and N. D. Goldberg, Regulation of cyclic GMP metabolism in toad photoreceptors. Definition of the metabolic events subserving photoexcited and attenuated states, *J. Biol. Chem.*, 1988, **263**, 8771–8785.

38. T. F. Walseth, R. M. Graeff and N. D. Goldberg, Monitoring cyclic nucleotide metabolism in intact cells by ^{18}O labeling, *Methods Enzymol.*, 1988, **159**, 60–74.

39. V. Saks, P. Dzeja, U. Schlattner, M. Vendelin, A. Terzic and T. Wallimann, Cardiac system bioenergetics: metabolic basis of the Frank-Starling law, *J. Physiol.*, 2006, **571**, 253–273.

40. I. Nobeli, H. Ponstingl, E. B. Krissinel and J. M. Thornton, A structure-based anatomy of the E-coli metabolome, *J. Mol. Biol.*, 2003, **334**, 697–719.

41. J. K. Nicholson, J. C. Lindon and E. Holmes, 'Metabonomics': understanding the metabolic responses of living systems to pathophysiological stimuli via multivariate statistical analysis of biological NMR spectroscopic data, *Xenobiotica*, 1999, **29**, 1181–1189.

42. K. P. Gartland, C. R. Beddell, J. C. Lindon and J. K. Nicholson, Application of pattern recognition methods to the analysis and classification of toxicological data derived from proton nuclear magnetic resonance spectroscopy of urine, *Mol. Pharmacol.*, 1991, **39**, 629–642.

43. H. Antti, T. M. D. Ebbels, H. C. Keun, M. E. Bollard, O. Beckonert, J. C. Lindon, J. K. Nicholson and E. Holmes, Statistical experimental design and partial least squares regression analysis of biofluid metabonomic NMR and clinical chemistry data for screening of adverse drug effects, *Chemometr. Intell. Lab. Syst.*, 2004, **73**, 139–149.

44. O. Cloarec, M. E. Dumas, A. Craig, R. H. Barton, J. Trygg, J. Hudson, C. Blancher, D. Gauguier, J. C. Lindon, E. Holmes and J. Nicholson, Statistical total correlation spectroscopy: An exploratory approach for latent biomarker identification from metabolic H-1 NMR data sets, *Anal. Chem.*, 2005, **77**, 1282–1289.

45. Y. Sekiguchi, N. Mitsuhashi, T. Kokaji, H. Miyakoda and T. Mimura, Development of a comprehensive analytical method for phosphate metabolites in plants by ion chromatography coupled with tandem mass spectrometry, *J. Chrom. A*, 2005, **1085**, 131–136.

46. S. P. Bessman, P. J. Geiger, T. C. Lu and E. R. McCabe, Separation and automated analysis of phosphorylated metabolic intermediates, *Anal. Biochem.*, 1974, **59**, 533–546.

47. T. Uehara, A. Yokoi, K. Aoshima, S. Tanaka, T. Kadowaki, M. Tanaka and Y. Oda, Quantitative Phosphorus Metabolomics Using Nanoflow Liquid Chromatography-Tandem Mass Spectrometry and Culture-Derived Comprehensive Global Internal Standards, *Anal. Chem.*, 2009, **81**, 3836–3842.

48. R. Alvarez, L. A. Evans, P. Milham and M. A. Wilson, Analysis of oxygen-18 in orthophosphate by electrospray ionisation mass spectrometry, *Int. J. Mass Spectrom.*, 2000, **203**, 177–186.

49. H. Fernando, S. Kondraganti, K. K. Bhopale, D. E. Volk, M. Neerathilingam, B. S. Kaphalia, B. A. Luxon, P. J. Boor and G. A. Ansari, H-1 and P-31 NMR Lipidome of Ethanol-Induced Fatty Liver, *Alcoholism Clin. Exp. Res.*, 2010, **34**, 1937–1947.

50. L. Coulier, R. Bas, S. Jespersen, E. Verheij, M. J. van der Werf and T. Hankemeier, Simultaneous quantitative analysis of metabolites using ion-pair liquid chromatography – Electrospray ionization mass spectrometry, *Anal. Chem.*, 2006, **78**, 6573–6582.

51. M. Smith and H. G. Khorana, Preparation of nucleotides and derivatives, in *Methods in Enzymology*, ed. S. P. Colowick and N. O. Kaplan, Academic Press Inc., New York, 1963, vol. 6, p. 645–669.

52. L. K. Olson, W. Schroeder, R. P. Robertson, N. D. Goldberg and T. F. Walseth, Suppression of adenylate kinase catalyzed phospho-transfer precedes and is associated with glucose-induced insulin secretion in intact HIT-T15 cells, *J. Biol. Chem.*, 1996, **271**, 16544–16552.

53. C. Perez-Terzic, A. M. Gacy, R. Bortolon, P. P. Dzeja, M. Puceat, M. Jaconi, F. G. Prendergast and A. Terzic, Structural plasticity of the cardiac nuclear pore complex in response to regulators of nuclear import, *Circ. Res.*, 1999, **84**, 1292–1301.

54. P. P. Dzeja and A. Terzic, Mitochondria-Nucleus Energetic Communication: Role for Phosphotransfer Networks in Processing Cellular Information, in *Handbook of Neurochemistry and Molecular Neurobiology*, ed. G. Gibson and G. Dienel, Springer, NY, 2007, vol. 5, Brain Energetics: Integration of Molecular and Cellular Processes, pp. 641–666.

55. G. F. Mason, K. F. Petersen, R. A. de Graaf, T. Kanamatsu, T. Otsuki and D. L. Rothman, A comparison of C-13 NMR measurements of the rates of glutamine synthesis and the tricarboxylic acid cycle during oral and intravenous administration of [1-C-13]glucose, *Brain Res. Protocol.*, 2003, **10**, 181–190.

56. N. Zamboni, E. Fischer and U. Sauer, FiatFlux – a software for metabolic flux analysis from C-13-glucose experiments, *BMC Bioinformatics*, 2005, **6**, 209.

57. T. W. Binsl, K. M. Mullen, I. H. M. van Stokkum, J. Heringa and J. H. G. M. van Beek, FluxSimulator: An R package to simulate isotopomer distributions in metabolic networks, *J. Stat. Software*, 2007, **18**, 1–17.

58. K. W. Saupe, M. Spindler, J. C. Hopkins, W. Shen and J. S. Ingwall, Kinetic, thermodynamic, and developmental consequences of deleting creatine kinase isoenzymes from the heart. Reaction kinetics of the creatine kinase isoenzymes in the intact heart, *J. Biol. Chem.*, 2000, **275**, 19742–19746.

59. P. P. Dzeja, R. J. Zeleznikar and N. D. Goldberg, Adenylate kinase: kinetic behavior in intact cells indicates it is integral to multiple cellular processes, *Mol. Cell. Biochem.*, 1998, **184**, 169–182.

60. M. Barany and P. P. de Tombe, Rapid exchange of actin-bound nucleotide in perfused rat heart, *Am. J. Physiol.*, 2004, **286**, H1394–1401.

61. D. M. Karl and P. Bossard, Measurement of Microbial Nucleic Acid Synthesis and Specific Growth Rate by PO(4) and [H]Adenine: Field Comparison, *Appl. Environ. Microbiol.*, 1985, **50**, 706–709.

62. A. Rossi, ^{32}P labelling of the nucleotides in alpha-position in the rabbit heart, *J. Mol. Cell Cardiol.*, 1975, **7**, 891–906.

63. C. M. Titman, J. A. Downs, S. G. Oliver, P. L. Carmichael, A. D. Scott and J. L. Griffin, A metabolomic and multivariate statistical process to assess the effects of genotoxins in Saccharomyces cerevisiae, *Molecular BioSystems*, 2009, **5**, 1913–1924.

64. E. J. Want, A. Nordstrom, H. Morita and G. Siuzdak, From exogenous to endogenous: The inevitable imprint of mass spectrometry in metabolomics, *J. Proteome Res.*, 2007, **6**, 459–468.

65. M. Barker and W. Rayens, Partial least squares for discrimination, *J. Chemometr.*, 2003, **17**, 166–173.
66. J. G. Xia, N. Psychogios, N. Young and D. S. Wishart, MetaboAnalyst: a web server for metabolomic data analysis and interpretation, *Nucleic Acids Res.*, 2009, **37**, W652–W660.
67. J. A. Westerhuis, H. C. J. Hoefsloot, S. Smit, D. J. Vis, A. K. Smilde, E. J. J. van Velzen, J. P. M. van Duijnhoven and F. A. van Dorsten, Assessment of PLSDA cross validation, *Metabolomics*, 2008, **4**, 81–89.
68. H. J. Atherton, M. K. Gulston, N. J. Bailey, K. K. Cheng, W. Zhang, K. Clarke and J. L. Griffin, Metabolomics of the interaction between PPAR-alpha and age in the PPAR-alpha-null mouse, *Mol. Sys. Biol.*, 2009, **5**.
69. A. J. Carrasco, P. P. Dzeja, A. E. Alekseev, D. Pucar, L. V. Zingman, M. R. Abraham, D. Hodgson, M. Bienengraeber, M. Puceat, E. Janssen, B. Wieringa and A. Terzic, Adenylate kinase phosphotransfer communicates cellular energetic signals to K-ATP channels, *J. Mol. Cell. Cardiol.*, 2001, **33**, A18–A18.
70. P. P. Dzeja, P. Bast, D. Pucar, B. Wieringa and A. Terzic, Defective metabolic signaling in adenylate kinase AK1 gene knock-out hearts compromises post-ischemic coronary reflow, *J. Biol. Chem.*, 2007, **282**, 31366–31372.
71. S. Neubauer, M. Horn, M. Cramer, K. Harre, J. B. Newell, W. Peters, T. Pabst, G. Ertl, D. Hahn, J. S. Ingwall and K. Kochsiek, Myocardial phosphocreatine-to-ATP ratio is a predictor of mortality in patients with dilated cardiomyopathy, *Circulation*, 1997, **96**, 2190–2196.
72. P. P. Dzeja, M. M. Redfield, J. C. Burnett and A. Terzic, Failing energetics in failing hearts, *Curr. Cardiol. Rep.*, 2000, **2**, 212–217.
73. Y. M. Cha, P. P. Dzeja, W. K. Shen, A. Jahangir, C. Y. Hart, A. Terzic and M. M. Redfield, Failing atrial myocardium: energetic deficits accompany structural remodeling and electrical instability, *Am. J. Physiol.*, 2003, **284**, H1313–H1320.
74. P. A. Bottomley, K. C. Wu, G. Gerstenblith, S. P. Schulman, A. Steinberg and R. G. Weiss, Reduced myocardial creatine kinase flux in human myocardial infarction: an *in vivo* phosphorus magnetic resonance spectroscopy study, *Circulation*, 2009, **119**, 1918–1924.
75. R. Ventura-Clapier, A. Garnier, V. Veksler and F. Joubert, Bioenergetics of the failing heart, *Biochim. Biophys. Acta*, 2011, **1813**, 1360–1372.
76. J. S. Ingwall, M. F. Kramer, M. A. Fifer, B. H. Lorell, R. Shemin, W. Grossman and P. D. Allen, The creatine kinase system in normal and diseased human myocardium, *New Engl. J. Med.*, 1985, **313**, 1050–1054.
77. M. Wyss, O. Braissant, I. Pischel, G. S. Salomons, A. Schulze, S. Stockler and T. Wallimann, Creatine and creatine kinase in health and disease–a bright future ahead?, *Sub Cell. Biochem.*, 2007, **46**, 309–334.
78. J. van Deursen, A. Heerschap, F. Oerlemans, W. Ruitenbeek, P. Jap, H. ter Laak and B. Wieringa, Skeletal muscles of mice deficient in muscle creatine kinase lack burst activity, *Cell*, 1993, **74**, 621–631.

79. A. J. de Groof, F. T. Oerlemans, C. R. Jost and B. Wieringa, Changes in glycolytic network and mitochondrial design in creatine kinase-deficient muscles, *Muscle Nerve*, 2001, **24**, 1188–1196.
80. E. Janssen, A. Terzic, B. Wieringa and P. P. Dzeja, Impaired intracellular energetic communication in muscles from creatine kinase and adenylate kinase (M-CK/AK1) double knock-out mice, *J. Biol. Chem.*, 2003, **278**, 30441–30449.
81. J. S. Ingwall, Transgenesis and cardiac energetics: new insights into cardiac metabolism, *J. Mol. Cell. Cardiol.*, 2004, **37**, 613–623.
82. R. Ventura-Clapier, A. Kaasik and V. Veksler, Structural and functional adaptations of striated muscles to CK deficiency, *Mol. Cell. Biochem.*, 2004, **256–257**, 29–41.
83. J. S. Ingwall, Energetics of the failing heart: new insights using genetic modification in the mouse, *Arch Mal. Coeur Vaiss.*, 2006, **99**, 839–847.
84. P. P. Dzeja and A. Terzic, Phosphotransfer networks and cellular energetics, *J. Exp. Biol.*, 2003, **206**, 2039–2047.
85. P. Dzeja, S. Chung and A. Terzic, Integration of adenylate kinase, glycolytic and glycogenolytic circuits in cellular energetics, in *Molecular System Bioenergetics: Energy for Life*, ed. V. Saks, Wiley-VCH, Weinheim, Germany, 2007, pp. 265–301.
86. V. Saks, C. Monge, T. Anmann and P. Dzeja, Integrated and organized cellular energetic systems: theories of cell energetics, compartmentation and metabolic channeling, in *Molecular System Bioenergetics: Energy for Life*, ed. V. Saks, Wiley-VCH, Weinheim, Germany, 2007, pp. 59–109.
87. A. Katz, D. C. Andersson, J. Yu, B. Norman, M. E. Sandstrom, B. Wieringa and H. Westerblad, Contraction-mediated glycogenolysis in mouse skeletal muscle lacking creatine kinase: the role of phosphorylase b activation, *J. Physiol.*, 2003, **553**, 523–531.
88. K. Nicolay, F. A. van Dorsten, T. Reese, M. J. Kruiskamp, J. F. Gellerich and C. J. van Echteld, *In situ* measurements of creatine kinase flux by NMR. The lessons from bioengineered mice, *Mol. Cell. Biochem.*, 1998, **184**, 195–208.
89. K. W. Saupe, M. Spindler, R. Tian and J. S. Ingwall, Impaired cardiac energetics in mice lacking muscle-specific isoenzymes of creatine kinase, *Circ. Res.*, 1998, **82**, 898–907.
90. B. Crozatier, T. Badoual, E. Boehm, P. V. Ennezat, T. Guenoun, J. Su, V. Veksler, L. Hittinger and R. Ventura-Clapier, Role of creatine kinase in cardiac excitation-contraction coupling: studies in creatine kinase-deficient mice, *FASEB J.*, 2002, **16**, 653–660.
91. M. Spindler, K. Meyer, H. Stromer, A. Leupold, E. Boehm, H. Wagner and S. Neubauer, Creatine kinase-deficient hearts exhibit increased susceptibility to ischemia-reperfusion injury and impaired calcium homeostasis, *Am. J. Physiol.*, 2004, **287**, H1039–1045.
92. M. R. Abraham, V. A. Selivanov, D. M. Hodgson, D. Pucar, L. V. Zingman, B. Wieringa, P. P. Dzeja, A. E. Alekseev and A. Terzic, Coupling of cell energetics with membrane metabolic sensing. Integrative signaling

through creatine kinase phosphotransfer disrupted by M-CK gene knock-out, *J. Biol. Chem.*, 2002, **277**, 24427–24434.

93. M. Spindler, R. Niebler, H. Remkes, M. Horn, T. Lanz and S. Neubauer, Mitochondrial creatine kinase is critically necessary for normal myocardial high-energy phosphate metabolism, *Am. J. Physiol.*, 2002, **283**, H680–687.

94. R. Ventura-Clapier, A. V. Kuznetsov, A. d'Albis, J. van Deursen, B. Wieringa and V. I. Veksler, Muscle creatine kinase-deficient mice. I. Alterations in myofibrillar function, *J. Biol. Chem.*, 1995, **270**, 19914–19920.

95. E. Boehm, R. Ventura-Clapier, P. Mateo, P. Lechene and V. Veksler, Glycolysis supports calcium uptake by the sarcoplasmic reticulum in skinned ventricular fibres of mice deficient in mitochondrial and cytosolic creatine kinase, *J. Mol. Cell. Cardiol.*, 2000, **32**, 891–902.

96. K. Steeghs, A. Benders, F. Oerlemans, A. deHaan, A. Heerschap, W. Ruitenbeek, C. Jost, J. van Deursen, B. Perryman, D. Pette, M. Bruckwilder, J. Koudijs, P. Jap, J. Veerkamp and B. Wieringa, Altered Ca^{2+} responses in muscles with combined mitochondrial and cytosolic creatine kinase deficiencies, *Cell*, 1997, **89**, 93–103.

Investigations of the Mechanisms of Action of Oral Healthcare Products using ^{1}H NMR-based Chemometric Techniques

C. J. L. SILWOOD[*a] AND MARTIN GROOTVELD[*b]

[a] Institute for Materials Research and Innovation, University of Bolton, Deane Road, Bolton, BL3 5AB, UK; [b] Leicester School of Pharmacy, De Montfort University, The Gateway, Leicester, LE1 9BH, UK
*Email: cjsilwood@gmail.com; mgrootveld@dmu.ac.uk

10.1 Introduction

Multicomponent bioanalytical technique-based investigations currently represent a novel approach to the rapid detection of biomarkers describing biometabolic dysfunctions arising from the induction, development and/or progression of clinical conditions,[1,2] or the administration of therapeutic agents which, for example, exert dose-dependent toxicological actions.[3] The technique involves the NMR- (predominantly ^{1}H) or alternative analytical technique (*e.g.* GC-MS)-based profiling of the metabolic status of biofluids coupled with selected multivariate (MV) analysis techniques in order to identify metabolites which serve as 'markers' of disease processes and, quantitatively, their severities (*i.e.* those with elevated or reduced

Issues in Toxicology No. 21
Metabolic Profiling: Disease and Xenobiotics
Edited by Martin Grootveld
© The Royal Society of Chemistry 2015
Published by the Royal Society of Chemistry, www.rsc.org

concentrations which correlate with physiological conditions).[4] Indeed, it serves to supply overall quantitative 'global' descriptors from the simultaneous multicomponent analysis of many endogenous biomolecules (typically >100) in biofluids such as urine, blood plasma and cerebrospinal fluid.[5–7]

Metabonomics represents the metabolism-based approach for the identification and monitoring of biomolecules which may reflect metabolic modifications arising from biological or xenobiotic challenges, phenomena which can be readily monitored by the NMR spectroscopic analyses of biofluids, tissues and tissue or cell culture extract samples.[4,8] Metabolomics is aligned with the *in vitro* cellular, microbial and plant science analytical disciplines, although a considered difference between the two descriptors is that metabonomic studies tend to be concerned with metabolic responses to biological stimuli, whereas a metabolomic investigation would normally require a comprehensive, global appraisal and quantification of all detectable components.[4,9] Of course, the '-omics' sphere of research activity also embraces genomics, transcriptomics, proteomics *etc.*, such that the aim of any approach that attempts to combine the information derived from these quite separate experimental disciplines will ultimately be to introduce a 'systems-biological' holistic understanding of cellular action, irrespective of whether it is attributable to development, ageing, health or disease. Indeed, within a molecular biological context, the optimum goal of such research is to integrate biomolecular datasets in order to represent the pertinent biology of a whole organism.[4]

Saliva represents an ideal biofluid medium for metabolomic investigations, particularly since when excreted from the salivary and mucous glands, it contains no invading bacteria and only very low concentrations of metabolic agents which may represent 'markers' of selected oral diseases. Indeed, in the oral environment, micro-organisms located in tooth plaque, gingival crevices or soft tissues such as gums conduct a range of metabolic functions linked to their growth and prevalence, and hence saliva contains many excreted catabolites (*e.g.* propionate, *n*- and *iso*-butyrates, *n*- and *iso*-valerates, *etc.*) which are unique and dependent on the infiltration, activity and preponderance of bacterial flora therein. Furthermore, elevated salivary concentrations of markers of inflammatory processes occurring within soft tissue would also be anticipated in, for example, periodontal diseases such as gingivitis (the 'type I' class of periodontal disease).

10.1.1 High-resolution NMR Analysis of Human Saliva

The multicomponent nature of high-resolution, high field proton (^{1}H) NMR spectroscopy permits the multicomponent detection and quantification of biomolecules present in a wide range of complex biofluids.[5–7] The technique offers many advantages over alternative time-consuming, labour-intensive analytical methods since (1) it permits the rapid, non-invasive and simultaneous study of a multitude of components present in biological samples (*e.g.* biofluids such as human saliva) and (2) it generally requires little or no knowledge of sample composition prior to analysis. Moreover, chemical shift values, coupling patterns and coupling constants of resonances present in the ^{1}H NMR spectra of such complex, multicomponent systems provide much valuable molecular information regarding both endogenous and exogenous chemical species detectable. The broad overlapping resonances which arise from macromolecules present are routinely suppressed by spin-echo pulse sequences, giving rise to spectra which contain many sharp, well-resolved signals attributable to a wide variety of low-molecular-mass (non-protein-bound) components and the mobile portions of macromolecules that are detectable at a sensitivity of $<5\times10^{-6}$ mol. dm^{-3} (at an operating frequency of 600 MHz) for samples not subjected to any form of pre-concentration techniques.

Grootveld and co-workers have employed a series of NMR techniques for the multicomponent analysis of human saliva and oral biopsies such as carious dentin.[10-17] For example, it has been found that, in addition to lactate, these specimens also contained high (*i.e.* mmol dm^{-3}) concentrations of pyruvate and formate, the corresponding acids of which are stronger than that of lactate. Hence, in view of its relatively high acid dissociation constant, pyruvate may play an important role in tooth demineralisation processes, and its removal by hydrogen peroxide (H_2O_2)-containing dentifrices (or those with alternative peroxo-adducts as tooth-whitening agents) *via* an oxidative decarboxylation process may suppress the development and progression of primary root caries lesions.[11]

Experimental spectral signal intensities obtained *via* the applications of this technique also offer the possibility of a wide range of statistical analyses, both univariate (*i.e.* the analysis of grouped single signal intensities) and multivariate (*i.e.* the simultaneous consideration of all metabolic resonance intensities). For example, in the case of salivary analyses, the capacity of high-resolution NMR techniques to (1) index salivary biomolecules and (2) supply valuable metabolic data regarding intra- and inter-subject variabilities in the concentrations of a range of readily detectable components has been studied in detail by Silwood *et al.*[10] The components detectable comprised organic acid anions and malodorous amines (experiments were conducted on 'whole' saliva samples collected from dental patients, either randomly throughout their daily activities, or, for investigations involving the quantification of salivary metabolites, immediately after they awoke in the morning). Results acquired revealed the ability of these NMR techniques to simultaneously detect >70 endogenous biomolecules, together with a series of agents arising from dietary, oral healthcare products and pharmaceutical sources. Moreover, highly significant 'between-subject' differences in the a.m. awakening salivary biomolecule concentrations were also found.

NMR spectroscopy also facilitates a full biomolecular appraisal of the consequences of the presence of metallic ions within biological matrices. Again, taking saliva as an example, the technique can probe the complexation of, for example, Ca^{2+} ions by low-molecular-mass salivary biomolecules; indeed, a study conducted by Silwood *et al.* has revealed that the organic acid anion citrate acts as a powerful oxygen-donor complexant for this metal ion.[17] Indeed, accurate determination of its resonances' chemical shift values and AB spin-system coupling pattern can be successfully employed to estimate its degree of saturation with Ca^{2+} (binding constant computer modelling investigations indicated that lactate represents an alternative, competing Ca^{2+}-complexing biomolecule present in human saliva). Moreover, the level of Ca^{2+} chelation by salivary citrate was found to be markedly influenced by the use of dentifrices containing this complexant. The detection of dentifrice-derived agents serves as but one example of the possibilities for the detection of exogenous components. To this extent, a range of NMR studies have considered, for example, the consequences of treatment with an oral rinse formulation containing a 'stabilised' form of the free radical species chlorine dioxide ($ClO_2^{\bullet}$) [chlorite anion (ClO_2^{-})], for both its oxidative[12] and microbicidal activities,[13] the oxidising actions of a peroxoborate-containing tooth-whitening dentifrice formulation[14] and the ability of dentifrices containing a 'smart' bioactive glass to deliver calcium ions to saliva.[15]

10.1.2 Applications of Multivariate (MV) Statistical Techniques to the Interpretation of Salivary ^{1}H NMR Profiles

Extension of the above ^{1}H NMR analytical methods to the inclusion of, for example, MV statistical techniques also offers the possibility of the detection of hitherto undetectable spectral components, the intensity of which is altered by treatment, in addition to the more obvious salivary biomarker detection (with the ideal experimental proviso that a significant change in signal intensity does/do not occur; should that be the case, then necessary steps must be taken to allow for any new, unique spectral regions). A single metabolite variable can be monitored as a function of possible sources of variation represented as 'factors' incorporated into an experimental design (*e.g.* study participants, their health status, drug dosages, time-points and, where appropriate, their selected interactions).

The methodologies involved in the discovery of biomarkers from spectroscopic datasets have recently been re-appraised.[4] Indeed, since there is an experimental (and theoretical) detection limit to biomedical NMR analysis, in recent years complementary information has been sought from inherently highly sensitive mass spectrometric (MS) techniques. Nevertheless, metabolic pathways can assist with the identification of 'missing' components from ^{1}H NMR spectral profiles; concomitantly NMR analysis can inform us with important information regarding the unique isomeric form of a hitherto novel MS-detected species.[4] All such techniques furnish information on structure, quantity, pathway and implied organism status ascribable to genetics, environment or any interactions between the two.[4] It is also usually the case that separation science techniques such as gas chromatography (GC), high-performance liquid chromatography (HPLC) and ultra-high-performance liquid chromatography (UPLC) can be used as a means of separating out, *i.e.* time-resolve, mixture components that otherwise confound an analysis through undesirable signal overlap (NMR spectra of biological samples can yield signals arising from up to 200 components, but MS could highlight >1000).[4] However, the introduction of such extra 'degrees of freedom' by increases in the dimensionality of a derived data matrix also requires the use of highly developed MV statistical procedures for data reduction purposes. Richards *et al.* have recognised that global sequestration of -omics datasets with the objective of determining generalised mechanisms can involve (a) integration of the results from multiple experiments (conceptual integration), (b) the simultaneous MV analysis of datasets arising from multiple sources (statistical integration) and (c) the construction of predictive models from multiple experiments (model-based integration).[4] The MV statistical approach presents a series of challenges ranging from the type of mathematical method employed (and in particular whether any assumptions about linear relationships are required), to the adoption of appropriate scaling procedures in order to allow for inter- and intra-sample variations in signal intensities, and also for permitting differing dimensionalities emanating from each -omic platform (*e.g.* metabolites *versus* genes).[4] Typically, data matrices are constructed with patients' or participants' experimental data as rows, the various responses as columns and a possible third dimension (designated 'tubes') representing the multiple experiments, be it differing types of spectroscopy, sample type or -omics technique employed.[4]

The MV statistical analysis of such multicomponent analytical data is generally based on either (1) supervised pattern recognition or (2) exploratory data analysis (EDA). The first of these classes of analyses involves a variety of methods for discriminant analysis (DA),[18] which are employed both to determine whether there is sufficient information to classify samples into pre-defined groups using MV hypothesis tests and, subsequently, to determine which are the potential 'marker' variables (*i.e.* metabolite concentrations, spectral 'bucket' intensities or further measures related to biomolecule levels) with regard to their classification into different groups [*e.g.* control *vs.* disease group(s), or control (untreated) *vs.* treatment-receiving group classes]. This technique requires prior knowledge of their group membership in a 'training' set to form a model that is tested on an independent 'test set'. EDA, however, involves approaches such as principal component analysis (PCA) and self-organising maps (SOMs), which can be unsupervised and permit the use of inter-related descriptors.[19–22]

A recent study by Lloyd *et al.*[21] has been concerned with a further important aspect of MV analysis, namely the determination of which specific compound signals or spectral regions constitute 'biomarkers'. A symbiotic relationship may exist between optimised biomarker spectral parameters and particular class memberships.[23] Such classes could be diseased/control,[24] male/female[25,26] or genetic groupings.[27,28] Numerical values may therefore be obtained in order to determine how well variables discriminate between such classes. Despite the possibility that the variable could be discriminating for, *e.g.*, gender sub-groups, the preliminary process of determining the relative variable significance is nevertheless still an essential one for establishing a list of 'candidate' markers for two or more groups.[21] Principal Component Analysis (PCA) and PLS-DA pattern recognition techniques performed in combination with supplementary genetic algorithm calculation calculations in order to improve spectroscopic variable selection have been applied by Ramadan *et al.* to the MV analysis of ^{1}H NMR spectroscopic data of saliva samples of healthy human volunteers in order to probe for genetic differences.[29] Another ^{1}H NMR salivary study carried out by Takeda *et al.* (and also employing PCA and PLS-DA) has searched for metabolic differences contained within samples describing both 'resting' and 'stimulated' saliva, smokers and non-smokers, as well as further differences in an effort to describe the 'salivary metabolome'.[30] ^{1}H NMR spectroscopy, PCA and PLS-DA were also used by De Laurentiis *et al.* in order to compare and contrast metabolomic differences between saliva and exhaled breath condensate (EBC) from healthy subjects with those from laryngectomised and chronic obstructive pulmonary disease patients.[31] This same research group have recently reviewed their findings in the context of EBC, which displays biomarkers for respiratory conditions such as airways inflammation and endothelial dysfunction.[32]

Univariate statistical methods such as ANOVA (involving the F variance-ratio statistic) often assume normal distributions for potentially discriminating variables in order to yield significance values, but most metabolomic/metabonomic datasets fail traditional 'normality' tests, and also do not allow for variable interaction.[22] Moreover, further assumptions regarding variance homogeneity and additivity of individual metabolite datasets are also often violated. PLS-DA[33–35] is, of course, a popular method, particularly when there are only two groups (otherwise a one *versus* all binary decision is faced, which is a much more complicated situation[22]), and PLS weights and regression coefficients can be employed as diagnostic criteria.[36,37] However, this method is limited, since both classification and variable information assume an equivalent significance. This is quite often undesirable, and the implementation of supervised SOMs by Wongravee *et al.* (*vide-infra*) has allowed the attachment of relative significance to both the classifiers and the experimental dataset.[22]

SOMs (Self Organising Maps) were predominantly an unsupervised 'learning' method using artificial neural networks[38,39] in order to visualise differential patterns in datasets and to determine relationships between experimental measurements and samples. They were first described over 20 years ago,[40–42] but only a limited number of general analytical or metabolomic studies have appeared.[43,44] Such machine learning-type methods have been difficult to implement for various reasons.[21] For example, there is a paucity of readily available, user-friendly software. Furthermore, there was a requirement for significant (but not commercially viable) computing power in the past, case studies have appeared primarily in the machine-learning literature and method descriptions have not been expressed in a form readily understandable by chemometricians.[21] This situation has now, of course, been revised so that the methodologies can be performed in real time with cost-effective desktop computers.[21]

The use of SOMs can be a very powerful approach for graphical visualisations for the purpose of establishing inter-class relationships without formal assignments.[21,22,38,45,46] For example, unlike PCA they make use of the entire data space and do not require an optimum graphical combination of extracted principal components (PCs), and are not overly influenced by 'outliers'.[22] It is also the case that PCA is a linear method which does not allow for any data non-linearities.[22] SOMs have traditionally been employed in an unsupervised manner as an EDA technique but may also be implemented in a supervised manner for classification strategies.[22,42,45,47,48] Whereas unsupervised SOMs generally highlight sample similarities, it may be of interest to examine minor variations such as those attributable to donors, experimental apparatus, sampling date, *etc.*[22] The supervised variant allows the study of such factors by maximising their influence on the maps, particularly when this influence can be controlled.[22] Since maps can probe variable significance, the differing factors can also be employed to weight maps, highlighting the most significant variables for each source of variation.[22]

This chapter will therefore examine the effect of the *in vitro* treatment of healthy human salivary supernatant specimens with a particular oral healthcare product (OHCP) through the discovery of appropriate spectroscopic signals and the subsequent ability of MV statistical techniques to facilitate the recognition of group membership such as control/treatment (as a more direct variant of healthy/diseased status), patient population and diurnal variation.

10.2 Case Study: ^{1}H NMR-based Multivariate Statistical Analyses of Human Saliva Samples before and after Treatment with an Oxyhalogen Oxidant-containing Oral Rinse Product

Two recent studies by Lloyd *et al.*[21] and Wongravee *et al.*[22] have described novel SOM variable selection methods through the employment of salivary ^{1}H NMR spectroscopic datasets that highlighted the differences between samples before and after the *in vitro* addition of an oral rinse predominantly containing the oxyhalogen oxidant chlorite (ClO_2^-) [but also with trace levels of chlorine dioxide ($ClO_2^{\bullet}$)].[12] The experimental design ensured identical numbers of samples (48) in the treated and control groups, with an equivalent number of samples from each donor included in each group.

The methods were validated employing two simulated datasets of similar size to the NMR dataset, either with or without discriminatory variables (the latter being a null dataset).[21]

10.2.1 Materials and Methods

10.2.1.1 Collection of Human Saliva Samples

A series of non-medically compromised participants ($n = 16$) without any form of active periodontal disease or active dental caries were recruited to the study. To avoid interferences arising from the introduction of exogenous agents into the oral environment, participants were requested to collect all saliva available, *i.e.* ('whole') saliva expectorated from the mouth, into a plastic universal tube immediately after waking in the morning on three separate sampling days. Each participant was also requested to refrain completely from oral activities (*i.e.* eating, drinking, tooth-brushing, oral rinsing, smoking, *etc.*) during the short period between awakening and sample collection (*ca.* 5 min.). Each collection tube contained sufficient sodium fluoride (15 µmol.) in order to ensure that metabolites are not generated or consumed *via* the actions of micro-organisms or their enzymes present in whole saliva (or their supernatants) during periods of sample preparation and/or storage. Saliva specimens were transported to the laboratory on ice and then centrifuged immediately (3500 rpm for 15 min.) on their arrival to remove cells and debris, and the resulting supernatants were stored at $-70\ ^\circ$C for a maximum duration of 18 hr prior to analysis. The pH values of each supernatant were determined prior to ^{1}H NMR analysis.[21]

10.2.1.2 Preparation of Human Salivary Supernatant Samples for ^{1}H NMR Analysis

Following the collection of all 48 samples, each sample was divided into two 0.60 ml portions. The first portion was treated with a 3.0 ml volume of oral rinse, whilst the second portion was retained as a control to which an equivalent volume of HPLC-grade water was added. Portions containing the oral rinse are referred to as class A samples, whilst the controls are referred to as class B ones. The resulting $16 \times 3 \times 2 = 96$ samples were then thoroughly rotamixed to ensure a homogenous mixture, and then equilibrated at $37\ ^\circ$C for a period of 30 s. Samples were prepared by adding 0.05 ml of deuterium oxide (^{2}H$_2$O, providing a field frequency lock) and 0.05 ml of a 5.0×10^{-3} mol dm^{-3} solution of sodium 3-trimethylsilyl [2,2,3,3-^{2}H$_4$] propionate [TSP, chemical shift reference ($\delta = 0.00$ ppm) and internal quantitative ^{1}H NMR standard] in ^{2}H$_2$O to a 0.60 ml volume of each sample mixture examined.[21]

10.2.1.3 ^{1}H NMR Measurements and Spectral Editing

^{1}H NMR measurements were conducted on a Bruker Avance AX-600 spectrometer operating at 600.13 MHz for ^{1}H, with a probe operating temperature of 20 °C. Pulsing conditions for single-pulse spectra were: sweep width 8389 Hz; pulse width 8.5 ms (pulse angle 70°); acquisition time 3.9 s; pulse delay 2 s; 32 768 (subsequently zero-filled to 65 536) data points; 64 transients. An exponential function corresponding to a line-broadening of 0.30 Hz was applied to Free induction delays (FIDs) prior to Fourier transformation. Where present, the methyl group resonances of acetate (s, $\delta = 1.920$ ppm), alanine (d, $\delta = 1.487$ ppm) and/or lactate (d, $\delta = 1.330$ ppm) served as secondary chemical shift references. The identities of ^{1}H NMR signals were routinely assigned *via* considerations of chemical shift values, coupling patterns and coupling constants, and also comparisons with established literature values (where required, making allowances for salivary supernatant pH values and the pH-dependence of biomolecule resonances).[10] The total intensities of each 'intelligently defined' spectral bucket region were determined by electronic integration *via* application of the ACD/Labs 1D NMR software suite as outlined below. The spectral regions of 1.03–1.35, 1.88–1.94, 2.42–2.79, 3.35–3.38 and >7.92 ppm were removed from all salivary ^{1}H NMR profiles in view of the presence of signals from the added oral rinse agents in these regions [including acetate ($\delta = 1.92$ ppm), citrate (centred at 2.66 ppm) and formate ($\delta = 8.46$ ppm)]. The selected edited region for the AB coupling pattern of citrate, a buffering agent present in the added oral rinse at a high level (relative to those of acetate and formate), was wide in view of the presence of its readily detectable ^{13}C satellite resonances in all spectra acquired on oral rinse-treated salivary supernatants. The 4.62–4.94 ppm region was also removed in view of the presence of the broad, residual H_2O/HOD signal in this spectral region.[21]

10.2.1.4 *'Intelligent Bucketing' of Spectra Acquired and Further Data Processing Prior to MV Statistical Analysis of ^{1}H NMR Data*

The relatively large dataset matrix (96 spectra×146 buckets) was generated through the application of macro procedures for line-broadening, zero filling, Fourier-transformation and phase and baseline corrections, followed by the application of a separate macro for the 'intelligent bucketing' processing sub-routine;[49] all procedures were performed within the ACD/Labs 1D NMR Manager software package (version 9.0, ACD/Labs, Toronto, Ontario, Canada M5C 1T4). Before commencing the bucketing procedure, all spectra were examined visually for any inherent distortions and manually corrected, if necessary. The experimental strategy employed involved the addition of all spectra acquired into one common file in which the 'intelligent bucketing algorithm' examined all spectra simultaneously and focused on the 'bucket limits' of commonly observed resonance (peak) intensity areas, where and if possible. Buckets were selected through the employment of an algorithm designed to make critical divisional decisions, *i.e.* those which precisely define the loci of bucket divisions with regard to an optimised selection of 'resonance-specific' ones. It was assumed that any bucket containing less than 1% of the maximum summed intensity would contain primarily noise, and hence these buckets were removed, leaving a 96×49 data matrix that was then imported as a text file into MS Excel for further manipulation.[21]

10.2.1.5   Data Simulations

In the unsupervised SOM study of Lloyd *et al.*, in order to test the ability of the variable selection method for the detection of known discriminators, two types of simulated datasets were generated.[21] The first type was a null dataset, containing randomly generated variables; a correlation structure was induced between the variables in order to better resemble experimental multivariate data matrices. For this dataset, half the samples were randomly assigned to one class (A), and half to the other (B). In addition, the second type was simulated on similar principles, but contained discriminatory variables that were able to distinguish between the two classes. The procedure for generating the simulated datasets was outlined[21] as follows: (a) a 96×J data matrix of random numbers was generated using a normal distribution with mean value equal to 0 and standard deviation equivalent to 1. Uniform random noise of values between -0.2 and $+0.2$ was then added to the matrix; (b) in order to introduce correlation, variables from the matrix were divided into sub-sets containing successive variables from the original matrix, the variable number being a randomly selected integer from a uniform discrete distribution between 1 and 5 for each sub-set; (c) the variables in each sub-set were replaced by the first variable multiplied (with an underlying probability of 0.50) by ± 1. This created further sub-sets of variables that were all exactly positively or negatively correlated within each sub-set, but not necessarily between sub-sets. The now correlated sub-sets of variables were then replaced in the major matrix; (d) to weaken the perfect correlation structure in the dataset, a noise matrix was generated, specifically (i) for each column a value was chosen from a uniform distribution between 0 and 5 and (ii) each column of the noise matrix was then generated from a uniform distribution. The noise matrix was then added to the perfectly correlated main matrix in order to weaken the correlation structure before randomly permuting the order of the variables to give the appearance of a real dataset.[21]

The correlated data matrix was thereafter referred to as dataset 1a. For the purposes of variable selection a class vector was still required, even though the dataset contained no discriminatory variables. It was therefore assumed that the first 48 samples belonged to class A, and the last 48 belonged to class B for this dataset. To simulate a discriminatory dataset, steps (a)–(c) were followed exactly to produce a second data matrix containing perfectly correlated sub-sets. To generate the discriminatory variables, samples 1 to 48 were assumed to be from class A, and the remainder from class B, and the following steps were then conducted: (e) 20 variables were randomly selected to be discriminatory and removed from the data matrix to form a 96×20 sub-matrix. For each discriminatory variable the samples in class A had their value increased by a random number chosen from a uniform distribution between −1.5 and 1.5, and samples in class B had their value decreased by the same amount; (f) the discriminatory variables were then replaced into the correlated matrix and step (d) was performed as for dataset 1a in order to weaken the perfect correlations present; (g) the order of variables was randomly permuted to give the appearance of a real dataset, although the position of the discriminatory variables was still known. This correlated matrix with discriminatory variables was thereafter referred to as dataset 1b.[21]

10.2.1.6 Data Preprocessing

The simulated datasets were standardised before performing variable selection. For the ^{1}H NMR dataset, in order to reduce the influence of large outlier peaks, the data was square-root transformed[35] prior to the application of further spectral preprocessing steps. Two different types of processing were then performed, which depended on the variable selection method being employed. For the PLS-DA regression coefficient extraction procedures, the data matrix was standardised as for the simulated datasets. For the SOM methods, only centring was required, since the scaling of component planes (to be described below) is achieved in automatic fashion when comparing variables.[21] With the supervised SOM method, the dataset was split into training set and test sets for classification. Centring was performed on only training set samples; test sets were in fact centred according to mean training set-derived parameters in order to ensure that the test set samples do not influence the model.[22]

10.2.1.7 Software

The constructed data matrix was then imported from MS Excel into Matlab 2008a (The Mathworks Inc., Natick, Massachusetts, USA), where in-house routines for the removal of low intensity buckets and all preprocessing procedures were subsequently employed.[21] Custom routines were also written in Matlab to generate the simulated datasets, to perform the described variable selection methods and for all graphical SOM representations. All variable selection calculations were subsequently performed on the entire dataset using autopredictive methods, without the prerequisite for separating into test or training sets.[35]

10.2.1.8 Variable Selection Methods

10.2.1.8.1 Self Organising Maps. A Self Organising Map (SOM) is a form of Artificial Neural Network that usually employs unsupervised learning to produce low-dimensional representations of the training samples, whilst preserving the topological properties of the input space.[21,38,42,50] Therefore, SOMs are suitable for visualising data possessing a large number of variables. A SOM can be considered to be a grid of map units of regular spacing, each map unit containing a weight for each dataset variable. Sample vectors are then compared to the map unit weight vectors, and the map unit with the most similar weight vector is declared the Best Matching Unit, or BMU.[21] The BMU and neighbouring map units are then updated in order to resemble the sample more closely. A learning rate controls the amount the units can 'learn' to represent the input sample, decreasing monotonically with each algorithm iteration, as well as a neighbourhood weight, which decreases with distance from the BMU. As the learning process proceeds, the samples gradually become restricted to the most similar region of the map with consolidation of samples that are also close together in the overall high-dimensional input space.[21] The number of map units can be chosen according to problem complexities. For example, in the study of Wongravee *et al.* the number of map units was set as 15×20, *i.e.* 300 in total,[22] approximately three times the number of data samples (96 samples).

SOM variables can be interpreted by examining weights for a selected variable for each map unit. These so called 'component planes' correspond to single variable-specific layers of the map, and can identify strongly corresponding variables for a particular sample.[42] By shading the map according to the weight intensity, and comparing with a map with the sample BMUs labelled, this can be visualised in a facile manner. For a small number of variables this is usually sufficient, since the number of layers in the map is equivalent to the number of variables. The visualisation process has to be replaced by an automated method for a large number of variables. Such a method, the SOM Discrimination Index (SOMDI), was developed by Lloyd *et al.*, and represents an overall summed ratio describing the amount a significant variable is present in map units representing a particular variable classification.[21]

If required, class information can also be visualised with SOMs whilst not influencing the map. This can be achieved by excluding class information when locating the BMU by calculating the Euclidean sample distance without class variables to the map weights, then including class weights and variables during learning in order to update the map weights.[21] As the map becomes more representative of the samples, its class weights update and approach 1 if the sample belongs to that class, and closer to 0 if not. In the study of Wongravee *et al.*, an adjustable class weight was introduced instead.[22] This necessarily meant that a low value weight essentially results in a map that is close to an unsupervised operation, whereas a high value may result in data overfit.[22] The most useful variables are, of course, those that distinguish between classes, so it is more appropriate to examine relative variable magnitudes within a class rather than considering its magnitude across the entire map. Since both the class and variable weights are scaled to be between 0 and 1, the product of the two for a single map unit will be close to 1 if the map unit is strongly associated with both specified classes and variables, and a value close to 0 if not.[21]

The SOMDI (SOM Discrimination Index) of Lloyd *et al.*[21] was established just for two-class unsupervised SOMs. This, in fact, can be generalised, and subsequently combined with supervised SOMs. There are often more than two groups within a dataset to consider, and this has also been allowed for in the study of Wongravee *et al.*,[22] for example an additional set of weight vectors is employed to contain class information, which therefore also updates the map.[22] The other important aspect of the study was the introduction of the facility to incorporate several factors that could influence metabolic profiles. Maps can be independently 'trained' for a different factor, and it therefore seeks and provides markers for each of the factors independently.[22] In the study by Wongravee *et al.*, the major source of variation was expected to be treatment with oral rinse, although a relatively minor source of variation might have been expected 'between-donors', but sampling day was not considered to be a factor having a major influence (since it was donor-dependent), and was therefore treated as a null factor.[22] Unsupervised SOM markers can, of course, only find the most dominant factors that influence map appearance. The study of Wongravee *et al.* also introduced a strategy for determining variable significance based on how many times they are selected over 100 iterative reformulations of the SOM map.[22]

The weight factor was further defined in the supervised SOM study of Wongravee *et al.* as the 'variable weight vector' (VWV).[38] To initialise the map unit, weight vectors were randomly generated by a uniform distribution between the maximum and minimum values of each data variable.[22] This was extended for supervised SOMs, such that the weight vector dimensions changed from $(1 \times J)$ to $[1 \times (J + K)]$, where K is the number of classes in the dataset.[22] The weight vector of such units was termed a 'supervised weight vector' (SWV), which further incorporated a 'class weight vector' (CWV) containing class membership information; the higher the constituent values, the more likely the map unit is describing a particular class membership. Sample input vectors used for supervised SOM map training contains two parts comprising the preprocessed 'variable sample vector' (VSV) in the data and the vector containing class information, *viz.* the 'class sample vector' (CSV), of dimensions $(1 \times K)$.[22] The combination of the sample vector and the scaled class vector yielded the 'supervised sample vector' (SSV).[22] The dimensions of the CSV are dependent on the number of dataset classes. For example, if the dataset contains three classes A, B and C, then the CSV can be described as $[\omega\ 0\ 0]$, $[0\ \omega\ 0]$, $[0\ 0\ \omega]$ for samples that are members of classes A, B and C, respectively, where ω is a scaling value.[22] If a particular class membership has little influence on 'learning', then the factor of interest would not be the primary cause of variation.[22] Hence, supervised SOM map training involves the use of an SSV, in contrast to unsupervised SOMs, which employ only the VSV. The SSV was then compared to each SOM map unit, and the unit whose SWV is most similar was assigned as the Best Matching Unit (BMU). After the learning process has been completed, samples that display similarities based on a consensus of class membership in measured variables should be assigned to similar map regions.[22]

Supervised SOMs can also be used to determine the class of an unknown sample, by locating the BMU of the unknown sample using only the VWV weights for each unit, and assigning the sample to the class in the CWV of the BMU that has the largest value.[22] The larger ω, the higher the risk of an overfit since the SOM may not be able to successfully classify test set samples, yet will force the training set into pre-defined groups, requiring the optimisation of ω and validation of the classifier.[22] Validation is usually performed by dividing the dataset into training and test sets.[22] In the study of Wongravee *et al.*, two-thirds of the samples in each class were randomly selected for the training set and the remainder assigned to the test set.[22] The map was trained using the SSVs of the training set samples to provide a map of VWVs that were then used to classify the test set samples. The procedure was repeated 100 times using different randomly selected training and test sets, in order to ensure that the map was not unduly influenced by outliers or typical samples from the training set.[22]

For the optimisation of ω, the VSV and corresponding VWV of the BMU were calculated for each test set sample, and the Euclidean distance between them computed.[22] The more similar the VSV and VWV values, the lower the computed value. The average computed value over the test set was used as a measure of how different the VSV and VWV values were after training. In view of separate test set calculations, this presents the possibility of different optimal values of ω. If too high, it will exert too great an influence on the final map because of overfit, and the BMU is actually a poor representation despite an apparent good class separation.[22] The optimum value of ω yielded a map with the most representative weights, and this was determined using the 'golden search' method.[51,52]

In the investigation conducted by Wongravee *et al.*, the SOM discrimination index (SOMDI) algorithm was extended to determine significant variables in datasets containing more than two classes, and thereafter combined with supervised SOMs.[22] To identify the markers for any specific group, 'in-group' *versus* 'out-group' comparisons were employed. For example, class A was defined as the 'in-group', and all other classes as the 'out-group' using a two-class one *vs.* all comparison.[35] Therefore, for K groups in the data, K such comparisons were made (where $K = 16$ individuals, 2 treatments and 3 sampling days). A SOMDI was calculated for both the 'in-group' and 'out-group', and usually the 'in-group' corresponded to a small proportion of the samples, *e.g.* an individual donor, so the calculations highlighted whether a specific variable was more often found for the 'in-group' compared with the rest of the samples. Markers would only be useful if the 'in-group' SOMDI was greater than the 'out-group' equivalent.[22]

The overall scheme for validating a supervised SOM map and classifying an unknown sample is shown in Figure 10.1. The Percent Correctly Classified (%CC)[35,53] is obtained by a majority vote,[35] which involves assigning a sample to a class for a maximum number of times in order to assess the classifier performance.[22] The %CC value can be calculated for both training and test sets. Although a high %CC could feasibly be obtained for both training and test sets, it is still possible to obtain a high %CC for the training set and a low %CC for the test set, a situation indicative of an overfit.[22]

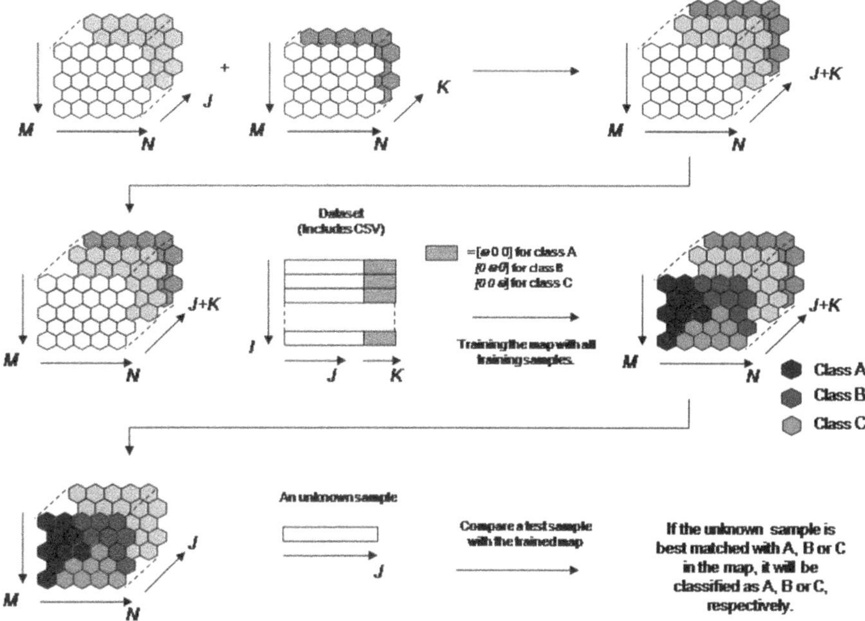

Figure 10.1   Scheme for classification of an unknown sample using supervised SOMs for K classes and J variables.
Reprinted with permission from K. Wongravee, G. R. Lloyd, C. J. Silwood, M. Grootveld and R. G. Brereton, Supervised Self Organizing Maps (SOMs) for classification and variable selection: illustrated by application to NMR metabolomic profiling, *Anal. Chem.*, 2010, **82**, 628–638. Copyright 2010 American Chemical Society.

10.2.1.8.2 Partial Least Squares Regression Coefficients. Partial Least Squares (PLS)[34,54,55] is a common supervised linear modelling technique. The technique performs a dimensionality reduction of the spectroscopic data X matrix, but also relates x variances to y variances contained in a Y response matrix. The matrices are simultaneously decomposed, exchanging respective scores information such that the technique maximises their covariance in this case. In a similar manner to PCA, components are generated that successfully maximise any remaining covariance, the optimum number defining the model dimensionality. A PLS-DA analysis involves a Y matrix containing class information, the NMR spectroscopic buckets (^{1}H or otherwise, X matrix) being related to nominal categorical codes (Y column dummy matrix) by an equivalent correlation matrix B [eqn (1)]

$$Y = XB \tag{1}$$

and the analysis can therefore maximise the correlation (or covariance) between X and Y. The PLS1 algorithm was employed in the study of Lloyd *et al.*, which considers only single column categorical Y matrices.[21] In their study the values of the elements in Y were set as $+1$ for samples that are members of group A and -1 for members of group B, and hence Y was automatically centred since there were equivalent numbers of samples in each class. The X and Y matrices are converted in the analysis to eqns (2) and (3)

$$X = TP^T + E \tag{2}$$

$$Y = UQ^T + F \tag{3}$$

T and P representing the scores and loadings matrices for X, respectively, U the corresponding Y scores matrix, Q^T the y weighting matrix, and E and F the residual matrices, accommodating information not related to the X/Y correlation. The x weights w (describing the variation in x correlated to the y class information, *i.e.* through their covariance, as well as containing information on the variation in x not related to y) are also used for calculating T [eqn (4)],

$$T = XW^* \tag{4}$$

the W^* matrix being transformed from the original W matrix so that it is PLS component-independent since the x scores T are linear combinations of the x variables, and hence when multiplied by P they will essentially return the original variables (with small E values). Equations (2) and (4) are then combined to yield eqn (5),

$$Y = XB + E \tag{5}$$

i.e. a modified form of eqn (1) which allows for residuals, to set up the regression model according to eqn (6)

$$B = W^*Q^T \tag{6}$$

B can be estimated from eqn (7)

$$B = W(P^TW)^{-1}Q^T \tag{7}$$

Significant ^{1}H NMR spectral bucket variables therefore appear in the regression coefficient matrix B. The larger the coefficient magnitude for a variable, the more likely it is to be a significant marker. The sign of the regression coefficient can also be employed to determine which group the variable is a marker for.[21] A range of further output parameters can be generated from PLS analytical packages, including goodness-of-fit parameters such as the fraction of the mean-corrected sum-of-squares (SS) of the Y codes explained for each generated PLS component, *i.e.* R^2 in eqn (8)

$$R^2 = (1 - RSS/SS) \tag{8}$$

where RSS is the fitted residual sum of squares, *i.e.* the sum of the squared difference between the observed and fitted y values in eqn (9)

$$RSS = \Sigma(y_{fitted} - y_{actual})^2 \tag{9}$$

The presence of many, potentially correlated, x variables indicates the possibility of data overfit, and hence there is a requirement to test the model's predictability for each PLS component. Model validation through deduction of the number of significant PLS components was determined in this study by a bootstrap procedure as described elsewhere.[21,56–58]

10.2.2 Results

10.2.2.1 ^1H NMR Spectra

The ^{1}H NMR investigations of the consumption of salivary biomolecules by agents in the oral rinse investigated revealed a wide range of further ^{1}H NMR-detectable salivary biomolecules which are predominantly consumed (scavenged) by oxyhalogen oxidants present therein, together with those of agents generated as products from such redox equilibria or scavenging processes. Indeed, such treatment-mediated spectral modifications were not necessarily directly observable by direct visual inspection of treatment-matched spectra acquired. The biomolecular species shown to be consumed in this manner comprise the amino acids valine, leucine, isoleucine, alanine, arginine, lysine, ornithine, glutamate, glutamine, proline, 4-hydroxyproline, methionine, aspartate, phenylalanine, tyrosine, histidine and taurine, together with the α-keto acid anions pyruvate and 2-oxoglutarate, carbohydrates including the α- and β-anomers of glucose, α-galactose, sucrose and mannitol, the molecularly mobile carbohydrate side-chains of 'acute-phase' glycoproteins, hyaluronate and oligosaccharides derived from this glycosaminoglycan [*via* the actions of bacterial hyaluronidase or phagocytically generated reactive oxygen species (ROS)], 3-D-hydroxybutyrate, glycolate, trimethylamine (TMA), creatinine, choline, phosphorylcholine, triacylglycerols (particularly polyunsaturated fatty acids), and possibly the exogenous agents chlorhexidine, thymol, dodecyl sulfate, caffeine, propane-1,2-diol, methyl paraben, aspartame, salicylate and paracetamol [although all participants instructed to refrain from all oral activities during the brief (*ca.* 5 min.) period between awakening and sample collection, it does, of course, remain a possibility that one or more of them did not follow this essential trial pre-requisite criterion; however, such agents could also remain detectable in the a.m. awakening saliva specimens *via* their consumption during the previous evening, and the detection of chlorhexidine in a series of appropriate intelligently selected buckets is conceivably explicable by its substantivity]. Agents which were found to increase in concentration subsequent to treatment with the oral rinse included (1) *n*-butyrate, *n*-valerate, *iso*-caproate and γ-aminobutyrate, this presumably representing a consequence of their mobilisation from positively charged protein binding-sites by the large excess of negatively charged oral rinse ClO_2^- present (or, alternatively, its ability to oxidatively damage such sites), and (2) treatment-elevated signal intensities in the conjugated diene vinylic proton regions of spectra (*ca.* 5.2–6.0 ppm), an observation arising from the possible generation of conjugated hydroperoxydienes (CHPDs) from the ClO_2^-- and/or $ClO_2^\bullet$-mediated oxidation of polyunsaturated fatty acids (PUFAs). Moreover, methionine sulfoxide (singlet -SO-CH$_3$ group resonance, $\delta = 2.725$ ppm) was also generated from the oxidation of methionine by ClO$_2^-$.

Figure 10.2 exhibits the expanded 0.670–1.460, 1.394–2.182, 2.031–3.962, 3.896–4.685, 5.000–6.760 and 6.749–8.510 ppm regions of the 600 MHz ^{1}H

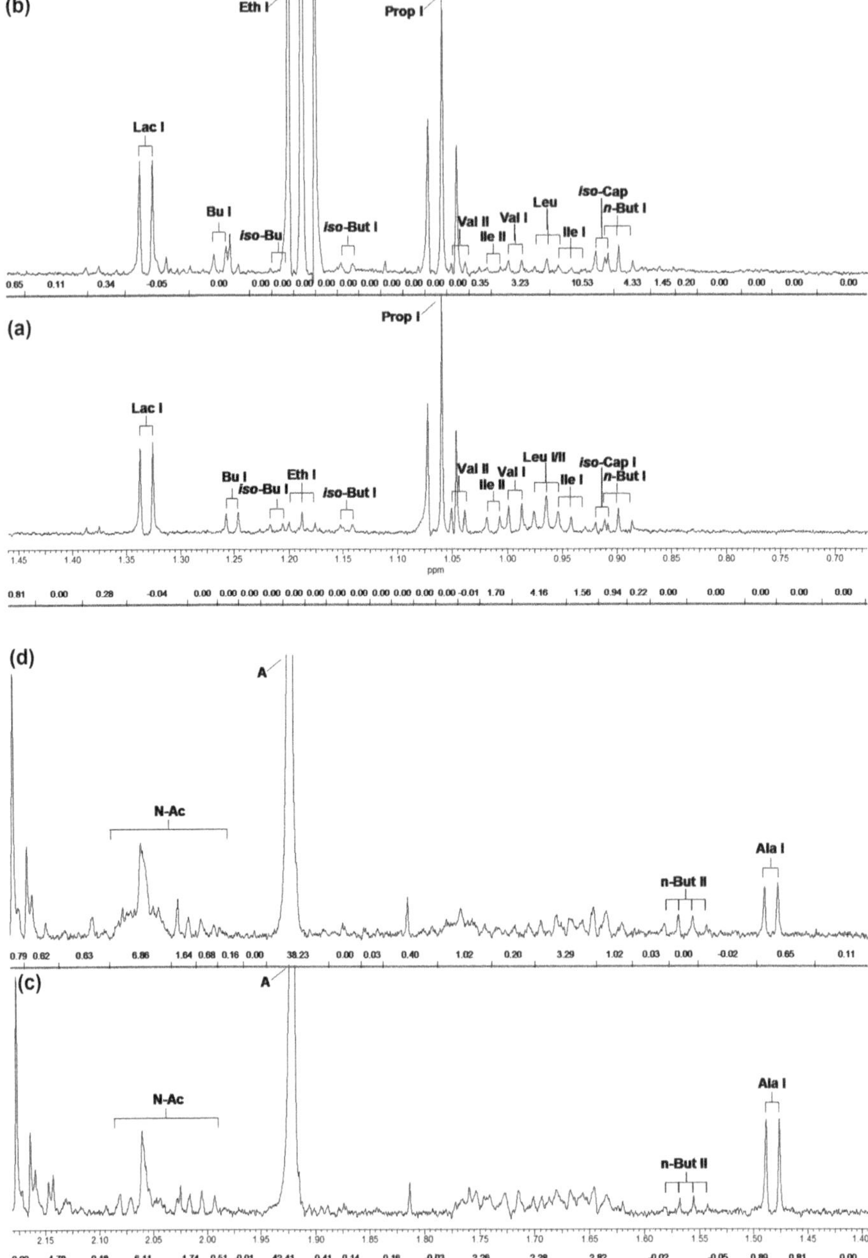

Figure 10.2

NMR spectra of a typical human salivary supernatant sample acquired prior and subsequent to equilibration with the oral rinse formulation. These typical spectral expansions clearly show a range of the oral rinse-dependent biomolecular modifications, for example the oxidative consumption of alanine, taurine, methionine and pyruvate. Indeed, particularly notable is the development of a multiplet resonance (with an apparent dt coupling pattern) centred at 4.272 ppm on oral rinse treatment. Clearly, the species responsible for this signal (which may serve as a unique, specific salivary 'marker' of the use of the oral rinses containing ClO_2^- by dental patients) may

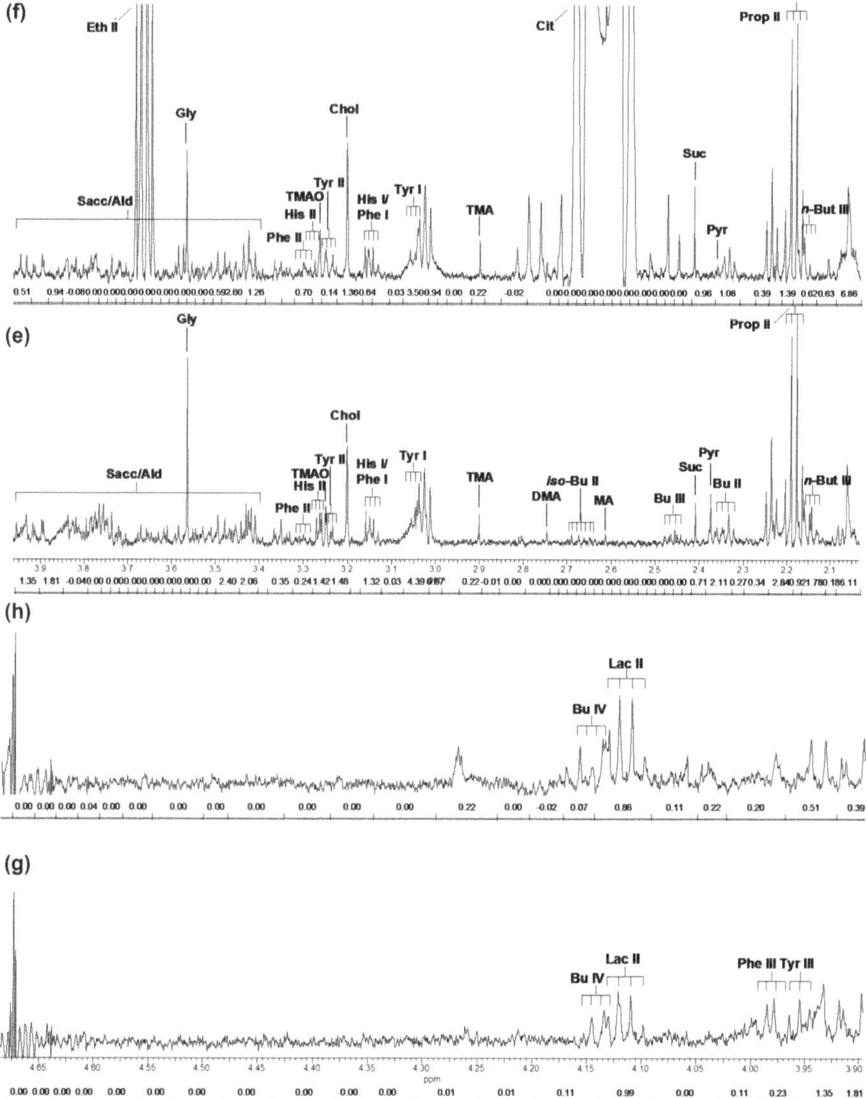

Figure 10.2 (Continued)

represent an agent arising from the reaction of ClO_2^- and/or, to a lesser extent, $ClO_2^\bullet$ with a selected salivary biomolecule.

Interestingly, the α-C$\underline{H}$ resonance of cysteine sulfinate (X proton of ABX system), an intermediate product expected from the reaction of ClO_2^- with cysteine in view of those arising from the attack of this oxyhalogen oxidant on the *N*-acetyl derivative of this amino acid,[59] is partially incorporated by

(j)

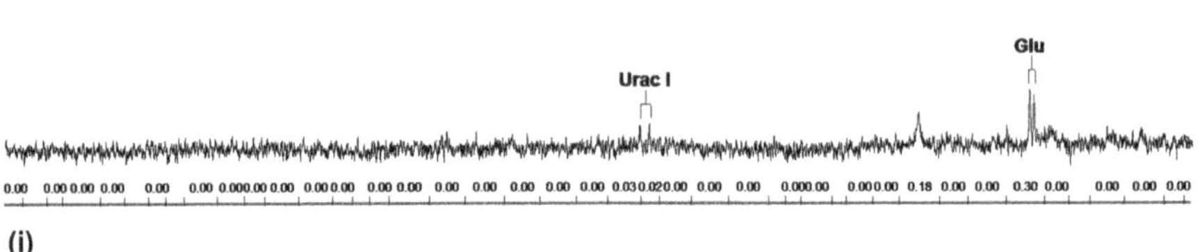

(i)

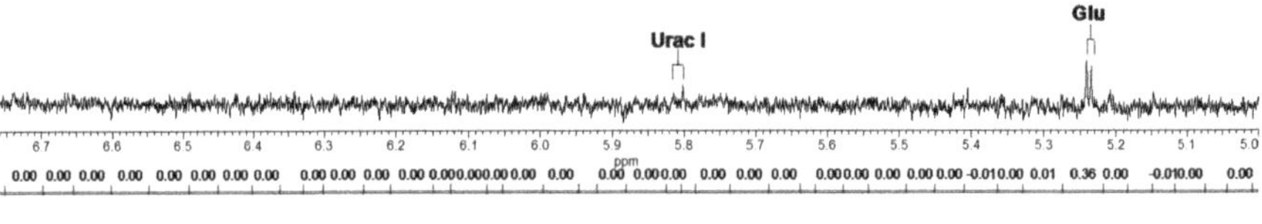

(l)

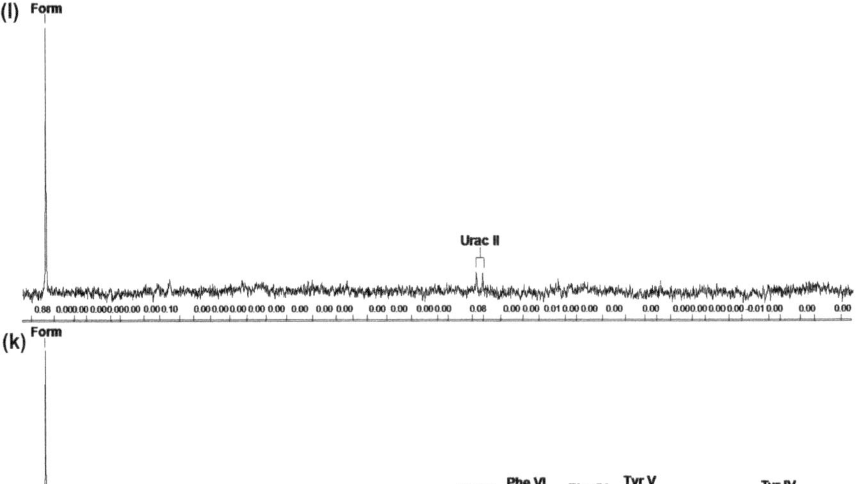

Figure 10.2   (Continued)

this bucket region (chemical shift value $\delta = 4.31$ ppm).[60] Moreover, the β-CH_2SH resonance of cysteine sulfonate (the final product derived from the reaction of *N*-acetylcysteine with ClO_2^{-}[59] occupies two of the 'intelligently selected' bucket regions (3.17–3.19 and 3.24–3.28 ppm), although it should be noted that the authors of the reference from which these chemical shift values (both cysteine sulfinate and sulfonate) were obtained[60] are unclear regarding the pH values of the aqueous solutions utilised in their NMR experiments. However, for each of these cysteine sulfonate-β-CH_2SH group bucket regions, significant decreases rather than increases in their resonance intensities were noted, an observation indicating either that treatment-induced decreases in those of alternative signals therein were more substantial than that of cysteine sulfonate production (the mean total salivary thiol level was found to be only 33 μmol dm^{-3} in one investigation[61]) or, alternatively, that this oxidation product is consumed by $HOCl/OCl^{-}$ generated as an intermediate in selected reactions of ClO_2^{-} with amino acids[62] (a process firstly producing N^{α}-monochloroamines which then degrade to corresponding aldehydes, NH_4^{+} ion and CO_2 at physiological temperatures,[63] as delineated in the Discussion section). Although further agents present in the oral rinse formulation (including citrate, acetate and formate) will also indicate the use of this product when detectable at

---

Figure 10.2 (a), (c), (e), (g), (i), (k) and (b), (d), (f), (h), (j), (l), expanded 0.670–1.460, 1.394–2.182, 2.031–3.962, 3.896–4.685, 5.000–6.760 and 6.749–8.510 ppm regions of the 600.13 MHz single-pulse ^{1}H NMR spectra of a human salivary supernatant specimen (pH value 6.78) before and after treatment with the oral rinse, respectively. Typical spectra are shown. Abbreviations: A, Acetate-C$\underline{H}_3$; Ala I and II, alanine-C$\underline{H}_3$ and –C$\underline{H}$ group protons, respectively; Bu I, II, III and IV, 3-D-hydroxybutyrate γ-C$\underline{H}_3$, α/α'-C$\underline{H}_2$ and β-C$\underline{H}$ protons, respectively; *iso*-Bu, 3-hydroxy-*iso*-butyrate-C$\underline{H}_3$ group protons; *iso*-But I and II, *iso*-butyrate-C$\underline{H}_3$ and – C$\underline{H}$ group protons, respectively; *n*-But I, II and III, *n*-butyrate γ-C$\underline{H}_3$, β- and α-C$\underline{H}_2$ protons, respectively; *iso*-Cap, *iso*-caproate δ-C$\underline{H}_3$ group protons; Chol, choline-N$^+$(C$\underline{H}_3$)$_3$; Cit, Citrate-*AB*-C$\underline{H}_2$-C-C$\underline{H}_2$; DMA, dimethylamine-C$\underline{H}_3$ group protons; Eth I and II, ethanol-C$\underline{H}_3$ and - C$\underline{H}_2$ group protons, respectively; Form, formate-$\underline{H}$; Glu, α-glucose H1; Gly, glycine-C$\underline{H}_2$; His I and II, histidine $\underline{AB}$X protons; His III and IV, histidine imidazole ring protons; Ile I and II, isoleucine-C$\underline{H}_3$ and β-C$\underline{H}_3$ group protons, respectively; Lac I and II, lactate-C$\underline{H}_3$ and –C$\underline{H}$ protons, respectively; Leu, leucine γ-C$\underline{H}_3$'s; N-Ac, spectral region for acetamido methyl groups of *N*-acetyl sugars; Phe I, II, III phenylalanine $\underline{AB}$X β-C$\underline{H}_2$ and AB$\underline{X}$ α-CH protons, respectively; Phe IV, V and VI, phenylalanine aromatic ring protons; Prop I and II, propionate-C$\underline{H}_3$ and –C$\underline{H}_2$ group protons, respectively; Pyr, pyruvate-CH_3; Sacc/Ald, saccharide/alditiol ring proton fingerprint region; Suc, succinate-C$\underline{H}_2$; Tau I and II, TMA, trimethylamine-C$\underline{H}_3$'s; TMAO, trimethylamine oxide ON(C$\underline{H}_3$)$_3$ group protons; Tyr I and II, tyrosine $\underline{AB}$X β protons; Tyr III, tyrosine AB$\underline{X}$ α proton; Tyr IV and V, tyrosine aromatic ring protons; Urac I and II, uracil H5 and H6 ring protons, respectively; Val I and II, valine-C$\underline{H}_3$ group protons; *n*-Val I and II, *n*-valerate δ-CH_3 and γ-CH_2 protons, respectively.

elevated levels in human saliva (especially citrate), such components are also commonly found in alternative dentifrice products, and hence are unspecific markers.

In this manner, a series of further biomolecular transformations were also notable, specifically (1) for the spectrally edited 1.03–1.35 ppm region, substantial reductions in the intensities of the lactate-CH_3 doublet resonance located at 1.330 ppm (an observation fully supporting the statistically detected decrease in that of its -CH signal) and significant decreases in the 3-D-hydroxybutyrate-CH_3 group doublet ($\delta = 1.240$ ppm), processes which presumably reflect the oxidation of these biomolecules to pyruvate (and subsequently acetate and CO_2) and acetoacetate (and subsequently acetone and CO_2), respectively, by oral rinse oxyhalogen oxidants; the (less reproducible) generation of a complex series of low intensity signals in this chemical shift range was also noted; (2) for the edited acetate-$C\underline{H}_3$ group region ($\delta = 1.88$–1.94 ppm), the generation of 2 or more low intensity singlet resonances was observed; (3) for the edited citrate-$C\underline{H}_2CO_2^-$ region ($\delta = 2.42$–2.79 ppm), no major modifications were observed, although it should be noted that one of the ^{13}C satellite lines of the 'A' proton of the citrate-$C\underline{H}_2CO_2^-$ system was significantly broadened on addition of the oral rinse to salivary supernatants (when expressed relative to that of the further 'A' proton lines of this component), a phenomenon that may reflect its complexation of salivary metal cations and/or binding to salivary macromolecules, *e.g.* proteins; (4) for the edited 3.50–3.85 ppm region, substantially and highly reproducible decreases in the intensity of the salivary glycine-α-CH_2 resonance ($\delta = 3.59$ ppm), as also noted for the alanine-CH_3 group signal. For the edited high-frequency region (7.78–7.92 ppm), no oral rinse-mediated modifications were observed.

Further highly notable oral rinse treatment-induced spectral differences observable *via* direct examination included marked decreases in the intensities of taurine, lysine, creatine, creatinine, phenylalanine and tyrosine ^1H NMR resonances, with a smaller but nevertheless highly reproducible reduction in those of the histidine resonances.

10.2.2.2 MV Statistical Techniques

Fully trained unsupervised SOM maps generated for the simulated datasets demonstrated several 'disjoint' regions for each class for the null dataset 1a, whilst the map for the discriminatory datasets 1b consisted of two main regions, one for each class.[21] It was found that variables that strongly distinguished between samples facilitated the learning process when organising the map, and hence variables that were constantly assigned a good rank predominantly contributed to the organising process. A comparison was made with regression coefficients generated from an equivalent set of PLS-DA calculations on the same null dataset.[21] The variables ranked in the top 20 by the SOM method were similar to their regression coefficient equivalents, but the actual rank of the variables varied considerably. However, for dataset 1b there was a good agreement between the SOM and PLS

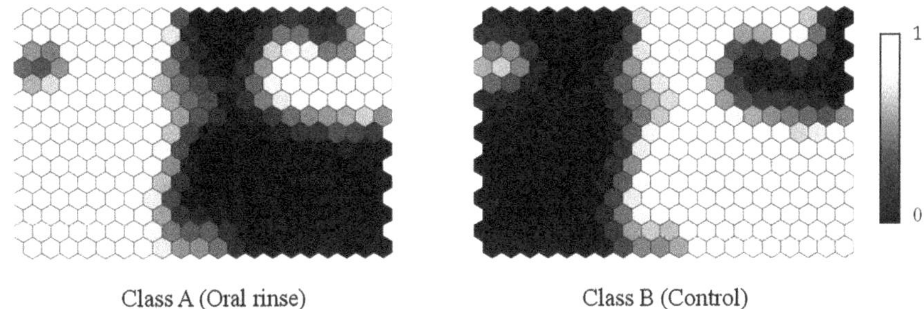

Figure 10.3   Class component planes for a trained unsupervised SOM for the human saliva dataset. The map units are shaded from highly representative (light) to non-representative (dark) of the two classes.
Reprinted from G. R. Lloyd, K. Wongravee, C. J. L. Silwood, M. Grootveld and R. G. Brereton, Self Organising Maps for variable selection: Application to human saliva analysed by nuclear magnetic resonance spectroscopy to investigate the effect of an oral healthcare product, *Chemometr. Intell. Lab. Syst.*, **98**, 149–161, Copyright (2009), with permission from Elsevier.

methods and the variable ranks demonstrated shared 'top 3' biomarkers, the two methods both being relatively successful at correctly identifying the true discriminatory variables.

An example SOM map for the human saliva dataset in the study by Lloyd *et al.*[21] is shown in Figure 10.3, where the units have been shaded according to the class component planes. These planes did display some overlap between the classes (this would be expected when examining biological samples subject to environmental factors), but nevertheless distinct control/oral rinse regions were evident.[21] The top 20 computed variables for dataset 2 (human saliva) using both SOM and PLS-DA methods are listed in Table 10.1, and from this it can be noted that there are a number of resonances that are ranked equally highly, in particular variables 45 (4.24–4.29 ppm) and 48 (7.20–7.22 ppm), which were given the highest (best) ranks (1st or 2nd) by both methods.[21] Variable 48 is assignable to the H3,H5 aromatic resonances of the oxidisable amino acid tyrosine. One advantage of the SOM technique to have emerged from the study is that one variable may only present with a significant magnitude for a small number of map units. Yet the equivalent PLS computations might rank this variable highly even though it would appear to be a significant one for a small number of samples, and hence is actually not such a good discriminatory marker for discriminating between the two classes. This arises from the mean or standard deviation not being utilised with SOM methodologies when estimating a variable's significance, and can be repeated many times even for autopredictive datasets such as those employed in the study of Lloyd *et al.*[21] This may not, however, indicate that such PLS-selected variables are indeed not significant, but this situation may require, for example, Monte Carlo simulation approaches[58] in order to probe their actual usefulness.[21]

Table 10.1 The top 20 ranked ^{1}H NMR buckets for dataset 2 using both PLS-RC and SOMDI (all buckets ranking within the top 20 using either method are listed). Reprinted from G. R. Lloyd, K. Wongravee, C. J. L. Silwood, M. Grootveld and R. G. Brereton, Self Organising Maps for variable selection: Application to human saliva analysed by nuclear magnetic resonance spectroscopy to investigate the effect of an oral healthcare product, *Chemometr. Intell. Lab. Syst.*, **98**, 149–161, Copyright (2009), with permission from Elsevier.

Variable	SOMDI rank	PLS-RC rank	Class	ppm	Tentative assignment
48	1	2	B (control)	7.20–7.22	Part protein tyrosine residue-Ar-H2,H6
45	2	1	A (oral rinse)	4.24–4.29	Part cysteine-sulfinate-α-CH
26	3	3	B (control)	2.35–2.38	Pyruvate-CH$_3$; glutamate-γ-CH$_2$; proline-β-CH$_2$
5	4	8	B (control)	0.99–1.03	Isoleucine-β-CH$_3$; valine-CH$_3$s
4	5	14	B (control)	0.96–0.99	Leucine-γ-CH$_3$s; valine-CH$_3$s
13	6	11	B (control)	1.72–1.78	Lysine-δ-CH$_2$
12	7	17	B (control)	1.70–1.72	Leucine-β- and γ-CH$_2$s; arginine-γ-CH$_2$
24	8	7	A (oral rinse)	2.28–2.31	γ-aminobutyrate-α-CH$_2$
17	9	15	B (control)	1.99–2.01	Isoleucine-β-CH
33	10	4	B (control)	3.24–3.28	Taurine-CH$_2$NH$_3$$^+$; betaine-$^+$N(CH$_3$)$_3$; arginine-$\delta$-CH$_2$; β-glucose-H2; phenylalanine-β-CH$_2$; trimethylamine-*N*-oxide-(CH$_3$)$_3$NO; histidine-β-CH$_2$; myo-inositol-H2
35	11	16	B (control)	3.32–3.35	Caffeine-NCH$_3$ (C3)?
20	12	12	B (control)	2.09–2.15	Methionine-S-CH$_3$ and -β-CH$_2$, glutamate-β-CH$_2$ glutamine-β-CH$_2$
7	13	9	B (control)	1.45–1.51	Alanine-CH$_3$; isoleucine-γ-CH$_2$; pyruvate hydrate-CH$_3$
16	14	21*	B (control)	1.86–1.88	γ-aminobutyrate-β-CH$_2$
42	15	5	A (oral rinse)	4.02–4.07	Phosphorylethanolamine-*O*-CH$_2$
37	16	19	B (control)	3.40–3.45	Taurine-$^-$O3SCH$_2$; proline-δ-CH$_2$NH-
49	17	26*	B (control)	7.38–7.43	Phenylalanine-Ar-H4; phenylalanine-Ar-H3,H5
30	18	10	B (control)	3.02–3.08	Lysine-ϵ-CH$_2$; creatinine-N-CH$_3$; creatinine-N-CH$_3$; cysteine-CH$_2$; ornithine-δ-CH$_2$; phenylalanine-β-CH$_2$
1	19	6	A (oral rinse)	0.84–0.86	n-valerate-CH$_3$; fatty acid-CH$_3$
36	20	28*	B (control)	3.38–3.40	Proline-δ-CH$_2$NH-; β-glucose-H4; methanol-CH$_3$
2	27*	13	A (oral rinse)	0.86–0.91	Fatty acid-CH$_3$; n-butyrate-CH$_3$; iso-caproate-δ-CH$_3$s
23	37*	18	B (control)	2.24–2.28	Valine-β-CH; acetone-CH$_3$
47	31*	20	A (oral rinse)	5.39–5.44	Unsaturated fatty acid vinylic >CH=CH<

*Variables not within the top 20 for the particular method are indicated by an asterisk. The class relates to the group in which the bucket has highest intensity, as determined by the sign of the coefficient (PLS-RC and SOMDI agree in all cases).

Table 10.2 Overall percentages correctly classified (%CC) over 100 iterations of training and test sets using the majority vote criterion with optimal scaling values $(\omega)^a$. Reprinted with permission from K. Wongravee, G. R. Lloyd, C. J. Silwood, M. Grootveld and R. G. Brereton, Supervised Self Organizing Maps (SOMs) for classification and variable selection: illustrated by application to NMR metabolomic profiling, *Anal. Chem.*, 2010, **82**, 628–638. Copyright 2010 American Chemical Society.

	% correctly classified		
Factor	Training set	Test set	Random
Treatment	94.72	70.79	50
Sampling day	92.36	38.19	33.33
Donor	89.26	19.53	6.25

[a]The far right-hand column represents the classification level that would be achieved if data were randomly assigned to each classification.

The %CC of the training and test sets for supervised SOMs in the study of Wongravee *et al.* using the majority vote system are shown in Table 10.2.[22] This table also lists the %CC that would have been achievable if samples had been randomly assigned to each group.[22] In essence, the training set %CC indicates how well the model is optimised.[22] It is clear from the table that the training set %CC for all cases is around 90% and higher than the test set %CC for all the designated cases, implying that the maps have been well organised and successfully classified the training set sample data when the optimum computed values of ω had been employed. Test set %CC values for the oral treatment regimen factor is *ca.* 70%, suggesting that this is a major source of variation. However, the donor variable test set %CC (19.53%) is also high when compared to that of the random model prediction (6.25%), whilst for sampling day (38.19%) is very close to the random model (33.33%). Therefore, a change in donor has a small effect on the dataset, and, as expected, the sampling day within-participants exerts an even more limited effect.

Unsupervised SOMs were also employed in the study of Wongravee *et al.* in order to examine any variation caused by the three factors.[22] The maps are shown in Figure 10.4, shaded according to their sources of variation. As with the previous study of Lloyd *et al.*, it is evident that there are very distinct regions for oral rinse/control classes. Conversely, there are not such good separations for the donor and sampling day analyses, observations suggesting that these factors contribute only a relatively small amount. Concomitantly, it might be expected for there to be some discrimination between donors in view of their differing habits (*e.g.* diet and exercise regimens), together with physical characteristics (*e.g.* age, gender, body mass index *etc.*) but these are likely to be indistinct when employing unsupervised methods.[22] Consequently, supervised SOMs using optimal scaling values were constructed (these are also shown in Figure 10.4). It is evident that there is an improved separation between groups for all cases, particularly for the minor factors, although it should also be noted that the sampling day effect is a random one and also donor-dependent, although there remains

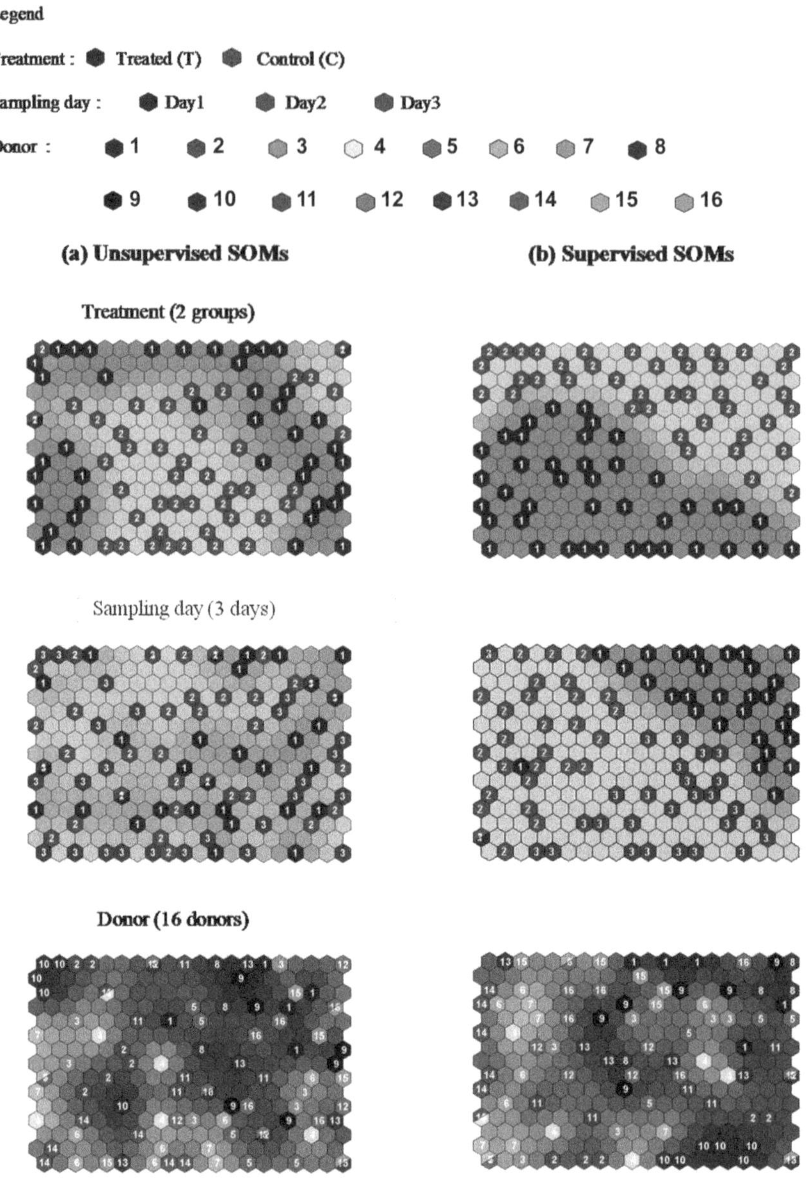

Figure 10.4 Unsupervised SOMs (left) and supervised SOMs (right) of the three factors, which are treatment, sampling day and donor, for the intelligently bucketed dataset. The optimal scaling values (ω) for each factor were employed to obtain the supervised SOMs. Reprinted with permission from K. Wongravee, G. R. Lloyd, C. J. Silwood, M. Grootveld and R. G. Brereton, Supervised Self Organizing Maps (SOMs) for classification and variable selection: illustrated by application to NMR metabolomic profiling, *Anal. Chem.*, 2010, **82**, 628–638. Copyright 2010 American Chemical Society.

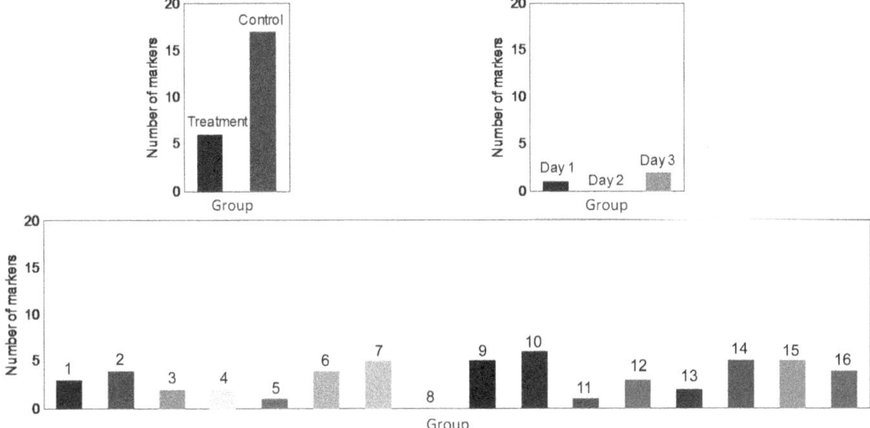

Figure 10.5　Number of significant variables found for each group and source of variation.
Reprinted with permission from K. Wongravee, G. R. Lloyd, C. J. Silwood, M. Grootveld and R. G. Brereton, Supervised Self Organizing Maps (SOMs) for classification and variable selection: illustrated by application to NMR metabolomic profiling, *Anal. Chem.*, 2010, **82**, 628–638. Copyright 2010 American Chemical Society.

at least some 'fixed' synchronous nature to this variable in terms of their sequential nature.[22]

In concentrating on the more powerful aspect of the use of supervised SOMs for finding discriminatory variables with respect to the three different sources of variation, the number of significant variables found in each case is shown in Figure 10.5 for the study of Wongravee *et al.*[22] There are 23 treatment-specific variables (^{1}H NMR spectral buckets) that pass this criterion for either the oral rinse- or H_2O-treated control groups. There are only 6 'markers' for the oral rinse group (T) and there are 17 markers corresponding to the H_2O control group (C); this is not unexpected since chlorite anion ($ClO_2{}^-$) present in the oral rinse reacts with and/or oxidatively consumes many salivary biomolecules.[12]

Markers were then sought for the other sources of variation (the calculations necessarily involving the use of multiclass methods).[22] Unlike the treatment/control factor, only 3 markers were found for the sampling days within-participants component of variance, of which none relate to sampling day 2; for sampling days 1 and 3, 1 and 2 variables, respectively, were found. These findings are as anticipated. In the case of the donor (participant) variable, only the samples from two donors were well clustered (ω, of course, could have been increased to provide an improved clustering within samples belonging to individual donors, but this may have arisen from data overfit). It transpired that both these donors possessed several characteristic variables according to the established criteria:[22] indeed, 29 characteristic variables were detected, and these 29 variables were found to be characteristic of at least one individual. In this case, biomarkers may be characteristic of several donors instead of one, *e.g.* personal habits or genetic make-up.[22]

It was found that 7 out of the 49 NMR spectral buckets did not act as markers for any of the factors, 29 for only one of the factors and 13 for more than one factor.[22] Results will, of course, be affected by the fact that signals contained within an intelligently selected bucket do not necessarily originate from a single compound, so the experimental observations of Wongravee *et al.* are not unexpected, although there is still a degree of selectivity, especially for the dominant factors.[22] This implies that ^{1}H NMR spectral regions may be influenced by more than two unrelated factors. As a potentially useful observation, six bucket regions serve as significant markers for oral rinse treatment and do not appear to be influenced by either of the other factors.[22]

10.2.3 Discussion

Although a series of the oral rinse-induced differences observed in the ^{1}H NMR profiles of human saliva are observable *via* direct inspection of the spectra acquired, many of the modifications detected *via* the *prima facie* application of MV analytical techniques are not. Moreover, direct examination of spectra acquired of course fails to provide critical information (diagnostic or otherwise) regarding the statistical significance of such modifications. In principle, the approach employed here could be extended to studies targeted at discriminating between human subjects receiving and not receiving a particular (or particular class of) oral healthcare product (OHCP). Alternatively, the proposed methodology is readily applicable to investigations targeted at distinguishing between subjects with selected periodontal diseases (*e.g.* gingivitis, *etc.*), and those with an acceptably high standard of oral health. Results acquired on the consumption of (relatively) simple amino acids such as glycine, alanine and taurine by the added oral rinse formulation are explicable by previous investigations conducted on the kinetics and mechanisms of the reactions of such biomolecules by oxyhalogen oxidants (including ClO_2^-) as outlined below.

Of much significance to the marked level of salivary taurine consumption by the oral rinse investigated are the experiments reported by Chinake and Simoyi[62] on the oxidation of this β-amino acid by ClO_2^- (at neutral to acidic pH values, *i.e.* those which are relevant to the oral environment). Indeed, the stoichiometry of this reaction system was found to involve the consumption of 3 molar equivalents of ClO_2^- per mole of taurine to generate 1 of taurine's *N*-monochloroamine [$Cl(H)NCH_2CH_2SO_3H$] and 2 of $ClO_2^{\bullet}$ (the production of *N*-monochlorotaurine is rapid when expressed relative to that of $ClO_2^{\bullet}$ accumulation); at the lower pH values investigated, *N*-monochlorotaurine disassociated to taurine and *N*-dichlorotaurine. An important characteristic of this reaction system involves a significant induction period in which both HOCl and the reactive intermediate $H(OH)NCH_2CH_2SO_3H$ are produced, a process leading to the formation of *N*-chlorotaurine and $ClO_2^{\bullet}$ auto-catalytically. As expected for redox reactions involving ClO_2^-, this auto-catalysis is mediated by a Cl_2O_2 intermediate species and, interestingly, taurine's C–S bond is not cleaved, despite the availability of the powerful oxidant HOCl.

Hence, these previously reported studies clearly explain the substantial ^1H NMR-detectable reductions in salivary taurine observed on treatment of human salivary supernatant specimens with oral rinse-containing ClO_2^-. They also indicate that the oral rinse-induced oxidative consumption of a range of α-amino acids present in this biofluid also detected in this investigation also proceed *via* this mechanism. However, since many N^{α}-monochloroamines generated in this manner are unstable at physiological temperature (37 °C),[63] and deteriorate to corresponding aldehydes [eqn (10)], further investigations focused on the detection and quantification of such species corresponding to the side-chains of α-amino acids (*e.g.* formaldehyde from salivary glycine, acetaldehyde from alanine, *etc.*) are required in order to demonstrate this.

$$Cl(H)N\text{-}CHR\text{-}CO_2^- + H_2O \rightarrow RCHO + NH_3 + CO_2 + Cl^- \qquad (10)$$

Interestingly, it is well known that aldehydes act as potent microbicidal agents, and hence those derived from the above processes may also exert this activity in the oral environment. Indeed, a 2.0% (w/v) solution of glutaraldehyde is frequently employed as a disinfecting agent.[64]

Similarly, the oxidative consumption of γ-aminobutyrate noted here is likely to proceed *via* a similar mechanism. However, the amino acids cysteine, methionine and tyrosine, each with redox-active side-chains, can, of course, also be oxidatively modified by ClO_2^- (and also $ClO_2^{\bullet}$ and HOCl/ OCl^- produced *via* its reaction with these and/or further α-amino acids, together with taurine) to cysteine sulfonate (and cystine), methionine sulfoxide [eqn (11)] and a tyrosine-derived quinone species, respectively.

$$H_3N^+CH(CH_2CH_2SCH_3)CO_2^- + ClO_2^- \rightarrow H_3N^+CH(CH_2CH_2SOCH_3)CO_2^- + OCl^- \qquad (11)$$

The simplest representation of the oxidative consumption of pyruvate, together with further α-keto acid anions such as 2-oxoglutarate, is that given in eqn (12).

$$2CH_3COCO_2^- + ClO_2^- \rightarrow 2CH_3CO_2^- + 2CO_2 + Cl^- \qquad (12)$$

Therefore, it should be noted that the production of reactive HOCl/OCl$^-$ during an induction period observed during the reaction of ClO_2^- with the β-amino acid taurine[62] (and also presumably the salivary α-amino acids and γ-aminobutyrate consumed on reaction with oral rinse ClO_2^-) will also serve to further reduce the amino acid concentrations of human saliva. Indeed, even if this mechanistic process only proceeds for the reactions of selected free amino acids with ClO_2^- (or those located at the N-termini of salivary proteins), the

HOCl/OCl$^-$ generated will, of course, be available to react with a much wider range of such HOCl/OCl$^-$ 'scavenger' species in a (relatively) unselective manner to form N$^\alpha$-monochloro- and dichloroamines, together with N$^\varepsilon$-monochloro- and -dichloroamines in lysine residues (either free or protein-incorporated). As noted above, specific aldehydes arising from the decomposition of their parent amino acid N$^\alpha$-monochloroamine precursors will serve as valuable indicators of the activity of HOCl/OCl$^-$ arising from these reaction systems (RCHO, where R represents an amino acid side-chain moiety).

Aldehydes produced from the interaction of HOCl/OCl$^-$ with salivary α-amino acids and the decomposition of the primary N$^\alpha$-monochloroamine products can also react with ClO_2^-, and the oxidation of formaldehyde (HCHO) by this oxyhalogen oxidant was critically examined by Chinake *et al.*[65] in both mildly acidic and alkaline media. This reaction gave rise to CO_2 and $ClO_2^\bullet$ as products, the latter in virtually quantitative yield, and was autocatalytic with respect to hypochlorous acid/hypochlorite (HOCl/OCl$^-$). Indeed, the primary phase of the process generated HOCl, which catalysed the production of $ClO_2^\bullet$ and the additional oxidation of formic acid/formate (HCO_2H/HCO_2^-); $ClO_2^\bullet$ rapidly accumulated in view of its (relative) lack of reactivity towards both HCHO and HCO_2H/HCO_2^-. Although with excess HCHO the stoichiometry of this process was determined to be $3ClO_2^- + HCHO \rightarrow HCO_2H + 2\ ClO_2^\bullet{}_{(aq.)} + Cl^- + 2H_2O$, when large excesses of ClO_2^- were present [as, of course, is expected in the case of 5 : 1 (v/v) mixtures of oral rinse: human salivary supernatant], the stoichiometric profile involved in the consumption of 6 molar equivalents of ClO_2^- per mole of HCHO to generate 4 of $ClO_2^\bullet$, 2 of Cl^- and 1 of CO_2.

With regard to the oral rinse-mediated decrease in the intensities of salivary cysteine resonances observed here (and also in previously conducted chemical model studies,[12] Darkwa *et al.*[59] investigated the oxidative consumption of *N*-acetylcysteine by ClO_2^-, and found that the final product generated from this reaction system was *N*-acetylsulfonate and that the process had a stoichiometry of $3ClO_2^- + 2RSH \rightarrow 3Cl^- + 2RSO_3H$; as expected, there was no evidence for the production of *N*-chloroamine derivatives. This oxidation process proceeds *via* a mechanism involving a stepwise S-oxygenation involving the consecutive generation of sulfenic and sulfonic acid adducts. Intriguingly, a notable characteristic of the reaction is the rapid, immediate formation of chlorine dioxide ($ClO_2^\bullet$) without a monitorable induction period, since oxidation of the thiol by this oxyhalogen free radical species is sufficiently slow for it to accumulate without such a time lag, which, in general, represents a characteristic of the oxidation of organosulfur compounds by ClO_2^-. A full description of the 'global' dynamics of this system involves eight reactions in a truncated mechanism.

10.3 Conclusions

Evidence provided in these ^{1}H NMR spectroscopy-linked metabolomic investigations clearly demonstrated that the generation of $ClO_2^\bullet$ from ClO_2^-

in the oral environment is not entirely dependent on entry of the latter into acidotic environments therein [eqns (13) and (14)], the pK_a value of the $ClO_2^-/HClO_2$ system being 2.31.[12] Although the mean pH value of this biofluid is *ca.* 7 when unchallenged with oral stimuli (*i.e.* 'resting'), the consumption of relatively large volumes of beverages of lower pH value (*ca.* pH 4) can clearly exert a significant influence on this parameter. However, it should also be noted that the pH value of primary root caries lesions can approach a limit of 4.5, and therefore this represents an environment in which there are expected to be marked elevations in the level of $HClO_2$ generated (*i.e.* from 0.0020% at pH 7.00 to 0.64% of total available oxyhalogen oxidant at pH 4.50), a value computed in this work, although it should be noted that, in view of the pK_a value of the $ClO_2^-/HClO_2$ couple, this value still remains very low when expressed relative to the total amount of oxyhalogen oxidant available (the remainder being ClO_2^- in the absence of an alternative means of producing $ClO_2^\bullet$, or $HOCl/OCl^-$, from the reaction of ClO_2^- with α-, β- and γ-amino acids available). Of course, from the stoichiometry of eqn (14), 2 molar equivalents of $ClO_2^\bullet$ are generated per 4 of $HClO_2$, and hence the above figures for $HClO_2$ generation represent double that of the total $ClO_2^\bullet$ producible (*i.e.* maximum percentages of 0.0010 and 0.32% of total oxyhalogen oxidant at pH values of 7.00 and 4.50, respectively). Clearly, the rate of $ClO_2^\bullet$ generation from $HClO_2$ should also be considered in view of the short oral rinse-salivary supernatant equilibration time involved in our studies (the half-life of this process is extremely slow!).

$$ClO_2^- + H^+ \rightarrow HClO_2 \; (pK_a = 2.31) \tag{13}$$

$$4HClO_2 \rightarrow 2ClO_2^\bullet + ClO_3^- + Cl^- + H_2O \tag{14}$$

The MV statistical studies conducted by Lloyd *et al.*[21] and Wongravee *et al.*[22] have demonstrated that both unsupervised and supervised SOM methodologies present a valuable alternative to more established methods used for biomarker discovery, and the modelling of class memberships. Indeed, SOMs facilitate the visualisation of the main trends within a dataset. Despite being part of the Machine-Learning literature for some time, until the above studies were conducted they have been rarely employed in metabolomic/metabonomic studies. SOMs can be considered as a viable alternative to the established MV techniques, although ideally they should be combined with alternative MV techniques for comparative purposes.[21] However, this problem has now largely been remedied through the availability of affordable computer equipment with suitable processing power.

References

1. J. C. Lindon, E. Holmes, M. E. Bollard, E. G. Stanley and J. K. Nicholson, Metabonomics Technologies and Their Applications in Physiological Monitoring, *Biomark.*, 2004, **9**, 1.

2. J. C. Lindon, E. Holmes and J. K. Nicholson, Metabonomics and its Role in Drug Development and Disease Diagnosis, *Exp. Rev. Mol. Diag.*, 2004, **4**, 189.

3. J. C. Lindon, E. Holmes and J. K. Nicholson, Systems Biology in Pharmaceutical Research and Development, *Curr. Opin. Mol. Ther.*, 2004, **6**, 265.

4. S. E. Richards, M.-E. Dumas, J. M. Fonville, T. M. D. Ebbels, E. Holmes and J. K. Nicholson, Intra- and inter-omic fusion of metabolic profiling data in a systems biology framework, *Chemometr. Intell. Lab Syst.*, 2010, **104**, 121.

5. J. C. Lindon, J. K. Nicholson, E. Holmes and J. R. Everett, Metabonomics: Metabolic Processes Studied by NMR Spectroscopy of Biofluids, *Conc. Mag. Reson.*, 2000, **12**, 289.

6. W. B. Dunn, N. J. Bailey and H. E. Johnson, Measuring the Metabolome: Current Analytical Technologies, *Analyst*, 2005, **130**, 606.

7. W. B. Dunn and D. J. Ellis, Metabolomics: Current Analytical Platforms and Methodologies, *Trend. Anal. Chem.*, 2005, **24**, 285.

8. J. K. Nicholson, J. C. Lindon and E. Holmes, 'Metabonomics': Understanding the Metabolic Responses of Living Systems to Pathophysiological Stimuli via Multivariate Statistical Analysis of Biological NMR Spectroscopic Data, *Xenobiot.*, 1999, **29**, 1181.

9. O. Fiehn, Metabolomics – The Link Between Genotypes and Phenotypes, *Plant Mol. Biol.*, 2002, **48**, 155.

10. C. J. L. Silwood, E. Lynch, A. W. D. Claxson and M. C. Grootveld, ^{1}H and ^{13}C NMR Spectroscopic Analysis of Human Saliva, *J. Dent. Res.*, 2002, **81**, 422.

11. C. J. L. Silwood, E. J. Lynch, S. Seddon, A. Sheerin, A. W. D. Claxson and M. C. Grootveld, ^{1}H NMR Analysis of Microbial-Derived Organic Acids in Primary Root Carious Lesions and Saliva, *NMR Biomed.*, 1999, **12**, 345.

12. E. Lynch, A. Sheerin, A. W. D. Claxson, M. D. Atherton, C. J. Rhodes, C. J. L. Silwood, D. P. Naughton and M. Grootveld, Multicomponent Spectroscopic Investigations of Salivary Antioxidant Consumption by an Oral Rinse Preparation Containing the Stable Free Radical Species Chlorine Dioxide $(ClO_2^{\bullet})$, *Free Rad. Res.*, 1997, **26**, 209.

13. M. Grootveld, D. Gill, C. J. L. Silwood and E. Lynch, Evidence for the Microbicidal Activity of a Chlorine Dioxide-Containing Oral Rinse Formulation *in vivo*, *J. Clin. Dent.*, 2001, **12**, 67.

14. E. Lynch, A. Sheerin, C. J. Silwood and M. Grootveld, Multicomponent Evaluations of the Oxidising Actions and Status of a Peroxoborate-Containing Tooth-Whitening System in Whole Human Saliva Using High Resolution Proton NMR Spectroscopy, *J. Inorg. Biochem.*, 1999, **73**, 65.

15. M. Grootveld, C. J. L. Silwood and W. T. Winter, High-Resolution ^{1}H NMR Investigations of the Capacity of Dentifrices Containing a Smart Bioactive Glass to Influence the Metabolic Profile of and Deliver Calcium Ions to Human Saliva, *J. Biomed. Mater. Res. B: App. Biomat.*, 2009, **91B**, 88.

16. C. J. L. Silwood, E. Lynch, A. W. D. Claxson and M. C. Grootveld, ^{1}H and ^{13}C NMR Spectroscopic Analysis of Human Saliva, *J. Dent. Res.*, 2002, **81**, 422 (appendices).

17. C. J. L. Silwood, M. C. Grootveld and E. Lynch, ^{1}H NMR Investigations of the Molecular Nature of Low-Molecular-Mass Calcium Ions in Biofluids, *J. Biol. Inorg. Chem.*, 2002, **7**, 46.

18. C. J. Huberty, *Applied Discriminant Analysis*, Wiley-Interscience (Wiley Series in Probability and Statistics), New York, 1994.

19. M. A. Constantinou, E. Papakonstantinou, D. Benaki, M. Spraul, K. Shulpis, M. A. Koupparis and E. Mikros, Application of Nuclear Magnetic Resonance Spectroscopy Combined with Principal Component Analysis in Detecting Inborn Errors of Metabolism Using Blood Spots: A Metabonomic Approach, *Anal. Chim. Acta*, 2004, **511**, 303.

20. M. A. Constantinou, E. Papakonstantinou, M. Spraul, S. Sevastiadou, C. Costalos, M. A. Koupparis, K. Schulpis, A. Tsanttili-Kakoulidou and E. Mikros, ^{1}H NMR-Based Metabonomics for the Diagnosis of Inborn Errors of Metabolism in Urine, *Anal. Chim. Acta*, 2005, **542**, 169.

21. G. R. Lloyd, K. Wongravee, C. J. L. Silwood, M. Grootveld and R. G. Brereton, Self Organising Maps for Variable Selection: Application to Human Saliva Analysed by Nuclear Magnetic Resonance Spectroscopy to Investigate the Effect of an Oral Healthcare Product, *Chemometr. Intell. Lab. Syst.*, 2009, **98**, 149.

22. K. Wongravee, G. R. Lloyd, C. J. Silwood, M. Grootveld and R. G. Brereton, Supervised Self Organizing Maps (SOMS) for Classification and Variable Selection: Illustrated by Application to NMR Metabolomic Profiling, *Anal. Chem.*, 2010, **82**, 628.

23. K. Hollywood, D. R. Brison and R. Goodacre, Metabolomics: Current Technologies and Future Trends, *Proteomics*, 2006, **6**, 4716.

24. M. M. W. B. Hendriks, S. Smit, W. L. M. W. Akkermans, T. H. Reijmers, P. H. C. Eilers, H. C. J. Hoefsloot, C. M. Rubingh, C. G. de Koster, J. M. Aerts and A. K. Smilde, How to Distinguish Healthy from Diseased? Classification Strategy for Mass Spectrometry-Based Clinical Proteomics, *Proteomics*, 2007, **7**, 3672.

25. S. J. Dixon, Y. Xu, R. G. Brereton, H. A. Soini, M. V. Novotny, E. Oberzaucher, K. Grammer and D. J. Penn, Pattern Recognition of Gas Chromatography Mass Spectrometry of Human Volatiles in Sweat to Distinguish the Sex of Subjects and Determine Potential Discriminatory Marker Peaks, *Chemom. Intell. Lab. Syst.*, 2007, **87**, 161.

26. D. J. Penn, E. Oberzaucher, K. Grammer, G. Fischer, H. A. Soini, D. Wiesler, M. V. Novotny, S. J. Dixon, Y. Xu and R. G. Brereton, Individual and Gender Fingerprints in Human Body Odour, *J. R. Soc. Interface*, 2007, **4**, 331.

27. H. K. Choi, Y. H. Choi, M. Verberne, A. W. M. Lefeber, C. Erkelens and R. Verpoort, Metabolic Fingerprinting of Wild Type and Transgenic Tobacco Plants by 1H-NMR and Multivariate Analysis Technique, *Phytochem*, 2004, **65**, 857.

28. S. Zomer, S. J. Dixon, Y. Xu, S. P. Jensen, H. Wang, C. V. Lanyon, A. G. O'Donnell, A. S. Clare, L. M. Gosling, D. J. Penn and R. G. Brereton, Consensus Multivariate Methods in Gas Chromatographic Mass Spectrometry and Denaturing Gradient Gel Electrophoresis : MHC-Congenic and Other Strains of Mice can be Classified According to the Profiles of Volatiles and Microflora in their Scent-Marks, *Analyst*, 2009, **134**, 114.

29. Z. Ramadan, D. Jacobs, M. Grigorov and S. Kochhar, Metabolic Profiling using Principal Component Analysis, Discriminant Partial Least Squares and Genetic Algorithms, *Talant.*, 2006, **68**, 1683.

30. I. Takeda, C. Stretch, P. Barnaby, K. Bhatnager, K. Rankin, H. Fu, A. Weljie, N. Jha and C. Slupsky, Understanding the Human Salivary Metabolome, *NMR Biomed.*, 2009, **22**, 577.

31. G. De Laurentiis, D. Paris, D. Melck, M. Maniscalco, S. Marisco, G. Corso, A. Motta and M. Sofia, Metabonomic Analysis of Exhaled Breath Condensate in Adults by Nuclear Magnetic Resonance Spectroscopy, *Eur. Respir. J.*, 2008, **32**, Article ID 1175.

32. M. Sofia, M. Maniscalco, G. De Laurentiis, D. Paris, D. Melck and A. Motta, Exploring Airway Diseases by NMR-Based Metabonomics: A Review of Application to Exhaled Breath Condensate, *J. Biomed. Biotechnol.*, 2011, Article ID 403260.

33. P. Geladi and B. R. Kowalski, Partial Least-Squares Regression: A Tutorial, *Anal. Chim. Acta*, 1986, **185**, 1.

34. H. Martens and T. Naes, *Multivariate Calibration*, Wiley, Chichester, UK, 1989.

35. R. G. Brereton, *Chemometrics for Pattern Recognition*, Wiley, Chichester, UK, 2009.

36. E. Sanchez and B. R. Kowalski, Tensorial Calibration: I. First-Order Calibration, *J. Chemom.*, 1988, **2**, 247.

37. O. Cloarec, M. E. Dumas, J. Trygg, A. Craig, R. H. Barton, J. C. Lindon, J. K. Nicholson and E. Holmes, Evaluation of the Orthogonal Projection on Latent Structure Model Limitations Caused by Chemical Shift Variability and Improved Visualization of Biomarker Changes in ^{1}H NMR Spectroscopic Metabonomic Studies, *Anal. Chem.*, 2005, 77, 517.

38. G. R. Lloyd, R. G. Brereton and J. C. Duncan, Self Organising Maps for Distinguishing Polymer Groups Using Thermal Response Curves Obtained by Dynamic Mechanical Analysis, *Analyst*, 2008, **133**, 1046.

39. T. Kohonen, S. Kaski and H. Lappalainen, Self-Organized Formation of Various Invariant-Feature Filters in the Adaptive-Subspace SOM, *Neur. Comput.*, 1997, **9**, 1321.

40. T. Kohonen, *Construction of Similarity Diagrams for Phenomes by a Self-Organising Algorithm*, Helsinki University of Technology, Espoo, Finland, 1981.

41. T. Kohonen, Self-Organized Formation of Topologically Correct Feature Maps, *Biol. Cybernet.*, 1982, **43**, 59.

42. T. Kohonen, *Self-Organizing Maps*, Springer, Berlin, 2000.

43. F. Marini, A. L. Magrìa, R. Buccia and A. D. Magrìa, Use of Different Artificial Neural Networks to Resolve Binary Blends of Monocultivar Italian Olive Oils, *Anal. Chim. Acta*, 2007, **599**, 232.

44. T. Murtola, M. Kupiainen, E. Falck and I. Vattulainen, Conformational Analysis of Lipid Molecules by Self-Organizing Maps, *J. Chem. Phys.*, 2007, **126**, 054707.

45. Y. D. Xiao, A. Clauset, R. Harris, E. Bayram, P. Santago II and J. D. Schmitt, Supervised Self-Organizing Maps in Drug Discovery. 1. Robust Behavior with Overdetermined Data Sets, *J. Chem. Inf. Model.*, 2005, **45**, 1749.

46. W. J. Melssen, J. R. M. Smits, G. H. Rolf and G. Kateman, Two-Dimensional Mapping of IR Spectra Using a Parallel Implemented Self-Organising Feature Map, *Chemom. Intell. Lab. Syst.*, 1993, **18**, 195.

47. U. Siripatrawan, Self-Organizing Algorithm for Classification of Packaged Fresh Vegetable Potentially Contaminated with Foodborne Pathogens, *Sensor. Actuat. B-Chem.*, 2008, **128**, 435.

48. W. Melssen, B. Ustun and L. Buydens, SOMPLS: A Supervised Self-Organising Map-Partial Least Squares Algorithm for Multivariate Regression Problems, *Chemom. Intell. Lab. Syst.*, 2007, **86**, 102.

49. B. Lefebvre, *Intelligent Bucketing for Metabonomics*, ACD/Labs Technical Note, 2004. http://www.acdlabs.com/publish/publ04/enc04_intelli_bucket.html.

50. T. Kohonen, J. Hynninen, J. Kangas and J. Laaksonen, *SOM_PAK: The Self-Organizing Map Program Package*, Technical Report A31, Helsinki University of Technology, Laboratory of Computer and Information Science, FIN-02150 Espoo, Finland, 1996.

51. G. E. Forsythe, M. A. Malcolm and C. B Moler, *Computer Methods for Mathematical Computations*, Prentice-Hall, New Jersey, USA, 1976.

52. R. P. Brent, *Algorithms for Minimization without Derivatives*, Prentice-Hall, New Jersey, USA, 1973.

53. K. Wongravee, N. Heinrich, M. Holmboe, M. L. Schaefer, R. R. Reed, J. Trevejo and R. G. Brereton, Variable Selection Using Iterative Reformulation of Training Set Models for Discrimination of Samples: Application to Gas Chromatography/Mass Spectrometry of Mouse Urinary Metabolites, *Anal. Chem.*, 2009, **81**, 5204.

54. P. Geladi, Notes on the history and nature of partial least squares (PLS) modelling, *J. Chemom.*, 1988, **2**, 231.

55. R. G. Brereton, Introduction to Multivariate Calibration in Analytical Chemistry, *Analyst*, 2000, **125**, 2125.

56. R. Wehrens and W. E. Van Der Linden, Bootstrapping Principal Component Regression Models, *J. Chemom.*, 1997, **11**, 157.

57. R. Wehrens, H. Putter and L. Buydens, The Bootstrap. A Tutorial, *Chemom. Intell. Lab. Syst.*, 2000, **54**, 35.

58. Y. Xu, S. J. Dixon, R. G. Brereton, H. A. Soini, M. V. Novotny, K. Trebesius, I. Bergmaier, E. Oberzaucher, K. Grammer and D. J. Penn, Comparison of Human Axillary Odour Profiles Obtained by Gas

Chromatography Mass Spectrometry and Skin Microbial Profiles Obtained by Denaturing Gradient Gel Electrophoresis Using Multivariate Pattern Recognition, *Metabolomics*, 2007, **3**, 427.

59. J. Darkwa, R. Olujo, O. Olagunju, A. Otoihan and R. H. Simoyi, Oxyhalogen-Sulfur Chemistry: Oxidation of N-Acetylcysteine by Chlorite and Acidic Bromate, *J. Phys. Chem.*, 2003, **107**, 9834.
60. J. Darkwa, C. Mundoma and R. H. Simoyi, Antioxidant Chemistry, *J. Chem. Soc. Faraday Trans.*, 1998, **94**, 1971.
61. J. Grigor and A. J. Roberts, Reduction in the Levels of Oral Malodor Precursors by Hydrogen Peroxide: *In-Vitro* and *in-vivo* Assessments, *J. Clin. Dent.*, 1992, **III**, 111.
62. C. R. Chinarke and R. H. Simoyi, Oxidation of Taurine by Chlorite in Acidic Medium, *J. Phys. Chem. B*, 1997, **101**, 1207.
63. S. L. Hazen, A. d'Avignon, M. M. Anderson, F. F. Hsu and J. W. Heinecke, Mechanistic Studies Identifying Labile Intermediates along the Reaction Pathway, *J. Biol. Chem.*, 1998, **273**, 4997.
64. R. A. Follente, B. J. Kovacs, R. M. Aprecio, H. J. Bains and J. D. Kettering, Efficacy of High-Level Disinfectants for Reprocessing GI Endoscopes in Simulated-Use Testing, *Gastro. Endosc.*, 2001, **53**, 456.
65. C. R. Chinarke, O. Olojo and R. H. Simoyi, Oxidation of Formaldehyde by Chlorite in Basic and Slightly Acidic Media, *J. Phys. Chem. A*, 1998, **102**, 606.

Metabolomics Investigations of Drug-induced Hepatotoxicity

WEI TANG[a] AND QIUWEI XU*[b]

[a] Department of Drug Metabolism and Pharmacokinetics, Merck Research Laboratories, Rahway, NJ 07065, USA; [b] Department of Safety Assessment, Merck Research Laboratories, West Point, PA 19486, USA
*Email: qiuwei_xu@merck.com

11.1 Introduction

Pharmacotherapy has contributed tremendously to improving human health for the last century, and it has improved quality of life and prolonged life span. These therapeutic agents range from remedies for alleviating headaches to those for treating devastating life-threatening infections and cancers. The effective management of human immunodeficiency virus (HIV) infection illustrates a remarkable achievement in the history of modern medicine.[1,2] More than 25 anti-viral drugs are currently available, and treatment with various combinations of these drugs enable HIV patients carrying the once-deadly virus to live relatively normal lives. However, pharmacotherapy can cause adverse effects (AEs) in spite of recent advancements in pharmacology and toxicology. Some of the AEs can be severe, or even fatal. For example, drug-induced liver injury (DILI) accounts for ~25–50% of clinical cases of liver failure.[3] DILI is also a leading cause in the discontinuation of drug development, or removal of therapeutic agents from the market. For example, a COX-2 selective inhibitor Lumiracoxib caused serious treatment-related hepatotoxicity, and resulted in liver transplantations and fatalities.[4] In consequence, its marketing authorisation was

Issues in Toxicology No. 21
Metabolic Profiling: Disease and Xenobiotics
Edited by Martin Grootveld
© The Royal Society of Chemistry 2015
Published by the Royal Society of Chemistry, www.rsc.org

revoked by the regulatory agencies in the European Union, Australia and Canada, and it was never approved in the USA. Lumiracoxib is a structural analogue of diclofenac that is known to induce liver injury. Although some therapeutic agents that carry warnings of hepatotoxicity may remain on the market, their clinical application can be severely limited. For example, aspirin is often excluded from treating children, since this popular painkiller may induce the so-called Reye's syndrome in that particular patient population.[5,6] Antibiotic flucloxacillin is reserved narrowly for the exclusive treatment of serious staphylococcal infections due to high risk of cholestasis. This chapter is intended to provide a brief description of DILI, with a focus on reviewing current understanding of likely underlying mechanisms. Metabolomics will be discussed in the context of its applications that aim to (1) understand the pathogenesis of hepatotoxicity and (2) help identify biochemical markers that may be utilised for the early identification, diagnosis, prevention, and management of DILI.

11.2   Drug-Induced Liver Injury (DILI)

The occurrence of DILI is estimated between 1/10 000 and 1/100 000 of treated patients; and females and the elderly often are more susceptible.[7,8] The term DILI is a broad description encompassing various hepatic responses arising from insults by agents with different indications and chemical structures.[9,10] Therefore, the so-called 'risk factors' can be specific to particular drugs in question. For example, liver injuries induced by aspirin and valproic acid occur more often in young children in comparison to adults.[5,11] DILI often appears in resemblance to such liver diseases as viral hepatitis, chronic hepatitis, and cholestatis. Typical clinical symptoms are abdominal pain, fever, elevated serum transaminase levels, and jaundice. Among pathological findings are usually inflammation, steatosis, granulomas, cholestasis, bile duct vanishing syndrome, and cell necrosis.[12,13] Cholestatis appears to be common in older patients, but hepatocellular damage often occurs in younger subjects.[10,14] Patients suffering hepatocellular injury are at a high risk of acute liver failure, thence leading to high mortality. Patients experiencing cholestatis-related liver injury tend to develop chronic liver disease. Histopathological analyses of liver biopsies can help clarify the patterns of tissue injuries, and are valuable to aiding mechanistic investigations of DILI. For example, microvesicular steatosis can be indicative of impairment to mitochondrial β-oxidation; cholestasis indicates likely disruption of canalicular biliary transport functions; and centrilobular necrosis suggests possible insults by reactive chemical species formed during drug metabolism.[13] The formation of reactive drug metabolites can cause tissue injury due to either blatant cytotoxicity or immune-mediated reactions. An immune reaction to DILI usually is characterised by skin rash, eosinophilia, and circulating autoantibodies. Although the majority of patients with DILI are often expected to recover following discontinuation of insulting drugs, up to 10% of those may progress to acute liver failure, requiring liver transplantation or leading to death.[7] According to the so-called 'Hy's Law', severe hepatotoxicity can usually be predictable by elevated serum bilirubin levels higher than two times the upper limit of a normal (ULN) value, and a concomitant increase in serum aminotransferase concentration higher than three times the ULN.[15] A large fraction of DILI incidents reported in the USA and the EU involves acetaminophen (also known as paracetamol, or APAP) and antimicrobials. In part, this is attributed to their widespread use.[16,17]

DILI can result from drug overdose, and it is dose dependent. There is always a threshold below which a drug is deemed safe, or poses a low risk. Such a kind of hepatotoxicity is likely to be associated with direct full-blown cytotoxicity caused by an offending drug and/or its metabolites. As an over-the-counter analgesic, APAP is usually well tolerated and effective within the recommended dose range. However, severe hepatocellular damage can occur if a large quantity of APAP is administered in a single dose or over a short interval. It can progress to liver failure, and in consequence a likely fatality.[18,19] Its hepatotoxicity often exhibits symptoms with a rapid onset within 48 hours after drug ingestion. Elevation of liver enzymes and hyperbilirubinemia occur between 12 and 36 hours, and hepatic abnormality reaches a peak on the third day. There is a reported dose-response curve for the DILI: liver injury is rare at dose levels <125 mg kg^{-1}, severe in $\sim 50\%$ of individuals administered at 250 mg kg^{-1}, and present in almost all dosed at 350 mg kg^{-1}.[20] A recent study reported that ~ 30–40% of healthy adults who ingested APAP at the maximum recommended daily dose of 4 g for 14 days experienced an elevation (>3 times the ULN) of serum alanine aminotransferase (ALT), whereas none in the placebo group did.[21] This lends additional evidence to APAP-induced hepatotoxicity. Its pathological characteristics include extensive centrilobular necrosis, suggesting a toxic reactive metabolite in the pathogenesis of cellular damage. Patients suffering from APAP-induced liver injury usually have a 60–80% chance of survival without liver transplant when *N*-acetylcysteine (NAC) is administered promptly.[22] Since NAC is capable of scavenging chemically reactive electrophiles and free radical chemicals, its effectiveness as an antidote for managing APAP overdose strongly supports that the formation of reactive species during APAP metabolism is likely a culprit responsible for DILI.

DILI can also be 'idiosyncratic'. It is characterised by a lack of a clear dose-response relation and a delayed onset of clinical symptoms after weeks to months of exposure to the offending drug. Hepatotoxicity of this type usually escapes detection by pre-clinical and clinical safety assessments, and it often occurs in a small sub-set in spite of tolerability by the majority of treated patients. For example, diclofenac, a non-steroidal anti-inflammatory drug (NSAID), is associated with rare but severe hepatotoxicity; its clinical symptoms include jaundice and elevated serum transaminase activity. However, these symptoms of liver toxicity are often not evident until after 1–3 months into diclofenac treatment, and the rate of incidences ranges from 1 to 5 per 100 000 exposed patients, with a fatality at $\sim 10\%$.[23,24] Liver biopsies revealed extensive cell necrosis, which resembled viral hepatitis or chronic hepatitis accompanied by cholestasis. Less than 25% of reported cases exhibited skin rash, fever, and eosinophilia; whereas re-challenge of patients with diclofenac rarely resulted in rapid adverse responses.[25,26] Its DILI is hepatocellular injury attributed to metabolic instead of immune idiosyncrasy.[27] Putative toxic metabolites were identified as reactive acyl glucuronide and quinone imine chemicals.[28] Another example is flucloxacillin, a beta-lactam antibiotic. Its utilisation sometimes leads to cholestatic liver injury with symptoms of pruritus (skin itch) and prolonged jaundice.[29,30] DILI is estimated in ~1 per 15 000 treatment courses, and the delayed onset of clinical symptoms can take up to 45 days. Hence, the toxicity may not show until after the cessation of the antibiotic for several weeks. Two risk factors are identified: 'age over 55' and 'duration of therapy over two weeks'. Pathological findings include canalicular cholestasis, eosinophilia, inter-lobular biliary epithelial tissue degeneration, severe bile duct damage, and minimum to absence of hepatocyte necrosis.[29] The adverse reactions to flucloxacillin were immediate in patients re-challenged inadvertently with the antibiotic, which suggests an immune-based mechanism in the pathogenesis of flucloxacillin-induced hepatotoxicity.[31] Idiosyncratic DILI may lead to acute liver failure that usually exhibits poor prognosis, ranging from 20 to 40% of that measured by transplant-free survival.[14]

The high frequency of DILI relative to other types of pharmacotherapy-related AEs can be attributable, in part, to liver physiological function including secretory and metabolic capacity. Following oral administration, an entire load of absorbed drug passes through the liver before entering into the systemic circulation. In the liver, drug molecules are subject to metabolism and/or biliary excretion. Those molecules that survive the first-pass hepatic extraction circulate back to the liver for ultimate elimination if other organs such as the kidney do not participate in clearance of the drug. Drug metabolism sometimes produces reactive chemicals capable of modifying hepatic proteins and/or nucleic acids, and interactions of parent drugs and/or metabolites with hepatic transporters may interfere with hepatocellular uptake and efflux. All those processes bear toxicological signs.

11.3   Possible Mechanisms Underlying DILI

11.3.1   Drug Metabolism and Elimination

After absorption, drug molecules undergo biotransformation such as hydroxylation of aliphatic/aromatic hydrocarbons, epoxidation of alkenes, nitrogen/sulfur oxygenation, *N*-, *O*- and *S*-dealkylations and dehydrogenation.[32,33] These oxidation reactions take place primarily in the liver as hepatocytes are rich in drug-metabolising enzymes such as cytochrome P450s, peroxidases, amine oxidase, and flavin-containing monooxygenase. Metabolites are the reaction products. They are usually more hydrophilic than the parent drug, and therefore are good candidates for either excretion directly *via* bile/urine or subsequent conjugation reactions. Although the majority of drug metabolites are inert pharmacologically and toxicologically at their exposure levels in the human body, some are chemically reactive and capable of alkylating proteins or nucleic acids *via* nucleophilic addition or substitution, causing structural and functional changes to those biological macromolecules. Biotransformation that leads to the formation of reactive species is termed metabolic bioactivation. Among the resulting electrophiles are α,β-unsaturated carbonyls (*e.g.* extremely reactive α,β-unsaturated aldehydes), epoxides, isocyanates, isothiocyanates, nitrogen and sulfur mustards, quinones, quinone imines, and quinone methides (Figure 11.1, panel A). Quinones and quinone imines also are capable of causing cellular 'oxidative stress' by participating in redox cycling in hepatocytes. A representative reductive bioactivation reaction is the conversion of nitro-substituted aryl compounds to *N*-hydroxy arylamines; the chemical reactivity of *N*-hydroxy arylamines is sometimes mediated by *O*-acetylation or sulfation.

In addition to oxidation and reduction, drug molecules and their metabolites can be subject to further biotransformations *via* conjugation reactions. For example, glucuronidation, acetylation, and sulfation are facilitated by the drug-metabolising enzymes UDP-glucuronosyltransferases (UGT), *N*-acetyltransferases, and sulfotransfcrascs (SULT), respectively; all those enzymes are abundant in hepatocytes. Acyl glucuronides are believed to be reactive since they can react with proteins through transacylation, in which a glucuronate moiety is displaced by cysteine, tyrosine, or lysine in proteins.[34] Alternatively, the aglycone of an acyl glucuronide can undergo migration to the 2-, 3- or 4-position on pyranose. An aldehyde can form by a scission of a hemiacetal ether bond on pyranose, and it can modify proteins and nucleic acids *via* the formation of an imine Schiff base linkage. Further intra-molecular *Amadori* rearrangement of the imine intermediate yields a stable 1-amino-2-keto protein or DNA adduct (Figure 11.1, panel B). Alkylation of biomolecules by acetylate and sulfate conjugates can be mediated by cationic nitrenium or carbenium species, with acetate or sulfate as leaving groups, respectively (Figure 11.1, panel C).[35,36]

Another conjugation reaction with toxicological importance is that between electrophilic metabolites and glutathione (GSH), catalysed by glutathione-*S*-transferases (GST). The resulting GSH adducts undergo further hydrolysis to generate the corresponding cysteinylglycine and cysteine adducts by the catalytic enzymes γ-glutamyltranspeptidase and dipeptidase, respectively. *N*-acetylation of cysteine adducts leads to mercapturic acids (*N*-acetylcysteine adducts). GSH conjugation usually represents a detoxification process under normal circumstances; however, extensive conjugation risks exhaustion of GSH in hepatocytes, in consequence increasing cellular oxidative stress and exposing critical biomolecules to reactive species.[37,38] In addition, certain GSH adducts can serve as 'carriers' in delivering electrophiles that otherwise are short-lived to distal tissues or compartments. The proposed underlying mechanism is reformation of reactive species through

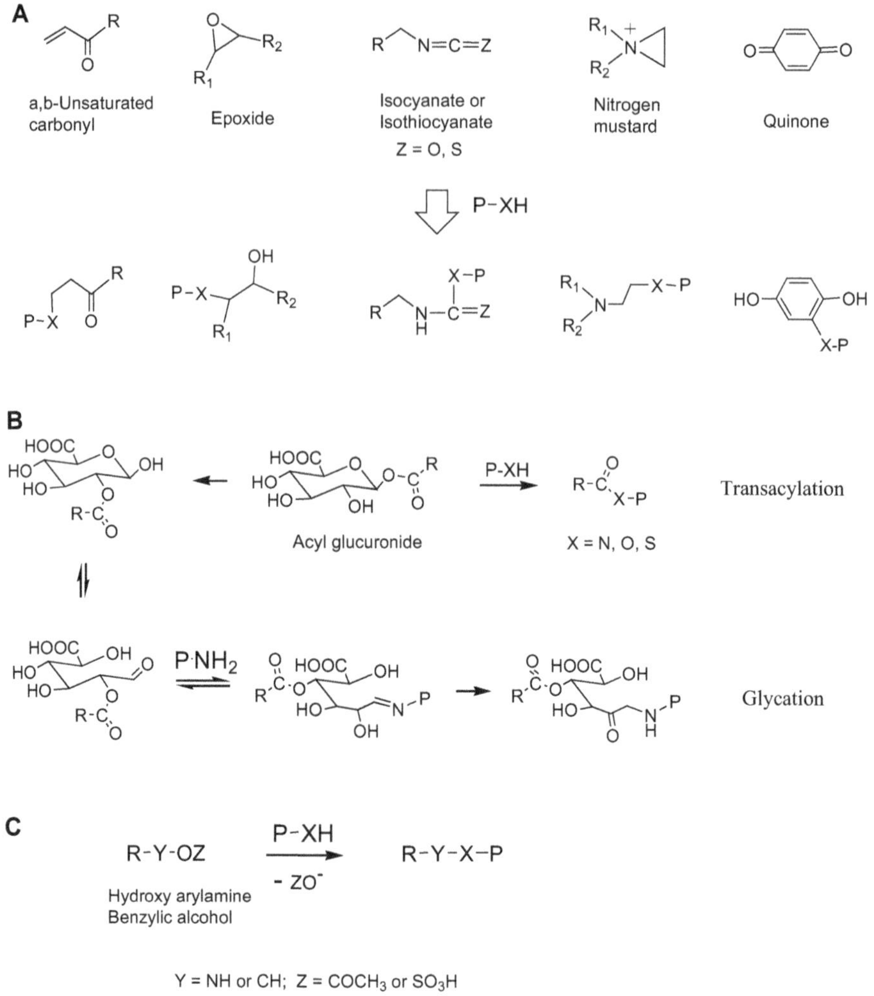

A

B

C

Figure 11.1 Examples of reactive species formed during drug metabolism (A: α,β-unsaturated carbonyls, epoxides, isocyanates and isothiocyanates, nitrogen mustards, quinones; B: acyl glucuronides; and C: acetyl or sulfate conjugate of hydroxyarylamines and their reactions with biological macromolecules (P-XH).

the reverse of the reversible GSH conjugation reaction, or further bioactivation of thiol adducts *via* the cysteine conjugate β-lyase pathway.[39,40]

Elimination of drug molecules and their metabolites usually happens by active or passive excretion through hepatobiliary and renal transporters. Biliary efflux often relies on transporters such as bile salt export pump (BSEP), multidrug resistance-associated proteins (MRP), or multidrug resistance proteins (MDR).[41] Transporter proteins reside in the canalicular membrane of hepatocytes. They were initially identified as pump proteins responsible for drug resistance developed by cancer cells during chemotherapy. MRP, MDR, and BSEP regulate hepatic removal of cytotoxic wastes generated by hepatocytes, in addition to the disposition of endogenous substances such as bile acids, cholesterol, and lipids. Active transporters are saturable, and subject to competitive or non-competitive inhibitions by drug molecules and metabolites.

Alkylation of hepatic proteins and nucleic acids, by reactive chemicals formed during drug metabolism, modifies chemical structures of biomolecules, and may disrupt their important biochemical functions. This process can lead to 'oxidative stress', mitochondrial dysfunctions, reduced hepatobiliary transport capacity, or immune responses in the liver. Impairment to biliary efflux can be reversible inhibition of hepatic transporters by drug molecules and metabolites; it gives rise to an increased deposit of bile acids that are usually cytotoxic to hepatocytes. These biochemical processes can trigger DILI *via* direct cytotoxicity or immune-mediated pathogenesis, a further discussion ensues in the following section. Depending on particular biochemical processes and degrees of cellular injury, subsequent death of hepatocytes follows either apoptotic or necrotic pathways. A delayed onset of idiosyncratic DILI with a clinical manifestation is likely attributable to time needed for sufficient accumulation of continuous tissue damages, or development of immune responses.

11.3.2 Direct Cytotoxicity and Immune-mediated Reactions

Cytotoxicity exerted by drug molecules and their metabolites is sometimes attributable to their insult on mitochondria.[42,43] Mitochondrial organelle consists of outer and inner membranes, and a matrix. The outer and inner membranes contain ion channels, fatty acid transporters, and electron-transport complexes; the matrix contains high concentrations of proteins, and is slightly alkaline (high pH) and negatively charged. These structural properties predispose mitochondria to the attraction and concentration of lipophilic or cationic toxic chemicals capable of interfering in mitochondrial oxidative respiration, or fatty acid β-oxidation. Both oxidative respiration and fatty acid β-oxidation are important for mitochondria to generate high-energy metabolites such as ATP. Any disruptions to oxidative respiration or fatty acid β-oxidation can result in limited generation of bioenergetics, and in consequence ATP depletion within impaired cells. Inhibition to fatty acid β-oxidation can lead to excessive free fatty acids that can be deleterious too. The pathological observation of hepatic steatosis often suggests an abnormal accumulation of triglycerides secondary to the impairment of fatty acid β-oxidation. Uncontrolled mitochondrial oxidative respiration produces excessive reactive oxygen species (ROS); their neutralisation relies on superoxide dismutase (SOD), GSH, and glutathione reductase (GR). 'Oxidative stress' arises in mitochondria and cells when neutralisation and detoxification processes are insufficient. They can be attributable to, for example, loss of the enzymatic activities of SOD and GR, or GSH depletion, or the enzymes can be overwhelmed by surges of ROS. It inevitably causes mitochondrial membrane permeability transition and perturbed cellular calcium homeostasis. Drug molecules and their metabolites can sometimes directly interfere with mitochondrial DNA (mtDNA) replication or transcription. It impairs the transcription or translation of mtDNA-encoded respiratory polypeptides, and thus the respiratory reaction. ATP depletion often accompanies mitochondrial dysfunction and hepatocyte necrosis.[44] Rupture of plasma membrane and leakage of bioactivated intra-cellular constituents, in consequence, can induce immunogenic responses to liver by the innate immune system. Subsequent release of cytokines and chemokines from activated immune cells such as Kupffer, NK and NKT cells can further exacerbate hepatic tissue injuries.[45,46]

Mitochondrion itself can regulate the programmed cell death (*i.e.*, apoptosis), and act as a mediator in the cascade of cellular damage. Drugs or their metabolites can impair or trigger the uptake, metabolism, or efflux of accumulated cytotoxic bile acids and free fatty acids. For example, inhibition to the bile acid efflux pump BSEP leads to the retention of bile acids in hepatocytes; and activation of Pregnane X receptor (PXR) and liver X receptor (LXR) up-regulates the expression of the uptake transporter Cd36, and results in an increased influx of fatty acids into hepatocytes.[47,48] Excessive bile acids and fatty acids in liver cause interaction of the death receptors with Fas ligand (FasL), TNF-related apoptosis-inducing ligand (TRAIL) and tumour necrosis factor-α (TNF-α). Subsequent activation of cysteine-dependent aspartate specific protease-8 (caspase-8) mediates the cleavage of Bid in the BH3 family into a truncated Bid (tBid). Translocation of the tBid to the surface of mitochondria provokes mitochondrial membrane leakage, cytochrome c efflux and inevitable apoptosis. An alternative apoptosis pathway is facilitated by fatty acid. It induces the response of the stress-related unfolded protein in endoplasmic reticulum, and the release of the cathepsin proteases in lysosome. Both routes lead to mitochondrial membrane permeability transition, and consequently hepatocyte apoptosis. Apoptotic cells also promote inflammation *via* the activation of Kupffer cells, and further aggravate liver tissue injury such as hepatic fibrosis. Extensive apoptosis can sometimes stimulate mitogenesis of hepatocytes, and likely lead to the development of liver cancer.

Much of the current understanding of DILI derives from the studies of APAP-induced hepatotoxicity. CYP2E1 catalyses the primary oxidative metabolism of APAP, and results in the formation of *N*-acetyl-*p*-benzoquinone imine (NAPQI).[49] NAPQI is a highly reactive electrophile, and its detoxification relies on a conjugation reaction with GSH. APAP overdose can lead to depletion of GSH in both cytosolic and mitochondrial compartments. When NAPQI overwhelms cellular reduction potential, it reacts with hepatic proteins, particularly those in mitochondria.[50] In addition, an *ipso* GSH adduct of NAPQI is reversible, and may likely serve as a carrier to facilitate the delivery of the reactive species from its original site to a different sub-cellular compartment, such as mitochondria.[51] Modified proteins were identified as aldehyde dehydrogenase and ATP synthase's α-unit.[52]

Depletion of GSH and the subsequent arylation of proteins in mitochondria lead to disruption of the oxidative respiratory chain reactions and fatty acid β-oxidation, and it results in impairment to mtDNA replication and transcription, and thus reduced ATP generation. The inhibition to hepatic fatty acid β-oxidation by APAP is likely attributable to the suppression of several pro-β-oxidative genes that are regulated by peroxisome proliferator-activated receptor-α (PPARα). Protein modification and GSH depletion due to APAP treatments can generate mitochondrial oxidative stress with increased levels of reactive oxygen and peroxynitrite species. These reactive chemicals can induce lipid peroxidation, lead to membrane permeability transition, and collapse the mitochondrial membrane potential. Injured mitochondria can release calcium and proteins such as endonucleases and apoptosis-inducing factor; and they cause a disturbance to cellular calcium homeostasis, nuclear DNA fragmentation, and ultimate hepatocyte necrosis. Alternatively, 'oxidative stress' can sometimes initiate the c-Jun NH_2-terminal kinase (JNK)-dependent intrinsic cell death process, and consequent necrotic hepatocyte death can result from mitochondrial dysfunction and ATP depletion.

The CYP2E1-mediated metabolism of APAP can also contribute directly to cellular 'oxidative stress', since the enzyme generates hydrogen peroxide (H_2O_2), superoxide anion (O_2^-), and hydroxyl radical ($^\bullet OH$) as by-products.[53,54] The formation of NAPQI initiates inflammatory responses, and activates the innate immune system in the liver; and it eventually triggers necrotic cell death. Subsequent secretion of cytokines and Fas ligand from Kupffer and natural killer cells propagates APAP-induced tissue injury in liver, thus leading to potentially fulminant liver failure.[55,56]

Alkylation of hepatic proteins by reactive metabolites can also trigger antigen production, in addition to altering protein function. This is the so-called 'hapten hypothesis': it is deemed necessary for small-molecule drugs and their metabolites to form protein adducts in order to invoke immune responses.[57,58] Drug-protein adducts are taken up and processed by antigen-presenting cells, and the resulting immunogenic adducts are presented in association with the major histocompatibility complex (MHC). It leads to proliferation of $CD4^+$ or $CD8^+$ T-cells that are cytotoxic to hepatocytes when self-tolerance is overwhelmed. Immunological reactions are dictated likely by both the extent of protein modification and the specific protein modified. Alternatively, drug molecules and metabolites can interact directly with MHC and T-cell receptors to sensitise the immune system and activate T-cells.[59,60] DILI of an immunological nature usually exhibits clinical symptoms of fever, rash, and eosinophilia.

Although characterised by a delayed onset of responses upon initial exposure to an offending drug, such DILI occurs rapidly upon re-challenge with either the implicated agent or a different drug molecule that is metabolised to generate a structurally similar reactive species.[61,62] Antibodies recognising drug-related antigens are sometimes detectable in patients with liver injury. For example, the anaesthetic agent halothane induces immune-mediated hepatotoxicity that occurs more frequently in individuals with a history of exposures to the anaesthetic. The rate of DILI increased from 1/35 000 in subjects treated for the first time with halothane to 1/3700 in those who had previously been exposed to the anaesthetic.[63] A reported unusual case study involves a patient who developed severe liver toxicity after receiving desflurane.[64] It transpired that the patient was treated previously with halothane. Antibodies were detected in blood serum of the patient, and they cross-reacted with proteins isolated from rats pre-dosed with halothane. This case study demonstrates that patients can be 'sensitised' by their prior exposures to halothane, and become susceptible to other fluorinated inhalation anaesthetics. The underlying mechanism of sensitisation is possibly attributable to a commonly generated putative reactive species, *i.e.*, trifluoroacetic chloride that is capable of alkylating proteins, and invokes the immune system.

Given the mediator function of MHC in the cascade leading to immunity, genetic differences and protein expression of MHC appear to predispose certain patient populations to DILI. Flucloxacillin treatment causes liver injury in a small sub-set of subjects. Its β-lactam moiety is essential for its irreversible inhibition to the synthesis of bacterial cell walls, but is also likely to alkylate hepatic proteins and generate antigens that provoke immune responses. Protein adducts were detected in both rats and humans treated with flucloxacillin.[65,66] Clinical case studies show that patients carrying the HLA-B*5701 allele of the MHC are more susceptible to liver injury with flucloxacillin therapy than controls.[67] A similar genomic investigation also revealed an association between HLA-DRB1 alleles and hepatotoxicity induced by anticoagulant ximelagatran.[68] Ximelagatran likely activates MHC directly and produces its immune-mediated AEs, since a search failed to detect any reactive metabolites.[69] In addition to genetic predisposition, environmental factors such as inflammation and poly-pharmacotherapy can possibly tip the balance from tolerance to immune response, leading to the idiosyncrasy of immune-mediated hepatitis.

Accumulation of bile acids in liver can lead to cholestasis and DILI. A possible pathogenesis by drug or metabolites is the disruption of ATP-binding cassette transporters localised at the hepatocyte canalicular membrane, and impacted transporters can be BSEP, MDR1, and MRP2. These transporters are a part of the constituent regulating and governing bile components and flow-rates. When activities of transporters are compromised, bile secretion is impaired. Excessive bile acids are cytotoxic to hepatocytes. Translocation of the cytoplasmic death receptor Fas to the plasma membrane happens in response to elevated bile acids, and interaction of the death receptor with a Fas ligand initiates cell death. Subsequent caspase-8 activation promotes a mitochondrial membrane permeability transition, cytochrome c release, the activation of caspase-3/7, and ultimate hepatocyte apoptosis.[70,71]

Bile acids may also interfere directly with mitochondrial respiration, and it results in 'oxidative stress', mitochondrial dysfunction, and consequently cytotoxicity.[72] If drugs or metabolites are competitive substrates to bile acid transporters, their inhibition to biliary bile acid efflux is reversible. Irreversible inhibition occurs if drug metabolites are reactive and alkylate transport proteins for bile acid efflux. Anti-fungal agent terbinafine causes rare but severe hepatotoxicity with a frequency of the incidents estimated at ~1 in 50 000 treated patients. Pathological findings of the DILI include persistent cholestasis accompanied by the so-called 'vanishing duct syndrome'.[73] A GSH adduct was identified in the studies of terbinafine metabolism in liver microsomes, and it was formed *via* a Michael addition of the thiol to a reactive allylic aldehyde species.[74] This conjugation reaction is reversible, and allyic aldehyde can be regenerated by retro-Michael conversion of the GSH adduct under physiological conditions. In addition, the GSH adduct is an electrophile with chemical features of an α,β-unsaturated carbonyl structure. There are two possible scenarios accounting for terbinafine-induced liver injury. Terbinafine molecule undergoes metabolic bioactivation to form allylic aldehyde that reacts readily with GSH. The resulting GSH adduct is capable of traversing the hepatocyte canalicular membrane and alkylates bile acid transport proteins. Alternatively, the allylic aldehyde directly reacts with bile acid transporters. In consequence, structurally modified transport proteins lose their activities and fail to facilitate bile acid secretion. Accumulation of bile acids in hepatocytes is an explanation for the DILI characteristic of cholestasis.

11.3.3 Ambiguous Nature of Current Understandings of DILI

In spite of recent progress in unveiling the pathogenesis of DILI, many of its mechanistic details remain poorly defined. Despite the detection of reactive metabolites, few protein targets have actually been identified. Even when modified proteins are identified, their roles in the etiology of hepatotoxicity often remain enigmatic. This can be illustrated by a rat model of diclofenac hepatotoxicity, and it serves to demonstrate a convoluted pattern and time course of cytotoxicity and protein covalent modification.[75] Pathological examination of rat liver revealed swollen and apoptotic hepatocytes on day 1 of the diclofenac treatment. Cellular injury progressed from days 2 to 4 post-dosing. It eventually led to focal necrosis with reduced levels of bile flow and bile acid secretion on day 5. Protein modification was initially related to an adduct of 110 kDa, and it was later identified in a separate study as dipeptidyl peptidase IV.[76] The concentration of this adduct was significant on day 1, but diminished to an undetectable level on day 5 with appearance of other predominant adducts of molecular weights 85 and 96 kDa.[75] The protein adducts are likely related to reactive acyl glucuronide and quinone imine of diclofenac metabolites.[28]

Another example is anti-diabetic troglitazone. It was withdrawn from the market due to severe DILI featuring hepatocyte necrosis, bile duct proliferation, and cholestasis. Many reactive metabolites were identified in the studies of troglitazone biotransformation, and they likely contributed to its

DILI. Several reactive species resulted from the P450-mediated oxidation of the hydroxy chromane and thiazolidinedione moieties.[77,78] Troglitazone is also subject to extensive SULT1A3-catalysed sulfation that leads to the formation of a sulfate conjugate with its plasma exposure 7–10 times that of the parent drug.[79,80] Both troglitazone and its sulfate conjugate inhibit BSEP *in vitro*, exhibiting IC_{50} values comparable to or lower than their respective plasma concentrations with therapeutic doses. Impairment to BSEP can occur in susceptible patients upon troglitazone treatment. It results in the accumulation of toxic bile acids in hepatocytes, and leads to cholestatic liver injury. However, there is a lack of clarity with regard to what extent the parent drug or its metabolites contribute to the DILI. Evidence so far is insufficient to convince us of an unambiguous 'culprit' in the troglitazone DILI pathogenesis. There may be a number of contributing factors such as patient genetic predisposition and disease states; DILI can arise from accumulative and collective injuries. Therefore, the detection and monitoring of signals as markers of various biochemical reactions at the onset of drug treatment and hepatotoxicity can aid DILI mechanistic understanding.

The inability to predict the likelihood of drug susceptibility to DILI severely hinders and frustrates the effort of producing safe molecules in drug discovery and development. In the current paradigm, potential drug candidates are subject to extensive pre-clinical and clinical safety assessments prior to submissions for approvals by regulatory agencies. The pre-clinical safety studies constitute experiments that usually require two different animals: a rodent and a non-rodent species; and they 'predict' $\sim$70% of toxicities that may occur in humans.[81,82]

There are several obvious deficiencies in the current pre-clinical safety assessments. For example, laboratory animals are homogeneous, but patient populations are heterogeneous in terms of their genetic make-up and environmental milieu. Laboratory animals are free from disease, but patients are often compromised by other illnesses. These issues can prevent a simple and straightforward extrapolation of pre-clinical findings to humans, and they often have to be addressed in subsequent clinical trials assessing their human safety and tolerability. These clinical trials are usually divided into three phases, with the number of subjects in each phase increasing from a dozen or so to a few thousand. In spite of these safety precautions and approaches, certain drugs approved for marketing can still show hepatotoxicity after their clinical application is extended to even a larger patient population. DILI of this type has been termed idiosyncratic in view of its elusive nature. It is characterised by the absence of an apparent dose-response relationship, and occurs in only a small sub-set of the treated patients after prolonged exposure ranging from days to months.

The estimated rate of idiosyncratic DILI is 1 per 10 000 to 100 000 treatments, and such a low incidence can contribute to the 'miss' of detection in early clinical trials with 'flawed' drug molecules. Hence, the number of human subjects selected for clinical trials has to increase significantly in order to detect such a low probability of these adverse events according to the so-called 'rule of three', *i.e.*, an investigation involving 30 000 or more subjects is necessary for the detection of DILI that takes place in 1 out of every 10 000 treated patients.[83,84] An alternative approach is to check drug safety in 'humanised' animal models that reflect the most sensitive patient population. Such an approach requires vigorous validation; otherwise, results can present many difficulties for interpretation.[85,86]

'Adaption' to idiosyncratic DILI can sometimes happen in certain individuals during their therapy when DILI resolves itself spontaneously despite continuing or reinstating treatment with the offending drug. For example, the anti-tuberculosis agent isoniazid (also known as isonicotinylhydrazine or INH) causes hepatotoxicity featuring focal necrosis accompanied sometimes by cholestasis. The DILI exhibits an incidence rate of 0.1–0.6%, and is linked to INH metabolism that generates acetylhydrazine.[87] Acetylhydrazine can further undergo CYP2E1-catalysed oxidative bioactivation. It produces reactive species that are capable of modifying hepatic proteins, and hence are likely responsible for INH-induced hepatotoxicity. About 20% of the patients initially experienced an elevation of serum liver enzymes, but they were able to complete INH therapy without progressing to severe DILI.[88] Re-challenge with the drug has also been a common practice since few options are available for the treatment of tuberculosis and its latent form.

It can be hard to predict whether a drug may cause DILI in a patient, or foretell whether a patent can become adapted after DILI and remain free from any adverse reaction. Given the difficulties of detecting DILI in preclinical and clinical safety studies, searching toxicological biomarkers with good understanding of underlying biochemistry is an alternative but promising approach. Biomarkers based on sound biochemical understanding are invaluable not only to post-marketing surveillance, but also to a proper selection of patients that can benefit from a suitable 'personalised' therapy. It is conceivable that medications can be much safer under the auspice of effective biomarkers that enable the distinction of patients who can or cannot tolerate a drug, especially for a lifesaving treatment.

Traditional biomarkers for detecting DILI include serum aminotransferase and bilirubin levels.[89,90] Serum aminotransferase reflects hepatocyte damage that leads to the leakage of cellular enzymes such as alanine and aspartate aminotransferases (ALT and AST) into the blood circulation, and serum bilirubin monitors hepatobiliary excretory function. These assays are routinely accessible, but they do not necessarily predict whether an individual is able to tolerate a specific pharmacotherapy. In addition, elevation of serum enzyme levels sometimes may arise from non-hepatic injury. Pathological data are important for the characterisation of tissue and cellular injuries, and aid an elucidation of the mechanisms underlying DILI. However, liver pathology is not always practically accessible. A limited few options for tissues to respond to various insults of distinctive natures can lead to different interpretation of a particular pattern of histological presentation.[13]

The advance of technology, such as nuclear magnetic resonance (NMR) spectroscopy and mass spectrometry (MS), provides a broad and quantitative profiling of both endogenous and exogenous metabolites in biofluids (*e.g.* blood and urine) of patients. They are expected, in principle, to provide specific molecular 'fingerprints' of individual patients, and enable phenotyping and monitoring of patients before, during and after drug treatments. Metabolomics with molecular imaging (*e.g.* MRS and PET) may represent a promising platform allowing for an examination of multi-organ functional integrity without a need for invasive tissue sampling.

11.4 Metabolomics

Metabolomics and metabonomics refer to studies and profiling of endogenous metabolites, although their original definitions were different. Metabolomics aims at covering the entirety of the metabolome in a biological system, and metabonomics intends to investigate the perturbation of cellular metabolism in response to pathophysiological stimuli.[91,92] However, these two terms have become increasingly interchangeable, and we will use metabolomics in this chapter. A recent proposal recommended that metabolomics should encompass xenobiotics such as those derived from drugs and environmental chemicals in order to provide a comprehensive description of the human metabolome in the milieu of surrounding environments.[93] A broad metabolome coverage can handle an exogenous metabolite that is metabolised in cells and becomes conjugated with an endogenous metabolite such as glucuronic acid, GSH, and amino acids. In addition, endogenous and drug metabolites are often present in the same sample source (*e.g.*, plasma, urine *etc.*); and their identification and qualification are frequently based on the same or very similar instrumentation and data analysis software. Metabolomics in this chapter describes metabolite profiling regardless of their origin.

The utility of metabolomics in studies of DILI is based on two premises: (1) the profiles of endogenous metabolites correspond to responses of a biological system to internal or external stimuli, (2) drug metabolites represent overall exposures by patients to drug-related substances, including those that are chemically reactive and capable of modifying proteins and nucleic acids.

Metabolites are the products of biochemical reactions that are often carried out by protein enzymes and controlled by protein receptors. Those proteins are, in turn, dependent on gene expressions at the transcriptional, translational, and post-translational levels. Therefore, quantitative profiling of endogenous metabolites can render molecular phenotypes to be an important link between pathological outcomes and biochemical response to either internal or external stimuli. External stimuli can be, for example, diet, environmental stress, disease, and medication. Internal stimuli are cellular perturbations such as enzyme inhibition and induction, receptor agonism and antagonism, and gene up- or down-regulation. The profiling of drug metabolites can reveal potential toxins formed during drug metabolism, and relate a patient's DILI susceptibility to drugs and their metabolites with respect to concentrations and exposure time. Collected data are usually analysed in light of systems biology.

A complete coverage of both endogenous and exogenous metabolites presents many difficulties for a single type of instrumentation. Chemicals in the whole metabolome often exhibit vastly different physicochemical properties, and are present in a wide range of concentrations. Therefore, there is a need for two or more complementary analytical approaches such as MS and NMR. They can be coupled to chromatographic separations such as gas chromatography (GC), or high- and ultra-performance liquid chromatography (HPLC and UPLC).

NMR is quantitative, non-discriminative, and non-destructive, although less sensitive.[94,95] Its non-discriminative nature is attributed to the same response factor to a particular nucleus (*e.g.*, proton) in all molecules for a given sample. Nuclei ^{1}H, ^{13}C, ^{31}P, and ^{15}N often found in biological molecules are detectable by NMR. Molecules are separated on an NMR spectrum based on different frequencies of their constituent nuclei, and the differences in frequency expressed in ppm are called chemical shifts. The chemical shift values often provide ample structural information on analytes of interest. NMR peak areas are proportional to metabolite concentrations; therefore, quantitative analysis is readily achievable when a single reference compound of known concentration is included in test samples. NMR is non-destructive, and analysed samples can be 'reused' for either different NMR analyses or other analytical methods. Sample preparation for NMR experiment is minimal, but often requires deuterium oxide or deuterated organic solvents in order to help keep magnetic field strength constant during sample analyses by 'locking' the field with respect to the ^{2}H signal. One major drawback of NMR is its relatively low sensitivity. It is capable of measuring most small molecules in the µM to mM concentration range. The sensitivity increases with increasing magnetic field strength, and a strong magnetic field improves the signal/noise ratio and peak dispersion.

An alternative but equally important analytical approach is the MS-based platform that detects ionised metabolites or their fragment ions formed following collision-induced dissociation.[96] Molecule ionisation relies on electronic, chemical, electrospray, or matrix-assisted laser desorption. MS measures mass according to the mass-to-charge ratio (*m/z*) of charged parent and fragment ions. Quantification by MS is based on comparisons of the signal strength of an analyte in biological samples with a standard curve constructed using a set of naive samples 'spiked' with the analyte of known amounts. MS sensitivity can reach low femtomolar. MS usually offers much improved sensitivity, but it often lacks essential structural details. High-resolution and accurate-mass MS make it easy for chemical composition determination, and thus enabling post-acquisition data-mining for metabolite profiling.

The identification of all individual metabolites in a mixture can be a challenge to either NMR or MS. NMR resolves metabolites based on chemical shifts exhibited by nuclei (*e.g.*, protons) in different compounds.

However, the resolution may not be sufficient when a sample containing hundreds of metabolites displays thousands of nucleus peaks in a limited chemical shift range. MS distinguishes metabolites according to their mass-to-charge (m/z) ratios with an optimal resolution at m/z ratios differing by 4 units. However, hundreds of small-molecule metabolites often share limited numbers of m/z. For example, more than 1000 metabolites can have molecular weights between 30 and 850 Da with a single ionisation charge. Competition for ionisation gives rise to suppression of ion signals in MS. These problems can be resolved by separating chemicals using a chromatographic column before detection by NMR or MS. In principle, LC-NMR can alleviate severe peak overlap on an NMR spectrum. However, NMR detection can be limited by low NMR sensitivity on a small quantity of eluted chemicals with limited acquisition time. A sensitive probe, such as a cryogenic probe with small diameter, can help improve signal-to-noise ratios. The combination of GC, HPLC or UPLC with mass spectrometry has been very successful; GC-MS, HPLC-MS or UPLC-MS enable qualitative and quantitative analyses of a wide range of metabolites following a single injection of a biological sample with minimal sample preparation.

Metabolite identification and quantity are important to metabolomics. Metabolic pathway analyses need identity of metabolites, and analyses of dynamic changes to metabolites require quantification of metabolites produced or consumed. The advantages of NMR and MS for metabolomics are their ability to separate chemicals and identify molecular structures. NMR and MS analyses are often routinely and iteratively applied to chemical structural elucidation, for example in drug metabolism. Synergy of these two analytical platforms is perfect for metabolomics. In order to reduce the time required for *de novo* chemical structure determination on all chemicals in every sample, reference libraries are often collected beforehand, and set up as an NMR or a LC/GC-MS database for chemical identification.

After data collection from NMR and MS, one commonly used statistical method is multivariate (MV) analysis for pattern recognition. In order to reduce data size, principal component analysis (PCA) reduces overwhelming and intricate raw data into a smaller and more manageable size of variables known as principal components.[95,97] The original dataset in a matrix is transformed and represented by two new small matrices, namely the scores and loading matrices. A scores plot can reveal a relationship between samples, and a loadings plot depicts variables that drive the separation of samples in scores plots. The scores matrix is often plotted with a limited number of principal components of decreasing importance in order to 'capture' significant changes or variances amongst the original dataset. The most significant principal component (usually the first one) is dominated by metabolites showing the largest changes in magnitude in samples. In a beneficially succinct way, examination of distribution patterns in the scores plot can often tell whether there are likely clusters that correspond to drug treatment or pathological changes.

PCA is an 'unsupervised' method, since the 'clustering' in a scores plot is not 'forced' or 'biased' by pre-defined groupings or pathological scores. In principle, this should lead to an unbiased discovery of 'important' metabolites. However, group clustering observed in scores plots can be affected by the manner in which the raw data are scaled. Therefore, PCA is often recommended for use in conjunction with other data analysis tools, and it can avoid an incorrect trend of metabolite profiles. Ranking important metabolites and understanding their biological roles are two parallel but interactive processes. They combine statistical analyses with biological interpretation, and help generate new working hypotheses for further validation. A synergistic integration of the two processes often is important to the success of metabolomics studies and investigations.

11.5 Application of Metabolomics in Studies of DILI

11.5.1 Mechanistic Investigation

Low-molecular-mass drugs associated with DILI may undergo metabolic bioactivation that produces reactive species capable of covalently modifying hepatic proteins or mitochondrial DNA. Reactive metabolites were detected in laboratory animals and patients administered with hepatotoxic drugs such as APAP, diclofenac, troglitazone, valproic acid, terbinafine and fialuridine. Identification and quantification of drug metabolites are the mainstays of drug development under regulatory requirements. Reactive metabolites can often be inferred only from their conjugates with GSH, peptides, cyanide, and nucleus bases *in vivo* or *in vitro*, due to their transient lives.[98,99] For example, lumiracoxib causes severe liver injury, and its metabolism in human liver preparations produces a quinone imine species that is identified by LC-MS/MS as an *N*-acetylcysteine adduct.[100]

Drug metabolite profiling can be assisted by PCA, and this is illustrated in the studies of fenofibrate metabolisms in rats and monkeys.[101,102] Fenofibrate, an agonist of PPARα, is prescribed in control of hypercholesterolemia and hypertriglyceridemia. This drug causes hepatocellular carcinoma in rodents but not in monkeys and humans. In these reported studies, multivariate analyses of LC-MS/MS data helped identify previously uncharacterised metabolites, including two taurine conjugates and an unusual metabolite B that likely resulted from *C*-decarboxylation of fenofibrate. The studies illustrate the benefit of the metabolomics methodology in profiling drug metabolites. These studies did not find differences of fenofibrate metabolisms between rats and monkeys. Identification of drug metabolites, especially reactive chemicals, is valuable to understanding DILI mechanisms.

The benefit of analytical techniques such as LC-MS and NMR is their capability of simultaneous detection of drug and endogenous metabolites. Identification of these two groups of molecules should provide complementary aspects of biochemistry in the investigation and elucidation of any possible underlying DILI mechanisms. Pathophysiological injuries incurred by drug treatment can result in metabolic perturbation. The detection and quantification of exogenous and endogenous metabolites enable holistic understanding to the biochemical pathways and cellular networks under drug-induced cellular stress. The merit of metabolomics to drug toxicity investigation is its systems biology approach; it is able to associate seemingly unrelated DILI cases based on common impaired metabolic pathways. For example, metabolic profiling revealed reduced levels of citrate, 2-oxoglutarate and succinate in the urine of rats or mice administered with the hepatotoxins such as APAP, bromobenzene, aroclor 1254 (polychlorinated biphenyls; PCBs) and 2,3,7,8-tetrachlorodibenzo-*p*-dioxin (TCDD).[103–105] The decreased concentrations in urine were accompanied by increased concentrations of 3-hydroxybutyrate, glucose, pyruvate, acetate, and lactate in plasma. The decreases of TCA intermediate metabolites suggest a possible link of impairment by these drugs to mitochondrial TCA or oxidative phosphorylation. It can result in poor utilisation of glycolysis and fatty acid β-oxidation by mitochondria to generate ATP. Mitochondria in these cases can be either a direct target or a mediator during the pathogenesis of DILI.

Metabolomics has the potential to monitor toxicity development in 'real time' without resorting to invasive tissue sampling, and it can be a powerful tool when combined with other 'omics' approaches. For example, in the case of APAP-induced liver injury in mice at a toxic dose of 500 mg kg^{-1}, the plasma concentrations of lactate, acetate, 3-hydroxybutyrate and lipids elevated within 15 to 240 minutes, whereas the plasma pyruvate level decreased initially between 15–30 minutes before it increased from 60 to 240 minutes.[106,107] These changes mirrored the proteomic and genomic profiles of APAP hepatotoxicity in mice. Decreases were noticed as early as 15 minutes post-dosing in the expression of ATP synthase sub-units and proteins for the fatty acid β-oxidation, and mitochondrial protein expression proceeded gene transcriptional regulation response to APAP treatment.[108] It suggests that gene response is less likely due to direct insults, but more likely exacerbation of toxicity or mediation to cell recovery.

It helps confirm DILI mechanisms by profiling endogenous and drug metabolites in animal models deficient in particular enzymes supposedly related to drug toxicity. For example, APAP-related hepatotoxicity has been attributed to reactive NAPQI. CYP2E1 is believed to be mainly responsible for NAPQI formation, with minor contributions from CYP1A2 and 3A. Comparison of metabolic profiles of *CYP2e1*-null and wild-type mice provides additional details to likely toxicity mechanisms.[109] The profiles showed that the concentrations of NAPQI-related thiol conjugates in urine were significantly higher in wild-type than in *CYP2e1*-null mice at 10 mg kg^{-1} of APAP, a dose below the rodent toxicity dosage. However, urinary levels of thiol conjugates were similar in both strains of mice following a dose of APAP (400 mg kg^{-1}) toxic to wild-type mice but not to *CYP2e1*-null mice. The increased thiol conjugate levels in *CYP2e1*-null mice suggest a switch from

CYP2E1- to the CYP1A2/3A-catalysed formation of NAPQI. In addition, double transgenic mice (humanised for PXR and CYP3A4) showed significantly increased serum aminotransferase activities and APAP dimer concentrations when pre-treatment of Rifampicin up-regulated CYP3A4 by activating human PXR.[110] These data suggest that the severity of APAP-induced 'oxidative stress' and toxicity, for example following drug overdose, relates to hepatic CYP3A4 activity in the liver. CYP3A represents up to ~60% of the total cytochrome P450 in human liver, and its contribution to clinically observed APAP hepatotoxicity warrants further investigation.

'Oxidative stress' generated by CYP2E1-mediated APAP metabolism is another important contributor to APAP cytotoxicity. Following a toxic dose of APAP (400 mg kg^{-1}), oxidative byproducts such as the APAP dimer and a benzothiazine derivative were excreted into wild type mouse urine. Hepatic GSH was significantly depleted within 1–2 hours in both wild-type and *Cyp2e1*-null mice.[109] GSH in *Cyp2e1*-null mice quickly 'rebounded', but it did not return to the pre-treatment level until 16 hours post-dosing in wild type mice. The recovery of GSH is driven by metabolic homeostasis.[112] Prolonged depletion of GSH in the liver of wild-type mice indicates 'oxidative stress', and is susceptible to tissue injury by further oxidative reactions. The dynamic changes to systemic GSH remind us of the importance of metabolic profiling over time to capture insults and responses by resilient biological systems, and temporal profiles can be informative to understanding DILI mechanisms.

APAP treatment related 'oxidative stress' is also indicated by the elevation of ophthalmate in mouse liver and serum.[111] Ophthalmate is synthesised from 2-aminobutyrate *via* a pathway catalysed by γ-glutamylcysteine synthetase (GCS) when cysteine is in short supply.

Hepatic fatty acid β-oxidation is also inactivated by APAP treatment. According to PCA, a scores plot of serum metabolites over a 24-hour period showed a separation of wild-type from *Cyp2e1*-null mice administered with APAP at 400 mg kg^{-1}, a dose toxic to wild-type but not *Cyp2e1*-null mice.[113] The separation was noticeably driven by four molecular ions, identified by LC-MS/MS, as palmitoylcarnitine, myristoylcarnitine, oleoylcarnitine, and palmitoleoylcanitine. Serum acylcarnitine concentrations increased and remained elevated for at least 24 hours in wild-type mice. In contrast, serum acylcarnitine changes in *Cyp2e1*-null mice were transient, and returned quickly to baseline concentrations. Excessive acylcarnitines in wild-type mice suggest problems associated with hepatic fatty acid β-oxidation, and it is supported by elevated triglycerides and free fatty acids in wild-type, but absent in *Cyp2e1*-null mice.

The APAP interference to fatty acid β-oxidation appears to be associated with the suppression of PPARα activation. It is supported by the accumulation of acylcarnitines in serum of fasted mice with knockout PPARα (*Pparα*-null).[113] Hepatocytes switch to fatty acid β-oxidation as an alternative energy supply during fasting. An inactive PPARα fails to up-regulate the fatty acid β-oxidation pathway, and in consequence leads to improper mitochondrial

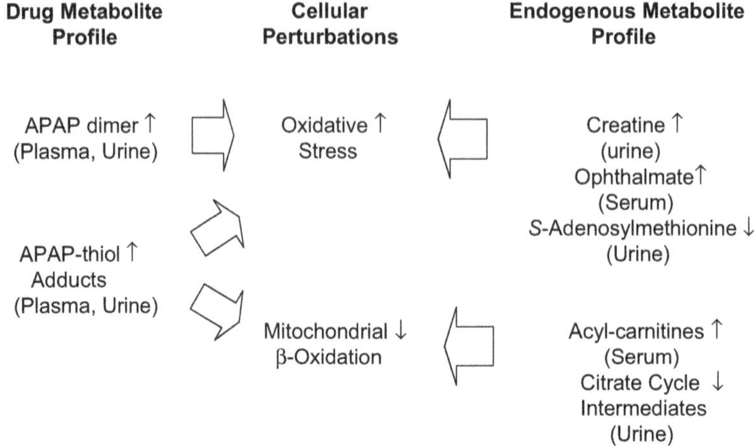

Figure 11.2 Cellular events upon APAP challenge: readouts from the analysis of exogenous and endogenous metabolites.

functioning. PPARα regulates four mouse liver genes, *i.e.* carnitine palmitoyltransferase 1 (*Cpt1*), carnitine palmitoyltransferase 2 (*Cpt2*), acyl-CoA thioesterase 1 (*Acot1*), and P450 4A10 (*Cyp4a10*). APAP induced strong and sustained expressions of these four genes in *Cyp2e1*-null mice, but the expressions were weak and transient in the wild-type strain. Therefore, stimulated PPARα activation in response to APAP challenge was normal in *Cyp2e1*-null mice, but suppressed in the wild-type. This may explain that wild-type mice were impaired by APAP treatment in comparison to *Cyp2e1*-null mice.

Metabolomics investigation of APAP hepatotoxicity using various mouse models is valuable to revealing and confirming its likely DILI mechanisms such as CYP2E1-mediated bioactivation, 'oxidative stress', and inactivation of fatty acid β-oxidation due to suppressed PPARα activation (Figure 11.2).

11.5.2 Searching for Biomarkers

Accessible biomarkers capable of detecting underlying biochemical changes preceding the onset of overt DILI toxicities are valuable to both pre-clinical and clinical safety assessment. Such 'idealised' biomarkers should also be able to differentiate patients who may or may not develop a tolerance to potential toxicity. Elevated transaminase activities in serum have traditionally served for the diagnosis of DILI, and it is based on the understanding that these enzymes abundant in liver leak into the blood circulation following hepatocyte injury. However, serum transaminases can be limited by two ambiguities. One is a lack of specificity; in addition to liver, they are present in heart and skeletal muscle, though as different isoforms.[114] The other is species dependence: for example, the basal ALT activity in human serum is mostly attributable to ALT1, but ALT2 gene expression is elevated in the fatty livers of obese mice.[115,116] These ambiguities can be improved by measuring additional enzymes or small molecules leaking from injured hepatocytes. Potential enzymes or molecules for improving the liver specificity include arginase I, GSTα, sorbitol dehydrogenase, glutamate dehydrogenase, malate dehydrogenase, bilirubin, and bile acids.[89,90] Although additional enzymes and small molecules improve tissue specificity, these molecular biomarkers lack etiological revelation of underlying mechanisms, and may not foretell patient progress.

Metabolomics has a promising potential to advance understanding of DILI mechanisms, and facilitate finding biomarkers with good sensitivity and specificity toward drug-related adverse effects. The advantage of metabolites is attributable to their direct participation in cellular anabolism and catabolism under toxicity insults or homeostasis. For example, some metabolites in hepatocytes may decrease due to substrate depletion or enzyme inhibition, but other metabolites may increase in response to the shunt of impaired metabolic pathways.

Small molecules are also capable of traversing the plasma membrane with the help of membrane transporters *via* either passive diffusion or active efflux. Their entry into the circulation in response to, for example, toxicity insults can precede ultimate cell necrosis and tissue damage. These kinds of metabolites detected in serum can serve as 'fingerprints' for the early indication of cellular toxicity or stress, and are good candidates for clinical trial surveillance.

Groups or types of low-molecular-mass molecules detected can also help stratify tissue toxicities according to underlying biochemical mechanisms, although morphological changes in injured tissues may appear similar. For example, an investigational drug for the HIV treatment caused hepatotoxicity in rats, and the histopathology found microvesicular steatosis, centrilobular hypertrophy, and both single-cell and focal necrosis.[117] Metabolomics investigation of urine samples from treated rats using NMR found decreased concentrations of intermediate metabolites (*i.e.*, citrate, 2-oxoglutarate and succinate) in the TCA cycle, but significant amounts of dicarboxylic acids such as sebacic and suberic acids. The appearance of dicarboxylic acids is attributed to elevated microsomal ω-oxidation of fatty acids as a result of the inhibition to mitochondrial β-oxidation. Histological detection of lipid accumulation in liver supported the likely inhibition to fatty acid β-oxidation. This example suggests that increased dicarboxylic acids, together with decreased TCA cycle metabolites, can serve as urinary molecular signatures for an early identification of hepatic β-oxidation dysfunction with an improved sensitivity and specificity.

Degradation of endogenous proteins or nucleic acids may accompany DILI. These degradation by-products can be nucleosides or amino acids, and the small molecules remain conserved across different species in spite of varying sequences of proteins and nucleic acids from rodents to humans. Such metabolites as biomarkers are likely translatable from pre-clinical *in vitro* or *in vivo* models to clinical observation. For example, GC-MS and

LC-MS were used to profile urinary metabolites in rats administered with valproic acid, and acquired data were subject to multivariate analysis.[118] The analyses indicated that increased concentrations of 8-hydroxy-2′-deoxyguanosine (8-OHdG) were accompanied by high levels of GSTα in rat serum, which clearly indicated DILI. In a separate study conducted in patients under valproic acid therapy, increases of 8-OHdG serum concentrations correlated with elevations of the traditional liver parameters AST, ALT, and γGT.[119] Therefore, 8-OHdG appears to be a molecular signature that is shared by both rats and humans due to valproic acid-induced liver injury. The precise mechanisms relating 8-OHdG to DILI requires further investigation; however, the metabolomic findings suggest that the valproic acid treatment induces 'oxidative stress' in mitochondria, and likely causes damage to mtDNA. 8-OHdG is often found as an oxidation product of nucleosides when DNA is exposed to 'oxidative stress'.

Metabolomics has been demonstrated to offer good utility for the provision of clues or the generation of hypotheses as a part of a pursuit of biomarkers. In a study of ximelagatran-induced liver injury, GC-MS analyses of patients' plasma revealed that those who developed hepatotoxicity started with low levels of circulating pyruvate, and their blood concentrations of pyruvate reduced further post-dose.[120] To test this hypothesis at the cellular level, hepatoma cells and hepatocytes were challenged with ximelagatran in media containing differing levels of pyruvate. The cells were found to be more sensitive to drug-induced cytotoxicity when pyruvate was absent. In addition, L-cysteine and L-glutamine were also relatively low in the plasma of those patients experiencing the DILI. Therefore, it appears that susceptible patients are predisposed to ximelagatran-induced liver injury when the drug treatment leads to a depletion of pyruvate and amino acids that are of importance to cellular survival.[120] This DILI model demonstrates that idiosyncratic hepatotoxicity can be caused by such factors as poor nutrition, disease, and possibly genetics.

Choline deficiency may cause liver damage, since this nutrient is a precursor of methionine and S-adenosylmethionine, which is vital for the methylation of nucleic acids and proteins. In a clinical study of nutrition, healthy human subjects were fed on a choline-adequate (550 mg per 70 kg^{-1} day^{-1}) diet for 10 days to establish a baseline level, and then on a choline-deficient (<50 mg per 70 $kg^{-1}day^{-1}$) diet for up to 42 days. Their blood was collected at the end of every treatment for the GC-MS and LC-MS analyses of plasma.[121] A low intake of dietary choline significantly decreased metabolites in the metabolic pathways of choline and methionine. The most significant finding was that ~52% of the participants developed fatty liver following a period of choline-deficient diet, and their plasma metabolite profiles during the early baseline period were already distinguishable from those who did not experience liver injury. Therefore, metabolic phenotypes in this case could predict hepatic dysfunction due to choline deficiency, and molecules in the metabolic pathways of choline, carnitine, keto acid, and amino acids can serve to discriminate between those two phenotypes.[121]

Metabolomics thus may be able to identify individual metabolic signatures that can predict the benefits or dangers of a drug treatment.

Metabolomics can serve personalised medicine or targeted treatments that promise a very effective yet safe pharmacotherapy, especially for those drugs that may elicit adverse effects in a small number of patients. In a rat model of APAP hepatotoxicity, ^{1}H NMR analyses profiled urine samples collected before and after drug treatments.[122] Multivariate analyses of the NMR spectral profiles showed a correlation between the ratio of APAP glucuronide to the parent drug and signal intensities in the 5.06–5.14 ppm regions of pre-dose spectra; revealing a connection between low degrees of liver tissue damage and the presence of high taurine concentrations in pre-dose urine. The concentration of urine taurine may reflect the availability of taurine to protect liver, and the availability of inorganic sulfate as a precursor essential for sulfation reactions. The sulfonate group of taurine can detoxify NAPQI by sulfation. In a follow-up study with human subjects administered with APAP, patients who experienced ALT elevation were designated as responders, and those who did not were classified as non-responders.[123] Analyses of urine samples using ^{1}H NMR spectroscopy could distinguish the responders from non-responders 4–5 days preceding the peak occurrence of serum ALT in responders. Increased urine glycine excretion suggested its decreased uptake by the liver for hepatic protection independent of glutathione. This example illustrates metabolomics for a post-dose surveillance using the combined profiles of endogenous metabolites with the drug and its metabolites. The combined profiles may become robust by including additional chemicals such as NAPQI thiol conjugates, glycine, methyl-histidine, and trimethylamine oxide.

A large quantity of NAPQI thiol conjugates in the urine of responders fit the known mechanism of APAP hepatotoxicity, in which NAPQI triggers the cascade of cytotoxic events. It is believed that responders with high levels of urinary glycine have problems in the hepatic uptake, and are vulnerable to DILI since glycine was shown to be hepatoprotective.[123] The connection remains unknown between responders and urine levels of methyl-histidine and trimethylamine oxide. 'Pattern recognition' is one of the 'omics' approaches. Sometimes, a response pattern may appear when some of the responding signals cannot be figured out for their chemical identities at an early stage; it can be used temporarily but needs to be properly validated later. Therefore, metabolomics has a promising potential helping the discovery of biomarkers for the identification of patients susceptible to DILI before the development of an overt clinical manifestation.

11.6   Summary and Closing Remarks

The application of metabolomics to DILI can serve mechanistic investigations and biomarker discovery. Cellular concentrations of endogenous metabolites usually are maintained as closely to steady-state values as possible for cell homeostasis.[124,125] Injury to cells or tissues can elicit transcriptional up- or down-regulations to counterbalance metabolic perturbations. However, metabolic re-adjustment may take place within a limited early time. Consequently, metabolite profiles shortly after a toxicity challenge may be helpful in capturing an attempt of hepatocytes to regain homeostasis before succumbing to DILI. Metabolite profiles exhibit signs of apoptosis or necrosis when cells experience overt cytotoxicity. Therefore, time-course profiles of metabolite changes in response to DILI can be valuable to an elucidation of the underlying mechanisms. Biomarkers indicative of underlying toxicity mechanisms can be very robust with respect to their performance in predicting and detecting drug-induced AEs.

Nutrition as an external factor directly influences metabolic homeostasis. Nutrient uptake supplements endogenous substrates, and can confer metabolic robustness fending off drug toxicities. It is known that nutrition deprivation can be a consequence of drug toxicity, but reduced nutrients such as during fasting can also influence the manifestation of drug toxicity. Lack of nutrition substrates can limit cellular metabolisms necessary for detoxification. Therefore, metabolomics, with the consideration of all important external contributors such as drugs and nutrients, should be preferred in the study of DILI. For example, excessive concentrations of dicarboxylic acids in urine have been taken to indicate inherited and acquired hepatic mitochondrial dysfunction.[126,127] Fasting, especially limited intake of carbohydrates, can aggravate the toxicity to fatty acid β-oxidation by diminishing the major alternative energy metabolism of glycolysis. It is noticed that dicarboxylic acids in urine showed up in significant amounts after fasting. Fasting itself is often not a direct cause of toxicity, but it can accentuate mitochondrial toxicities that sometimes are manifested as an idiosyncrasy. A broad scope of dynamic metabolite coverage by metabolomics makes it possible to track seemingly complicated DILI in toxicology studies, and avoid relying solely on limited terminal histopathological assessments.

The flux analysis of metabolites by metabolomics is another useful technique, although it is not covered in this review. It can facilitate the detection of DILI when detrimental metabolic pathways are pre-defined.[128] It is similar to the use of 2-fluoro(^{18}F)-2-deoxy-D-glucose (*i.e.*, FDG) for cancer diagnoses using positron emission tomography (PET). The flux approach can be readily adoptable simultaneously to both NMR and MRS (magnetic resonance spectroscopy). It was illustrated by examining ^{13}C-labelled pyruvate metabolisms in rat liver; hyperpolarised MRS imaging can monitor the production of ^{13}C-lactate and ^{13}C-alanine with greatly improved sensitivity.[129] Results showed that livers of fasted rats had unchanged lactate levels, but alanine decreased significantly. Such dynamic metabolic diagnoses can be readily transferrable from pre-clinical animals to clinical humans thanks to almost very similar metabolite space among animals and humans. With advancements in technology for sensitivity and reliable quantification, endogenous and drug metabolites can be profiled in the same biological matrices and with the same analytical instrument. This can provide a helpful connection between metabolite concentrations and biological responses when dealing with likely DILI.

Challenges to metabolomics remain considering different sample types, techniques or instrumentation. For example, the variation of metabolic profiles amounted to ~50% in 72 metabolites when serum were compared to plasma, despite their isolation from the same rat blood collection.[130] In a separate study, urine from rats dosed with isoniazid was analysed by two different MS instruments (QTRAP and Q-TOF) coupled with a single UPLC. The result from both machines discriminated treated rats from controls.[131] However, one MS instrument identified 34 up- and 59 down-regulated signature metabolites, and the other MS-based system detected 27 up- and 60 down-regulated signatures. Only 16 up- and 18 down-regulated signatures were shared between these two instruments. Therefore, a wide acceptance of metabolomics applications requires further improvements in managing data integrity and consistency.

Diagnoses based on metabolic pathways help bridge morphological phenotypes and genotypes of DILI. Together with genomics and proteomics, metabolomics should be routinely employed assisting drug toxicity investigations, and serving early diagnosis of DILI or other toxicities.

Acknowledgements

We thank Dr Frank Sistare for his critical review and insightful suggestions to this manuscript.

References

1. K. Bhaskaran, O. Hamouda, M. Sannes, F. Boufassa, A. M. Johnson, P. C. Lambert and K. Porter, Changes in the risk of death after HIV seroconversion compared with mortality in the general population, *J. Am. Med. Assoc.*, 2008, **300**, 51.
2. E. D. Clercq, Anti-HIV drugs: 25 compounds approved within 25 years after the discovery of HIV, *Inter. J. Antimicrob. Agents*, 2009, **33**, 307.
3. G. Abboud and N. Kaplowitz, Drug-induced liver injury, *Drug Saf.*, 2007, **30**, 277.
4. http://www.recalls.gov.au/content/index.phtml/itemId/953426; http://www.tga.gov.au/safety/alerts-medicine-lumiracoxib-070813.htm; http://www.bpac.org.nz/magazine/2007/september/docs/bpj8_lumiracoxib_page_24-25.pdf.
5. S. M. Hall, Reye's syndrome and aspirin: a review, *J. Royal. Soc. Med.*, 1986, **79**, 596.
6. E. Belay, J. S. Bresee, R. C. Holman, A. S. Khan, A. Shahriari and L. B. Schonberger, Reye's syndrome in the United States from 1981 through 1997, *New Eng. J. Med.*, 1999, **340**, 1377.
7. V. J. Navarro and J. R. Senior, Drug-related hepatotoxicity, *N. Engl. J. Med.*, 2006, **354**, 731.
8. L. N. Bell and N. Chalasani, Epidemiology of idiosyncratic drug-induced liver injury, *Semin. Liver Dis.*, 2009, **29**, 337.

9. N. Chalasani and E. Bjornsson, Risk factors for idiosyncratic drug-induced liver injury, *Gastroenterol.*, 2010, **138**, 2246.

10. M. I. Lucena, R. J. Andrade, N. Kaplowitz, M. García-Cortes, M. C. Fernandez, M. Romero-Gomez, M. Bruguera, H. Hallal, M. Robles-Diaz, J. F. Rodriguez-Gonzalez, J. M. Navarro, J. Salmeron, P. Martinez-Odriozola, R. Perez-Alvarez, Y. Borraz and R. Hidalgo, Phenotypic characterization of idiosyncratic drug-induced liver injury: the influence of age and sex, *Hepatology*, 2009, **49**, 2001.

11. F. E. Dreifuss, N. Santilli, D. H. Langer, K. P. Sweeney, K. A. Moline and K. B. Menander, Valproic acid hepatic fatalities: a retrospective review, *Neurology*, 1987, **37**, 379.

12. R. Ramachandran and S. Kakar, Histological patterns in drug-induced liver disease, *J. Clin. Pathol.*, 2009, **62**, 481.

13. J. M. Cullen and R. T. Miller, The role of pathology in the identification of drug-induced hepatic toxicity, *Expert Opin. Drug Metab. Toxicol.*, 2006, **2**, 241.

14. E. Bjornsson and R. Olsson, Outcome and prognostic markers in severe drug-induced liver disease, *Hepatology*, 2005, **42**, 481.

15. R. Temple, Hy's law: predicting serious hepatotoxicity, *Pharmacoepidemiol. Drug Saf.*, 2006, **15**, 241.

16. W. M. Lee, Acetaminophen and the U.S. Acute Liver Failure Study Group: lowering the risks of hepatic failure, *Hepatology*, 2004, **40**, 6.

17. M. Robles, E. Toscano, J. Cotta, M. I. Lucena and R. J. Andrade, Antibiotic-induced liver toxicity: mechanisms, clinical features and causality assessment, *Curr. Drug Saf.*, 2010, **5**, 212.

18. M. Black, Acetaminophen hepatotoxicity, *Ann. Rev. Med.*, 1984, **35**, 577.

19. A. M. Larson, J. Polson, R. J. Fontana, T. J. Davern, E. Lalami, L. S. Hynan, J. S. Reisch, R. V. Schiodt, G. Ostapowicz, A. O. Shakil and W. M. Lee, Acetaminophen-induced acute liver failure: results of a United States multicenter, prospective study, *Hepatology*, 2005, **42**, 1364.

20. L. F. Prescott and J. A. J. H. Critchley, The treatment of acetaminophen poisoning, *Ann. Rev. Pharmacol. Toxicol.*, 1983, **23**, 87.

21. P. B. Watkins, N. Kaplowitz, J. T. Slattery, C. R. Colonese, S. V. Colucci, P. W. Stewart and S. C. Harris, Aminotransferase elevations in healthy adults receiving 4 grams of acetaminophen daily: a randomized controlled trial, *J. Am. Med. Assoc.*, 2006, **296**, 87.

22. L. F. Prescott, Paracetamol overdosage. Pharmacological considerations and clinical management, *Drugs*, 1983, **25**, 290.

23. P. Purcell, D. Henry and G. Melville, Diclofenac hepatitis, *Gut*, 1991, **32**, 1381.

24. U. A. Boelsterli, H. J. Zimmerman and A. Kretz-Rommel, Idiosyncratic liver toxicity of nonsteroidal antiinflammatory drugs: molecular mechanisms and pathology, *Crit. Rev. Toxicol.*, 1995, **25**, 207.

25. S. M. Helfgott, J. Sandberg-Cook, D. Zakim and J. Nestler, Diclofenac-associated hepatotoxicity, *J. Am. Med. Assoc.*, 1990, **264**, 2660.

26. E. G. Breen, J. McNicholl, E. Cosgrove, J. McCabe and F. M. Stevens, Fatal hepatitis associated with diclofenac, *Gut*, 1986, **27**, 1390.
27. A. T. Banks, H. J. Zimmerman, K. G. Ishak and J. G. Harter, Diclofenac-associated hepatotoxity: analysis of 180 cases reported to the Food and Drug Administration as adverse reactions, *Hepatology*, 1995, **22**, 820.
28. W. Tang, The metabolism of diclofenac–enzymology and toxicology perspectives, *Curr. Drug Metab.*, 2003, **4**, 319.
29. R. P. Eckstein, J. F. Dowsett and M. R. Lunzer, Flucloxacillin induced liver disease: histopathological findings at biopsy and autopsy, *Pathology*, 1993, **25**, 223.
30. S. Russmann, J. A. Kaye, S. S. Jick and H. Jick, Risk of cholestatic liver disease associated with flucloxacillin and flucloxacillin prescribing habits in the UK: cohort study using data from the UK General Practice Research Database, *Br. J. Clin. Pharmacol.*, 2005, **60**, 76.
31. S. Lobatto, B. A. Dijkmans, H. Mattie and J. P. van Hooff, Flucloxacillin-associated liver damage, *Neth. J. Med.*, 1982, **25**, 47.
32. A. Parkinson and B. W. Ogilvie, Biotransformation of Xenobiotics, in *Casarett & Doull's Toxicology, The Basic Science of Poisons*, ed. C. D. Klaassen, McGraw-Hill, New York, 7th edn, 2008, p. 161.
33. W. Tang, Drug metabolite profiling and elucidation of drug-induced hepatotoxicity, *Expert Opin. Drug Metab. Toxicol.*, 2007, **3**, 407.
34. H. Spahn-Langguth and L. Z. Benet, Acyl glucuronides revisited: is the glucuronidation process a toxification as well as a detoxification mechanism?, *Drug Metab. Rev.*, 1992, **24**, 5.
35. H. Glatt, Sulfation and sulfotransferases 4: bioactivation of mutagens via sulfation, *FASEB J.*, 1997, **11**, 314.
36. P. E. Hanna, Metabolic Activation and Detoxification, *Curr. Med. Chem.*, 1996, **3**, 195.
37. L. D. DeLeve and N. Kaplowitz, Glutathione metabolism and its role in hepatotoxicity, *Pharm. Ther.*, 1991, **52**, 287.
38. L. Yuan and N. Kaplowitz, Glutathione in liver diseases and hepatotoxicity, *Mol. Asp. Med.*, 2009, **30**, 29.
39. T. A. Baillie and J. G. Slatter, Glutathion: a vehicle for the transport of chemically reactive metabolites in vivo, *Acc. Chem. Res.*, 1991, **24**, 264.
40. T. J. Monks and S. S. Lau, Glutathione conjugation as a mechanism for the transport of reactive metabolites, *Adv. Pharmacol.*, 1994, **27**, 183.
41. C. Klaassen and L. M. Aleksunes, Xenobiotic, bile acid, and cholesterol transporters: function and regulation, *Pharmacol. Rev.*, 2010, **62**, 1.
42. R. Scatena, P. Bottoni, G. Botta, G. E. Martorana and B. Giardina, The role of mitochondria in pharmacotoxicology: a reevaluation of an old, newly emerging topic, *Am. J. Physiol. Cell Physiol.*, 2007, **293**, C12.
43. G. Labbe, D. Pessayre and B. Fromenty, Drug-induced liver injury through mitochondrial dysfunction: mechanisms and detection during preclinical safety studies, *Fund. Clin. Pharmacol.*, 2008, **22**, 335.

44. H. Malhi, M. E. Guicciardi and G. J. Gores, Hepatocyte death: a clear and present danger, *Physiol. Rev.*, 2010, **90**, 1165.
45. D. H. Adams, C. Ju, S. K. Ramaiah, J. Uetrecht and H. Jaeschke, Mechanisms of immune-mediated liver injury, *Toxicol. Sci.*, 2010, **115**, 307.
46. B. V. Martin-Murphy, M. P. Holt and C. Ju, The role of damage associated molecular pattern molecules in acetaminophen-induced liver injury in mice, *Toxicol. Lett.*, 2010, **192**, 387.
47. J. Zhou, M. Febbraio, T. Wada, Y. Zhai, R. Kuruba, J. He, J. H. Lee, S. Khadem, S. Ren, S. Li, R. L. Silverstein and W. Xie, Hepatic fatty acid transporter Cd36 is a common target of LXR, PXR, and PPARgamma in promoting steatosis, *Gastroenterology*, 2008, **134**, 556.
48. M. Moya, M. J. Gomez-Lechon, J. V. Castell and R. Jover, Enhanced steatosis by nuclear receptor ligands: a study in cultured human hepatocytes and hepatoma cells with a characterized nuclear receptor expression profile, *Chem. Biol Interact.*, 2010, **184**, 376.
49. P. T. Manyike, E. D. Kharasch, T. F. Kalhorn and J. T. Slattery, Contribution of CYP2E1 and CYP3A to acetaminophen reactive metabolite formation, *Clin. Pharmacol. Ther.*, 2000, **67**, 275.
50. M. S. Rashed, T. G. Myers and S. D. Nelson, Hepatic protein arylation, glutathione depletion, and metabolite profiles of acetaminophen and a non-hepatotoxic regioisomer, 3′-hydroxyacetanilide, in the mouse, *Drug Metab. Dispos.*, 1990, **18**, 765.
51. W. Chen, J. P. Shockcor, R. Tonge, A. Hunter, C. Gartner and S. D. Nelson, Protein and nonprotein cysteinyl thiol modification by N-acetyl-p-benzoquinone imine via a novel ipso adduct, *Biochemistry*, 1999, **38**, 8159.
52. Y. Qiu, L. Z. Benet and A. L. Burlingame, Identification of the hepatic protein targets of reactive metabolites of acetaminophen in vivo in mice using two-dimensional gel electrophoresis and mass spectrometry, *J. Biol. Chem.*, 1998, **273**, 17940.
53. J. A. Hinson, A. B. Reid, S. S. McCullough and L. P. James, Acetaminophen-induced hepatotoxicity: role of metabolic activation, reactive oxygen/nitrogen species, and mitochondrial permeability transition, *Drug Metab. Rev.*, 2004, **36**, 805.
54. H. Jaeschke and M. L. Bajt, Intracellular signaling mechanisms of acetaminophen-induced liver cell death, *Toxicol. Sci.*, 2006, **89**, 31.
55. H. Jaeschke, Role of inflammation in the mechanism of acetaminophen-induced hepatotoxicity, *Expert Opin. Drug Metab. Toxicol.*, 2005, **1**, 389.
56. Z.-X. Liu and N. Kaplowitz, Role of innate immunity in acetaminophen-induced hepatotoxicity, *Expert Opin. Drug Metab. Toxicol.*, 2006, **2**, 493.
57. B. K. Park, M. Pirmohamed and N. R. Ketteringham, Role of drug disposition in drug hypersensitivity: a chemical, molecular, and clinical perspective, *Chem. Res. Toxicol.*, 1998, **11**, 969.

58. J. Uetrecht, New concepts in immunology relevant to idiosyncratic drug reactions: the "danger hypothesis" and innate immune system, *Chem. Res. Toxicol.*, 1999, **12**, 387.

59. M. P. Zanni, S. von Greyerz, B. Schnyder, K. A. Brander, K. Frutig, Y. Hari, S. Valitutti and W. J. Pichler, HLA-restricted, processing- and metabolism-independent pathway of drug recognition by human alpha beta T lymphocytes, *J. Clin. Invest.*, 1998, **102**, 1591.

60. M. J. Rieder, Immune mediation of hypersensitivity adverse drug reactions: implications for therapy, *Expert Opin. Drug Saf.*, 2009, **8**, 331.

61. J. P. Sanderson, D. J. Naisbitt and B. K. Park, Role of bioactivation in drug-induced hypersensitivity reactions, *AAPS J.*, 2006, **8**, E55.

62. J. Uetrecht, Immune-mediated adverse drug reactions, *Chem. Res. Toxicol.*, 2009, **22**, 24.

63. J. M. Neuberger, Halothane and hepatitis. Incidence, predisposing factors and exposure guidelines, *Drug Saf.*, 1990, **5**, 28.

64. J. L. Martin, D. J. Plevak, K. D. Flannery, M. Charlton, J. J. Poterucha, C. E. Humphreys, G. Derfus and L. R. Pohl, Hepatotoxicity after desflurane anesthesia, *Anesthesiology*, 1995, **83**, 1125.

65. M. A. Carey and F. N. A. M. Van Pelt, Immunochemical detection of flucloxacillin adduct formation in livers of treated rats, *Toxicology*, 2005, **216**, 41.

66. R. E. Jenkins, X. Meng, V. L. Elliott, N. R. Kitteringham, M. Pirmohamed and B. K. Park, Characterisation of flucloxacillin and 5-hydroxymethyl flucloxacillin haptenated HSA in vitro and in vivo, *Proteomics Clin. Appl.*, 2009, **3**, 720.

67. A. K. Daly, P. T. Donaldson, P. Bhatnagar, Y. Shen, I. Pe'er, A. Floratos, M. J. Daly, D. B. Goldstein, S. John, M. R. Nelson, J. Graham, B. K. Park, J. F. Dillon, W. Bernal, H. J. Cordell, M. Pirmohamed, G. P. Aithal and C. P. Day, HLA-B*5701 genotype is a major determinant of drug-induced liver injury due to flucloxacillin, *Nat. Genet.*, 2009, **41**, 816.

68. A. Kindmark, A. Jawaid, C. G. Harbron, B. J. Barratt, O. F. Bengtsson, T. B. Andersson, S. Carlsson, K. E. Cederbrant, N. J. Gibson, M. Armstrong, M. E. Lagerstom-Fermer, A. Dellsen, E. M. Brown, M. Thornton, C. Dukes, S. C. Jenkins, M. A. Firth, C. O. Harrod, T. H. Pinel, S. M. E. Billing-Clason, L. R. Cardon and R. E. March, Genome-wide pharmacogenetic investigation of a hepatic adverse event without clinical signs of immunopathology suggests an underlying immune pathogenesis, *Pharmacogenomics J.*, 2008, **8**, 186.

69. K. Kenne, I. Skanberg, B. Glinghammar, A. Berson, D. Pessayre, J.-P. Flinois, P. Beaune, I. Edebert, C. D. Pohl, S. Carlsson and T. B. Andersson, Prediction of drug-induced liver injury in humans by using in vitro methods: the case of ximelagatran, *Toxicology in Vitro*, 2008, **22**, 730.

70. H. Jaeschke, G. J. Gores, A. I. Cederbaum, J. A. Hinson, D. Pessayre and J. J. Lemasters, Mechanisms of hepatotoxicity, *Toxicol. Sci.*, 2002, **65**, 166.

71. D. P. Williams, Toxicophores: investigations in drug safety, *Toxicology*, 2006, **226**, 1.

72. C. M. Palmeira and A. P. Rolo, Mitochondrially-mediated toxicity of bile acids, *Toxicology*, 2004, **203**, 1.

73. N. Fernandes, S. A. Geller and T.-L. Fong, Terbinafine hepatotoxicity: case report and review of the literature, *Am. J. Gastroenterol.*, 1998, **93**, 459.

74. S. L. Iverson and J. P. Uetrecht, Identification of a reactive metabolite of terbinafine: insights into terbinafine-induced hepatotoxicity, *Chem. Res. Toxicol.*, 2001, **14**, 175.

75. N. Somchit, L. T. Wade, L. Ramsay, R. D. Goldin, J. G. Kenna and J. Caldwell, Hepatotoxicity and hepatic protein adduct formation in rats dosed ip with diclofenac, *Hum. Exp. Toxicol.*, 1997, **16**, 401.

76. S. J. Hargus, B. M. Martin, J. W. George and L. R. Pohl, Covalent modification of rat liver dipeptidyl peptidase IV (CD26) by the non-steroidal anti-inflammatory drug diclofenac, *Chem. Res. Toxicol.*, 1995, **8**, 993.

77. H. Yamazaki, A. Shibata, M. Suzuki, M. Nakajima, N. Shimada, F. P. Guengerich and T. Yokoi, Oxidation of troglitazone to a quinone-type metabolite catalyzed by cytochrome P-450 2C8 and P-450 3A4 in human liver microsomes, *Drug Metab. Dispos.*, 1999, **27**, 1260.

78. K. Kassahun, P. G. Pearson, W. Tang, I. McIntosh, K. Leung, C. Elmore, D. Dean, R. Wang, G. Doss and T. A. Baillie, Studies on the metabolism of troglitazone to reactive intermediates in vitro and in vivo. Evidence for novel biotransformation pathways involving quinone methide formation and thiazolidinedione ring scission, *Chem. Res. Toxicol.*, 2001, **14**, 62.

79. C. Funk, M. Pantze, L. Jehle, C. Ponelle, G. Scheuermann, M. Lazendic and R. Gasser, Troglitazone-induced intrahepatic cholestasis by an interference with the hepatobiliary export of bile acids in male and female rats. Correlation with the gender difference in troglitazone sulfate formation and the inhibition of the canalicular bile salt export pump (Bsep) by troglitazone and troglitazone sulfate, *Toxicology*, 2001, **167**, 83.

80. W. Honma, M. Shimada, H. Sasano, S. Ozawa, M. Miyata, K. Nagata, T. Ikeda and Y. Yamazoe, Phenol sulfotransferase, ST1A3, as the main enzyme catalyzing sulfation of troglitazone in human liver, *Drug Metab. Dispos.*, 2002, **30**, 944.

81. H. Olson, G. Betton, D. Robinson, K. Thomas, A. Monro, G. Kolaja, P. Lilly, J. Sanders, G. Sipes, W. Bracken, M. Dorato, K. van Deun, P. Smith, B. Berger and A. Heller, Concordance of the toxicity of pharmaceuticals in humans and in animals, *Regul. Toxicol. Pharmacol.*, 2000, **32**, 56.

82. P. Greaves, A. Williams and M. Eve, First dose of potential new medicines to humans: how animals help, *Nature Rev. Drug Discov.*, 2004, **3**, 226.

83. J. A. Hanley and A. Lippman-Hand, If nothing goes wrong, is everything all right? Interpreting zero numerators, *J. Am. Med. Assoc.*, 1983, **249**, 1743.

84. E. Eypasch, R. Lefering, C. K. Kum and H. Troidl, Probability of adverse events that have not yet occurred: a statistical reminder, *BMJ*, 1995, **311**, 619.

85. U. A. Boelsterli and C.-J. J. Hsiao, The heterozygous Sod2(+/ −) mouse: modeling the mitochondrial role in drug toxicity, *Drug Dis. Today.*, 2008, **13**, 982.

86. W. Tang and A. Y. H. Lu, Metabolic bioactivation and drug-related adverse effects: current status and future directions from a pharmaceutical research perspective, *Drug Metab. Rev.*, 2010, **42**, 225.

87. P. Preziosi, Isoniazid: metabolic aspects and toxicological correlates, *Curr. Drug Metab.*, 2007, **8**, 839.

88. C. M. Nolan, S. V. Goldgerg and S. E. Ruskin, Hepatotoxicity associated with isoniazid preventive therapy: a 7-year survey from a public health tuberculosis clinic, *J. Am. Med. Assoc.*, 1999, **281**, 1014.

89. D. E. Amacher, A toxicologist's guide to biomarkers of hepatic response, *Human Exp. Toxicology*, 2002, **21**, 253.

90. J. Ozer, M. Ratner, M. Shaw, W. Bailey and S. Schomaker, The current state of serum biomarkers of hepatotoxicity, *Toxicology*, 2008, **245**, 194.

91. J. K. Nicholson, J. Connelly, J. C. Lindon and E. Holmes, Metabonomics: a platform for studying drug toxicity and gene function, *Nat. Rev. Drug Discov.*, 2002, **1**, 153.

92. R. Kaddurah-Daouk, B. S. Kristal and R. M. Weinshilboum, Metabolomics: a global biochemical approach to drug response and disease, *Ann. Rev. Pharmacol. Toxicol.*, 2008, **48**, 653.

93. C. Chen, F. J. Gonzalez and J. R. Idle, LC-MS-based metabolomics in drug metabolism, *Drug Metab. Rev.*, 2007, **39**, 581.

94. E. Y. Xu, W. H. Schaefer and Q. Xu, Metabolomics in pharmaceutical research and development: metabolites, mechanisms and pathways, *Curr. Opin. Drug Discov. Dev.*, 2009, **12**, 40.

95. R. D. Beger, J. Sun and L. K. Schnackenberg, Metabolomics approaches for discovering biomarkers of drug-induced hepatotoxicity and nephrotoxicity, *Toxicol. Appl. Pharmacol.*, 2010, **243**, 154.

96. R. S. Plumb, C. L. Stumpf, J. H. Granger, J. Castro-Perez, J. N. Haselden and G. J. Dear, Use of liquid chromatography/time-of-flight mass spectrometry and multivariate statistical analysis shows promise for the detection of drug metabolites in biological fluids, *Rapid Commun. Mass Spectrom.*, 2003, **17**, 2632.

97. E. J. Want, A. Nordstrom, H. Morita and G. Siuzdak, From exogenous to endogenous: the inevitable imprint of mass spectrometry in metabolomics, *J. Proteome Res.*, 2007, **6**, 459.

98. W. Tang and R. R. Miller, In Vitro Drug Metabolism: Thiol Conjugation, in *Methods in Pharmacology and Toxicology Optimization in Drug*

Discovery: In Vitro Methods, ed. Z. Yan and G. W. Caldwell, Humana Press, Totowa, NJ, 2005, p. 369.

99. Z. Zhang and J. Gan, Protocols for Assessment of In vitro and In vivo Bioactivation Potential of Drug Candidates in *Drug Metabolism in Drug Design and Development: Basic Concepts and Practice*, ed. D. Zhang, M. Zhu and W. G. Humphreys, John Wiley & Sons, Hoboken, NJ, 2008, p. 447.

100. Y. Li, J. G. Slatter, Z. Zhang, Y. Li, G. A. Doss, M. P. Braun, R. A. Stearns, D. C. Dean, T. A. Baillie and W. Tang, In vitro metabolic activation of lumiracoxib in rat and human liver preparations, *Drug Metab. Dispos.*, 2008, **36**, 469.

101. A. Liu, Y. Chen, Z. Yang, Y. Feng, W. Rui, W. Luo, Y. Liu, F. J. Gonzalez and R. Dai, New metabolites of fenofibrate in Sprague-Dawley rats by UPLC-ESI-QTOF-MS-based metabolomics coupled with LC-MS/MS, *Xenobiotica*, 2009, **39**, 345.

102. A. Liu, A. D. Patterson, Z. Yang, X. Zhang, W. Liu, F. Qiu, H. Sun, K. W. Krausz, J. R. Idle, F. J. Gonzalez and R. Dai, Fenofibrate metabolism in the cynomolgus monkey using ultraperformance liquid chromatography-quadrupole time-of-flight mass spectrometry-based metabolomics, *Drug Metab. Dispos.*, 2009, **37**, 1157.

103. J. Sun, L. K. Schnackenberg, R. D. Holland, T. C Schmitt, G. H. Cantor, Y. P. Dragan and R. D. Beger, Metabonomics evaluation of urine from rats given acute and chronic doses of acetaminophen using NMR and UPLC/MS, *J. Chromatogr. B*, 2008, **871**, 328.

104. N. J. Waters, C. J. Waterfield, R. D. Farrant, E. Holmes and J. K. Nicholson, Integrated metabonomic analysis of bromobenzene-induced hepatotoxicity: novel induction of 5-oxoprolinosis, *J. Proteome Res.*, 2006, **5**, 1448.

105. C. Lu, Y. Wang, Z. Sheng, G. Liu, Z. Fu, J. Zhao, J. Zhao, X. Yan, B. Zhu and S. Peng, NMR-based metabonomic analysis of the hepatotoxicity induced by combined exposure to PCBs and TCDD in rats, *Toxicol. Appl. Pharmacol.*, 2010, **248**, 178.

106. M. Coen, E. M. Lenz, J. K. Nicholson, I. D. Wilson, F. Pognan and J. C. Lindon, An integrated metabonomic investigation of acetaminophen toxicity in the mouse using NMR spectroscopy, *Chem. Res. Toxicol.*, 2003, **16**, 295.

107. M. Coen, S. U. Ruepp, J. C. Lindon, J. K. Nicholson, F. Pognan, E. M. Lenz and I. D. Wilson, Integrated application of transcriptomics and metabonomics yields new insight into the toxicity due to paracetamol in the mouse, *J. Pharm. Biomed. Anal.*, 2004, **35**, 93.

108. S. U. Ruepp, R. P. Tonge, J. Shaw, N. Wallis and F. Pognan, Genomics and proteomics analysis of acetaminophen toxicity in mouse liver, *Toxicol. Sci.*, 2002, **65**, 135.

109. C. Chen, K. W. Krausz, J. R. Idle and F. J. Gonzalez, Identification of novel toxicity-associated metabolites by metabolomics and mass isotopomer analysis of acetaminophen metabolism in wild-type and Cyp2e1-null mice, *J. Biol. Chem.*, 2007, **283**, 4543.

110. J. Cheng, X. Ma, K. W. Krausz, J. R. Idle and F. J. Gonzalez, Rifampicin-activated human pregnane X receptor and CYP3A4 induction enhance acetaminophen-induced toxicity, *Drug Metab. Dispos.*, 2009, **37**, 1611.
111. T. Soga, R. Baran, M. Suematsu, Y. Ueno, S. Ikeda, T. Sakurakawa, Y. Kakazu, T. Ishikawa, M. Robert, T. Nishioka and M. Tomita, Differential metabolomics reveals ophthalmic acid as an oxidative stress biomarker indicating hepatic glutathione consumption, *J. Biol. Chem.*, 2006, **281**, 16768.
112. H. S. Buttar, A. Y. Chow and R. H. Downie, Glutathione alterations in rat liver after acute and subacute oral administration of paracetamol, *Clin. Exp. Pharmacol. Physiol.*, 1977, **4**, 1.
113. C. Chen, K. W. Krausz, Y. M. Shah, J. R. Idle and F. J. Gonzalez, Serum metabolomics reveals irreversible inhibition of fatty acid beta-oxidation through the suppression of PPARalpha activation as a contributing mechanism of acetaminophen-induced hepatotoxicity, *Chem. Res. Toxicol.*, 2009, **22**, 699.
114. R. A. Nathwani, S. Pais, T. B. Reynolds and N. Kaplowitz, Serum alanine aminotransferase in skeletal muscle diseases, *Hepatology*, 2005, **41**, 380.
115. P. Lindblom, I. Rafter, C. Copley, U. Andersson, J. J. Hedberg, A. L. Berg, A. Samuelsson, H. Hellmold, I. Cotgreave and B. Glinghammar, Isoforms of alanine aminotransferases in human tissues and serum–differential tissue expression using novel antibodies, *Arch. Biochem. Biophys.*, 2007, **466**, 66.
116. S. B. Jadhao, R. Z. Yang, Q. Lin, H. Hu, F. A. Anania, A. R. Shuldiner and D. W. Gong, Murine alanine aminotransferase: cDNA cloning, functional expression, and differential gene regulation in mouse fatty liver, *Hepatology*, 2004, **39**, 1297.
117. R. J. Mortishire-Smith, G. L. Skiles, J. W. Lawrence, S. Spence, S. A. W. Nicholls, B. A. Johnson and J. K. Nicholson, Use of metabonomics to identify impaired fatty acid metabolism as the mechanism of a drug-induced toxicity, *Chem. Res. Toxicol.*, 2004, **17**, 165.
118. M. S. Lee, B. H. Jung, B. C. Chung, S. H. Cho, K. Y. Kim, O. S. Kwon, B. Nugraha and Y.-J. Lee, Metabolomics study with gas chromatography-mass spectrometry for predicting valproic acid-induced hepatotoxicity and discovery of novel biomarkers in rat urine, *Inter. J. Toxicol.*, 2009, **28**, 392.
119. K. H. Schulpis, C. Lazaropoulou, S. Regoutas, G. A. Karikas, A. Margeli, S. Tsakiris and I. Papassotiriou, Valproic acid monotherapy induces DNA oxidative damage, *Toxicology*, 2006, **217**, 228.
120. U. Andersson, J. Lindberg, S. Wang, R. Balasubramanian, M. Marcusson-Stahl, M. Hannula, C. Zeng, P. J. Juhasz, J. Kolmert, J. Backstrom, L. Nord, K. Nilsson, S. Martin, B. Glinghammar, K. Cederbrant and I. Schuppe-Koistinen, A systems biology approach to understanding elevated serum alanine transaminase levels in a clinical trial with ximelagatran, *Biomarkers*, 2009, **14**, 572.

121. W. Sha, K.-A. da Costa, L. M. Fischer, M. V. Milburn, K. A. Lawton, A. Berger, W. Jia and S. H. Zeisel, Metabolomic profiling can predict which humans will develop liver dysfunction when deprived of dietary choline, *FASEB J.*, 2010, **24**, 2962.
122. T. A. Clayton, J. C. Lindon, O. Cloarec, H. Antti, C. Charuel, G. Hanton, J. P. Provost, J. L. Le Net, D. Baker, R. J. Walley, J. R. Everett and J. K. Nicholson, Pharmaco-metabonomic phenotyping and personalized drug treatment, *Nature*, 2006, **440**, 1073.
123. J. H. Winnike, Z. Li, F. A. Wright, J. M. Macdonald, T. M. O'Connell and P. B. Watkins, Use of pharmaco-metabonomics for early prediction of acetaminophen-induced hepatotoxicity in humans, *Clin. Pharmacol. Ther.*, 2010, **88**, 45.
124. E. Holmes, I. D. Wilson and J. K. Nicholson, Metabolic phenotyping in health and disease, *Cell*, 2008, **134**, 714.
125. G. Shinar and M. Feinberg, Structural sources of robustness in biochemical reaction networks, *Science*, 2010, **327**, 1389.
126. N. Shimizu, S. Yamaguchi and T. Orii, A study of urinary metabolites in patients with dicarboxylic aciduria for differential diagnosis, *Acta Paediatr. Jpn*, 1994, **36**, 139.
127. A. E. Vickers, Characterization of hepatic mitochondrial injury induced by fatty acid oxidation inhibitors, *Toxicol. Pathol.*, 2009, **37**, 78.
128. D. A. Schoeller, Uses of stable isotopes in the assessment of nutrient status and metabolism, *Food Nutr. Bull.*, 2002, **23**, 17.
129. S. Hu, A. P. Chen, M. L. Zierhut, R. Bok, Y. F. Yen, M. A. Schroeder, R. E. Hurd, S. J. Nelson, J. Kurhanewicz and D. B. Vigneron, In vivo carbon-13 dynamic MRS and MRSI of normal and fasted rat liver with hyperpolarized 13C-pyruvate, *Mol. Imaging Biol.*, 2009, **11**, 399.
130. L. Liu, J. Aa, G. Wang, B. Yan, Y. Zhang, X. Wang, C. Zhao, B. Cao, J. Shi, M. Li, T. Zheng, Y. Zheng, G. Hao, F. Zhou, J. Sun and Z. Wu, Differences in metabolite profile between blood plasma and serum, *Anal. Biochem.*, 2010, **406**, 105.
131. H. G. Gika, G. A. Theodoridis, M. Earll, R. W. Snyder, S. J. Sumner and I. D. Wilson, Does the mass spectrometer define the marker? A comparison of global metabolite profiling data generated simultaneously via UPLC-MS on two different mass spectrometers, *Anal. Chem.*, 2010, **82**, 8226.

Chemogenomics

VIRENDRA S. GOMASE,* AKSHAY N. PARUNDEKAR AND
ARCHANA B. KHADE

Department of Bioinformatics, Padmashree Dr. D. Y. Patil University,
Plot No-50, Sector-15, CBD Belapur, Navi Mumbai, 400614 (MS), India
*Email: virusgene1@yahoo.co.in

12.1   Introduction

In the post-genomic era, one of the key challenges for drug discovery is
making optimal use of the comprehensive genomic data available after the
elucidation of the human genome and others in order to identify effective
new medicines. To overcome this challenge, chemogenomics aims to identify systematically all ligands and modulators for all the gene products
expressed and allow the accelerated exploration of their biological function.
Chemogenomics aims towards the systematic identification of small molecules that interact with the products of the genome and to modulate their
biological function and the different knowledge-based strategies which are
followed, and outlines the challenges and opportunities that will impact
drug discovery.[1,2] The subject brings together diverse disciplines including
chemistry, genetics, chemo- and bioinformatics, structural biology and
biological screening in phenotypic and target-based assays.

Chemogenomics is defined as, in principle, the screening of the chemical
universe, *i.e.* all possible chemical compounds, against the target universe,
i.e. all proteins and their potential drug targets. The systematic screening of
libraries of congeneric compounds against members of a target family offers
unprecedented chances in the search for compounds with significant target
or sub-type specificity. Chemogenomics is a new strategy in drug discovery

Issues in Toxicology No. 21
Metabolic Profiling: Disease and Xenobiotics
Edited by Martin Grootveld

Published by the Royal Society of Chemistry, www.rsc.org

which, in principle, searches for all molecules which are capable of interacting with any biological target. Therefore, chemogenomics has been defined as the investigation of classes of compounds, *i.e.* libraries, against families of functionally related proteins. It deals with the systematic analysis of chemical-biological interactions. Whilst historically the approach has been based on efforts that systematically explore target gene families such as kinases, today additional knowledge-based systematisation principles are followed within early drug discovery projects, which aim to biologically validate the targets and identify starting points for chemical lead optimisation. Whilst the expectations of chemogenomics are very high, the reality of drug discovery remains quite sobering with a very high level of attrition in projects. This article summarises the different knowledge-based chemogenomics strategies that are followed and also outlines the challenges and potential opportunities that will impact drug discovery.[3–7] The most widely used definition of chemogenomics refers to the perturbation of biological systems by small molecules, thus gaining a holistic understanding of the interactions of such molecules with complex molecular systems. In this context, chemogenomics represents a sub-set of genomics in which the focus is on small molecules. Congeneric series of chemical analogues serve as probes to investigate their action on specific target classes, *e.g.* GPCRs, kinases, phosphodiesterases, ion channels, serine proteases and others. Such a strategy was developed in the pharmaceutical industry almost 20 years ago, and it is now more systematically applied in the search for target- and sub-type-specific ligands. The term 'privileged structures' has been defined for scaffolds, such as the benzodiazepines, which very often produce biologically active analogues in a target family, in this case in the class of G-protein-coupled receptors. The SOSA approach is a strategy utilised to modify the selectivity of biologically active compounds, generating new drug candidates from the side activities of therapeutically employed drugs.[8–11]

12.2 Privileged Structures

Many drugs have been derived from certain chemotypes, *e.g.* phenethylamines, tricyclics, steroids or benzodipines. However, others have certain common structural features *e.g.* diphenylamine or arylpiperazine groups, *e.g.* systematic variation of GABA agonist diazepam-1 produced not only tranquilisers, but also GABA antagonists, inverse agonists and the strong kappa-opiate receptor against trifluadom. However, certain privileged structures are capable of providing useful ligands for more than one receptor. Indeed, judicious modification of such structures could provide viable alternatives in the search for new agonists and antagonists. Matrix metalloproteinase (MMP) inhibitors have been pursued as clinical candidates since the first drug discovery program targeting this enzyme family began in the late 1970s. Targeted indications for them included cancer, arthritis, cardiovascular diseases and many others. However, the clinical development of most of the MMP inhibitors have been discontinued for

safety reasons, and so far only doxycycline hyclate, a non-specific MMP inhibitor, has been approved for periodontal diseases. Since they are of high
therapeutic potential, the development of MMP inhibitors continues (as
shown by several recent patents and research publications). Moreover, the
development of selective MMP inhibitors lacking serious side-effects such as
musculoskeletal syndrome is of high importance. Innovative approaches for
the design of selective MMP inhibitors include the integration of classical
medicinal chemistry structure-based properties, and also design features
into the emerging chemogenomics concept of target-family based drug
discovery. The research work approach, which includes privileged structures,
molecular frameworks, bioisosteric and bioanalogous-isofunctional modifications (known as the matrixinome approach), may lead to highly selective
MMP inhibitors in the future.[12–16]

G-protein coupled receptors (GPCRs) are promising targets for the
discovery of novel drugs. In order to identify novel chemical series,
high-throughput screening (HTS) is often complemented by rational chemogenomics lead-finding approaches. These sets of compounds are supplemented with novel libraries synthesised around proprietary scaffolds.
Such target-directed libraries are designed using the knowledge of privileged
fragments and pharmacophores to address specific GPCR sub-families, *e.g.*
chemokine-binding GPCRs. Research testing of the GPCR collection has
provided a novel chemical series for several GPCR targets including the
adenosine A1, the P2Y12 and the chemokine CCR1 receptors. Moreover,
GPCR sequence motifs linked to the recognition of GPCR ligands, *i.e.* chemoprints, are identified using homology modelling, molecular docking and
experimental profiling. These chemoprints can support the design and
synthesis of compound libraries tailor-made for a novel GPCR target.[9,17–19]

12.3 Drugs Arising from the Side-effects – SOSA Approach

Many drugs of the past have resulted from the experimental or clinical observation of side-effects. Diuretic, anti-hypertonic, anti-glaucoma and anti-diabetic
drugs were derived from the bacteriostatic sulfonamides; interestingly, the
mood-improving effect of iproniazid was discovered when it was tested as an
anti-tuberculous drug. Camille Wermuth[76] proposed selective optimisation
of this side-activities approach. Hence, whenever a side-effect of a drug is
observed, it might be possible to optimise the candidate to a selective
analogue with this further biological activity. For example, Wermuth demonstrated by his own research the optimisation of differing weak side-effects
of the anti-depressant minaprine to the nanomolar muscarinic M1 receptor
ligand and the reversible acetylcholinesterase inhibitor. A closely related
analogue of minaprine was optimised to the nanomolar 5-HT. Chemogenomics is mainly based on the 'master-key' concept of tailor-made privileged
structures. Starting from such 'master-keys', selective ligands can be derived,

either by classical medicinal chemistry, or alternatively by systematic structural variation in combinatorial libraries. For example, selective β1 and β2 agonists, as well as β-antagonists (β-blockers) derived from the mixed alpha/beta agonist epinephrine.[17–20]

12.4 Classification of Chemogenomics

Chemogenomics can be classified into three categories:

1. *Ligand-based Chemogenomics*: techniques pool together targets at the level of families or sub-families and a model for ligands is 'learned' at the level of the family. Such strategies could be facilitated by the design of libraries containing annotated ligands. The basic paradigm underlying ligand-based chemogenomic approaches is that molecules sharing enough similarity to existing biologically annotated ligands have an enhanced probability to share the same biological profile. Indeed, Novartis scientists linked chemical space to target space by merging fields from separate chemical and biological databases to provide a unified and searchable chemogenomic database.[21,22] The target's sequence is linkable to the ligand, and hence sequence-based similarity searches of ligands using the chemogenomic database for protein homologues of liganded 'targets' are feasible.[22] For proteins, spatial structure is more conserved in evolution than the primary amino acid sequence. The ligand-sensing cores of individual protein domains are grouped on the basis of structural similarities, and are irrespective of sequence similarities in order to generate a protein structure similarity cluster (PSSC). The structures of ligands that bind to one member of this cluster may be used for the development of novel ligands for other members of that particular cluster. Natural product-derived compound libraries are expected to yield comparatively high 'hit rates' at small library sizes.[23] The concept of ligand-based chemogenomics is slowly re-emerging nowadays, since it gives the opportunity to evaluate novel phenotypes and allows the discovery of novel targets that are currently poorly understood.
2. *Target-based Chemogenomics*: These methods approach cluster receptors based on ligand-binding site similarity, and again pool together known ligands for each cluster in order to infer shared ligands.[22] Target based-chemogenomics is classified into two types: sequence-based approaches that are intended to be employed for any class of target family, provided that a multiple alignment of all targets to compare is achievable. After the alignment of all sequences, key residues which are supposed to 'map' the binding-site of most non-peptide ligands can be extracted and concatenated into an ungapped sequence of a few residues, which can be later used to derive a distance matrix based on sequence identity, sequence similarity or physicochemical properties. Through cavity-based clustering techniques, this approach is applied in

'target-hopping', which involves the discovery of ligands for a particular receptor *via* a consideration of firstly the known ligands of closely related receptors. The second type, *i.e.* structure-based approaches, are used for target families with well-validated template structures wherein only ligand binding sites of related targets are compared. These structure-based comparisons can be performed using the following strategies:[22]

a) Comparisons of computed molecular interaction fields from the cavities. This is performed using the sc-pdb database.[24] It involves the use of grid-mapped knowledge-based potentials in order to rapidly 'cluster' proteins into sub-families according to similarities in the hydrophobic and polar fields of their ligand-binding sites. Regions of the binding site which are common within a protein family are then identified and analysed for the design of family-targeted libraries, or, alternatively, those which differ for the improvement of ligand selectivity.[25]

b) Comparisons of 3D protein co-ordinates to measure a distance between two 'targets'. Recently developed, effective methods represent an active-site of interest by pseudocentres (dummy atoms located along or close to every side-chain of interest), encoding physico-chemical properties such as H-bonding capacity, aromaticity, *etc.* of their cognate residues, and then the pseudocentres are linked together *via* edges and thus define a molecular 'graph'.[22]

c) Comparisons of proteins of the same family by the examination of packing defects. These packing defects are localised at the dehydrons (back-bone heavy atoms with unsatisfied H-bonding partners), which are good indicators of a protein's capacity to interact with potential ligands, and are predictable from the amino-acid sequences involved.[22]

3. *Target Ligand-based Chemogenomics:* This approach attempts to predict ligands for a given 'target' by leveraging binding information for other targets in a single step, that is, without first attempting to define a particular set of similar receptors.[22] Lapinsh *et al.* used a proteo-chemometrics strategy for the analysis of interactions of a range of proteins with series of ligands. They used this strategy for the modelling of the interaction of psycho-active organic amines with all the five known families of amine G protein-coupled receptors.[26] Erhan *et al.* suggested the use of collaborative filtering on a family of biological targets. Collaborative filtering techniques build predictive models that link multiple targets to multiple examples. Clearly, the greater the number of commonalities between the targets, the better the multi-target model built therefrom.[27]

12.5 Chemogenomics Screens

This approach was pioneered by Paul Elhrich and created a one-dimensional (1D) screen for investigating the action of 606 compounds against

Trepanoma palladium. In 2D screens, the first dimension is a chemical library, and the second dimension is a library of different cell types (*e.g.* yeast deletion strains, cancer cell lines, *etc.*) The resulting data structure is a two-dimensional matrix in which each data point has two coordinates (one chemical and the other genetic), and one specific associated value. The value of each data point represents a measurement of the phenotype of interest, such as viability, growth-rate or cell size and shape. Differing small-molecule libraries can also be used for chemogenomic screens. Two fundamentally different approaches for the design of small molecule libraries are commonly employed. One approach uses small-molecule libraries that show as much chemical diversity as possible, whilst the other draws its small molecule library from only a small fraction of the defined chemical space. Here, the range of small molecules is limited by a choice of those compounds that are likely to have some form of biological activity. For data interpretation and analysis, clustering methods are used, and hence the action(s) of untested molecules can be predicted using the available cluster dataset. The immediate purpose of a chemogenomic screen is to characterise the effect that a set of small molecules has at the gene or protein level. From a biotechnological point of view, such chemogenomic data can allow for the identification of proteins as novel drug targets.[28]

12.6 Haploinsufficiency Profiling

Haploinsufficiency is defined as a dominant phenotype in diploid organisms that are heterozygous for a loss-of-function allele.[29] Haploinsufficiency profiling (HIP) in *Saccharomyces cerevisiae* is commonly employed to identify genes that, when deleted, confer sensitivity on small molecules *in vivo*.[30] In haploinsufficiency profiling, lowering of the gene dosage from two copies (diploid yeast strain) to one copy (heterozygous deletion strain) results in a strain that is sensitised to compounds that inhibit the product of the heterozygous locus. Giaever *et al.* found that for several well-characterised compounds, some of the most sensitive heterozygous strains often carry a deletion in the gene whose product is known to interact with the test molecule. This occurs in view of the compound possibly inhibiting cellular proliferation *via* a reduction in the activity of the remaining gene product of the heterozygous locus, thereby mimicking a complete deletion. However, only ~3% of the *S. cerevisiae* genome displays HI under the standard growth conditions.[29,31]

Gene deletions render cells hypersensitive to specific drug identification pathways that 'buffer' the cell against the toxic effects of the drug, and thereby provide clues regarding both gene and compound functions. Moreover, compounds that show similar chemical-genetic profiles often perturb similar target pathways. Notwithstanding, gene dosage can be exploited to discover connections between compounds and their targets. Thus, HIP allows us to detect putative molecular mechanisms underlying the action of particular drugs.[30,32] Deming Xu *et al.* applied chemically induced HI on *Candida albicans* on a genomic scale (*C. albicans* fitness test or CaFT). The genes selected for the construction of heterozygous deletion strains in this pilot study were those which had orthologues in fungi and/or in higher eukaryotes. The CaFT strains were screened against anti-fungal agents. The MOA (Mechanism of Action) information obtained from CaFT and ScFT (*S. cerevisiae*) was extrapolated to a group of chemically related active compounds with unknown MOAs. These drugs were shown to inhibit growth by affecting microtubule function.[33] Many of the haploinsufficient mutations in humans are observed in transcription factors, including *TWIST* and *GATA3*.[31] Haploinsufficiency has also been implicated in cancer; in a minority of tumour suppressor genes such as SMAD-4, mutation in one of the two alleles is sufficient to initiate tumorigenesis.[34]

12.7 High-content Screening

High-content screening has emerged as a new and powerful technique for identifying small-molecule modulators of mammalian cell biology. The use of automated microscopes combined with digital imaging, machine learning and other analytical tools has enabled high-content screening (HCS) in a variety of experimental systems. A complementary set of biological descriptors that can be employed to monitor the effects of compounds on living cells are the read-outs of high-content screening (HCS) campaigns. HCS combines automated fluorescence microscopy with state-of-the-art image processing and quantification in order to generate a biological 'fingerprint' that is based on the quantity, activity and organisation of biomolecules within the spatial context of cellular milieu.[35–37] For the analysis of HCS data, recently integrated data-mining tools applicable to building biological networks have been developed by GeneGo, which are similar to those of MetaCore and MetaDrug.[38] In a screen employed by Wilson *et al.* to identify a molecule that induces mitotic arrest in a simple DNA stain (DAPI), a sensitive non-parametric statistical test was employed to identify compounds from an internal collection of approximately 13 000 high-quality lead-like small molecules. They identified an active compound, which is a quinazolinone originating from a natural product-like sub-set of the screened compounds, active in cells at concentrations of approximately 500 nM; it probably acts by inhibiting the polymerisation of tubulin.[35] Cgoi *et al.* performed a high-content screening of small molecules targeting the thioredoxin redox system, and identified glitoxin, a fungal metabolite with powerful antioxidant properties.[39] Perlman *et al.* screened a library of compounds of known activity, and they found unexpected effects on centrosome duplication by a number of drugs. From a 16 320-member library of uncharacterised small molecules, we identified five potent centrosome-duplication inhibitors that do not target microtubule dynamics or protein synthesis.[40] Mitchison and colleagues.[37] quantified more than 90 cytological features in Hela cells that described nuclear and cytoskeletal structure, signalling pathway activity and transcription factor localisation at a single-cell level using fluorescent stains and indirect immunofluorescence.

12.8 Mode of Action by Network Identification

Mode of action by network identification (MNI) does not require libraries of genetic mutants or fitness-based assays of drug response. In this approach (given by Bernandol and Thompson[41]), a network model of regulatory interactions in the organism of interest was reverse-engineered using a training dataset of whole-genome expression profiles. The model was then used to analyse the expression profile of compound-treated cells to determine the pathways and genes targeted by the compound. The reverse-engineered model was a 'directed graph' relating the concentrations of transcripts to each other (an 'edge' in the graph indicated that the activity of one gene product influences the transcription of another[41]). The MNI algorithm given by Xing and Gardener[42] uses a training dataset of hundreds of expression profiles in order to construct a statistical model of gene-regulatory networks in a cell or tissue. This model describes combinatorial influences of genes on one another. The algorithm then uses the model to filter the expression profile of a particular experimental treatment and thereby distinguish the molecular targets or mediators of the treatment response from hundreds of additional genes that also exhibit expression changes.[42] Iorio *et al.*[43] built a drug-similarity network starting from a public reference dataset containing genome-wide gene expression profiles (GEPs) following treatments with more than a thousand compounds. In this network, drugs sharing a sub-set of molecular targets are connected by an 'edge' or 'lie-in' the same community. The approach is based on a novel similarity distance between two compounds, and the distance is computed by combining GEPs *via* an original rank-aggregation method, followed by a gene-set enrichment analysis (GSEA) to compute similarities between pairs of drugs. The network is obtained by considering each compound as a node, and adding an 'edge' between two compounds if their similarity distance is below a given significance threshold.[43] Lauria *et al.* developed NIRest, a tool for gene network and mode of action inference. NIRest is Network Identification by multiple Regressions with a perturbation Estimate, and the approach behind the tool is based on ordinary differential equations (ODEs) model of the network, and on an assumption of linearity based on an equilibrium point of the cell machinery.[44]

12.9 Current Research in Chemogenomics

The advent of chemogenomics and its subsequent development has facilitated its application in other 'omics' fields, and has also added a new dimension to drug discovery research.

12.10 Bioinformatics

Annotated chemogenomics databases integrate chemical and biological domains and can provide a powerful tool to predict and validate new targets for compounds with unknown effects. The MDL Drug Data Report (MDDR) (Molecular Design Ltd, San Leandro, California) is one of the well-known and widely used databases that contains chemical structures and corresponding biological activities of drug-like compounds. Many biological activities reported in MDDR are very generic, *e.g.* anti-neoplastic, anti-hypertensive and anti-inflammatory. It covers information available in patent literature, journal articles, research conferences and meetings. (MDDR is updated annually.) WOMBAT (World of Molecular Bioactivity) is another consistently annotated chemogenomics database. This database follows a hierarchical scheme which allows one to seek the target family. The target families are based on the functional properties of targets. Targets are usually proteins, but can also be DNA or RNA. Moreover, WOMBAT is organised in a manner so that one compound may have more than one target.[45] The Comparative Toxicogenomics Database (CTD) is a curated database that promotes an understanding of the effects of environmental chemicals on human health. Biocurators at CTD manually curate chemical-gene interactions, chemical-disease relationships and gene-disease relationships from the literature. Over 350 000 gene-disease and 77 000 chemical-disease relationships can be inferred from this database.[46]

12.11 Kinase Activity

Protein kinases represent a large family of enzymes involved in regulating complex molecular machineries that control many cellular functions, from survival and proliferation to apoptosis. Abnormal protein kinase activity has been implicated in a variety of pathophysiologic states, including cancer, inflammatory and autoimmune disorders and cardiac diseases. Indeed, protein kinases have become one of the major therapeutical targets of the past ten years. The major problem associated with ATP-competitive kinase inhibition is target specificity, since many other enzymes, kinases and non-kinases alike all utilise ATP.[47]

Currently, structural information is available for relatively few of the protein kinases encoded in the human genome (7% of the estimated 518). Chemogenomics attempts to combine genomic and structural biological data, classical dendrograms and selectivity data to explore, define and classify the medicinally relevant kinase space. Exploitation of this information in the discovery of kinase inhibitors defines practical kinase chemogenomics (kinomics). Vieth and Higgs[48] presented the first dendrogram of kinases based entirely on small-molecule selectivity data. They found that the selectivity dendrogram differs from sequence-based clustering, mostly in the higher-level groupings of the smaller clusters, and remains very comparable for closely homologous targets. Highly homologous kinases are, on average, inhibited comparably by small molecules. This observation is very important to the process of target selection, since we would expect difficulty in achieving inhibitor selectivity for kinases that share a high sequence identity.[48] The Cyclic-AMP Response Element Binding (CREB) proteins (a family of transcription factors) plays an important role in learning and long-term memory (LTM) formation by coupling neuronal activity with changes in gene expression. CREB signalling occurs through activation of protein kinase A by cAMP, the event mediated by the action of neurotransmitter on a G-protein coupled receptor. Menghang Xia *et al.* screened 73 000 compounds for CREB enhancer activity using a cell-based CRE-β-lactamase reporter gene assay in qHTS mode. A structure-activity relationship (SAR) analysis on the active compounds was performed based on information obtained from qHTS. The enhancers of CREB screened from the 73 000 compounds will be useful for the study of long-term memory, and potentially lead to new clinical 'memory enhancers' for widely prevalent disorders such as Alzheimer's disease.[49]

12.12 Oncology

Cancer is a disease of genes. Multiple mutations get accumulated over a number of generations of a cell type, and this transforms a normal cell to a malignant one. Somatic rearrangements of transcription factors are common abnormalities in the acute leukaemias. With rare exception, however, the resultant protein products have remained largely intractable as pharmacological targets. In order to identify AML1-ETO modulators, Corsello *et al.* screened a small-molecule library using a chemical genomic approach. Gene expression signatures were used as surrogates for the expression *versus* loss of the translocation in AML1-ETO-expressing cells. The top classes of compounds that scored in this screen were corticosteroids and dihydrofolate reductase (DHFR) inhibitors. This work suggests a role for DHFR inhibitors and corticosteroids in treating patients with AML1-ETO-positive disease.[50] Tran[51] utilised a robust and high-resolution chemical genomics procedure to examine the pharmacological structure-activity relationships of dithiolethiones in the livers of male rats by microarray analyses. They identified 226 differentially expressed genes that were common to all treatments. Functional analysis identified the relationship of these genes to glutathione metabolism and the nuclear factor erythroid derived 2-related factor 2 pathway (Nrf2), which is known to regulate many of the protective actions of dithiolethiones. Thus, chemogenomics can help identify genes involved in cancer. Microtubules are a promising target for new therapeutic agents. In view of its dynamic characteristics, the microtubule cytoskeleton represents a suitable target for small molecules that rapidly diffuse in the cell cytoplasm. Chemogenomics and cell-based assays can therefore be of potential use in the discovery of new therapeutic compounds.[52]

12.13 Ligand-binding Study

Predicting interactions between small molecules and proteins represents a crucial step required to decipher many biological processes, and plays a critical role in drug discovery. When no detailed 3D structure of the protein target is available, ligand-based virtual screening allows the construction of predictive models by learning how to discriminate known, specific ligands from non-ligands. Using the chemogenomic approach one can attempt to screen the chemical space against whole families of proteins simultaneously. The lack of known ligands for a given target can then be compensated by the availability of known ligands for similar targets. This strategy has been tested on three important classes of drug targets, namely enzymes, G-protein-coupled receptors (GPCRs) and ion channels, and dramatic improvements in prediction accuracy have been reported over those of classical ligand-based virtual screening, in particular for targets with few or no known ligands.[53]

12.14 Metabolomics

Metabolomics is a newborn cousin to genomics and proteomics. Specifically, metabolomics involves the rapid, high-throughput characterisation of the small-molecule metabolites found in an organism.[54] The compound activity refers to inhibition or protein-binding measurements, but intra-cellular (or inter-cellular) signalling networks are not accounted for. This is possible by describing the effect of a compound on a cell by measuring gene expression (mRNAs) and changes in its nature arising from compound administration. One of the pioneering works in the area was performed by Covell *et al.* at the National Cancer Institute.[37]

The Connectivity Map is employed as a resource and tool to connect small-molecule drugs, genes and diseases. The main assumption behind the concept of a connectivity map is that a biological state, whether physiological, pathological or that induced with chemical or genomic perturbations, can be described in terms of a genomic signature.[55,56] Kutalik *et al.* developed a 'ping-pong' algorithm that predicts drug-gene associations. Using information from the DrugBank and the Connectivity Map databases, they proved that the 'ping-pong' algorithm predicts drug-gene associations significantly better than other methods.[57]

Ganter and Tugendreich[58] evaluated the utility of pairing clinical pathology assessments with gene expression data by using three anti-neoplastic drugs, carmustine, methotrexate and thioguanine, which had similar effects on the blood compartment, but also diverse effects on hepatotoxicity. They also demonstrated that gene expression events monitored in the liver can be used to predict pathological events occurring in that tissue, as well as in hematopoietic tissues.

12.15 Pharmacophore

The IUPAC definition of a pharmacophore is 'an ensemble of steric and electronic features that is necessary to ensure the optimal supramolecular interactions with a specific biological target and to trigger (or block) its biological response'. In modern computational chemistry, pharmacophores are used to define the essential features of one or more molecules with the same biological activity.[59] Chemogenomic approaches out-perform individual approaches, in particular in cases where very limited or no ligand information is available. Whilst 2D structures are known to be very competitive in ligand-based virtual screening for the identification of molecules presenting some given chemical, physical or biological properties, the protein-ligand recognition process takes place in the 3D space. Hence, descriptors representing the presence of potential 3-point pharmacophores are tested for the generation of ligands for GPCRs, using an *'in-silico'* chemogenomics approach. For this, a 3D pharmacophore kernel, which generalises 3D pharmacophore fingerprint descriptors, is employed.[60]

12.16   Cheminformatics

Cheminformatics is a generic term that encompasses the design, creation, organisation, management, retrieval, analysis, dissemination, visualisation and use of chemical information. Chemogenomic strategies, which involve the generation of small-molecule compounds that can be used both as tools to probe biological mechanisms, and also as leads for drug-property optimisation, provide a highly parallel, industrialised solution. A key to the success of this strategy is an integrated suite of cheminformatics applications that can permit the rapid and directed optimisation of chemical compounds with drug-like properties using 'just-in-time' combinatorial chemical synthesis. An effective embodiment of this process requires new computational and data-mining tools that cover all aspects of library generation, compound selection and experimental design, and work input on an effectively massive scale. Hence, the use of combinatorial chemistry is still a more effective approach for ligand searching.[61]

12.17   Pharmacogenomics

This form of 'omics' investigates an inherited basis for differences observed in drug response amongst individuals. Indeed, the information gleaned from such high-content molecular data has begun to augment traditional approaches to the assessment of drug safety. The optimal approach is a hybrid strategy employing chemogenomic data and gene expression-based biomarkers of drug efficacy and toxicity sought to supplement low content and insensitive methods for risk assessment and the mechanistic evaluation of drug candidates. Indeed, large reference databases of chemogenomic data are essential for the derivation and validation of accurate and predictive gene expression biomarkers. Transitional structural chemogenomics (TSCg) is employed to regulate gene expression by using ultrasensitive small-molecule drugs that target nucleic acids. By using chemicals to target transitional changes in the helical conformations of single-stranded (ss) and double-stranded (ds) DNA (*e.g.* B- to Z-DNA) and RNA (*e.g.* A- to Z-RNA), gene expression can be regulated (*i.e.* turning genes 'on or off' and variably controlling them). Alternative types of ds- and ssDNA and RNA (*e.g.* cruciform DNA), and other multistranded nucleic acids (*e.g.* triplex-DNA) are also targeted by this method.[62]

Glucocorticoids are the most effective anti-inflammatory drugs used in the treatment of chronic inflammatory diseases such as asthma. They act by binding to a specific glucocorticoid receptor that on activation translocates to the nucleus and controls the expression of responsive genes. The ability of the transcription factors AP-1 and NF-kappaB to induce gene transcription is attenuated by glucocorticoid receptors. Although only 5–10% of asthmatic subjects are glucocorticoid-insensitive, these subjects account for over 50% of the health-care costs for asthma. The development of small molecule therapies that interfere directly with AP-1 transcription may therefore be of benefit in corticosteroid-resistant airway disease. Recently, Seattle-based researchers used a chemogenomics approach to screen for small-molecule inhibitor(s) of AP-1 transcription. Using this approach, a small-molecule inhibitor (PNRI-299) that selectively inhibited AP-1 transcription without affecting NF-kappaB transcription was identified. This effect was suggested to involve the inhibition of redox factor 1 (Ref-1), a nuclear factor that regulates AP-1 transcription. PNRI-299 significantly reduced airway eosinophil infiltration, mucus hypersecretion, edema and IL-4 levels in a mouse asthma model. These data validate AP-1 as an important therapeutic target in allergic airway inflammation, and molecules such as PNRI-299 may therefore be of value in the treatment of asthma.[63]

12.18   Drug Safety

Today, the information gleaned from high-content molecular data has begun to augment traditional approaches to the assessment of drug safety. The optimal approach is a hybrid strategy employing chemogenomic data and gene expression-based biomarkers of drug efficacy and toxicity, in order to supplement low-content and insensitive methods for risk assessment and the mechanistic evaluation of drug candidates.[64]

12.19   Evaluating Complex Signalling Networks

Chemical genomics is a powerful method to complement more traditional genetic techniques (*i.e.* knockout mice, siRNA) for the dissection of complex signalling networks. A key step in Wnt activation of target genes is the nuclear translocation of beta-catenin, and the formation of a complex between it and members of the T-cell factor (TCF) family of transcription factors. Using a forward chemical genomics strategy, they identified ICG-001, a selective inhibitor of a sub-set of Wnt-beta-catenin-driven gene expression.[65]

12.20 Current Trends in Chemogenomics

12.20.1 Stem Cells

Understanding how survival is regulated in human embryonic stem cells (hESCs) could improve expansion of stem cells for the production of those for regenerative therapy. Damoiseaux[66] developed a high-content screening (HCS) approach with small molecules to examine hESC survival. These researchers identified novel small molecules that improve survival by inhibiting either Rho-kinase (ROCK) or Protein kinase C (PKC). Re-screening with stable hESCs that were genetically altered in order to improve survival enabled the identification of groups of pathway 'targets' that are important for modifying survival status.

12.20.2 Schistosomiasis

Schistosomiasis is a prevalent and chronic helminthic disease in tropical regions. Caffrey *et al.*[67] took a comparative chemogenomics approach utilising the putative proteome of *Schistosoma mansoni* compared to the proteomes of two model organisms, the nematode *Caenorhabditis elegans* and the fruit fly *Drosophila melanogaster*. They used the genome comparison software Genlight, and implemented two separate *in silico* work-flows to derive a set of parasite proteins for which gene disruption of the orthologues in both the model organisms yielded deleterious phenotypes. They identified 57 drug-responsive protein homologues, the further scrutiny of which selectively revealed 35 *S. mansnoni* sequences, which were homologous to proteins with 3D structures including co-crystallised ligands.[67]

12.20.3 Ligand-Enzyme Interaction

Strombergsson *et al.* built an interactive model based on local protein substructures generalised to the entire structural enzyme-ligand space. This model was trained on a dataset composed of all available enzymes co-crystallised with drug-like ligands. To evaluate the model, a comprehensive test set consisting of enzyme structures and ligands was manually created. The test set of enzymes were characterised by matching their entire structures to the local descriptor library constructed from the training set. Both the training and the test sets contained enzyme-ligand complexes from all major enzyme classes, and the enzymes spanned a large range of sequences and folds. The experimental binding affinities (pK_i) ranged from 0.5 to 11.9. This demonstrates that the use of local descriptors makes it possible to create approximate predictive models that can be generalised over a wide range of protein 'targets'.[68]

12.20.4 Cytoscape Plug-ins

DrugViz is a Cytoscape plug-in that is designed to visualise and analyse small molecules within the framework of the interactome. This plug-in can import drug-target network information in an extended SIF file format to Cytoscape, and display the two-dimensional (2D) structures of small molecule nodes in a unified visualisation environment. It can also identify small-molecule nodes by means of three different 2D structure-searching methods, specifically isomorphism, sub-structure and fingerprint-based similarity searches. Subsequent to selections, users can furthermore conduct a two-side 'clustering' analysis on drugs and targets, which allows for a detailed analysis of the active compounds in the network, and also elucidate relationships between these drugs and 'targets'.[69] BiNoM (Biological Network Manager) is a new Cytoscape plug-in that significantly facilitates the usage and analysis of biological networks in standard systems biology formats. BiNoM is able to work with huge BioPAX files such as whole pathway databases. In addition, BiNoM permits the analysis of networks created with CellDesigner software and their conversion into BioPAX format. It is supplied as a library and as a Cytoscape plug-in, which adds a rich set of operations to Cytoscape such as path and cycle analysis, clustering sub-networks, decomposition of networks into modules, clipboard operations and others.[70]

12.20.5 Novel Screening Technologies

Novartis has developed two novel high-throughput screening (HTS) technologies for that purpose: NanoScreen and SpeedScreen. NanoScreen is a highly miniaturised and fully automated HTS/uHTS test system with both confocal single-molecule and non-confocal detection capabilities, and is employed for functional screening in the range of 1–5 µl per sample. The integration of the single-molecule readout technologies into the system enables highly sophisticated biochemical test systems with multiparameter readouts for a very high level of data quality. SpeedScreen is a highly miniaturised and automated screening system for the high-throughput affinity-selection of compounds.[71]

12.20.6 Anti-HIV Drugs

Several phenolic compounds isolated and characterised from natural sources have been found to exhibit inhibitory effects against different stages of the HIV-1 life cycle. Hence, chemogenomic approaches can be useful for the rapid identification of promising new anti-HIV lead molecules without having any other unwanted or undesirable pharmacological effects.[72]

12.21 Discussion

Chemogenomics is a study of the intersection of biological and chemical spaces. It aims towards the systematic identification of small molecules that interact with the products of the genome and hence modulate their biological function. Chemogenomics requires expertise in biology, chemistry and computational sciences (bioinformatics, chemoinformatics, large-scale statistics and machine-learning methods), but it is more than the simple apposition of each of these disciplines.[73] The aim of this approach, to find possible drugs for all target families, is the reason why chemogenomics is continually affecting the drug-discovery process. Whilst historically the approach is based on efforts that systematically explore target gene families such as kinases, today additional knowledge-based systematisation principles are followed within early drug discovery projects which aim to biologically validate the targets, and also to identify starting points for chemical lead optimisation.[1] In order to realise the value of chemogenomics information, a contextual database is required to relate the physiological outcomes induced by diverse compounds to the gene expression patterns measured in the same species. Massively parallel gene expression characterisation, coupled with traditional assessments of drug candidates, provides additional, important mechanistic information, and therefore a means to increase the accuracy of critical decisions.[58] RNA interference is a conserved biological process that has evolved to specifically and efficiently silence genes. Genome-wide screens using RNA interference have proven powerful in elucidating components of functionally related pathways, and have therefore become integral for the development of new and improved therapeutic targets. Chemogenomics is undergoing changes and developing with RNA interference-based screening, and also shaping the discovery of new targeted therapies.[74,75] Since gene expression technologies are continually improving, biomarkers will achieve higher throughput, and become more cost-effective and increasingly accurate. This will elevate the value of chemogenomics in drug-development research programmes, shift attrition to earlier in the process and reduce the overall cost of drug development. Over the past two to three years, the transition of chemogenomics from a research tool to a decision-making one has begun, and regulatory agencies are anxiously awaiting implementation of this technology to accelerate and make more informed evaluations of potential drugs.[64]

12.22 Conclusion

This review presented various present and future applications, and also advances in the area of chemogenomics. It provides an insight into various approaches used to characterise novel targets and ligands, and seeks new metabolic pathways that further aid the gene discovery process. This review also provides a brief overview of how chemogenomics can play a vital role in further research areas such as cheminformatics, pharmacogenomics, ligand-binding studies, kinase activities *etc.* This technology can be valuable in diminishing the time required for drug design. The dual approaches of chemogenomics promises to add a fresh influx of knowledge to cheminformatics, gene discovery, gene regulation and molecular signalling, and hence supply directions for finding novel therapeutic agents against disorders such as a range of cancers. It offers hope to resolve the ethical issues regarding stem cells *via* a provision of safer and more practical methods for stem cell regeneration.

References

1. E. Jacoby, Chemogenomics: drug discovery's panacea?, *Mol. Biosyst.*, 2006, **2**(5), 218–220.
2. E. Jacoby, A. Schuffenhauer and P. Floersheim, Chemogenomics knowledge-based strategies in drug discovery, *Drug News Perspect.*, 2003, **16**(2), 93–102.
3. X. F. Zheng and T. F. Chan, Chemical genomics: a systematic approach in biological research and drug discovery, *Curr. Issues Mol. Biol.*, 2002, **4**(2), 33–43.
4. A. Sehgal, Drug discovery and development using chemical genomics, *Curr. Opin. Drug Discov. Devel.*, 2002, **5**(4), 526–531.
5. H. Kubinyi, Chemogenomics in drug discovery, *Ernst Schering Res. Found. Workshop*, 2006, **58**, 1–19.
6. J. Mestres, Computational chemogenomics approaches to systematic knowledge-based drug discovery, *Curr. Opin. Drug Discov. Devel.*, 2004, **7**(3), 304–313.
7. M. Murphy, Discovery on Target 2006 – CHI's fourth annual event. Chemogenomics: small molecules as biological probes, *IDrugs*, 2007, **10**(1), 30–32.
8. T. Klabunde and R. Jäger, Chemogenomics approaches to G-protein coupled receptor lead finding, *Ernst Schering Res. Found. Workshop*, 2006, **58**, 31–46.
9. W. Guba, Chemogenomics strategies for G-protein coupled receptor hit finding, *Ernst Schering Res. Found. Workshop*, 2006, **58**, 21–29.
10. A. Schuffenhauer, J. Zimmermann, R. Stoop, J. J. van der Vyver, S. Lecchini and E. Jacoby, An ontology for pharmaceutical ligands and its application for in silico screening and library design, *J. Chem. Inf. Comput. Sci.*, 2002, **42**(4), 947–955.
11. H. Tomioka and K. Namba, Development of antituberculous drugs: current status and future prospects, *Kekkaku*, 2006, **81**(12), 753–774.
12. G. Dormán, K. Kocsis-Szommer, C. Spadoni and P. Ferdinandy, MMP inhibitors in cardiac diseases: an update, Recent Patents Cardiovasc, *Drug Discov.*, 2007, **2**(3), 186–194.
13. J. T. Peterson, The importance of estimating the therapeutic index in the development of matrix metalloproteinase inhibitors, *Cardiovasc. Res.*, 2006, **15;69**(3), 677–687.
14. B. Pirard, Insight into the structural determinants for selective inhibition of matrixmetalloproteinases, *Drug Discov. Today*, 2007, **12**(15–16), 640–646.
15. J. W. Skiles, N. C. Gonnella and A. Y. Jeng, The design, structure, and clinical update of small molecular weight matrix metalloproteinase inhibitors, *Curr. Med. Chem.*, 2004, **11**(22), 2911–2977.

16. Q. X. Sang, Y. Jin, R. G. Newcomer, S. C. Monroe, X. Fang, D. R. Hurst, S. Lee, Q. Cao and M. A. Schwartz, Matrix metalloproteinase inhibitors as prospective agents for the prevention and treatment of cardiovascular and neoplastic diseases, *Curr. Top. Med. Chem.*, 2006, **6**(4), 289–316.

17. T. Klabunde and G. Hessler, Drug design strategies for targeting G-protein-coupled receptors, *Chembiochem*, 2002, **4**;3(10), 928–944.

18. R. M. Eglen, R. Bosse and T. Reisine, Emerging concepts of guanine nucleotide-binding protein-coupled receptor (GPCR) function and implications for high throughput screening, *Assay Drug Dev Technol.*, 2007, **5**(3), 425–451.

19. P. Jimonet and R. Jäger, Strategies for designing GPCR-focused libraries and screening sets, *Curr. Opin. Drug Discov. Devel.*, 2004, **7**(3), 325–333.

20. H. Tomioka and K. Namba, Development of antituberculous drugs: current status and future prospects, *Kekkaku*, 2006, **81**(12), 753–774.

21. L. Jacob, B. Hoffmann, V. Stoven and J.-P. Vert, Virtual screening of GPCRs: An in silico chemogenomics approach, *BMC Bioinformatics*, 2008, **9**, 363, DOI: 10.1186.

22. D. Rognan, Chemogenomic approaches to rational drug design, *Br. J. Pharmacol.*, 2007, **152**(1), 38–52SeptemberSeptember.

23. M. A. Koch, L. O. Wittenberg, S. Basu, D. A. Jeyaraj, E. Gourzoulidou, K. Reinecke, A. Odermatt and H. Waldmann, Compound library development guided by protein structure similarity clustering and natural product structure, *Proc. Natl Acad. Sci. USA*, 2004, **101**(48), 16721–16726.

24. E. Kellenberger, P. Muller, C. Schalon, G. Bret, N. Foata and D. Rognan, SC-PDB: an annotated database of druggable binding sites from the Protein Data Bank, *J. Chem. Inf. Model.*, 2006, **46**(2), 717–727.

25. C. Hoppe, C. Steinbeck and G. Wohlfahrt, Classification and comparison of ligand-binding sites derived from grid-mapped knowledge-based potentials, *J. Mol. Graph. Model.*, 2006, **24**(5), 328–340.

26. M. Lapinsh, P. Prusis, S. Uhlén and J. E. Wikberg, Improved approach for proteochemometrics modeling: application to organic compound–amine G protein-coupled receptor interactions, *Bioinformatics*, 2005, **21**(23), 4289–4296.

27. D. Erhan, P. J. L'heureux, S. Y. Yue and Y. Bengio, Collaborative filtering on a family of biological targets, *J. Chem. Inf. Model.*, 2006, **46**(2), 626–635.

28. A. Wuster and M. Madan Babu, Chemogenomics and biotechnology, *Trends Biotechnol.*, 2008, **26**(5), 252–258.

29. A. M. Deutschbauer, D. F. Jaramillo, M. Proctor, J. Kumm, M. E. Hillenmeyer, R. W. Davis, C. Nislow and G. Giaever, Mechanisms of haploinsufficiency revealed by genome-wide profiling in yeast, *Genetics*, 2005, **169**(4), 1915–1925.

30. P. Flaherty, G. Giaever, J. Kumm, M. I. Jordan and A. P. Arkin, Latent variable model for chemogenomic profiling, *Bioinformatics*, 2005, **21**(15), 3286–3293.

31. G. Giaever, P. Flaherty, J. Kumm, M. Proctor, C. Nislow, D. F. Jaramillo, A. M. Chu, M. I. Jordan, A. P. Arkin and R. W. Davis, Chemogenomic profiling: Identifying the functional interactions of small molecules in yeast, *Proc. Natl Acad. Sci. USA*, 2004, **101**(3), 793–798.

32. A. Lopez, A. B. Parsons, C. Nislow, G. Giaever and C. Boone, Chemical-genetic approaches for exploring the mode of action of natural products, *Prog. Drug Res.*, 2008, **66**(237), 239–271.

33. D. Xu, B. Jiang, T. Ketela, S. Lemieux, K. Veillette, N. Martel, J. Davison, S. Sillaots, S. Trosok, C. Bachewich, H. Bussey, P. Youngman and T. Roemer, Genome-wide fitness test and mechanism-of-action studies of inhibitory compounds in Candida albicans, *PLoS Pathog.*, 2007, **3**(6), e92.

34. P. Alberici, C. Gaspar, P. Franken, M. M. Gorski, I. de Vries, R. J. Scott, A. Ristimäki, L. A. Aaltonen and R. Fodde, Smad4 haploinsufficiency: a matter of dosage, *Pathogenetics*, 2008, **1**(1), 2.

35. C. J. Wilson, Y. Si, C. M. Thompsons, A. Smellie, M. A. Ashwell, J. F. Liu, P. Ye, D. Yohannes and S. C. Ng, Identification of a small molecule that induces mitotic arrest using a simplified high-content screening assay and data analysis method, *J. Biomol. Screen.*, 2006, **11**(1), 21–28.

36. F. J. Vizeacoumar, Y. Chong, C. Boone and B. J. Andrews, A picture is worth a thousand words: Genomics to phenomics in the yeast Saccharomyces cerevisiae, *FEBS Lett.*, 2009, **583**(11), 1656–1661.

37. A. Bender, D. W. Young, J. L. Jenkins, M. Serrano, D. Mikhailov, P. A. Clemons and J. W. Davies, Chemogenomic Data Analysis: prediction of small-molecule targets and the advent of biological fingerprints, *Combinatorial Chemistry & High Throughput Screening*, 2007, **10**, 719–731.

38. H. S. Choi, J. S. Shim, J. A. Kim, S. W. Kang and H. J. Kwon, Discovery of gliotoxin as a new small molecule targeting thioredoxin redox system, *Biochem. Biophys. Res. Commun.*, 2007, **359**(3), 523–528.

39. S. Ekins, Y. Nikolsky, A. Bugrim, E. Kirillov and T. Nikolskaya, Pathway mapping tools for analysis of high content data, *Methods Mol. Biol.*, 2007, **356**, 319–350.

40. Z. E. Perlman, T. J. Mitchison and T. U. Mayer, High-content screening and profiling of drug activity in an automated centrosome-duplication assay, *Chembiochem*, 2005, **6**(1), 145–151.

41. D. di Bernardo, M. J. Thompson, T. S. Gardner, S. E. Chobot and E. L. Eastwood, Chemogenomic profiling on a genomewide scale using reverse-engineered gene networks, *Nat. Biotechnol.*, 2005, **23**(3), 377–383.

42. H. Xing and T. S. Gardener, The mode-of-action by network identification (MNI) algorithm: a network biology approach for molecular target identification, *Nat. Protoc.*, 2006, **1**(6), 2551–2554.

43. F. Iorio, R. Tagliaferri and D. di Bernardo, Identifying network of drug mode of action by gene expression profiling, *J. Comput. Biol.*, 2009, **16**(2), 241–251.

44. M. Lauria, F. Iorio and D. di Bernardo, NIRest: a tool for gene network and mode of action inference, *Ann. NY Acad. Sci.*, 2009, **1158**, 257–264.

45. Nidhi, M. Glick, J. W. Davies and J. L. Jenkins, Prediction of biological targets for compounds using multiple-category Bayesian models trained on chemogenomics databases, *J. Chem. Inf. Model.*, 2006, **46**, 1124–1133.
46. A. P. Davis, C. G. Murphy, C. A. Saraceni-Richards, M. C. Rosenstein, T. C. Wiegers and C. J. Mattingly, Comparative Toxicogenomics Database: a knowledgebase and discovery tool for chemical-gene-disease networks, *Nucleic Acids Res. J.*, 2009, **37**, D786–792.
47. G. Scapin, Protein kinase inhibition: different approaches to selective inhibitor design, *Curr. Drug Targets*, 2006, **7**(11), 1443–1454.
48. M. Vieth, R. E. Higgs, D. H. Robertson, M. Shapiro, E. A. Gragg and H. Hemmerle, Kinomics-structural biology and chemogenomics of kinase inhibitors and targets, *Biochim. Biophys. Acta*, 2004, **1697**(1–2), 243–257.
49. M. Xia, R. Huang, V. Guo, N. Southall, M. H. Cho, J. Inglese, C. P. Austin and M. Nirenberg, Identification of compounds that potentiate CREB signaling as possible enhancers of long-term memory, *Proc. Natl Acad. Sci. USA*, 2009, **106**(7), 2412–2417.
50. S. M. Corsello, G. Roti, K. N. Ross, K. T. Chow, I. Galinsky, D. J. DeAngelo, R. M. Stone, A. L. Kung, T. R. Golub and K. Stegmaier, Identification of AML1-ETO modulators by chemical genomics, *Blood*, 2009 [Epub ahead of print].
51. Q. T. Tran, L. Xu, V. Phan, S. B. Goodwin, M. Rahman, V. X. Jin, C. H. Sutter, B. D. Roebuck, T. W. Kensler, E. O. George and T. R. Sutter, Chemical genomics of cancer chemopreventive dithiolethiones, *Carcinogenesis*, 2009, **30**(3), 480–486.
52. L. Lafanechere, Chemogenomics and cancer chemotherapy: cell-based assays to screen for small molecules that impair microtubule dynamics, *Combinatorial Chemistry & High Throughput Screening*, 2008, **11**(8), 617–623.
53. L. Jacob and J. P. Vert, Protein-ligand interaction prediction: an improved chemogenomics approach, *Bioinformatics*, 2008, **24**(19), 2149–2156.
54. The Human Metabolome Project (http://www.metabolomics.ca/).
55. M. Glick, J. W. Davies and J. L. Jenkins, Prediction of Biological Targets for Compounds Using Multiple-Category Bayesian Models Trained on Chemogenomics Databases, *J. Chem. Inform. Model.*, 2006, **46**(3), 1124–1133.
56. S.-D. Zhang and T. W Gant, A simple and robust method for connecting small-molecule drugs using gene-expression signatures, *BMC Bioinformatics*, 2008, **9**, 258.
57. Z. Kutalik, J. S. Beckmann and S. Bergmann, A modular approach for integrative analysis of large-scale gene-expression and drug-response data, *Nat. Biotechnol.*, 2008, **26**(5), 531–539.
58. B. Ganter, S. Tugendreich, C. I. Pearson, E. Ayanoglu, S. Baumhueter, K. A. Bostian, L. Brady, L. J. Browne, J. T. Calvin, G. J. Day, N. Breckenridge, S. Dunlea, B. P. Eynon, L. M. Furness, J. Ferng,

M. R. Fielden, S. Y. Fujimoto, L. Gong, C. Hu, R. Idury, M. S. Judo, K. L. Kolaja, M. D. Lee, C. McSorley, J. M. Minor, R. V. Nair, G. Natsoulis, P. Nguyen, S. M. Nicholson, H. Pham, A. H. Roter, D. Sun, S. Tan, S. Thode, A. M. Tolley, A. Vladimirova, J. Yang, Z. Zhou and K. Jarnagin, Development of a large-scale chemogenomics database to improve drug candidate selection and to understand mechanisms of chemical toxicity and action, *J. Biotechnol.*, 2005, **119**(3), 219–244.

59. Glossary of Terms Used in Medicinal Chemistry (IUPAC Recommendations 1998) (http://www.chem.qmul.ac.uk/iupac/medchem/ix.html#p7).

60. L. Jacob, B. Hoffmann, V. Stoven and J.-P. Vert, Virtual screening of GPCRs: An in silico chemogenomics approach, *BMC Bioinformatics*, 2008, **9**, 363, DOI: 10.1186/1471-2105-9-363.

61. D. K. Agrafiotis, V. S. Lobanov and F. R. Salemme, Combinatorial informatics in the post-genomics era, *Nat. Rev. Drug Discov.*, 2002, **1**(5), 337–346.

62. C. E. Gagna and W. C. Lambert, Cell biology, chemogenomics and chemoproteomics – application to drug discovery, *Expert Opin. Drug Discov*, 2007, **2**(3), 381–401.

63. Nguyen *et al.*, Chemogenomic identification of Ref-1/AP-1 as a therapeutic target for asthma, *Proc. Natl Acad. Sci. USA*, 2003, **100**(3), 1169–1173.

64. M. R. Fielden, C. Pearson, R. Brennan and K. L. Kolaja, Preclinical drug safety analysis by chemogenomic profiling in the liver, *American Journal of PharmacoGenomics*, 2005, **5**(3), 161–171.

65. M. McMillan and M. Kahn, Investigating Wnt signaling: a chemogenomic safari, *Drug Discov. Today*, 2005, **10**(21), 1467–1474.

66. R. Damoiseaux, S. P. Sherman, J. A. Alva, C. Peterson and A. D. Pyle, Integrated chemical genomics reveals modifiers of survival in human embryonic stem cells, *Stem Cells*, 2008, Dec 18 [Epub ahead of print].

67. C. R. Caffrey, A. Rohwer, F. Oellien, R. J. Marhöfer, S. Braschi, G. Oliveira, J. H. McKerrow and P. M. Selzer, A comparative chemogenomics strategy to predict potential drug targets in the metazoan pathogen, Schistosoma mansoni, *PLoS ONE*, 2009, **4**(2), e4413.

68. H. Strömbergsson, P. Daniluk, A. Kryshtafovych, K. Fidelis, J. E. Wikberg, G. J. Kleywegt and T. R. Hvidsten, Interaction model based on local protein substructures generalizes to the entire structural enzyme-ligand space, *J. Chem. Inf. Model.*, 2008, **48**(11), 2278–2288.

69. B. Xiong, K. Liu, J. Wu, D. L. Burk, H. Jiang and J. Shen, DrugViz: a Cytoscape plugin for visualizing and analyzing small molecule drugs in biological networks, *Bioinformatics*, 2008, **24**(18), 2117–2118.

70. A. Zinovyev, E. Viara, L. Calzone and E. Barillot, BiNoM: a Cytoscape plugin for manipulating and analyzing biological networks, *Bioinformatics*, 2008, **24**(6), 876–877.

71. L. M. Mayr, Tackling the chemogenomic space by novel screening technologies, Ernst Schering Res, *Found. Workshop*, 2006, **58**, 111–173.

72. M. T. Hassan Khan and A. Ather, Potentials of phenolic molecules of natural origin and their derivatives as anti-HIV agents, *Biotechnol. Annu. Rev.*, 2007, **13**, 223–264.
73. E. Marechal, Chemogenomics: a discipline at the crossroad of high throughput technologies, biomarker research, combinatorial chemistry, genomics, cheminformatics, bioinformatics and artificial intelligence, *Combinatorial Chemistry & High Throughput Screening*, 2008, **11**(8), 583–586.
74. A. Kourtidis, C. Eifert and D. S. Conklin, RNAi applications in target validation, *Ernst Schering Res. Found. Workshop*, 2007, **61**, 1–21.
75. L A. Gaither, Chemogenomics approaches to novel target discovery, *Expert Rev. Proteomics*, 2007, **4**(3), 411–419.
76. C. G. Wermuth, Selective optimization of side activities: the SOSA approach, *Drug Discov. Today*, 2006, **11**, 160–164.

Subject Index

adenylate kinase AK1 knockout hearts, 270–273

A–D (Anderson–Darling) test, 45–49

agglomerative hierarchal clustering (AHC) methods, 89–91

AHC (agglomerative hierarchal clustering) methods, 89–91

AID (automatic interaction detection) trees, 81

alanine aminotransferase (ALT), 325

ALT (alanine aminotransferase), 325

AnalyserPro (SpectralWorks) software, 175

analysis-of-covariance (ANCOVA), 93

analysis-of-variance (ANOVA), 50–58
    factorial/multifactorial models, 54–57
    fixed effects, 50–53
    and glucosinolates production, in *Brassicaceae oleracea,* 121
    hierarchical or 'nested' models, 54
    interaction components of variance in, 57–58
    random effects, 53–54
    simultaneous component analysis (ASCA), 57
        and glucosinolates production, in *Brassicaceae oleracea,* 121–125

analysis-of-variance simultaneous component analysis (ASCA), 57

ANCOVA (analysis-of-covariance), 93

Anderson–Darling (A–D) test, 45–49

ANOVA simultaneous component analysis (ASCA), 118
    and glucosinolates production, in *Brassicaceae oleracea,* 121–125

anti-HIV drugs, and chemogenomics, 371

APCI (atmospheric pressure chemical ionisation), 170, 171t

API (atmospheric pressure ionisation), 170

APPI (atmospheric pressure photo ionisation), 171t

ASCA (analysis-of-variance simultaneous component analysis), 57

ASCA (ANOVA simultaneous component analysis), 118

assumption of normality, and experimental design, 44–50

atmospheric pressure chemical ionisation (APCI), 170, 171t

atmospheric pressure ionisation (API), 170

atmospheric pressure photo ionisation (APPI), 171t

ATR-FTIR (attenuated total reflectance-Fourier transform infrared) spectroscopy, 203

attenuated total reflectance-Fourier transform infrared (ATR-FTIR) spectroscopy, 203

automatic interaction detection (AID) trees, 81

backward elimination, 144. *See also* stepwise backward selection
bagging (bootstrap aggregating), 158–159
Bayesian belief networks (BBNs), 205
BBNs (Bayesian belief networks), 205
BCAAs (branched-chain amino acids), 101
best matching unit (BMU), 298, 299
bias-variance trade-off, 102
bile salt export pump (BSEP), 328
BiNoM (Biological Network Manager), 371
bioinformatics, and chemogenomics, 365
biologically interpretable multivariate biomarkers, 156–160
identification of parsimonious biomarkers, 159–160
informative set of genes, 157–158
Modified Bagging Schema, 158–159
Biological Network Manager (BiNoM), 371
biomarkers searching, and DILI, 342–345
BioPAX plug-in, 371
blood serum, estrogens in, 227
BMI (body mass index), 130
BMU (best matching unit), 298, 299
body mass index (BMI), 130
Bonferroni correction, for multiple comparisons, 62–64
bootstrap aggregating (bagging), 158–159
bounded support vectors, 153
branch and bound method, 144
branched-chain amino acids (BCAAs), 101
Brassicaceae oleracea, glucosinolates production in, 119–125
and ANOVA, 121
and ASCA, 121–125
Bruker Avance AX-600 spectrometer, 294
BSEP (bile salt export pump), 328

canonical correlation analysis (CCorA), 75–80
case study, 76–80
capillary electrophoresis-mass spectrometry, 184
Capillary zone electrophoresis (CZE), 184
CART (classification and regression tree) analysis, 80–81
CBA (cost-benefit analysis), 30–31
CCorA. *See* canonical correlation analysis (CCorA)
CCR (correlated component regression), 104–110
case study, 106–110
central carbon analysis, and GSIST, 227–234
derivatisation and analytical conditions, 229–230
method evaluation and validation, 230–234
CHAID (chi-square automatic interaction detection), 81
charged residue model (CRM), 184
chemical ionisation (CI), 171t, 179
chemical shifts, 337
cheminformatics, and chemogenomics, 368
chemogenomics
and bioinformatics, 365
and cheminformatics, 368
classification of, 360–361
ligand-based, 360
target-based, 360–361
target ligand-based, 361
current research in, 364
current trends in, 370–371
anti-HIV drugs, 371
cytoscape plug-ins, 370–371
ligand-enzyme interaction, 370
novel screening technologies, 371
schistosomiasis, 370
stem cells, 370

description, 371–372
and drug safety, 369
and evaluating complex
  signalling networks, 369
and haploinsufficiency
  profiling, 362–363
and high-content screening,
  363–364
and kinase activity, 365–366
and ligand-binding study, 367
and metabolomics, 367
and mode of action by network
  identification (MNI), 364
and oncology, 366
overview, 357–358
and pharmacogenomics, 368–
  369
and pharmacophore, 367–368
and privileged structures, 358–
  359
screens of, 361–362
and selective optimisation of
  side-activities (SOSA)
  approach, 359–360
chemometric techniques, in
  metabolomics
  partial least squares-
    discriminatory analysis
    (PLS-DA), 18–33
    case study, 20–22
    cost-benefit analysis
      (CBA), 30–31
    and final calibration
      model, 28
    permutation testing, 22–23
    quality evaluation
      processes, 28–30
    validation and cross-
      validation of, 24–27
  principal component analysis
    (PCA), 2–18
    assumptions, 4–9
    case study, 13–15
    examination of wider
      range of components,
      15–16

interpretability criteria of,
  11–12
number and significance of
  explanatory variables, 9
number of extractable
  PCs, 9–10
sample size for, 10–11
suitability of MV datasets,
  17–18
total variance of dataset, 10
and Type I (false-positive)
  errors, 16–17
varimax rotation, 12–13
chi-square automatic interaction
  detection (CHAID), 81
chi-squared statistic, 18
ChromaToF (Leco) software, 175
CI (chemical ionisation), 171, 179
CIs (confidence intervals), 28
classification, of chemogenomics,
  360–361
  ligand-based, 360
  target-based, 360–361
  target ligand-based, 361
classification and regression tree
  (CART) analysis, 80–81
class sample vector (CSV), 299
class weight vector (CWV), 299
cluster analysis, 86–92
  agglomerative hierarchal
    clustering (AHC) methods,
    89–91
  case study, 91–92
comparative toxicogenomics
  database (CTD), 365
complete search strategy, 144
complex signalling networks, and
  chemogenomics, 369
component planes, 298
comprehensive GCxGC-MS, 180–181
COMSPARI software, 175
confidence intervals (CIs), 28
confusion matrix, 19
correlated component regression
  (CCR), 104–110
  case study, 106–110

correlation-based feature selection,
143
cost-benefit analysis (CBA), 30–31
creatine kinase M-CK knockout
hearts, 273–277
CREB (cyclic-AMP response element
binding), 366
CRM (charged residue model),
184
CSV (class sample vector), 299
CTD (comparative toxicogenomics
database), 365
curse of dimensionality, 58, 96, 137,
138, 140, 156
1CV (single cross-validation)
method, 25
CWV (class weight vector), 299
cyclic-AMP response element
binding (CREB), 366
cytoscape plug-ins, and
chemogenomics, 370–371
cytotoxicity, and DILI, 329–333
CZE (capillary zone electrophoresis),
184

DAD (diode array detection) system,
207
DART (direct analysis in real time),
172
data acquisition, and mass
spectrometry, 172–174
data normalisation, and
experimental design, 42–44
data normality assumption, and
experimental design, 44–50
data preprocessing
oxyhalogen oxidant-containing
oral rinse product treatment,
297
steps, in experimental design,
39–42
data processing, and mass
spectrometry, 174–175
data scaling, and experimental
design, 42–44
DBP (diastolic blood pressure), 77

dendograms
defined, 87
polar, 89
derivatisation strategy
for energy metabolism
analysis, 229–230
for estrogens, 223–224
and triterpenoid metabolomic
fingerprints, 236
DESI (desorption electrospray
ionisation), 172
desorption electrospray ionisation
(DESI), 172
DHFR (dihydrofolate reductase)
inhibitors, 366
diastolic blood pressure (DBP), 77
dihydrofolate reductase (DHFR)
inhibitors, 366
DILI. *See* drug-induced liver injury
(DILI)
dimensionality reduction, and
experimental design, 42–44
diode array detection (DAD) system,
207
direct analysis in real time (DART),
172
direct cytotoxicity, and DILI, 329–333
directed graph, 364
direct infusion mass spectrometry
(DIMS), 176–177
drug-induced liver injury (DILI),
324–326
mechanisms of, 326–336
and ambiguous nature,
333–336
and direct cytotoxicity,
329–333
drug metabolism and
elimination, 326–329
and immune-mediated
reactions, 329–333
and metabolomics, 336–339
mechanistic
investigation, 339–342
searching for biomarkers,
342–345

drug metabolism, and DILI,
    326–329
drug safety, and chemogenomics,
    369
DrugViz plug-in, 371
dynamic non-linear analysis, of
    polyphenols, 133

EBAM (empirical Bayesian approach
    modelling), 82
EBC (exhaled breath condensate),
    292
EI (electron impact), 171
EIT (electrical impedance
    tomography), 207
electrical impedance tomography
    (EIT), 207
electron impact (EI), 171
        ionisation, 179
electron multiplier (EM), 172
electrospray ionisation (ESI), 170,
    171t, 206
EM (electron multiplier), 172
embedded models, 143
empirical Bayesian approach
    modelling (EBAM), 82
energy metabolism analysis, and
    GSIST, 227–234
        derivatisation and analytical
            conditions, 229–230
        method evaluation and
            validation, 230–234
ensemble classifier, 155
epoch, 96
error analysis, 69
ESI (electrospray ionisation), 170,
    171t, 206
estrogens, and GSIST, 223–227
        in blood serum from breast
            cancer patients, 227
        derivatisation strategy for, 223–
            224
        and isotopic labelling, 225
        LC-MS analysis of, 224–225
        method validation in complex
            sample, 225–227

exhaled breath condensate (EBC),
    292
exhaustive search strategy, 143–144
experimental design
        analysis-of-variance (ANOVA),
            50–58
                factorial/multifactorial
                    models, 54–57
                fixed effects, 50–53
                hierarchical or 'nested'
                    models, 54
                interaction components
                    of variance in, 57–58
                random effects, 53–54
                simultaneous component
                    analysis (ASCA), 57
        applications of univariate
            approaches, 58–64
                Bonferroni correction for
                    multiple comparisons,
                    62–64
                and homogeneity of
                    variances, 60–62
                and homoscedasticity
                    assumptions, 60–62
                and statistical
                    assumptions, 60
        assumption of normality,
            44–50
        considerations for sample
            collection, 36–39
        data normalisation, scaling
            and dimensionality
            reduction, 42–44
        error analysis, 69
        overview, 35–36
        power (sample size)
            computations, 64–66
        raw data preprocessing steps,
            39–42
        sample size requirements,
            67–68
        statistical power computations
            for, 67–68
experimental/statistical
    proliferation. *See* error analysis

factorial/multifactorial models, ANOVA, 54–57
false discovery rate (FDR), 81
false-positive (Type I) errors, 16–17
family-wise error rate (FWER), 62
FDR (false discovery rate), 81
feature selection, and multivariate biomarkers, 142–145
  and random forests, 156
  search models, 143
    correlation-based feature selection, 143
    embedded models, 143
    filter models, 143
    hybrid models, 143
    shrunken centroid filters, 143
    wrapper models, 143
  search strategies, 143–144
    backward elimination, 144
    branch and bound method, 144
    complete search, 144
    exhaustive search, 143–144
    heuristic searches, 144
    heuristic sequential searches, 144
    hill-climbing strategies, 144
    stepwise backward selection, 144
    stepwise hybrid selection, 144
  stability of results, 144–145
  with T^2 and LDA, 149–150
filter models, 143
final calibration model, of PLS-DA, 28
FIT (quality-of-fit) model, 26
fixed effects model, ANOVA, 50–53
Fourier-transform infrared (FTIR) spectroscopy, 200
Fourier-transform ion cyclotron resonance (FTICR), 170, 174

frequent primary genes, 160
FTICR (Fourier-transform ion cyclotron resonance), 170, 174
FTIR (Fourier-transform infrared) spectroscopy, 200
FWER (family-wise error rate), 62

gamma-tocopherol metabolites, 243–245
GAs (genetic algorithms), 95–96
gas chromatography-mass spectrometry, 177–180
Gaussian graphical models (GGMs), 96–98
GC/MS analysis, of ^{18}O-assisted ^{31}P NMR and mass spectrometry, 262–264
gene expression profiles (GEPs), 364
gene-set enrichment analysis (GSEA), 364
genetic algorithms (GAs), 95–96
geometric trajectory analysis, 118
GEPs (gene expression profiles), 364
GGMs (Gaussian graphical models), 96–98
glucosinolates production, in *Brassicaceae oleracea*, 119–125
  and ANOVA, 121
  and ASCA, 121–125
GPCRs (G-protein coupled receptors), 359, 367
G-protein coupled receptors (GPCRs), 359, 367
greedy strategies, 144. *See also* heuristic sequential searches
group-specific internal standard technology (GSIST)
  applications of, 223–247
  basic principles of, 221–222
  central carbon and energy metabolism analysis, 227–234
    derivatisation and analytical conditions, 229–230
    method evaluation and validation, 230–234

determination of estrogens,
223–227
   in blood serum from
breast cancer patients,
227
   derivatisation strategy
for, 223–224
   and isotopic labelling,
225
   LC-MS analysis of,
224–225
   method validation in
complex sample,
225–227
and metabolites discovery,
240–247
   identification of gamma-
tocopherol
metabolites, 243–245
   structural determination
of identified ions, 245–
247
overview, 220–221
and triterpenoid metabolomic
fingerprints, 234–240
   and bioavailability study,
239–240
   and derivatisation
evaluation step, 236
   and ganoderic acids in
mushroom extracts,
236–239
GSEA (gene-set enrichment
analysis), 364
GSIST. *See* group-specific internal
standard technology (GSIST)

haploinsufficiency profiling (HIP),
362–363
hapten hypothesis, 331
hard-margin support vector
machines, 150
HCC (hepatocellular carcinoma),
206, 339
HCS (high-content screening), 363
HDL (high-density-lipoprotein), 101

hepatocellular carcinoma (HCC),
206, 339
hESCs (human embryonic stem
cells), 370
heuristic searches, 144
heuristic sequential searches, 144
hierarchical model, of ANOVA, 54
high-content screening (HCS), 363
   and chemogenomics,
363–364
high-density-lipoprotein (HDL), 101
high-performance liquid
chromatography (HPLC), 166, 233
   mass spectrometry, 181–184
high-throughput metabolomics
datasets analysis, 92–101
   Gaussian graphical models
(GGMs), 96–98
   genetic algorithms (GAs),
95–96
   independent component
analysis (ICA), 98–101
high-throughput screening (HTS),
371
HILIC (hydrophilic interaction
liquid chromatography),
182, 206
hill-climbing strategies, 144. *See also*
heuristic sequential searches
HIP (haploinsufficiency profiling),
362–363
hixels, defined, 39
^{1}H NMR-based multivariate (MV)
statistical analyses, of human
saliva. *See* multianalyte human
biofluid datasets
homogeneity of variances, and
experimental design, 60–62
homoscedasticity assumptions, and
experimental design, 60–62
HPLC. *See* high-performance liquid
chromatography (HPLC)
HTS (high-throughput screening),
371
human embryonic stem cells
(hESCs), 370

human urine, and polyphenols,
125–134
analysis of pooled samples,
130–133
dynamic non-linear analysis of,
133
multilevel PLSDA model,
128–129
multivariate consequence,
126–128
study setup, 130
HUSERMET project, 188
hybrid models, 143
hydrophilic interaction liquid
chromatography (HILIC), 182, 206
Hy's Law, 325

ICA (independent component
analysis), 98–101
ICP-MS (inductively coupled plasma-
mass spectrometry), 207
ICP-OES (inductively coupled
plasma-optical emission
spectrometry), 207
IEM (ion evaporation model), 171
immune-mediated reactions, and
DILI, 329–333
inborn errors of metabolism, 177
independent component analysis
(ICA), 98–101
inductively coupled plasma-mass
spectrometry (ICP-MS), 207
inductively coupled plasma-optical
emission spectrometry (ICP-OES),
207
informative set of genes, and
multivariate biomarkers, 157–158
infrared spectroscopy, 202–203
INH. *See* isoniazid (INH);
isonicotinylhydrazine (INH)
instrumentation, mass
spectrometry, 168–176
instrument control and data
processing, 174–175
ion detection and data
acquisition, 172–174

ion formation, 170
mass ion separation, 170–172
sample introduction, 169–170
instrument control, and mass
spectrometry, 174–175
intelligent bucketing, and oral rinse
product treatment, 295–296
interaction components of variance,
in ANOVA, 57–58
interpretability criteria, of PCA,
11–12
ion detection, and mass
spectrometry, 172–174
ion evaporation model (IEM), 171
ion formation, and mass
spectrometry, 170
isoniazid (INH), 335
isonicotinylhydrazine (INH), 335
isotopic labelling, and estrogens,
225

Kaiser–Meyer–Olkin (KMO)
measure, 17
kernels, defined, 154
kernel trick, 154
kinase activity, and chemogenomics,
365–366
KMO (Kaiser–Meyer–Olkin)
measure, 17
knockout hearts
adenylate kinase AK1, 270–273
creatine kinase M-CK, 273–277
Kolmogorov–Smirnov (K–S) curve
fitting algorithm, 44–49
K–S (Kolmogorov–Smirnov) curve
fitting algorithm, 44–49

laser-induced fluorescence (LIF), 207
LC-MS analysis, of estrogens,
224–225
LDA. *See* linear discriminant
analysis (LDA)
LDL (low-density-lipoprotein), 38
learning algorithms. *See also* specific
types
LDA, 146–149

nonparametric, 145
parametric, 145
random forests, 155
supervised (*See* supervised
learning algorithms, and
multivariate biomarkers)
SVM, 150–154
unsupervised, 140–142
LIF (laser-induced fluorescence), 207
ligand-based chemogenomics, 360
ligand-binding study, and
chemogenomics, 367
ligand-enzyme interaction, and
chemogenomics, 370
linear discriminant analysis (LDA),
145–150
feature selection with T^2,
149–150
learning algorithm, 146–149
linear ion trap (LIT), 170, 173t, 174
linear quadrupole (Q), 173
LIT (linear ion trap), 170, 173, 174
liver X receptor (LXR), 330
low-density-lipoprotein (LDL), 38
LXR (liver X receptor), 330

machine learning techniques, 83–86
random forests (RFs), 86
self-organising maps (SOMs),
83–85
support vector machines
(SVMs), 85–86
magnetic resonance spectroscopy
(MRS), 203
major histocompatibility complex
(MHC), 331
MALDI (matrix-assisted laser
desorption/ionisation), 168, 172,
206
mammalian metabolomes, 185–188
MAP (mean arterial pressure), 77
MarkerLynx (Waters) software, 175
MarkerView (AB Sciex) software, 175
MassHunter (Agilent) software, 175
mass ion separation, and mass
spectrometry, 170–172

mass spectrometry. *See also* specific
types
capillary electrophoresis, 184
comprehensive GCxGC-MS,
180–181
direct infusion mass
spectrometry (DIMS),
176–177
gas chromatography, 177–180
high performance liquid
chromatography, 181–184
instrumentation, 168–176
instrument control and
data processing,
174–175
ion detection and data
acquisition, 172–174
ion formation, 170
mass ion separation,
170–172
sample introduction,
169–170
and multivariate chemometric
profiling of cancer, 206–207
and ^{18}O-assisted ^{31}P NMR
(*See* ^{18}O-assisted ^{31}P NMR
and mass spectrometry)
overview, 162–168
terminologies and definitions
applied in, 163t–165t
mass-to-charge ratio, and mass ion
separation, 170–172
MathDAMP software, 175
matrix-assisted laser desorption/
ionisation (MALDI), 168, 172, 206
matrixinome approach, 359
matrix metalloproteinase (MMP)
inhibitors, 358–359
MDDR (MDL drug data report), 365
MDL drug data report (MDDR), 365
MDR (multidrug resistance
proteins), 328
mean arterial pressure (MAP), 77
mechanism of action (MOA), 363
mechanistic investigation, and DILI,
339–342

metabolites discovery, and GSIST,
240–247
  identification of gamma-
tocopherol metabolites,
243–245
  structural determination of
identified ions, 245–247
metabolomics
  and chemogenomics, 367
  chemometric techniques in
(*See* chemometric
techniques, in
metabolomics)
  and DILI, 336–339
    mechanistic
investigation, 339–342
    searching for biomarkers,
342–345
MetAlign software, 175
method validation/evaluation
  energy metabolism analysis,
230–234
  in estrogens sample, 225–227
MET-IDEA software, 175
M-fold cross-validation process, 102–
103
MHC (major histocompatibility
complex), 331
microchannel plate (MCP), 172
mid-infrared (MIR) spectroscopy,
200
MIR (mid-infrared) spectroscopy,
200
MMP (matrix metalloproteinase)
inhibitors, 358–359
MOA (mechanism of action), 363
model tuning, and MV regression
modelling, 102–103
mode of action by network
identification (MNI), 364
moderated t-statistic methods,
81–82
  empirical Bayesian approach
modelling (EBAM), 82
  significance analysis of
microarrays (SAM), 81–82

Modified Bagging Schema, 158–159
Monte-Carlo simulation, 45, 69, 309
MRP (multidrug resistance-
associated proteins), 328
MRS (magnetic resonance
spectroscopy), 203
MSFACTS software, 175
multianalyte human biofluid
datasets
  high-resolution NMR analysis
of, 288–290
  interpretation of salivary
profiles, 290–293
  overview, 287–288
  treatment with oxyhalogen
oxidant-containing oral
rinse product, 293–316
    data preprocessing, 297
    data simulations,
296–297
    description, 314–316
    ^{1}H NMR spectra, 303–308
    intelligent bucketing,
295–296
    measurements and
spectral editing,
294–295
    MV statistical techniques,
308–314
    partial least squares
regression coefficients,
301–303
    preparation of
supernatant samples,
294
    sample collection, 294
    self organising maps,
297–301
    software, 297
multidrug resistance-associated
proteins (MRP), 328
multidrug resistance proteins
(MDR), 328
multifactorial/factorial models,
ANOVA, 54–57
multilevel analysis, 128

multilevel PLSDA model, and
polyphenols, 128–129
multivariate biomarker discovery
biologically interpretable,
156–160
identification of
parsimonious
biomarkers, 159–160
informative set of genes,
157–158
Modified Bagging
Schema, 158–159
common misconceptions in,
138–142
univariate analysis,
139–140
using unsupervised
learning algorithms,
140–142
feature selection, 142–145
search models, 143
search strategies,
143–144
stability of results,
144–145
overview, 137–138
supervised learning
algorithms, 145–156
linear discriminant
analysis (LDA), 145–150
random forests, 155–156
support vector machines,
150–155
multivariate chemometric profiling
of cancer
infrared spectroscopy,
202–203
mass spectrometry, 206–207
nuclear magnetic resonance
spectroscopy, 203–206
overview, 199–201
multivariate consequence, and
polyphenols, 126–128
multivariate statistical analysis, and
^{18}O-assisted ^{31}P NMR and mass
spectrometry, 269–270

MV dataset analysis
canonical correlation analysis
(CCorA), 75–80
case study, 76–80
classification and regression
tree (CART) analysis, 80–81
cluster analysis, 86–92
agglomerative hierarchal
clustering (AHC)
methods, 89–91
case study, 91–92
high-throughput
metabolomics datasets
analysis, 92–101
Gaussian graphical
models (GGMs), 96–98
genetic algorithms (GAs),
95–96
independent component
analysis (ICA), 98–101
machine learning techniques,
83–86
random forests (RFs), 86
self-organising maps
(SOMs), 83–85
support vector machines
(SVMs), 85–86
moderated t-statistic methods,
81–82
empirical Bayesian
approach modelling
(EBAM), 82
significance analysis of
microarrays (SAM), 81–82
and multidimensional data
problems, 101–110
correlated component
regression (CCR),
104–110
and M-fold cross-
validation process,
102–103
model tuning and
optimisation, 102–103
partial least squares
regression (PLS-R), 104

MV dataset analysis (*continued*)
    principal component regression (PCR), 103–104
    regression regularisation, 102
    overview, 74–75
MV regression modelling, 101–110
    correlated component regression (CCR), 104–110
    and M-fold cross-validation process, 102–103
    model tuning and optimisation, 102–103
    partial least squares regression (PLS-R), 104
    principal component regression (PCR), 103–104
    regression regularisation, 102
MZMine software, 175
mzML software, 175

nanoelectrospray (nanoESI), 184
nested model, of ANOVA, 54
nonparametric bootstrap, 155
nonparametric learning algorithms, 145
non-steroidal anti-inflammatory drug (NSAID), 325
novel screening technologies, and chemogenomics, 371
NSAID (non-steroidal anti-inflammatory drug), 325
nuclear magnetic resonance spectroscopy, 203–206

^{18}O-assisted ^{31}P NMR and mass spectrometry
    GC/MS analysis of, 262–264
    methodology, 259–270
    and multivariate statistical analysis, 269–270
    ^{18}O metabolic labelling procedure, 261–270
    overview, 255–259

and phosphometabolite analysis, 267
phosphometabolomic platforms, 259–261
    and adenylate kinase AK1 knockout hearts, 270–273
    and creatine kinase M-CK knockout hearts, 273–277
    and transgenic animal models, 270–277
    and phosphoryl metabolites, 265–267
    and phosphotransfer fluxes, 267–269
ODEs (ordinary differential equations), 364
OHCP (oral healthcare product), 314
OLS (ordinary least squares) regression, 103
^{18}O metabolic labelling procedure, 261–262
    for cultured cells, 262
    heart perfusion and ^{18}O phosphoryl labelling, 262
    for isolated cardiomyocytes, 262
'omnibus' test, 45
oncology, and chemogenomics, 366
OOB (out-of-bag) samples, 155–156
OPLS (orthogonal-PLS) technique, 95
optimally weighted predictor, 9
optimisation, and MV regression modelling, 102–103
oral healthcare product (OHCP), 314
orbitrap, 174t
ordinary differential equations (ODEs), 364
ordinary least squares (OLS) regression, 103
orthogonal-PLS (OPLS) technique, 95
'out-of-bag' (OOB) samples, 155–156
overfitting phenomenon, 101
oxidative stress, 341

oxyhalogen oxidant-containing oral rinse product treatment, 293–316
 data preprocessing, 297
 data simulations, 296–297
 description, 314–316
 ^{1}H NMR spectra, 303–308
 intelligent bucketing, 295–296
 measurements and spectral editing, 294–295
 MV statistical techniques, 308–314
 partial least squares regression coefficients, 301–303
 preparation of supernatant samples, 294
 sample collection, 294
 self organising maps, 297–301
 software, 297

parametric learning algorithms, 145
Pareto-scaling, 43, 47, 130
parsimonious biomarkers, identification of, 159–160
partial least squares-discriminatory analysis (PLS-DA), 18–33
 case study, 20–22
 cost-benefit analysis (CBA), 30–31
 and final calibration model, 28
 multilevel model, and polyphenols, 128–129
 permutation testing, 22–23
 quality evaluation processes, 28–30
 validation and cross-validation of, 24–27
partial least squares regression (PLS-R), 104
 and oxyhalogen oxidant-containing oral rinse product treatment, 301–303
PCR (principal component regression), 103–104
PCs (principal components), 2, 338
percent correctly classified (%CC), 300

permutation testing, of PLS-DA, 22–23
pharmacogenomics, and chemogenomics, 368–369
pharmacophore, and chemogenomics, 367–368
phosphometabolite analysis, 267
phosphometabolomic profiling
 and knockout hearts
  adenylate kinase AK1, 270–273
  creatine kinase M-CK, 273–277
 and ^{18}O-assisted ^{31}P NMR and mass spectrometry, 259–261
 and transgenic animal models, 270–277
phosphoryl metabolites, 265–267
phosphotransfer fluxes, 267–269
'ping-pong' algorithm, 367
PKC (protein kinase C), 370
platykurtic distributions, 46
PLS-DA. *See* partial least squares-discriminatory analysis (PLS-DA)
PLS-R (partial least squares regression), 104
PNNs (probabilistic neural networks), 205
polar dendograms, 89
polyphenols, and human urine. *See* human urine, and polyphenols
power (sample size) computations, 64–66
 for high-dimensional metabolomic datasets, 67–68
PRC (principal response curves), 118
pregnane X receptor (PXR), 330
'prime' genes, 105
principal component analysis (PCA), 2–18
 assumptions, 4–9
 case study, 13–15
 examination of wider range of components, 15–16
 interpretability criteria of, 11–12

principal component analysis (PCA)
(*continued*)
   number and significance of
      explanatory variables, 9
   number of extractable PCs,
      9–10
   sample size for, 10–11
   suitability of MV datasets, 17–18
   total variance of dataset, 10
   and Type I (false-positive)
      errors, 16–17
   varimax rotation, 12–13
principal component regression
   (PCR), 103–104
principal components (PCs), 2, 338
principal response curves (PRC), 118
privileged structures, and
   chemogenomics, 358–359
probabilistic neural networks
   (PNNs), 205
protein kinase C (PKC), 370
protein structure similarity cluster
   (PSSC), 360
PSSC (protein structure similarity
   cluster), 360
PXR (pregnane X receptor), 330

Q (linear quadrupole), 173
QC-RLSC (quality control based loess
   signal correction), 187
QIT (quadrupole ion trap), 170, 173
QQQ (triple quadrupole), 173t
Q-TOF (quadrupole-time of flight),
   174
quadrupole ion trap (QIT), 170, 173
quadrupole-time of flight (Q-TOF),
   174
quality control based loess signal
   correction (QC-RLSC), 187
quality evaluation processes, for
   PLS-DA, 28–30
quality-of-fit (FIT) model, 26
QUEST (quick, efficient, statistical
   tree) approach, 81
quick, efficient, statistical tree
   (QUEST) approach, 81

random effects model, of ANOVA,
   53–54
random forests (RFs), 155–156
   feature selection with, 156
   learning algorithm, 155
   and MV dataset analysis, 86
raw data preprocessing steps, in
   experimental design, 39–42
reactive oxygen species (ROS),
   329
receiver operator characteristic
   (ROC) curve, 28, 33
recursive feature elimination,
   154–155
regression regularisation, and MV
   regression modelling, 102
relative standard deviations (RSDs),
   187
Reye's syndrome, 324
Rho-kinase (ROCK), 370
Ridge regression technique, 102
ROC (receiver operator
   characteristic) curve, 28, 33
ROCK (Rho-kinase), 370
ROS (reactive oxygen species),
   329
RSDs (relative standard deviations),
   187

salivary metabolome, 292
SAM (significance analysis of
   microarrays), 81–82
sample introduction systems,
   176–184
   capillary electrophoresis-mass
      spectrometry, 184
   comprehensive GCxGC-MS,
      180–181
   direct infusion mass
      spectrometry (DIMS),
      176–177
   gas chromatography-mass
      spectrometry, 177–180
   high performance liquid
      chromatography-mass
      spectrometry, 181–184

sample size (power) computations,
64–66
for high-dimensional
metabolomic datasets, 67–68
for PCA, 10–11
SAR (structure-activity relationship)
analysis, 366
SBP (systolic blood pressure), 77
scaling data process, and
experimental design, 42–44
schistosomiasis, and
chemogenomics, 370
screening, high-content, 363–364
screens, of chemogenomics, 361–362
search models, and feature
selection, 143
correlation-based, 143
embedded models, 143
filter models, 143
hybrid models, 143
shrunken centroid filters, 143
wrapper models, 143
search strategies, and feature
selection, 143–144
backward elimination, 144
branch and bound method, 144
complete search, 144
exhaustive search, 143–144
heuristic searches, 144
heuristic sequential searches,
144
hill-climbing strategies, 144
stepwise backward selection,
144
stepwise hybrid selection, 144
secondary ion mass spectrometry
(SIMS), 172
SELDI-TOF-MS (surface-enhanced
laser desorption/ionisation
time-offlight MS), 206
selective optimisation of side-
activities (SOSA) approach, 359–360
self-organising maps (SOMs), 83–85
and oxyhalogen oxidant-
containing oral rinse
product treatment, 297–301

sensitivity, defined, 28–29
Shapiro–Wilks (S–W) approach,
45–49
shrunken centroid filters, 143
SIEVE (ThermoScientific) software,
175
significance analysis of microarrays
(SAM), 81–82
SIMCA (soft independent modelling
of class analogy), 203
SIMS (secondary ion mass
spectrometry), 172
single cross-validation (1CV)
method, 25
soft independent modelling of class
analogy (SIMCA), 203
soft-margin support vector
machines, 150
software, for oral rinse product
treatment, 297
SOMDI (SOM discrimination index),
298, 300
SOM discrimination index (SOMDI),
298, 300
SOMs (self-organising maps), 83–85
statistical assumptions
of normality, and experimental
design, 44–50
and univariate approaches, 60
statistical/experimental
proliferation. *See* error analysis
stem cells, and chemogenomics,
370
stepwise backward selection, 144.
See also backward elimination
stepwise hybrid selection, 144
structure-activity relationship (SAR)
analysis, 366
supervised learning algorithms, and
multivariate biomarkers, 145–156
linear discriminant analysis
(LDA), 145–150
feature selection with T^2,
149–150
LDA learning algorithm,
146–149

supervised learning algorithms, and
multivariate biomarkers
(*continued*)
    random forests, 155–156
       feature selection with,
         156
       learning algorithm, 155
    support vector machines,
      150–155
       recursive feature
         elimination, 154–155
       SVM learning algorithms,
         150–154
support vector machines (SVMs),
  150–155
    and MV dataset analysis,
      85–86
    recursive feature elimination,
      154–155
    SVM learning algorithms,
      150–154
surface-enhanced laser desorption/
ionisation time-offlight MS
(SELDI-TOF-MS), 206
SVMs. *See* support vector machines
(SVMs)
S-W (Shapiro–Wilks) approach, 45–49
systolic blood pressure (SBP), 77

target-based chemogenomics,
  360–361
target ligand-based chemogenomics,
  361
T-cell factor (TCF), 369
TCF (T-cell factor), 369
thyroid-stimulating hormone (TSH),
  26
time-of-flight (TOF), 170, 173
TNF-related apoptosis-inducing
ligand (TRAIL), 330
TOF (time-of-flight), 170, 173
TRAIL (TNF-related apoptosis-
inducing ligand), 330
transgenic animal models, and
  phosphometabolomic profiling,
  270–277

triple quadrupole (QQQ), 173
triterpenoid metabolomic
  fingerprints, and GSIST, 234–240
    and bioavailability study,
      239–240
    and derivatisation evaluation
      step, 236
    and ganoderic acids in
      mushroom extracts,
      236–239
TSH (thyroid-stimulating hormone),
  26
2D-COSY (two-dimensional
  correlation spectroscopy), 41
2D-HRMAS (two-dimensional high-
  resolution magic angle spinning),
  41
two-dimensional correlation
  spectroscopy (2D-COSY), 41
two-dimensional high-resolution
  magic angle spinning
  (2D-HRMAS), 41
two-dimensional J-resolved
  spectroscopy (2D-JRES), 41
two-dimensional total correlation
  spectroscopy (2D-TOCSY), 41
2D-JRES (two-dimensional J-resolved
  spectroscopy), 41
2D-TOCSY (two-dimensional total
  correlation spectroscopy), 41
Type I (false-positive) errors, 16–17

UHPLC (ultra high performance
  liquid chromatography), 183
ultra high performance liquid
  chromatography (UHPLC), 183
ultra high performance liquid
  chromatography (UPLC), 183, 291
unbounded support vectors, 153
univariate analysis, and multivariate
  biomarkers, 139–140
univariate approaches applications,
  in experimental design, 58–64
    Bonferroni correction for
      multiple comparisons,
      62–64

and homogeneity of variances,
  60–62
and homoscedasticity
  assumptions, 60–62
and statistical assumptions,
  60
unsupervised learning algorithms,
  and multivariate biomarkers,
  140–142
unweighted clustering analysis
  (UPGMC), 90
UPGMC (unweighted clustering
  analysis), 90
UPLC (ultra high performance liquid
  chromatography), 183, 291

vanishing duct syndrome, 333
variable sample vector (VSV), 299
variable weight vector (VWV), 299
varimax rotation, and PCA, 12–13
VSV (variable sample vector), 299
VWV (variable weight vector), 299

Warburg effect, 206
WOMBAT (World of Molecular
  Bioactivity), 365
World of Molecular Bioactivity
  (WOMBAT), 365
wrapper models, 143

XCMS software, 175